IDEAS FOR 21ST CENTURY EDUCATION

T0173984

PROCEEDINGS OF THE ASIAN EDUCATION SYMPOSIUM (AES 2016), 22–23 NOVEMBER, 2016, BANDUNG, INDONESIA

Ideas for 21st Century Education

Editors

Ade Gafar Abdullah, Ida Hamidah, Siti Aisyah,
Ari Arifin Danuwijaya, Galuh Yuliani &
Heli S.H. Munawaroh
Universitas Pendidikan Indonesia, Bandung, Indonesia

Taylor & Francis Group

LONDON AND NEW YORK

First published 2017 by Routledge
2 Park Square, Milton Park, Abingdon, Oxon OX14 4RN
605 Third Avenue, New York, NY 10017

First issued in paperback 2020

Routledge is an imprint of the Taylor & Francis Group, an informa business

Typeset by V Publishing Solutions Pvt Ltd., Chennai, India

ISBN 13: 978-0-367-73590-6 (pbk)
ISBN 13: 978-1-138-05343-4 (hbk)

Ideas for 21st Century Education – Abdullah et al. (Eds)
© 2017 Taylor & Francis Group, London, ISBN 978-1-138-05343-4

Table of contents

Learning Teaching Methodologies and Assessment (TMA)

Other Areas of Education (OAE)

Pedagogy (PDG)

Ubiquitous Learning (UBL)

Ideas for 21st Century Education – Abdullah et al. (Eds)
© 2017 Taylor & Francis Group, London, ISBN 978-1-138-05343-4

Preface

Invited speakers, Distinguished Guests, Presenters, Participants, and Authors of Asian Education Symposium.

It is such an honor to have had you at the Asian Education Symposium (AES) 2016, organized by the School of Postgraduate Universitas Pendidikan Indonesia. The AES 2016 is an international refereed conference dedicated to the advancement of theories and practices in education. The AES 2016 promotes collaborative excellence between academicians and professionals in education. The conference aimed to develop a strong network of researchers and pioneers in education worldwide. The aim of AES 2016 was to provide an opportunity for academicians and professionals from various educational fields with cross-disciplinary interests to bridge the knowledge gap, promote research esteem and the evolution of pedagogy.

The AES 2016 main theme was Ideas for 21st Century Education. Education plays an important role in countries all over the globe. It will enable countries to achieve sustainable development goals by 2030. As for countries in the Asian region, education is a vehicle that can move people's mobility particularly in a time when we are welcoming the Asian Economic Community. It is without a doubt, there is a need to develop a strong collaboration and partnership among countries, both at regional and international levels. This symposium was one of our attempts to provide space for networking among academics and researchers in education. It is our hope that the symposium would contribute to the development of education as a distinct body of knowledge.

This symposium was a platform for us to disseminate and discuss our research findings. It is our expectation that the conversation from this symposium will inform policy and practices of education. It was also hoped that this symposium will open up future research on education, while at the same allowing all participants to expand their network. It is our hope that during this two-day symposium, all the participants had engaged in fruitful and meaningful discussions.

This AES 2016 proceedings contains papers that have been subjected to a double blind refereeing process. The process was conducted by academic peers with specific expertise in the key scopes and research orientation of the papers. It provides an opportunity for readers to engage with a selection of refereed papers that were presented during the symposium. The scopes of this symposium proceedings are: i) art education, ii) adult education, iii) business education, iv) course management, v) curriculum, research and development, vi) educational foundations, vii) learning/teaching methodologies and assessment, viii) global issues in education and research, ix) pedagogy, x) ubiquitous learning, and xi) other areas of education. We strongly believe that the selected papers published in the symposium proceedings will pay a significant contribution to the spread of knowledge.

We also would like to express our gratitude to all the keynote speakers from overseas who have travelled to our country to deliver and exchange their ideas. Our appreciation also goes to all the committee members who have worked hard to make this event possible. Once again, deepest gratitude for everybody's participation to the symposium as well as the proceedings.

Ade Gafar Abdullah,
Ida Hamidah,
Siti Aisyah,
Ari Arifin Danuwijaya,
Galuh Yuliani &
Heli S.H. Munawaroh
Universitas Pendidikan Indonesia, Bandung, Indonesia

Acknowledgments

Furqon—*Universitas Pendidikan Indonesia, Indonesia*
Asep Kadarohman—*Universitas Pendidikan Indonesia, Indonesia*
Edi Suryadi—*Universitas Pendidikan Indonesia, Indonesia*
Aim Abdulkarim—*Universitas Pendidikan Indonesia, Indonesia*
Didi Sukyadi—*Universitas Pendidikan Indonesia, Indonesia*
M. Solehudin—*Universitas Pendidikan Indonesia, Indonesia*
Takuya Baba—*Hiroshima University, Japan*
Christine C.M. Goh—*Nanyang Technological University, Singapore*
Allan L. White—*University of Western, Australia*
Tuğba Öztürk—*Ankara University & University of Philipines, Philipines*
Vasilis Strogilos—*NIE Nanyang Technological University, Singapore*
Tom Nelson Laird—*Indiana University, US*
Simon Clarke—*The University of Western Australia, Australia*
Diana Baranovich—*University of Malaya, Malaysia*
Taehee Kim—*Youngsan University, Busan South Korea*
Ikuro Yamamoto—*Kinjo Gakuin University Japan, Japan*
Numyoot Songthanapitak—*President of RAVTE*
Frank Bünning—*University of Magdeburg, Germany*
Margarita Pavlova—*UNESCO-UNEVOC Center, Hongkong*
Maizam Alias—*Universiti Tun Hussein Onn, Malaysia*
Takahashi Mitsuru—*Tohoku University, Japan*
Shahbaz Khan—*Director and Representative of UNESCO Indonesia, Indonesia*
Gumpanat Boriboon—*Srinakharinwirot University, Bangkok, Thailand*

Ideas for 21st Century Education – Abdullah et al. (Eds)
© *2017 Taylor & Francis Group, London, ISBN 978-1-138-05343-4*

Organizing committees

ADVISORS

Prof. Furqon
Prof. Asep Kadarohman
Dr. Edi Suryadi
Prof. Aim Abdulkarim
Prof. Didi Sukyadi
Dr. M. Solehuddin
Prof. Takuya Baba
Prof. Christine C.M. Goh
Prof. Allan L. White
Dr. Tuğba Öztürk
Prof. Tom Nelson Laird
Prof. Simon Clarke
Dr. Diana Baranovich
Prof. Taehee Kim
Prof. Ikuro Yamamoto
Assoc. Prof. Numyoot Songthanapitak
Prof. Frank Bünning
Dr. Margarita Pavlova
Prof. Maizam Alias
Prof. Takahashi Mitsuru
Prof. Dr. Shahbaz Khan
Gumpanat Boriboon, Ph.D

CONFERENCE CHAIR

Prof. Anna Permanasari

COMMITTEE

Dr. Ida Hamidah
Dr. Ade Gafar Abdullah
Vina Adriany, Ph.D
Dr. Siti Nurbayani
Dr. Ana
Dr. Vanessa Gaffar
Dr. Dian Budiana
Dr. Siti Aisyah
Didin Wahyudin, Ph.D
Ari Arifin Danuwijaya, M.A.

Adult Education (ADE)

Ideas for 21st Century Education – Abdullah et al. (Eds)
© 2017 Taylor & Francis Group, London, ISBN 978-1-138-05343-4

Practicing critical thinking through extensive reading activities

N. Husna
Universitas Islam Negeri Syarif Hidayatullah, Jakarta, Indonesia

ABSTRACT: To become creative, people must evolve critical thinking skills. The skills in Critical Thinking (CT) are not developed unplanned. Those skills can be taught and practiced as part of an Extensive Reading (ER) program. The present study was to find out whether the activities in ER influence L2 learners' CT skills as well as their other language skills. The program was implemented at the English Education Department, Faculty of Tarbiyah and Teachers Training, UIN Jakarta. The results show that the majority (more than 60%) of the students stated that not only were their language skills improved, but their CT was also developed. They also said that this program not only helped them in doing other subjects, but also influenced their way of thinking or seeing things.

1 INTRODUCTION

Since information and innovation are the key factors in the global world, intensive knowledge is highly needed (Carnoy, 2005). Therefore, ways of transforming the knowledge will play a crucial role in adjusting to new technology. Higher education is one factor which undergoes and experiences pressures with those changes.

It has been common knowledge that education is one of the main factors which determines the development of a country. Education plays a vital role in preparing the human resources (Carnoy, 2005). Moreover, it can also characterize a nation and a civilization.

Like other countries, Indonesia has been very concerned in preparing its human resources to be able to compete with those from other countries. One of the first steps to create high quality human resources is to improve the quality of education.

In the report from the Ministry of National Education (Kurniawan, 2003), Indonesia in terms of the quality of education, is in twelfth place in Asia after Vietnam. It is still debatable whether the low quality of human resources is related to the quality of education given at the higher education level or to the whole system of education. The fact remains that many of Indonesia's future employees lack creativity, and as a result, fail to get the job opportunities.

To become creative, people must evolve Critical Thinking (CT) skills. In a sense, CT is a way in which people put all efforts into thinking deeply and quietly in an even-handed way before they make up their minds (Ennis, 2001). To educate the students to become critical thinkers is very crucial, both for themselves and the continuity of the society (Facione, 1990).

To bring up the students to become critical thinkers means helping them change into people with natural curiosity who are knowledgeable, trustful, open- and fair-minded, adaptable, honest, careful, and sensible in making decisions (Facione, 1990). They also have good organization in complex matters, are diligent in searching for related information and persistent in trying to find out the legitimate solutions. However, although it is definitely not easy to reach those ideals of critical thinkers in a very limited time and condition, it does not mean it is totally impossible. Those characteristics mostly require cognitive work. In fact, the specialists portray certain cognitive skills as *fundamental* or the *essence* of CT skills (Facione, 1990).

The skills in CT are not developed unplanned. They need practice and experience. Those skills can be taught and practiced from the very beginning level of study and in various subjects. Some simple and practical reading activities in encouraging the skills in CT can be applied in Indonesian higher education classrooms.

Some core skills of CT like interpretation, analysis, evaluation, inference, explanation, and self-regulation, can be conducted in simplicity (Facione, 2007) in Reading subjects. Of the various subjects that are taught in Indonesian formal education, reading is the crucial one, because, based on the 2004 Curriculum, skills in reading are to be emphasized.

Critical reading practices will be one of the ways that can be used by Indonesia with its limited influence to empower students' CT ability. Through both intensive and extensive practice of reading critically, they will learn how to stimulate their thoughts to see various aspects of the discourse.

Critical reading is assumed to be one of the effective ways to improve students' CT ability in this study.

Based on the need to improve students' CT skills through critical reading while the time is very limited, Extensive Reading (ER) (as one of the Reading subjects in the English Education Department (EED) in the Faculty of Educational Sciences, UIN Jakarta) was chosen as the shelter subject to practice it.

Therefore, the objective of this study is to find out whether the implementation of ER activities can improve students' CT skills. Students' opinions about ER and CT, both before and after the class, will be part of the data to support the findings.

2 LITERATURE REVIEW

2.1 General conception of CT

The teacher should drive the learners to have a higher order of thinking by creating their classroom to become a place for the thinker community (Davidson & Dunham, 1996).

The skills in CT are merely just an outcome. CT skills should also be observed during the process (Garrison et al., 2001). The judging will come from an individual perspective, where the teacher's responsibility is to see that the students acquire a deep and meaningful understanding as well as content-specific critical inquiry abilities, skills, and dispositions. Even though the CT skills as a product are difficult to assess, because as a cognitive process CT is very complex and can only be accessed indirectly, the skills still can be acknowledged through individual assessments.

To be recognized as a person who possesses CT ability does not mean that the individual must be skillful in every aspect of CT (Facione, 1990). Even though it is ambiguous to assess, still, five upper-levels of Bloom's Taxonomy (that is, analysis, synthesis, evaluation, comprehension, and application) were offered as guidelines in CT practices (Ennis, 1993). Ennis (1993) also stated that experts still try to get the best way to analyze the assignment with CT elements. Elaboration is suggested as a way to assess CT existing in students' assessment, to see the comprehensiveness of the CT assignment. According to Ennis (1993), the failure in assessing CT is usually because the assessor or teacher is less open-minded and fails to judge the credibility of the sources used in the assignment.

One of the very first definitions about CT is what was called Reflective Thinking (RT) by Dewey (1933) who stated that RT is 'Active, persistent, and careful consideration of a belief or supposed form of knowledge in the light of the grounds which support it and the further conclu-

sions to which it tends' (as cited in Rodgers, 2002). However, both CT and RT have their own focuses. CT is a kind of thinking that is purposeful, reasoned and goal-directed. Meanwhile RT is referring specifically to the processes of analyzing and making judgments about what has happened.

It can be concluded that CT involves a wide range of thinking skills leading towards interesting products while RT focuses more on the process of making and analyzing judgments about what has happened.

Borrowing ideas from Dewey, another definition on CT was given by Glaser (1941) as cited in (Fisher, 2011):

- an attitude of being disposed to consider in a thoughtful way the problems and subjects that come within the range of one's experience;
- knowledge of the methods of logical enquiry and reasoning;
- some skill in applying those methods. CT calls for a persistent effort to examine any belief or supposed form of knowledge in the light of the evidence that supports it and the further conclusions to which it tends.

It is clearly seen that Glaser's (1941) definition on CT was much influenced by RT from Rodgers (2002). The effort to scrutinize the topic before delivering the opinion was emphasized as part of CT skills.

The CT skills that will be emphasized in this present study are analyzing, inference, elaborating, and reasoning. All those skills should be recognized within their assignments, including when asking questions during the presentations.

2.2 Understanding extensive reading

The initial characteristic of ER is, as the name implies, the large amount of reading compared with the amount that readers would read in different types of reading programs. However, ER is not just a matter of submerging students in a bath of print (Bamford & Day, 1998): it takes superior materials, clever teachers who love to read themselves, time, and effort to develop the reading habit (Harris & Sipay, 1990).

ER is an approach to the teaching and learning of second-language reading in which learners read large quantities of books and other materials that are well within their linguistic competence (Bamford & Day, 1998). This program will help L2 learners to acquire their L2 language because they try to understand the material they read. This was in line with what has been stated by Krashen (1982) about comprehensible input. While learning to understand the text, the program of ER is expected to help the L2 learners develop good reading habits while building up their knowledge of vocabulary and structure (Richards & Schmidt, 2013).

With the various practices of ER, this program has developed several approaches in its implementation. This decreases the use of silent reading class as an additional assignment. However, Bamford and Day have provided ten guidelines to put ER into practice:

- The reading material is easy.
- A variety of reading material on a wide range of topics must be available.
- Learners choose what they want to read.
- Learners read as much as possible.
- The purpose of reading is usually related to pleasure, information and general understanding.
- Reading is its own reward.
- Reading speed is usually faster rather than slower.
- Reading is individual and silent.
- Teachers orient and guide their students.
- The teacher is a role model of a reader.

In addition, Richard (1998) presents the objectives of giving ER to L2 learners:

- to improve positive attitude towards the language being learned;
- to enhance confidence in reading;
- to build high motivation in reading;
- to develop the ability to avoid looking up the dictionary too many times;
- to develop word recognition ability;
- to build the ability to read based on need and purpose.

This study will apply what is known as Integrated ER, where other skills such as the four language skills and CT are embedded in the program. The study that ER can be integrated with other programs has been widely researched, such as with writing (Stevens et al., 1987), reading (Sheu, 2004), with software implementation such as the Moodle course management system (Robb & Kano, 2013), and language skills (Lituanas et al., 1999). Many also used ER to improve the English or literacy programs (Macalister, 2008; Judge, 2011; Yu, 1993; Wanzek & Vaughn, 2007).

3 RESEARCH METHODS

This study was a qualitative research. Most data was taken based on observation and written interview. The observation was started for 120 students from the 2007 academic year who were in the 5th semester in 2009 and finished in 2014 or for students from the 2012 academic year. The first data collected was the reading speed of each student,

which was taken three times in each semester. The need to get a particular speed was to ensure they would not meet too many difficulties in doing the ER subject next semester.

ER is the last part of the Reading course in EED's curriculum. It is given to the 5th semester students who have passed their Reading 4. The ER subject was started in 2009 with two credits, or a 100-minute session each week for each class. It was started with the obligation to read eight novels and seven textbooks. In 2014, the number of the materials increased. The students had to read ten English novels which consisted of a minimum of 300 pages, ten articles in English from journals, proceedings or anthology, and eight English text books in the form of references, instruction, workbook, and biography.

The ER subject in EED is an Integrated ER, where other skills are embedded to get the maximum benefit of ER. All language skills are used as part of ER activities, and within those skills CT plays an important part as the acknowledgement of their ER projects.

In this subject, the students chose any type or genre from the listed novels in the EED Reading Corner library. One title of the novels could be read by a maximum of two students, who had to be from different classes. After reading, they were given a test as shown in Figure 1.

To get the maximum points, they had to be able to explore their answer by giving evidence, analyzing the story, and correlating it with their own thought or opinion. Meanwhile, for textbooks reading, the requirement was in different projects of summary based on book types. They were required to write bibliography for each article.

To support their speaking and listening skills, two of the materials were presented in front of the

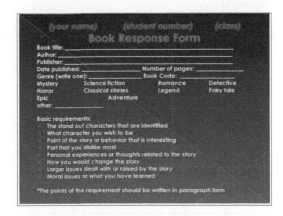

Figure 1. Book response form for novel review.

classroom, where the audience were obliged to ask critical questions, for which they received a score.

Additionally, the participants were also given questionnaires. The questions were given before and after (in the last meeting) the program started and only given to students from the 2011 and 2012 academic years. This means, they had their ER subject in the 2013 and 2014 academic years. In those academic years, there were three classes each year, consisting of 40 students. The questions were given to 20 students from each class, totaling 60 students, for each academic year, and 120 for two academic years.

4 RESULTS AND DISCUSSION

Time is needed to see whether ER has benefited students' CT skills within 2–5 years. The length of time needed in conducting this research was due to the necessities of understanding the effects of the whole program (Grabe & Stoller, 2001).

The first data was the students' reading speed that was recorded since they were in the 1st semester. In each semester their reading speed was tested using a speed-reading test formula three times, except in the 5th semester when they were tested once only at the end of the semester. The results are shown in Table 1.

From the Table it can be seen that all students' reading speed improved in their 12th rapid-reading test in semester and kept improving in their 5th semester. Of course, reading cannot take the ultimate honor as the one that influences that improvement. Bias can happen since the students also received skills input from other subjects.

Meanwhile, the responses from 120 students of the open-ended question on CR and ER are as follows:

Before the class

Question 1. What do you know about ER class? (The answers have been grouped based on the similarity.)

Table 1. Students' average reading rate.

Average reading rate	2009	2010	2011	2012	2013	2014
1st*		65	64	100	90	112	100
12th*		280	300	330	300	345	360
After ER class		300	300	340	320	350	360
Number of Novels		8	10	10	10	10	10
Number of Text-books		7	8	8	8	8	8
Number of Articles		0	0	8	8	10	10

*WPM (words per-minute).

- Terrifying class (38)
- Time consuming (29)
- Strict rules (17)
- Afraid (10)
- Difficult process (10)
- Curious (8)
- High pressure (8)

Question 2. You have heard about CT briefly, do you think you have those skills in CT?

- Yes (24)
- No (57)
- Not sure (16)
- Do not know at all (23)

After the class

Question 1. What do you know about ER class? (The answers have been grouped based on the similarity.)

- Good effect on language skills and components (63)
- Challenging (24)
- Time management (14)
- Time consuming (12)
- High pressure (7)

Question 2. You have practiced CT skills in ER, what do you learn?

- Thinking management (35)
- Think before act or say (33)
- Improve curiosity (19)
- Learn to be objective (14)
- Learn to elaborate (11)
- Being sensitive (8)

The comparison of the responses between before and after classes has shown that the students had mixed feelings, considered being negative about ER before the class, then changed by the end of the semester. The reason for their negative thoughts on ER was because they were informed by their seniors about ER and it seems that the seniors' explanation frightened them.

Before the class started, CT skills were explained, embedded in the explanation of the requirements. Many of them did not think that they had CT skills ability; in fact, most of them did not know about CT skills. From the observation, many of them looked worried when it was explained that the CT skills were to become part of the assessment in their ER assignment. The questions about how to do those skills emerged. Many of them worried that inability to perform CT skills well would affect their grades.

However, by the end of the semester, their answers had changed. The majority believed that ER had improved their language skills as well as their language components' mastery. All activities in ER had forced them to use all those skills consecutively. They felt more confident to read in English and they said that they had no problem in reading journal articles which all this time had become a burden for them in their study.

The activities in ER had also forced them to be more creative, including in managing time, due to the workload while their time is limited and they still need to do other assignments from other subjects. 20% of the participants said that they were challenged by the reading and workload in ER in a positive way. They felt very motivated and were urged to conquer those challenging tasks.

However, 15.8% of the participants still felt under pressure with ER activities. They admitted that they had difficulties in managing the time. It was difficult for them to fulfill the requirement in the assignment. The workload, limited time, and limited materials, made them feel so pressured with ER.

Nevertheless, with regard to CT skills, no negative effects had been detected from their answers. 29% of participants said that they got used to managing the way they thought after ER class. They knew which one need to think deeper and which one that they can think later. They also said that they learned to be more careful in giving their opinions. Their curiosity also improved because they always had to give logical reasons for their opinions.

5 CONCLUSIONS

From the findings, it can be said that ER has improved students' language skills (slightly more than 50%), making them be challenged by the assignments as well as getting them used to managing their time to meet the demands of the study. Moreover, they were satisfied with their new understanding about their ability in CT. They also mentioned what kind of CT ability they thought they are able to perform.

Even though some negative feedbacks still persist, it can still be said that the program was successful in developing students' ability in CT and therefore, to achieve better understanding on whether the findings can be applied in a general sense, the continuation of the program is strongly suggested.

However, this study still needs other studies to support the findings. For further research, the possibility to use an English standardized test such as Test of English as a Foreign Language (TOEFL) or International English Language Testing System (IELTS) will empower the findings of language skills improvement. Questionnaires and other types of assessments with specific rubrics can

also be used to measure the improvement of the students' CT skills.

REFERENCES

Bamford, J. & Day, R.R. (2004). *Extensive reading activities for language teaching.* Cambridge: Cambridge University Press.

Bamford, J. & Day, R.R. (1998). Teaching reading. *Annual Review of Applied Linguistics, 18,* 124–141.

Carnoy, M. (2005). *Globalization, educational trends and the open society.* Budapest, Hungary: Open Society Institute Education.

Davidson, B.W. & Dunham, R.L. (1996). *Assessing EFL student progress in critical thinking with the Ennis-Weir Critical Thinking essay test.* Nagoya, Japan: Association for Language Teaching.

Ennis, R. (2001). An outline of goals for a critical thinking curriculum and its assessment. In A. Costa (Ed.), *Developing minds: A resource book for teaching thinking.* Alexandria, VA: Association for Supervision and Curriculum Development.

Ennis, R.H. (1993). Critical thinking assessment. *Theory into practice, 32*(3), 179–186.

Facione, P.A. (1990). *Critical thinking: A statement of expert consensus for purposes of educational assessment and instruction. Research findings and recommendations.* Fullerton: American Philosophical Association.

Facione, P.A. (2007). Critical thinking: What it is and why it counts. *Insight Assessment, 200,* 1–23.

Fisher, A. (2011). *Critical thinking: An introduction.* Cambridge: Cambridge University Press.

Garrison, D.R., Anderson, T. & Archer, W. (2001). Critical thinking, cognitive presence, and computer conferencing in distance education. *American Journal of Distance Education, 15*(1), 7–23.

Grabe, W. & Stoller, F. (2001). *Teaching and researching reading.* Harlow, England: Pearson Education.

Harris, A.J. & Sipay, S.R. (1990). *How to improve reading ability: A guide to developmental remedial methods.* White Plains, NY: Longman.

Judge, P.B. (2011). Driven to read: Enthusiastic readers in a Japanese high school's extensive reading program. *Reading in a Foreign Language, 23*(2), 161–186.

Kurniawan, K. (2003). *Transformasi perguruan tinggi menuju Indonesia baru (University transformation toward new Indonesia).* [Online] Available at: http://www.depdiknas.go.id/jurnal/41/Khaerudin.htm.

Krashen, S. (1982). *Principles and practice in second language acquisition.* London, UK: Pergamon Press Inc.

Lituanas, P.M., Jacobs, G.M. & Renandya, W.A. (1999). A study of extensive reading with remedial reading students. *Language Instructional Issues in Asian Classrooms,* 89–104.

Macalister, J. (2008). Integrating extensive reading into an English for academic purposes program. *The Reading Matrix, 8*(1), 23–34.

Richards, J.C. & Schmidt, R.W. (2013). *Longman dictionary of language teaching and applied linguistics.* London: Routledge.

Robb, T. & Kano, M. (2013). Effective extensive reading outside the classroom: A large scale experiment. *Reading in a Foreign Language, 25*(2), 234–247.

Rodgers, C. (2002). Defining reflection: Another look at John Dewey and reflective thinking. *Teachers college record, 104*(4), 842–866.

Sheu, S.P. (2004). The effects of extensive reading on learners' reading ability development. *Journal of National Taipei Teachers College, 17*(2), 213–228.

Stevens, R.J., Madden, N.A., Slavin, R.E. & Farnish, A.M. (1987). Cooperative integrated reading and composition: Two field experiments. *Reading Research Quarterly*, 433–454.

Wanzek, J. & Vaughn, S. (2007). Research-based implications from extensive early reading interventions. *School Psychology Review, 36*(4), 541–561.

Yu, V. (1993). Extensive reading programs – How can they best benefit the teaching and learning of English? *TESL Reporter, 26*(1), 1–9.

Ideas for 21st Century Education – Abdullah et al. (Eds)
© *2017 Taylor & Francis Group, London, ISBN 978-1-138-05343-4*

Teaching–learning sequence: Designing ionic bonding concept through model of educational reconstruction

E. Nursa'adah, L. Liliasari & A. Mudzakir
Universitas Pendidikan Indonesia, Bandung, Indonesia

ABSTRACT: Ionic bonding is a topic of inorganic chemistry that has many contextual applications. Ionic bonding and structures are highly abstract. Therefore, students must make extra effort to comprehend the concept to avoid misconceptions. Furthermore, the importance of teaching content knowledge to students is intended to make them able to solve their contextual problems. A teaching-learning sequence for ionic bonding has been designed, adopted from the Model of Educational Learning (MER). Reconstruction focuses on the ionic bonding content knowledge, in order to help learners' competences and make connections between scientists' and students' conceptions. The aim of this study is to describe a teaching-learning sequence ionic bonding context-based MER. Reconstruction began by analyzing scientists' conception to produce a concept map that describes ionic bonding comprehensively. Next, the analysis learner's conception uses interviews and also a concept map. There are three types of learner's conception that occur—proper for chemists, misconception, and incompetence. Based on these criteria, a teaching-learning sequence produced.

1 INTRODUCTION

1.1 *Ionic bonding and characteristics*

Inorganic chemistry focuses on all elements of the periodic table and their compounds except hydrocarbon and its derivatives (Lee, 2014). A study of inorganic chemistry is more concerned with applying the concept of analyzing structures, properties and reactions that are used to solve problems relating to the structure and properties of inorganic compounds. Understanding structures v. properties leads us to know the advantages of compounds in our life, and to have a basic model for students in studying inorganic chemistry.

Ionic Bonding is a topic of inorganic chemistry which has many contextual applications. The type of this topic has an abstract critical attribute with real daily life contexts. In addition, learning about chemical bonding allows the learner to make predictions and give explanations about the physical and chemical properties of substances (Nahum et al., 2010; Uce, 2015; Daniel et al., 2010).

Chemical bonding is one of key concepts in chemistry and the most fundamental one. Ionic bonding and structures are highly abstract; we must make extra effort to comprehend the concept to avoid misconceptions (Gudyanga & Madambi, 2014). There are many misconceptions about ionic bonding. They only focus on ionic bonding as the electron transfer, they do not understand the difference between ionic bonding and ionic bond,

they also assume that covalent bonds are weaker than ionic bonding, they suppose that NaCl is a molecule and also do not understand the three-dimensional nature of ionic bonding in solid chloride (Gudyanga & Madami 2014; Nahum et al., 2010; Daniel et al., 2010; Tan & Treagust, 1999; Barke & Yitbarek, 2009).

The ionic bonding concept must be reconstructed into various compounds of ionic modeling when it will be taught to the students. This process will help students comprehend the content knowledge to solve their contextual problem in daily life.

1.2 *Model of educational reconstruction*

The Model of Educational Reconstruction (MER) is the German didactic tradition. It has been developed as a theoretical framework for studies, whether or not a concept area is possible to teach (Duit et al., 2012). Educational reconstruction is a design to make science contents that are simple and accessible to students (Viiri & Savinainen, 2008). However, it also provides significant guidance for planning science instruction at every level of education.

There are several research studies about MER in designing learning chemistry concepts and also designing laboratory research in the course-based context. The results show that MER is successful in improving student conception, contextual knowledge, and also intelligent thinking (Sam et al., 2015; Doloksaribu et al., 2015; Reinfrieda et al., 2015).

Based on background, the aim of this research is in designing a teaching-learning sequence of ionic bonding concept through MER. As a result, a research question in this research is: what teaching-learning sequence would be employed to improve the teaching of the ionic bonding concept?

2 RESEARCH METHODS

2.1 *Research design*

This study used a descriptive method. The writer intended to describe the teaching-learning sequence of the ionic bonding concept using MER. The study used three relevant phases of MER: (1) clarification of science conceptions, (2) investigation in students' conception, (3) development of teaching-learning sequence as described in Figure 1.

Figure 1 shows the design for a teaching-learning sequence of ionic bonding, based on scientists' and students' beliefs. This reconstruction makes scientific content simple and accessible to students (Virii & Savinainen, 2008) and also helps students to comprehend the content knowledge required to solve a contextual problem. In order to make a balance between scientists' and students' conceptions, Chemie im Kontext (Chemistry in Context) (Parchmann et al., 2001) interventions were adopted in the implementation of a teaching-learning sequence. Phases of Chemistry in Context are cyclical. Students can investigate content with context, or investigate context with content.

2.2 *Research subject*

Research subjects are: (1) library research about scientists' conception of ionic bonding, (2) 12 students of pre-service chemistry and teachers' conceptions of ionic bonding.

2.3 *Instruments*

The instruments of this study were: (1) list of propositions and a scientists' concept map, (2) students' conceptions described in a students' concept map, (3) students' interviews regarding their concept maps of ionic bonding.

The students' and scientists' conceptions were analyzed by qualitative content analysis (Mayring cited Sam et al., 2015) using the following steps: scanning the student statement/mapping concept, editing to improve readability, rearrangement of the statement, interpretation, and designing a teaching-learning sequence to make a balance between the students' and scientists' conceptions.

3 RESULTS AND DISCUSSION

3.1 *Scientists' conceptions*

The conceptions that scientists and students have on ionic bonding are presented from the analysis of a section of selected data. The scientists' ionic bonding conception focuses on a described ionic compound from an ionic bonding process, its structures, thermodynamic aspect, its properties, and the application of the ionic bonding concept. Knowing structures v. properties makes students able to predict and explain about the implementation of daily phenomena.

The areas of the scientists' ionic bonding conception are described in Figure 2.

Figure 2 explains that when a metal atom binds with a non-metal atom, electrons are transferred from the metal atom, turns cation to the non-metal atom and anion. It is caused by the metal atom having low ionization energies and its tendency to lose electrons. The electrostatic attraction of cations and anions results in ionic bonding. Most of the earth's solid ionic compounds consist of ions held together by ionic bonding. An example of the simplest structure

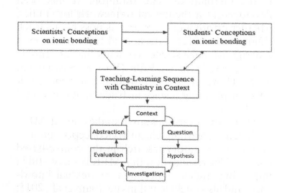

Figure 1. Research design adopted using MER.

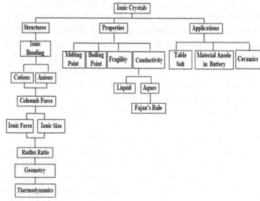

Figure 2. Scientists' conception of ionic bonding.

10

of ionic compound is the NaCl ionic compound; each Na^+ ion touches six Cl^- ions and each Cl^- in turn touches six Na^+ ions. This arrangement is assigned its own name of the rock salt structure. The strength of ionic bonding depends on these attractions and repulsions, and it is described by Coulomb's Law (Silberberg et al., 2009; Gilbert et al., 2009).

Lattice energies play a critical role in the formation of ionic compounds. They indicate the strength of ionic interactions and influence melting point, boiling point, hardness, solubility and other properties.

3.2 *Students' conception*

Students expressed their conception in various ways: orally, and/or figuratively. Therefore, students were expected to translate their conception about ionic bonding in a concept map form and also in interviews.

The students' conception is described in Figures 3a and 3b.

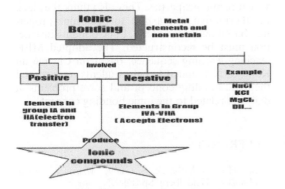

Figure 3a. Students' conception about ionic bonding.

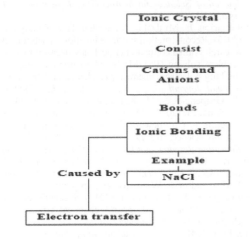

Figure 3b. Students' conception about ionic bonding.

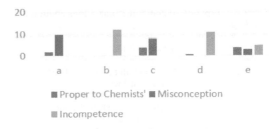

Figure 4. Quantity of students' conception about ionic bonding.
Bonding
a = ionic bonding process, b = thermodynamic aspect, c = conductivity, d = solubility, e = boiling point

Figure 3 shows that almost all students did not focus on the structure and properties of ionic compounds. The students' focus of ionic bonding is the electron transfer event. This result is in line with Nahum et al. (2010), who found that students knew well the covalent bond as electron sharing and the ionic bond as electron transfer. Furthermore, they also thought that sodium chloride exists as molecules and that these molecules were held together in the solid state by covalent bonds.

These misconceptions indicated that almost all ionic bonding learning did not emphasize on the correct conceptual language. For example, the ionic sodium chloride compound is represented as NaCl which is very similar to covalent hydrogen chloride, so the students might have the idea that one particle of sodium is bonded to one particle of chlorine, just as one atom of hydrogen is bonded to one atom of chlorine (Daniel et al., 2010). Figure 4 shows the quality of students' conception about ionic bonding.

3.3 *Teaching-learning sequence of ionic bonding*

There is no doubt that the ionic bonding concept is abstract, difficult, and most students have misconceptions about it (Unal et al., 2010). Table 1 describes the types of students' misconceptions about ionic bonding.

Table 1 describes that students' misconceptions in explaining the ionic bonding process. They suggest that a cation is a proton and that an anion is an electron. Students thought that metal is similar to metal atom, and also that ionic bonding always influences electron transfer. Taber cited Daniel et al. (2010) also gives as an example that when ionic material is formed through precipitation, ionic bonds can form, even no electron transfer is involved.

In this study, most of the students lacked an understanding about the structure, and thermodynamic aspects and properties, of ionic

Table 1. Types of students' misconceptions about ionic bonding.

Concept	Misconception
Ionic Bonding	Ionic bonding consists of metal and non-metal. There is electron transfer. There is electrostatic between proton and electron.
Thermodynamic Aspects	No answer.
Conductivity	Ionic compound, especially NaCl, can conduct electricity in solid state.
Solubility	All ionic compounds dissolve in water.
Boiling Point	Ionic bonding is stronger than covalent bonding, because ionic compounds have a greater boiling point than covalent molecules.

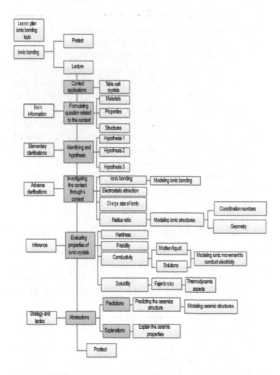

Figure 5. Teaching-learning sequence of ionic bonding.

compounds that they can find in their daily life. Uce (2015) suggests constructing the model or teaching-with-analogies model (Pabuccu & Geban, 2012) for teaching chemical bonding, to increase the students' understanding of the chemical bonding concept.

According to MER, teaching and learning sequence of ionic bonding is designed based on modeling and critical thinking skills (Duit et al., 2012). *Chemie im Kontext* (Parchmann et al., 2001) was adopted to find a balance between the students' and scientists' conceptions.

Figure 5 shows the teaching-learning sequence for the ionic bonding concept.

Figure 5 describes the phases of ionic instruction. Every phase of learning is adopted from the Chemie im Kontext. In every phase a critical thinking skills indicator was developed. The modeling skills of students were developed to help the students comprehend ionic bonding, structures, properties, and its applications.

4 CONCLUSIONS

There are three categories of students' conception about the ionic bonding concept proper for chemists, misconception, and incompetence. Most students have misconceptions about ionic bonding structure and properties. They also think they lack the thermodynamic aspect to determine properties. Based on these criteria, ionic bonding instruction must be reconstructed. The adopted MER teaching-learning sequence, based on Chemie im Kontext, is designed to help students comprehend the ionic bonding concept and solve problems in daily life related to the ionic bonding concept.

REFERENCES

Barke, H.D. & Yitbarek, S. (2009). *Misconceptions in chemistry*. Heidelberg: Springer.

Daniel, K.C., Khang, T.N., Sai, G.L., Kwen, H. & Boo. (2010). Alternative conception of chemical bonding. *Journal of Science and Mathematics Education in SE Asia*, XXIV(2).

Doloksaribu, F., Mudzakir, A., Solihin, H. & Sudargo, F. (2015). Reconstruction model education of laboratory research course context chemical clay decision making problem solving (PSDM) based to improve research thinking skills from chemistry teacher candidates. *International Journal of Science and Research*, 4(4). Duit, R., Gropengiesser, H., Kattman, U., Komorek, M. & Parchman, I. (2012). The model of educational reconstruction – a framework for improving teaching and learning science. In J. Doris & D. Justin (Eds.), *Science Education Research and Practice in Europe*: (pp. 13–47).

Gilbert, T.R., Kirss, R.V, Foster, N. & Davies, G. (2009). *Chemistry: The science in context* (2nd ed.). New York-London: W.W Norton & Company.

Gudyanga, E. & Madambi, T. (2014). Student's misconceptions about bonding and chemical structure in chemistry. *The Dyke Journal*, 8(1).

Lee, J.P. (2014). *Engaging students in the inorganic chemistry classroom with well-defined group activities and literature discussions*. Washington DC: American Chemical Society.

Nahum, T.L., Mamlok-Naaman, R., Hofstein, A. & Taber, K.S. (2010). Teaching and learning the concept of chemical bonding. *Journal of Chemical Education*, 179–207.

Pabuccu, A. & Geban, O. (2012). Students' conceptual level of understanding on chemical bonding. *International Journal of Educational Science 4*(3), 563–580.

Parchmann, I., Demuth, R., Ralle, B.E.R.N.D., Paschmann, A., & Huntemann, H.E.I.K.E. (2001). Chemie im Kontext-Begrundung und Realisierung eines Lernens in sinnstiftenden Kontexten. *Praxis der Naturwissenschaften Chemie in der Schule,* 50(1), 2–7.

Reinfrieda, S., Aeschbacherb, U., Kienzlerc, P.M., & Tempelmannd, S. (2015). The model of educational reconstruction – a powerful strategy to teach for conceptual development in physical geography: the case of water springs. *International Research in Geographical and Environmental Education*, *24*(3), 237–257.

Sam, A., Niebert, K., Harson, R. & Twumasi, A.K. (2015). The model of educational reconstruction: scientists and students conceptual balance to improve teaching of coordination chemistry in higher education. *International Journal of Academic Research and Reflection*, *3*(7), 67–77.

Silberberg, M.S., Durn, R., Haas, C.G. & Norman, A.D. (2009). *Chemistry: The molecular nature of matter and change*. Mc Graw Hill Higher Education.

Tan, D.K.C. & Treagust, D.F. (1999). Evaluating students' understanding of chemical bonding. *School Science Review*, *81*(294).

Uce, M. (2015). Constructing models in teaching of chemical bonding: ionic bond, covalent bond, double and triple bond, hydrogen bond on molecular geometry. *Educational Research and Review*, *10*(4), 491–500.

Unal, S., Costu, B. & Aya, A. (2010). Secondary school students' misconceptions of covalent bonding. *Journal of Turkish Science Education*, *7*(2).

Viiri, J. & Savinainen, A. (2008). Teaching-learning sequences: A comparison of learning demand analysis and educational reconstruction. *Lat. Am. J. Phys. Educ.*, *2*(2).

Art Education (ADE)

Ideas for 21st Century Education – Abdullah et al. (Eds)
© *2017 Taylor & Francis Group, London, ISBN 978-1-138-05343-4*

Design-based research to explore Luk Keroncong as vocal technique exercise

R. Milyartini
Universitas Pendidikan Indonesia, Bandung, Indonesia

ABSTRACT: The art of singing Keroncong is commonly taught informally, directly by a Keroncong maestro. This makes the development of Keroncong as a vocal art highly dependent on the availability of Keroncong maestros. Vocal exercise books are rarely found in the library. It creates a limited access for young generations to learn Keroncong. This research aimed to produce a learning material by exploring the benefits of Luk Keroncong. Using a design-based research method, we analyzed the characteristics of Luk Keroncong, constructed vocal exercises and implemented them on a vocal class. Research participants were music education students from Indonesia University of Education. Responses and comments from students were collected, and used to refine the vocal learning materials. It was indicated that luk is somewhat similar to portamento. Vocal technique exercises using Luk Keroncong were able to facilitate students to master breath support, *passagio* and head voice.

1 INTRODUCTION

Keroncong is one type of Indonesian traditional music that is less developed although some songs—one of which is Bengawan Solo by Gesang—are known worldwide. How to sing Keroncong is commonly learned informally via a maestro and is limited to their community. Almost no research or books describe how to sing Keroncong, or explore Keroncong singing techniques. That is probably why Keroncong is not well developed.

The artistic point of singing Keroncong lies in how to use ornamentation. Ornaments in Keroncong have a historical linkage with Coração in the tradition of fado singing in Portuguese. Coração, according to Ganap (2006), is an expression that comes from the heart of the singers. According to Fiksianina (2014), there are three types of ornaments in Keroncong, including *luk*, *gregel* and *cengkok*.

Using ornaments as a basic material to develop vocal technique is commonly found in Western tradition (Vacai, N., 1923; Panofka, H., 1970; Spicker, M. & Lutgen, 1987; Austin, 2013). These references motivate us to explore luk as an exercise for building vocal technique. We assumed that luk can be used to develop vocal technique.

The availability of vocal learning materials (etude) based on Luk Keroncong gives opportunity among the young to learn the artistic values of Keroncong, and its benefits for the development of vocal technique. Three research questions are delivered to produce vocal technique etude based on luk Keroncong. First, what are the characteristics of luk Keroncong, second, what is the characteristic of vocal exercise (etude) based on luk Keroncong, and third, what is the impact of Luk Keroncong etude on the students. This research objective is to contribute to the sustainability and development of the Keroncong singing tradition.

2 LITERATURE REVIEW

Research from Ayunda et al. (2013) and Fiksianina (2014) have different perspectives about luk Keroncong. According to Ayunda et al. (2013), Luk is similar to portamento, but according to Fiksianina (2014) luk in Keroncong is similar to *appoggiatura.*

Portamento is a vocal technique used to tie two distant notes by gliding between them lightly and smoothly (Stark, 1999; Elliott, 2007). There are two ways of expressing the voice: by anticipating and postponing. Pilotti (2009) explains that portamento, specifically in the Bel Canto vocal technique, helps to achieve the position of voice from lowest to highest pitch. This happens because the portamento allows the cavity to remain open which makes similar sound, color, and fuse. This means that there is relationship between portamento and register.

Aldrich (2011. p. 14) explained that there are two perspectives about register, from vocal teacher and scientist. The vocal teacher tends to define register as a sensation of sound by changing circumstances in the body. Mastering the register means being able to sing notes from the lowest to highest smoothly, although the mechanism of sound production is changed. The scientist says that the register is the terminology associated with the function of the larynx while singing. They have investigated the action mechanism of the larynx and the muscles around the vocal cords. The thyroarytenoid and cricothyroid are two muscles that affect the process of setting the length, thickness and tension of the vocal cords. These muscles have an important role in generating the fundamental frequency. So portamento can be used to strengthen the mechanism of the vocal organ when producing *passagio* – the passage between two registers.

Appoggiatura is ornaments sung by leaning on another note, adding dissonance before resolving the tension to the intended note. It usually occurs in a strong beat (Chung-Ahn, 2015. p. 28). Luk is similar to *appoggiatura* as an ornament, because it is a dissonant note added before the intended notes; however, sound production is similar to portamento.

3 RESEARCH METHODS

Design-based research was used to produce vocal technique exercises based on luk Keroncong. 'Design based research addressed theoretical questions about the nature of learning in context, ... in real situations and the need to derive research findings from formative evaluations' (Collin in Latukefu, 2010. p. 24). 'In this genre of research, educational product development serves as a case of that which is being studied.' (McKenney & Visscher-Voerman, 2013. p. 3).

Research steps began with an analysis of the character of Luk Keroncong through document analysis such as preliminary research by Fiksianina (2014), Ganap (2006), Ayunda et al. (2013), and Darini (2012), as well as analysis of Keroncong discography from Sundari Sukoco, Waljinah, Gesang and Mus Mulyadi.

Based on this finding, a vocal learning etude was constructed. There were four numbers of exercises with accompaniment in Keroncong style. We used two exercise versions for one woman and two men.

Participants in this research were music education students from Indonesia University of Education (UPI). The responses and comments of the student were collected and were implemented in an individual vocal class. Three students as participants, aged 18 and 19 years old, joined this research. Observation in class was conducted to clarify the improvement of student competences in breathing technique, head voice, *passagio* and stability of tone production. A rubric was used to analyze the data (Table 1).

Using a qualitative approach, comments of students, document analysis and field notes were analyzed and utilized for reflection, and refinement of the etude.

4 RESULTS AND DISCUSSION

4.1 *Characteristics of luk as vocal techniques materials*

Luk is a vocal ornament, added to decorate the melody. Audio analysis and review of previous research explained that singing luk needs little pressure at the first note. Luk appears second major, or minor before the intended note in the original melody. This distinguishes luk from *appogiatura*. Luk can be sung move up or down, but move up is commonly used.

Luk move up trains thyroarytenoid muscles little by little. Austin (2013) said that thyroarytenoid will make the vocal cords become shorter, thus serving to raise the pitch. Conversely, when luk rides down, the cricothyroid functioned. Cricothyroid muscles move when the tone goes down.

A luk exercise that moves up and down trains the mechanism of the thyroarytenoid and cricothyroid muscles. Activity of these two muscles combined with breathing management, can help students improve their vocal technique.

Table 1. Rubric for observation.

Indicator	Score 30	Score 20	Score 10
Breathing technique	Using diaphragm and intercostal consistently	Sometime using diaphragm and intercostal while breathing	Not using diaphragm and intercostal while breathing
Head voice	All the high tones singing with ringing and relaxed	Most of the high tones singing with ringing and relaxed	Several high tones singing with ringing and relaxed
Passagio	All *the passagio* is sung smoothly	Most of *passagio* section is sang smoothly	*Passagio* section is not sung correctly
Stability of tone	All of the tone is sung on pitch	Most of the tone is sung on pitch	Most of the tone is sung unpitched

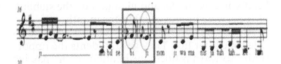

Figure 1. Luk move up and down (Fiksianina, 2014. p.82).

4.2 Characteristics of vocal exercises using luk Keroncong

Etude no. 1 was sung in andante, that is designed to help students improving their vocal technique (Figure 2). At the end of the phrase, there was a silence, to allow students to take a breath. Through this exercise, students learned voice placement and breath control. This exercise involved singing in several tonalities consecutively, so the students could learn the *passagio*.

Etude no. 2 was more difficult, especially in breathing (Figure 3). The exercise began with the same two-tone repetition and continued with luk towards the same tone. All the tone was moved in a second major up and down. This exercise was aimed at utilizing luk as a medium to maintain the stability of tone production. Tone stability can be achieved if singers use good breath support.

The length of melodic phrase was three and a half bars. Furthermore, the range of voice was wider than in the first exercise. The range of the first exercise was a sixth, while the second exercise was one octave (including ringing luk).

The third exercise was basically a series of notes in a major scale, but with variations in rhythm (Figure 4). When sung without luk, vocal chords actually became tense. The appearance of luk for a four count, tended to relax the vocal fold in a moment. Luk was expected to control the balancing between the mechanism of diaphragm and voice placement in supporting the long-sustained tone. The difficult point of this exercise was the ability to organize breath while supplying the energy to maintain tone.

Exercise no. 4 consisted of a sequence motif melodic line that moves up three times, followed by a retrograde motif with descending sequence for four times (Figure 5). There was a sound territorial expansion that started from the middle towards upper registers, and slowly downward to the lower registers. There was a melodic motion like climb up and down the stairs.

These exercises were equipped with audio samples in several tonalities. Each etude was repeated in three tonalities. There was also a specific exercise for each voice: soprano, mezzo, alto, tenor, baritone and bass.

Figure 2. Etude Luk Keroncong no. 1.

Figure 3. Etude Luk Keroncong No. 2.

Figure 4. Etude Luk Keroncong No. 3.

Figure 5. Etude Luk Keroncong No. 4.

Audio samples for each ornament were presented in two versions. The first version contained an exercise sung by models with musical accompaniment. The second version contained only musical accompaniment.

4.3 Implementation of the materials

Students involved in the implementation consisted of three people: two men (initials R and A), and one woman (identified as Y). Vocal characteristics of each student are described in Table 2.

'R' had a bass sound character and had not mastered diaphragmatic breathing properly. In the production of high pitch his sound was stiff and unpitched. 'A' was a student with a baritone voice and had a better vocal technique than 'R'. He had mastered the breathing technique. The sound quality was more rounded and stable. Occasionally he was still a bit tense in producing leap upward-moving tone from middle to upper registers, so the sound colors were not similar.

The third student 'Y' had problems in stability of the tone and ability to capture the tone. She had not been able to organize breath. Usually she was often out of breath before the melodic phrase of song ended. Her voice was clear, and had a good echo and volume.

Implementation took one month, and two times measurements on 16 and 30 March 2016. The results are tabulated in Figure 6 and Table 3.

Table 3 describes the implication of vocal exercise using Luk Keroncong to the student competence in singing. Breathing technique was the first aspect that gained improvement after half a month. The other technique including mastering

Table 2. Basic condition.

Student	Breathing technique	Head voice	*Passagio*	Stability of tone	Score
R	10	10	10	20	50
A	30	20	20	30	100
Y	20	20	20	20	80

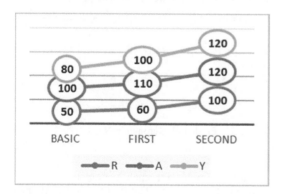

Figure 6.   Student improvement.

Table 3.   Improvement of each aspect.

St	Breathing technique			Head voice			Passagio			Stability of tone		
	bsc	1st	2nd	bsc	1st	2nd	bsc	1st	2nd	bsc	1st	2nd
R	10	20	30	10	10	20	10	10	20	20	20	30
A	30	30	30	20	30	30	20	20	30	30	30	30
Y	20	30	30	20	20	30	20	30	30	20	20	30

head voice, *passagio*, and maintaining the stability of tone, were improved after one month.

All the Luk Keroncong exercises used sustainable tone after luk. They gave students the opportunity to learn a fine coordination of the passive and active forces in breath control. As a result, students can master the breath support, with an indication that the stability of tone improves. Additionally, students also improved in mastering the *passagio* and head voice. These findings supported what Cleveland (2001) said about the interdependence of the vocal fold and breath management system.

Singing etude with Keroncong accompaniment motivated students to learn and pay attention to the sound quality. It also encouraged them learn to sing Keroncong. Their appreciation to Keroncong was expressed in interview: 'I like this lesson, now I know little bit about Keroncong and I can sing it' (A student interview March 30, 2015). 'Thanks mam, it's nice to be able to sing Keroncong' (R student interview on 2 April 2015). 'Actually singing Keroncong is not difficult mam' (Y student, interview on 30 April 2015). The students also gave positive response to the use of learning materials based on luk Keroncong.

5   CONCLUSIONS

Luk Keroncong exercises helped students manage the breath step by step, and at the same time also prepared voice transition between registers. Mastering of breath, as well as registers, gave an impact to good voice placement. As a result, sound production became clear and more relaxed.

Learning vocal technique using luk Keroncong, motivated students to sing Keroncong. It also provided the potential for Keroncong to sustain and develop. Appreciation and competence of the pre-service teacher in singing Keroncong gave an opportunity to transform Keroncong for the next generation.

REFERENCES

Aldrich, N. (2011). *Teaching registration in the mixed choral rehearsal: Physiological and acoustical consideration*. College Park: Faculty of the Graduate School of The University of Maryland.
Austin, F. (2013). Building strong voices – Twelve different ways. *Journal of Singing*, 69(3), 345–352.
Ayunda, P.R., Gustina, S. & Virgan, H. (2013). Gaya menyanyi musik keroncong tugu. *Antologi Departemen Pendidikan Musik UPI*, 1(3).
Chung-Ahn, G.C.Y. (2015). *An introduction to the art of singing Italian baroque opera*. Los Angeles: Escholarship University of California.

Cleveland, T.F. (2001). Constructing exercises that enhance the interdependence of the vocal folds and breath management in singing. *Journal of Singing*, *58*(1), 81–62.

Darini, R. (2012). Keroncong dulu dan kini. *Mozaik, 6*, 19–31. Retrieved Juni 23, 2015, from http://journal.uny.ac.id/index.php/mozaik/article/view/3875.

Edhi, K. (2009). *Tehnik nyanyian keroncong asli.* [Online] Available at: http://pushtaka.upsi.edu.my/web/guest/hom.

Elliott, M. (2007). *Singing in style: A guide to vocal performance practices.* London: Yalle Univerity Press.

Fiksianina, E. (2014). *Analisis ornamen keroncong sebagai sumber bahan ajar teknik vokal.* Bandung: UPI.

Ganap, V. (2006). Pengaruh Portugis pada musik keroncong. *Harmonia Jurnal Pengetahuan dan Pemikiran Seni,* VII(2).

Latukefu, L. (2010). *The constructed voice—A socio-cultural approach to teaching and learning singing.* University of Wolongong.

McKenney, S. & Visscher-Voerman, I. (2013). *Formal education of curriculum and instructional designer.* Educational Designer, 1–20.

Panofka, H. (1970). *24 vocalizzi.* Milano: Ricordi.

Pilotti, K. (2009). *The road to bell canto.* Sweden: Master's Thesis at the Academy of Music, Orebro University.

Spicker, M. & Lutgen, B. (1987). *Vocalises vol 1 for high voices.* New York: Schirmers Library of Musical Classic.

Stark, J. (1999). *Bel canto: A history of vocal pedagogy.* Toronto: University of Toronto Press.

Vaccai, N. (1923). *Practical Italian vocal method.* New York: G. Schirmer, Inc.

Business Education (BED)

Ideas for 21st Century Education – Abdullah et al. (Eds)
© *2017 Taylor & Francis Group, London, ISBN 978-1-138-05343-4*

The effect of psychological contract in improving university effectiveness

A.L. Kadiyono, R.A. Sulistiobudi & M. Batubara
Universitas Padjadjaran, Bandung, Indonesia

ABSTRACT: Change is undeniable. One of those who cannot run from the change is the university. As an educational institution under the Ministry of Indonesia, there are many changes associated with the system of governance rules, remuneration system, leadership system, as well as the performance evaluation system in the organization. This study was conducted to examine whether the Psychological Contract (PC) can be managed to increase the organizational effectiveness. The PC is an individual's belief regarding the terms and conditions of a reciprocal exchange agreement between that individual and another party. A PC emerges when one party believes that a promise of future return has been made, a contribution has been given, and thus an obligation has been created to provide future benefits. The study was conducted on 166 lecturers at the State University. The sampling technique is simple random sampling. Based on research, it is found that the PC mapping is a reciprocal relationship that occurs between lecturers and institutions. This gives the variation perception among lecturers towards the organization that affects their performance.

1 INTRODUCTION

Change is a fact that must be faced and managed properly for an organization to move forward and compete towards globalization. Change must be managed to make a development impact on the organization. Implementation of a change of course is a positive action, but it cannot be denied that there must be a member of the organization who gave negative responses or even rejection. Therefore, in the context of organizational change, it is important to map the perception perceived by members of the organization, because they have a relationship with the organization.

Relatively little is known about organizational development in universities and few models or frameworks for organizational development in universities exist. Since forces are present in universities, there are unique challenges associated with organizational development in universities that are not well understood. What is the impetus or catalyst for adopting organizational development-based change in a university? What strategies are associated with the development and expansion of organizational development-based change? Why do some organizational development initiatives succeed while others fail? To date these problems have not been addressed by research. The purpose of this article is to gain a better understanding of how planned change can be initiated in higher education using the models of the Psychological Contract (PC).

With the new status stated by the Ministry of Education and Culture to change its status, the State University has the authority to have more autonomy, including (1) academic freedom and autonomy of science, (2) greater autonomy in setting policy substance and management of research and dedication to the community, (3) governance and making decisions independently, (4) the optimization of physical assets, (5) independent and responsible financial management, (6) changes in work culture that fits the criteria of quality of work, and (7) the authority established as a business entity. It is implemented in the establishment of new systems such as the remuneration system, the leadership system, organizational structure changes, and the performance appraisal system.

Institutions have been cushioned from some of the external environmental pressures that businesses encounter, but are also subject to their own unique pressures, both external and internal. But the external environment can no longer be ignored because of several critical trends in higher education, including changes in their ultimate markets (the businesses and other institutions that hire their graduates) and changes in technology that have fostered a new way of thinking about the design and delivery of learning services. In addition to differences in external environments, colleges and universities are using a more consensual model of governance that involves faculties, administrators, trustees, and (sometimes) students, their parents, and the communities served by the university. Organizational changes must be made, considering the challenges, by initiating planned change in universities.

One of the most exposed to this change is a lecturer. Lecturers became the core resources and the determinants of educational achievement of organizational goals. The university make an effort so that lecturers can carry out their duties and responsibilities well, by changing their remuneration and administration obligation. Lecturers who can do many things for the organization, work harder and be loyal, would expect to get in return from the organization as a form of reciprocity as appreciation, respect, training, promotion and security. This relationship is referred as a psychological contract. The PC is the associated perception of mutual obligations between the employer and the employee (Rousseau, 1990). The PC sets lecturers to work in order to achieve the expected performance results. By managing what happens to the lecturers' PC, the university can manage to plan effectively and improve their effective performance in boosting university productivity.

There are previous studies that suggested that understanding an employee's perception of mutual obligations may be at least as important as creating a contractual relationship with one particular set of terms by showing how employer violations can affect employees' perception of their contracts with employers. This shows that there is an important relationship between manager's actions and employees' perception and fulfillment of their obligations (Robinson, Kraatz & Rousseau, 1994). Other previous research reveals, confirms and extends existing empirical evidence concerning the basis of employee reciprocity. But in Indonesia, there is still less research about the PC, even though it has a deep impact on organizational development. Indonesia is a collectivism society, so we have to see how to manage organizational development with social structure between the individual and organization, which have many forms and contracts.

This study will fulfill the gap, to investigate the PC in improving organizational productivity in state organizations in Indonesia. According to those phenomena, the purposes of the present study are: 1) the examination of the lecturer's psychological contract, and 2) the examination of the PC's dimensions, which are relational contracts and transactional contracts.

2 PSYCHOLOGICAL CONTRACT

PE research has been identified as a useful concept for understanding employees' relationships with their employers and the subsequent consequences including work attitudes and performance (Robinson, Kraatz & Rousseau, 1994; Shore & Tetrick, 1994; Turnley & Feldman, 2000). The PC is generally defined in the academic literature as the implicit and explicit promises two parties make to one another (Rousseau & Tijoriwala, 1998).

The origins of the PC construct dates back to the early 1960s. Argyris (1960) used the term psychological work contract to describe the mutual respect he observed between foremen and workers, and that he gathered from interview conversations. The foremen supported their employees' informal culture norms that they too had experienced before being promoted to their foremen positions.

In an organizational context, the norm of reciprocity within exchange relationships has been extensively used as a framework for understanding employee attitude and behavior. In particular, social exchange theory underlies much of the research in this area. As described by Blau (1964), social exchange entails unspecified obligations; where an individual does another a favor, there is an expectation of some future return. The future return is based on an individual trusting the other party to fairly discharge their obligations over the long run (Holmes, 1981). Falling within the domain of social exchange is the psychological contract defined by Rousseau (1989) – '...an individual's beliefs regarding the terms and conditions of a reciprocal exchange agreement between the focal person and another party. Key issues here include the belief that a promise has been made and a consideration offered in exchange for it, binding the parties to some set of reciprocal obligations'.

The norm of reciprocity represents the key explanatory mechanism that underlies psychological contract theory. Rousseau (1989) argues that in the exchange relationship, there is a belief 'that contributions will be reciprocated and that... the actions of one party are bound to those of another'.

The idea of reciprocation draws from the work of Blau (1964) who argues that the exchange partner will strive for balance in the relationship and if imbalance occurs, attempts will be made to restore the balance.

Morrison and Robinson (1997) divides the psychological contract into two types, transactional contracts and relational contracts. The essence of the transactional contract is the hope to build a relationship in terms of economic exchange, and therefore relationships are built not in the form of loyalty and in the long term (Suryanto, 2008). Relationships within the framework of the economy can be seen from the number of hours given to employees and wages paid by the company. A transactional contract refers to short-term tasks or single situation (Lee & Liu, 2009). Relational contracts involve loyalty and stability, and the employees with relational contracts have a greater desire to work, helping other employees in the work, and supporting the changes in the organization. Zagenczyk et al.

(2011) describe the relational and transactional dimensions in more detail. According to them, the relational dimension of the PC is specialized training, professional development, fair treatment and job security given by the organization in exchange for employee commitment and their desire to undertake tasks outside their job description. While at the same time, the dimensions of the PC captures transactional organizations providing adequate compensation, working conditions, job security and a reasonable short-term agreement in exchange for the fulfillment of contractual obligations of the employee. Restubog et al. (2008) offer a clear distinction between the types of relational and transactional PC: relational contracts are matters relating to socio-emotional needs, while transactional contracts represent the material interests of employees. Robinson et al. Millward and Hopkins (in Nelson et al., 2006) state that the transactional and relational aspects are interconnected in reverse; so that, '…the higher the relational orientation of employees, the lower the transactional orientation, and vice versa'. So, we can conclude that the transactional and relational dimension represents the material and socio-emotional needs of employees in compiling a phenomenon of PC.

3 RESEARCH METHODS

This research approach was a non-experimental research approach, descriptive and a verifiable research method. The verification method was done to test the hypothesis by using statistical tests (Wilcoxon Signed Rank Test). Then, we measured the discrepancies of obligation and fulfillment of the PC perceived by the lecturers as employees in ongoing changes at the university. In this study, data collection was conducted by using a questionnaire, consisting of three different sections. The first section included questions regarding individual profile (gender, age, tenure, status, position). The second was the main PC questionnaire. We also used qualitative supporting data through open-ended questions to broaden the discussion of the research.

3.1 Participants

To determine the selected sample, researchers used a non-probability sampling technique. The population of this research are the lecturers in the State University with a minimum length of work of around two years. The sampling technique used is simple random sampling. A total of 166 lecturers responded to the questionnaire. The background characteristic of the samples is described in demographic data. Altogether, 57.8% of participants

were male. Some 88.6% of participants were employed as government civil servants, and 38% were 31–40 years old. About 42.8% had a tenure of more than 15 years. This means that they actually felt the condition before the change in the university system and the time period of this change.

3.2 Instrument

The PC was measured with 15 items assessing employer obligations as a part of the employee's PC. The employees were asked to what extent the listed obligations were perceived in their present employment relationship. The Psychological Contract Scale (PCS) was developed by Coyle-Shapiro (2005). The answer was given using a five-point scale ranging from 1 = not at all responsible, to 5 = completely responsible. For fulfillment, 1 = not fulfilled at all, ranging to 5 = very much fulfilled. Analysis of reliability, measured by Cronbach Alpha, gives a result of 0.734 for 15 items, that reflects the level of high-moderate reliability.

4 RESULTS AND DISCUSSION

The study was conducted on 166 lecturers at the State University in West Java which is undergoing organizational changes. In addition to assessing whether there is agreement between the two parties regarding their obligations and the fulfillment of those obligations, the lecturer's perspective would allow researchers to examine how the exchange relationship operates; there is consensus in the norm of reciprocity in the governance relationship. The aim of this study is to examine reciprocity from the employer's and lecturer's perspectives to determine whether mutuality exists in how the relationship operates. Specifically, we explore the extent to which lecturers reciprocate lecturers' perceived obligations and the fulfillment of those obligations by adjusting their own obligations and the extent to which they fulfilled of those obligations.

The result of descriptive statistic for PC (Table 1) show that the mean of obligation in PC compared to the mean of fulfillment is relatively different (4.4222 and 3.0759). Most of the participants perceived the numbers for obligation of employer as high, and vice versa, the fulfillment was not yet equal. Both median and standard deviation show the same result too. The calculation of median in PC-Obligation is 4.4000, whereas the PC-Fulfillment is around 3.0. Score of standard deviation is 0.32 and 0.26 for PC-Obligation and Fulfillment.

Perception of obligations is quite different (Figure 1). A total of 98.8% of participants feel a high level of perception of the obligation to be fulfilled by institutions, and on the contrary, as much

Table 1. Descriptive statistic of PC.

Attribute	Mean	SD	Median
PC-Obligation	4.4222	0.32754	4.4000
PC-Obligation-Transactional	4.3916	0.40021	4.4000
PC-Obligation-Relational	4.4373	0.33977	4.4000
PC-Fulfillment	3.0759	0.26094	3.0667
PC-Fulfillment-Transactional	3.2241	0.44683	3,2000
PC-Fulfillment-Relational	3.0018	0.30381	3.0000

Table 2. Statistic test result.

		Z	Sign (2 tailed)
Pair 1	PC Fulfillment-Obligation	−11.156	.00
Pair 2	Transactional Fulfillment-Obligation	−10.968	.00
Pair 3	Relational Fulfillment-Obligation	−11.177	.00

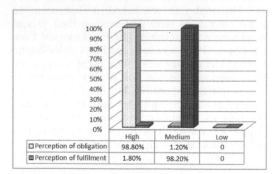

Figure 1. Total comparison of perception of obligation and fulfillment of PC.

Table 3. Coefficient d and effect size on PC.

	Cohen's d	r
Fulfillment of Obligation	4.546	0.915
Fulfillment of Obligation (Transactional)	2.752	0.809
Fulfillment of Obligation (Relational)	4.454	0.912

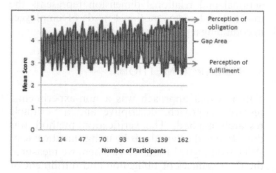

Figure 2. Gap mean score of participant.

as 98.2% of participants perceived its fulfillment only in the moderate level. In addition, not one of participants perceived the obligation and its fulfillment by the employer in low level.

Then, we calculated a comparison between obligation and its fulfillment of the PC. To comprehend our result, we also measured each pair of dimensions of the PC, both the fulfillment and obligation in transactional contract, and the fulfillment and obligation in the relational contract. As indicated in Table 2, the result of the measurements is significantly negative for each pair of obligation and fulfillment of the PC perception (two tailed).

Then, the results show significant differences in the statistical tests; calculations are also performed to calculate the effect size. As indicated in Table 3, results of the effect size (r) are at a large level (≥ 0.8) for each dimension of the PC which compared fulfillment of obligation. It shows that participants perceived large differences to obligation compared to its fulfillment by the university for their lecturers, for both transactional and relational contracts.

Through the mean score, we describe the discrepancies between an obligation and its fulfillment which is perceived by lecturers as indicated in Figure 2. The gap of the mean score ranges from around −1 and above. It has been confirmed by the result of effect size on the PC (Table 3), which means quite large differences.

The results in Table 2 indicate that there were significant negative discrepancies between the perceptions about institutional obligations towards its fulfillment. This means that the lecturers feel that the obligation has not been fulfilled optimally by institutions. In addition, the results of calculation of the gap score in each dimension of the PC (transactional and relational contract) show the same. The lecturers feel the negative gap between the fulfillment of institution obligations, both of transactional and relational contract.

Consistent with research adopting a social exchange framework, the PC examines employee reciprocity based on the behavior of the lecturer. Therefore, the PC framework expands our conceptualization of reciprocity by incorporating a cognitive dimension, that is, what lecturers feel are their obligations to their employer. With this in mind, the PC means reciprocity in its focus on perceived obligations and the fulfillment of those obligations.

Regarding the transactional contract, lecturers feel that promises that have been made such as remuneration, career opportunities, and so on, are not given by the university. There is no general consensus between the two parties about what the exchange appears to be between them. Based on relational contract, there is a low connection between the two parties.

Meanwhile, through explicit and implicit relational contract promises, the lecturers feel that the fulfillment is still not yet in accordance with the high expectations they perceived relating to employer obligation. The results showed that there was a large gap on the aspect of perception about support from the university so that the higher performance was hard to be achieved.

The results of this study have several implications. For organizations, they may consider factors that cause employee perceptions of the PC violations, especially about the rights that are not optimally met by the organization so that mutual perceptions of the PC violation are met. This knowledge is expected to assist organizations in preventing and avoiding negative prolonged effects.

The PC is not a formal contract. However, it actually exists in the minds of employees and the company. Attempts to analyze it will provide an overview on organizational development efforts in institutions becoming better organizations in the future.

5 CONCLUSIONS

- There were significant negative differences between the perceptions about institutional obligations towards its fulfillment. This means that the lecturers felt that the obligation has not been fulfilled optimally by institutions.
- The emerging conclusion therefore, is that employees reciprocate treatment by the employer by adjusting their attitudes and behavior accordingly.
- Perceived obligation sets the parameters of the exchange while fulfillment of obligations captures behavior within the exchange.

REFERENCES

Argyris, C. (1960). *Understanding organizational behavior*. Homewood: IL: Dorsey Press.
Blau, P. (1964). *Exchange and power in social life*. New York: Wiley.
Conway, N. & Coyle-Shapiro, J.A-M. (2011). The reciprocal relationship between psychological contract fulfilment and employee performance and the moderating role of perceived organizational support and tenure. *Journal of Occupational and Organizational Psychology, 85*, 277–299.
Coyle-Shapiro, J.A-M. & Kessler, I. (2000). Consequences of the psychological contract for the employment relationship: A large scale survey. *Journal of Management Studies, 37*, 903–930.
Coyle-Shapiro, J.A-M. (2002). Exploring reciprocity through the lens of the psychological contract: Employee and employer perspectives. *European Journal of Work and Organizational Psychology, 11*(1), 1–18.
Coyle-Shapiro, J.A-M. & Conway, N. (2005). Exchange relationship: Examining psychological contract and perceived organizational support. *Journal of Applied Psychology, 90*, 774–781.
Coyle-Shapiro, J.A-M. & Parzefall, M. (2008). Psychological Contracts. In: C.L. Cooper & J. Barling (Eds.), *The SAGE handbook of organizational behavior*. London, UK: SAGE Publication.
Guest, D. (1998). Is the psychological contract worth taking seriously? *Journal of Organizational Behavior, 19*, 649–664.
Holmes, J.G. (1981). The exchange process in close relationships: Microbehavior and macromotives. In M.J. Lerner & S.C. Lerner (Eds.), *The justice motive in social behavior*: 261–284. New York: Plenum.
Lee, H.W., & Liu, C.H. (2009). The relationship among achievement motivation, psychological contract and work attitudes. *Social Behavior and Personality: an international journal, 37*(3), 321–328.
McDonald, D.J. & Makin, P.J. (2000). The psychological contract, organizational commitment and job satisfaction of temporary staff. *Leadership & Organization Development Journal, 21*, 84–91.
Morrison, E.W. & Robinson, S.L. (1997). When employees feel betrayed: A model of how psychological contract violation develops. *The Academy of Management Review, 22*(1), 226–256.
Nazir, M. (2005). *Metode penelitian Bisnis*. Bogor: Ghalia Indonesia.
Nelson, L., Tonks, G. & Weymouth, J. (2006). The psychological contract and job satisfaction: Experiences of a group of casual workers. *Research and Practice in Human Resource Management, 14*(2), 18–33.
Restubog, S.L.D., Hornsey, M.J., Bordia, P., & Esposo, S.R. (2008). Effects of psychological contract breach on organizational citizenship behavior: Insights from the group value model. *Journal of Management Studies, 45*(8), 1377–1400.
Robinson, S.L., Kraatz, M.S., & Rousseau, D.M. (1994). Changing obligation and the psychological contract: A longitudinal study. *Academy of Management Journal, 37*(1), 137–152.
Robinson, S.L., Rousseau, D.M. (1994). Violating the psychological contract: Not the exception but the norm. *Journal of Organizational Behavior, 15*(3), 245–259.
Robinson, S.L. & Morrison, E.W. (1995). Psychological contract and OCB: The effect of unfulfilled obligations on civic virtue behavior. *Journal of Organizational Behavior 16*(3), 289–298.
Rousseau, D.M. (1989). Psychological and implied contracts in organizations. *Employee Responsibilities and Rights Journal, 2*(2), 121–138.

Rousseau, D.M. (1990). New hire perceptions of their own and their employer's obligations. *Journal of Organizational Behavior*, *11*(5), 389–400.

Rousseau, D.M. & Parks, J.Mc.L. (1993). The contracts of individuals and organizations. *Journal of Organizational Behavior, 15*, 1–43.

Rousseau, D.M. & Tijoriwala, S.A. (1998). Assessing psychological contracts: Issues, alternatives and measures. *Journal of Organizational Behavior*, *19*, 679–695.

Rousseau, D.M. (2000). *Psychological contract inventory*. Pittsburg, Pennsylvania: Carnegie Mellon University.

Ruokolainen, M., Mauno, S., Diehl, M.R., Tolvanen, A., Mäkikangas, A., & Kinnunen, U. (2016). Patterns of psychological contract and their relationships to employee well-being and in-role performance at work: Longitudinal evidence from university employees. *The International Journal of Human Resource and Management*. 1–24.

Shore, L.Mc.F. & Tetrick, L.E. (1994). The psychological contract as an explanatory framework in the employment relationship. *Trends in Organizational Behavior*, *1*, 91–109.

Sparrow, P.R. (1996). Transitions in the psychological contract: Some evidence from the banking sector. *Human Resource Management Journal*, *6*(4), 75–92.

Suryanto. (2008). Sistem Informasi Akuntansi Penjualan Dan Persediaan. *CommIT,* 2 (2), 106–110.

Turnley, H.W. & Feldman, D.C. (2000). Re-examining the effects of psychological contract violation: Unmet expectation and job dissatisfaction as mediators. *Journal of Organizational Behavior*, *21*(1), 25–42.

Wocke, A. & Sutherland, M. (2008). The impact of employment equity regulations on psychological contracts in South Africa. *International Journal of Human Resource Management*.

Zagenczyk, T.J., Gibney, R., Few, W.T., & Scott, K.L. (2011). Psychological contracts and organizational identification: The mediating effect of perceived organizational support. *Springer Science and Business Media, 32*, 254–281.

Zhao, H., Wayne S.J., Glibkowski, B.C. & Bravo J. 2007. The impact of psychological contract breach on work related outcomes: A meta-analysis. *Personnel Psychology* 60: 547–680.

Ideas for 21st Century Education – Abdullah et al. (Eds)
© *2017 Taylor & Francis Group, London, ISBN 978-1-138-05343-4*

Event as a means to educate youth through the volunteers program

D.R. Erlandia & I. Gemiharto
Universitas Padjadjaran, Bandung, Indonesia

ABSTRACT: This paper examined the volunteers program managed by the organizer of the Kreative Independent Clothing Kommunity Festival (Kickfest). This was a great nationwide, repetitive event which involved a lot of young people as volunteers. The event promoted, displayed, performed and sold products made by the Indonesian people, in the forms of clothing, music, and cuisines. The organizer had a program named the Volunteers Program, aimed at accommodating young people who wanted to learn how to organize a big event. The purposes of this study were to determine: how the organizer of Kickfest managed the volunteers program, starting from the pre-event, event execution and post-event, and how the organizer of Kickfest educated volunteers to organize an event. The method used in this study was qualitative with a case study approach, which was to examine the volunteers program in the event, in an intensive, in-depth, detailed, and comprehensive way. The uniqueness of the Kickfest event was that the organizer involved various communities in Indonesia. Then, the event was organized professionally and profitably but still upheld ideals. Besides, it focused on communicating and selling products created by Indonesian people. In general, the results of this study showed that the organizer of Kickfest managed the volunteers program through the process of informing, selecting, discussing, checking the venue, rehearsing, executing the event, evaluating, and following up. The organizer of Kickfest educated volunteers through sharing and discussing, observing, involving, practicing, trying and implementing.

1 INTRODUCTION

1.1 Background

In large-scale activities such as big events, there will be a lot of human resources involved. Core personnel, temporary employees, contracted personnel, and certainly volunteers, will be required.

The existence of volunteers in mega events is unavoidable. It is difficult to run a big event successfully without the involvement of volunteers. Gratton & Taylor (2000) in Kolar et al. (2016) explained that volunteers have had major influences on events and organizations in recent decades and continue to be essential for the success of a variety of organizations. McNamara (2012) stressed that volunteers can help out with tasks that the management team may not be able to do. So, what is a volunteer? Nassar and Tallat (2010) cited by Yen defined volunteer as, 'An individual who offers his or her time and service to an organization, cause, or event with no expectation of a monetary reward or other tangible compensation.' An organization which organizes major events will necessarily require volunteers. So, the organization needs to know the reasons why volunteers are interested to be involved in its program. Noordegraaf (2016) explained that a volunteer might join an event for several reasons, such as learning and growing, helping others, cultivating friendships, using existing skills and learning new skills, gaining work experience, repaying a debt to society, and using leisure time more effectively.

The reasons for the people to be volunteers, as explained above, should become the opportunities for the organizations to give 'a room' for them where mostly youngsters learn and create things. One of the big events that always involves young volunteers is the Kreative Independent Clothing Kommunity Festival (Kickfest). This event is routinely runs every year. Kickfest originated from a community of young people who were members of the Kreative Independent Clothing Kommunity (KICK) whose ambition was to enhance the creativity of young people, especially in the field of clothing (Handiman, 2013). In Kickfest 2013, the event involved many volunteers and communities. Volunteers involved in the event totaled as many as 200 people. They were selected from 1,500 applicants. Pamungkas (2016a), director of Kickfest, explained that there were several objectives in involving so many young people to become volunteers in the event, namely: (1). to provide a 'room' for young people to learn how to organize an event; (2). to be together (synergy) to love and promote Indonesian products created by young people; (3). to instill the values of nationalism and a sense of belonging in Indonesia.

Having known that Kickfest had become an event that provided the opportunities for volunteers to learn, the author was interested in researching Kickfest in 2013, because the number of registrants to be volunteers in Kickfest 2013 was the biggest, as many as 1,500 persons. The author was interested to know how the organizer managed and educated the volunteers in how to organize an event. Qualitative method with case study approach was used to examine the volunteers program in Kickfest, in an intensive, in-depth, detailed, and comprehensive way.

1.2 Research purposes

The purposes of this study were described in the research questions below:

1. How did the organizer of Kickfest manage the volunteers program starting from pre-event, event execution and post-event?
2. How did the organizer of Kickfest educate volunteers to organize the event?

1.3 Research method

This study used a qualitative approach with case study method. The case study was used to answer the questions of 'how' and 'why', and at a certain level was also used to answer the question of 'what' in research activities (Yin in Bungin, 2003). This method gives an opportunity to researchers to study the case intensively, in-depth, in detail, and comprehensively (Faisal, 1995). The case study is also flexible with regard to the data collection method used. It can reach the real dimension of the topic being investigated and can be practically implemented in many social environments, offers the opportunity to test the theory, and the cost spent for the research depends on the range and type of data collection techniques used (Black & Dean, 2001).

This research studied how the organizer managed the volunteers program starting from pre-event, event execution and post-event, and found out how the organizer educated volunteers to organize an event. The research process was done through several steps. First, the data collection was done through literature study, field research, interviews, observation and review of visual documentation. Second, the data analysis was carried out in three steps, in accordance with the concept of Milles and Huberman in Denzin and Lincoln (2000), which was through data reduction, data display, and data verification (conclusion). Third, a data validity test was applied through the process of extending the participation of researchers in the field, perseverance and constancy of observation, a

triangulation process, discussions with colleagues, analyzing the negative cases, checking members involved in the study, outlining the data in detail, and a data auditing process (Moleong, 2006).

2 LITERATURE REVIEW

Volunteers are individuals who choose to contribute their time and effort for no monetary reward (Nassar, 2010). Meanwhile, volunteering is as an activity involving a person's time and effort which is not compensated by regular payment or monetary reward, but is freely undertaken and produces goods and services for organizations, and by extension, for other individuals (Wilson & Musick, 1997) in Yen (2015). Further, Cnaan et al. (1996) explained that volunteering is ultimately associated or defined according to: the degree of freedom involved, the degree of remuneration involved, the structure of volunteering, and the intended beneficiaries of the volunteer act. For students, Astin and Sax (1998) in Cnaan (Repository.upenn.edu) found that volunteering can enhance students' academic development, personal skills development, and sense of civic responsibility. Sheard (1994) in Kolar et al. (2016) said that the specifics of what volunteering involves vary between countries and cultures, but there is some consensus regarding the basic tenets of volunteerism: volunteering is not low- or semi-paid work, work that is compulsorily coerced (i.e. by government or court order), or informal assistance for friends or family, but rather individuals donating their time or services for no financial gain.

3 RESULTS AND DISCUSSION

3.1 Research results

After going through the process of interview, observation and literature study, the following results were obtained:

3.1.1 The ways the organizer managed the volunteers program starting from pre-event, event execution and post-event

The organizer of Kickfest had some stages in managing the volunteers program, as follows:

1. Informing

This stage is the process which informs about volunteer vacancies, which is delivered via social media, such as Twitter, FB, Path and YouTube. Social media was used because the target of the message was young people. The message conveyed in the social media was about the event in general, such as calls to join, how to register, contact person and email address. Registrants could send their CV via email or bring their CV directly at the time of the call. Here there was a process of communication between prospective volunteers and the committee on any matters which had to be understood. Information about openings for volunteers were delivered one month before the event implementation. There were about 1,500 people who signed up to be a volunteer at this event. Kickfest provided an explanation of what should be written in the CV, so as to find out various things about the registrants.

2. Selecting

All applicants were called to join the selection process. Pamungkas (2016b) stated that in 2013, the registrants to be volunteers in Kickfest reached about 1,500 people, so that the applicants were called in groups for an interview on different days. Interview materials had more emphasis relating to motivation to join the event, areas of interest, and a commitment to follow all activities scheduled by the event organizer.

After completion of the interview process, the committee selected from all participants who had attended an interview and decided to receive 200 people. The registrants who failed to be accepted into a volunteer team were especially those who did not attend the interview process or who were unable to follow the commitments required by the committee. From the interview, it was found that there were three favorite fields that the candidates like to be involved with, namely Liaison Orgenizer (LO), Backstage, and Areas.

3. Discussing (Brainstorming)

All received applicants were then summoned to the secretariat of the committee. Here there was the time for an interaction process between volunteers, or between volunteers and the committee. On the first day of the meeting, the organizer had a briefing with the volunteers about the event in general. Then they were grouped according to the desired field. Each division had a coordinator who gave directions on each task. Each volunteer was given a production pack which consisted of the event that must be learned and understood, and discussed together. Explanation of the duties and authority of each volunteer was discussed in detail. The volunteers had to understand what became their task, and they were required to know about the event in general to avoid overlapping with another team. In Kickfest, senior volunteers were also assigned to recruit new volunteers. They were given authority to recruit people whom they thought deserved to be volunteers. They were also given the opportunity to educate those new volunteers.

4. Checking Venue

Three days before the event (D-3), all volunteers were invited to get to know the venue. They certainly needed to know all of the area to be used for the event, in accordance with the floor plan that had been given by the organizer. Here they had to explore all areas that would be used during the event, and also the area that would be their task when the event was running. They had to really understand the technical work, tools or equipment required, coordination between the committee, and be able to find solutions when problems arose. Volunteers could learn what were the steps that had to be done when there were problems. Also, they were able to see first-hand how to design and manage the venue. In the Kickfest, the venue consists of a variety of areas, including: the creative area (stage, community attraction, etc.), clothing and food area (booths), parking area (cars and motorcycles), the audience area, and the committee area.

5. Rehearsing

This phase was done the day before the event. Each division made the simulation. Here was the opportunity for all volunteers to practice how much they had understood the material they had learned beforehand. If they had difficulties, they could ask the coordinator of each division. Experimental activities were essential considering that for most of the volunteers this was their first time in joining Kickfest. Battle (1988) and McDuff (1995) in Noor (2009), said that there are some effective steps to building a team for the volunteers, namely: volunteers and staff give each other encouragement.

6. Executing the Event

Kickfest lasted three days. On the D day, all volunteers had to come a minimum of one hour before the execution. One hour prior to implementation, all the personnel had a briefing and checked the readiness of each division. Then all of the committee and the volunteers moved to each area. They carried out duties in accordance with the procedures and protocols that had been studied and discussed previously. Briefing before starting to work lasted 15–30 minutes, as to the job description of each. At the time of carrying out tasks in the field, volunteers were accompanied by the coordinator. Each volunteer was taught to be independent and brave, and to make decisions when there were problems when on duty, but if they could not decide, they could communicate with the coordinator. Each volunteer had been given ways to solve the problems before the implementation time. The organizer of Kickfest always maintained unity and enthusiasm of volunteers. The organizer managed the volunteers work in shifts, kept their health by providing vitamins, and also provided food with a varied menu.

7. Evaluating

This stage was carried out after the completion of the event execution. The Kickfest organizer invited all volunteers who had been involved in the event. The evaluation was conducted in each division first, then at all of the committee levels. The results of the evaluation in each division were not discussed in forum to maintain the credibility of each volunteer. But all problems were noted by the organizer for the next volunteer recruiting process. In the evaluation phase, the organizer also held a thank-God moment for the success of the event execution. They also conveyed the gratitude for all the committee, including volunteers. In addition, the organizer provided financial rewards for volunteers. Besides, they provided the documents needed by volunteers, such as certificates, information letters, and documentation.

8. Following Up

The organizer of Kickfest kept in touch with former volunteers. They were the best people chosen through a rigorous selection. They often had various ideas for the event, needed by the organizer. Communication was done through social media owned by the volunteer group. The Independent Network as the organizer of Kickfest always supported the activities carried out by volunteers, in the name of the Independent Community (IC). The organizer hoped that the ex-volunteers could join again at the next events.

3.1.2 *The organizer of Kickfest educated volunteers on how to organize an event*

1. Sharing and discussing

Sharing and discussing are the activities that were frequently performed by the organizer and volunteers. These activities provided a wide opportunity for volunteers to learn more about how to manage an event. The activities were carried out before the event, during the event, as well as after the event. Before the event, sharing and discussing things were carried out during the event introduction, the discussion of job description, technical implementation, and so on. During the event, the activities were conducted at briefings before the event began. Meanwhile, after the event, the activities were performed at the time of evaluation. The volunteers could get exposure and important knowledge about details of the event.

2. Involving

Involving volunteers in various activities before the event, during the event and after the event was the organizer's way to educate them. Before the event, volunteers were involved in several activities, such as promotion, ticket sales, and the recruitment process for senior volunteers. During the event, volunteers were involved with the committees. After the event, the organizer involved volunteers in the evaluation process. Clearly, by being involved in those activities, volunteers could gain knowledge, experience, and network.

3. Observing

Volunteers were given the opportunity to observe how an event was carried out. The organizer let volunteers watch how the committee completed a job, such as arranging the venue, and managing the team. Volunteers could learn the process of carrying out an event here.

4. Trying

After understanding each task, volunteers were asked to try it on the field. Previously, all volunteers were given production packs that had to be discussed, understood and then tested in the field. Volunteers were divided into groups, depending on the responsibility. Each group/division was guided by a coordinator who was always ready to help when volunteers were having problems. This activity was done three days before the main event. Here, volunteers could learn practical knowledge.

5. Practicing

All volunteers were to practice their tasks (rehearsal). This activity was carried out one day before the execution. All volunteers had to be present and practiced in accordance with their respective fields. They made a simulation. This step was done to avoid confusion during the event execution. Besides that, the volunteers also practiced coordinating in the division and among the divisions. Here, the volunteers could get new experience.

6. Implementing

Volunteers applied the knowledge gained beforehand. This implementation was performed during the main event. All volunteers were trusted to do their work independently. They were also given the authority to decide when there was a problem. The volunteers learned to be responsible for their work.

3.2 *Discussion*

The existence of volunteers is very important in a variety of great activities. We cannot imagine the activities of the Olympics, the Asian-African conference or a political campaign without the involvement of volunteers. They are the ones who will help the success of an activity. The volunteers are existing everywhere in a variety of activities, ranging from political, social to commercial activities. Basically, they are the ones who are ready to give the time and provide services to organizations without expecting some reward financially. However, they also expect that there is something that

Table 1. The results of the research.

No.	Research question	Results
1.	How did the organizer of Kickfest manage the volunteers program starting from pre-event, event execution and post-event?	Informing, selecting, discussing, checking the venue, rehearsing, executing the event, evaluating, and following up.
2.	How did the organizer of Kickfest educate volunteers to organize the event?	Sharing and discussing, observing, involving, practicing, trying, and implementing.

Figure 1. The volunteers in Kickfest.

they can get from their involvement. There are reasons which will motivate them to participate and be loyal to the organization. Therefore, the organizer must know what motivates of volunteers to join. Ringuet (2007) describes some of the things that make them reluctant to return as volunteers at an organization or event, which include: the overall workload, a lack of appreciation of their contribution, problems with how the event was organized, wanting more free time for other activities, a lack of 'sense of community' among volunteers, family responsibilities, the inability to make decisions regarding their own position, a dislike for some of their responsibilities, and lack of remuneration.

Independent Network Indonesia, as the organizer of Kickfest in 2008, 2009, 2010, 2011, 2012, and 2013, held the values in managing and educating volunteers, among them being: 1). Nationalism. The organizer upholds this value highly. The organizer sells and displays only products made in Indonesian Youth. With a very strong brand of Kickfest, it is not surprising that a lot of products from other countries wanting to join this event were rejected, because the vision of the Kickfest event is to promote the products of the nation. Nationalism was instilled in young people who joined the team of volunteers. 2). Education. The organizer wanted to give 'room' to young people to get the knowledge and experience in organizing an event. The era of after image made people want to see products they love, both goods and services directly. It was not enough to just watch on the mass media. Therefore, the organizer created several activities that volunteers could learn from. 3). Independent. For the organizer of Kickfest, everyone has his/her own uniqueness and advantages, which should be developed. Youngsters as independent people are free to be creative and to create things. Therefore, the organizer encouraged the advancement of Indi(pendent) companies to join. These values were embedded in the volunteer's mind, although not directly. 4). Togetherness & Synergy. This value is the power to create things together. Everyone must focus on each passion and expertise. If everyone with each expertise moves together, it will produce great work: move forward together in making work for the nation. The organizer was very open to feedback and new breakthroughs raised by volunteers. 5). Appreciation. The organizer treats the volunteers, who have been accepted through a rigorous selection, as a team. Note that what motivated volunteers to join in the event at the interview became material for knowing what he/she wanted. So, they were placed according to their desired field. If they had to be placed in another division, there was a process of persuasion. 6). Communication. The organizers communicated with volunteers before the event, during the event and after the event. The volunteers were individuals who had contributed to the success of the activity, so the organizer of Kickfest already found it helpful. After the event, the organizer kept communicating with volunteers by joining the media owned by the IC. IC was the community of Kickfest volunteers.

4 CONCLUSIONS

It can be concluded that volunteers are an important source of manpower in carrying out big events. The organizations which need volunteers to be involved in their programs must know the reasons why volunteers are willing to join. The organizer of Kickfest had managed the volunteers program through the process of informing, selecting, discussing, checking the venue, rehearsing, executing the event, evaluating, and following up. Then, volunteers were educated through sharing and discussing, observing, involving, practicing, trying and implementing. There were several values that were upheld by the organizer relating to volunteers, namely nationalism, education, independence, togetherness & synergy, appreciation, and communication. The relationship between the organizer and volunteers had mutual benefits and mutual needs.

REFERENCES

Beaudry, A. (1986). *The guide to organizing your campaign.* New York: The Free Press.

Black, J.A. & Dean J.C. (2001). *Methods and issues in social research.* New Jersey: John Wiley & Sons.

Bungin, B. (2003). *Analisis data penelitian kualitatif.* Jakarta: Raja Grafindo Persada, PT.

Cnaan, R.A., Smith, K.A., Holmes, K., Haski-Leventha, D., & Handy, F. (2010). Motivations and benefits of student volunteering: Comparing regular, occasional, and non-volunteers in five countries. *Canadian Journal of Nonprofit and Social Economy Research, 1*(1) 65–81.

Denzin, N.K. & Lincoln, Y.S. (2000). *Handbook of qualitative research.* Translated by Dariyatno in 2009. Yogyakarta: Pustaka Pelajar.

Faisal, S. (1995). *Format-format penelitian sosial.* Jakarta: Rajagrafindo Persada, PT.

Handiman, D.P. (2013). Revised in 2015. *Mendongkrak kreatifitas anak muda bangsa.* http://m.kompasiana.com/dimashandiman/kickfest-sukses-mendongkrak-kreatifitas-anak-muda-bangsa.

Kolar, D., Skilton, S., & Judge, L.W. (2016). Human resource management with a volunteer workforce. *Journal of Facility Planning, Design, and Management, 4*(1), 5–12.

McNamara, M. (2012). *The political campaign desk reference: A guide for campaign manager, professionals, and candidates running for office.* USA: Outskirts Press, Inc.

Moleong, L.J. (2006). *Metode penelitian kualitatif.* Bandung: Remaja Rosdakarya.

Nassar, N.O., & Talaat, N.M. (2010). Motivations of young volunteers in special events. *Tourismos: An International Multidisciplinary Journal of Tourism, 4*(1), 145–152.

Noor, A. (2009). *Manajemen event.* Bandung: CV Alfabeta.

Noordegraaf, M.A. (2016). *Volunteering: Is it a waste of time or best experience ever?* Turkey: The Sportjournal.org.

Ramli, N., Ghani, W.S.W.A., Bahry, N.S., & Rohaidah, A. (2014). Evaluating volunteer motivation and satisfaction at special event. *Tourismos: An International Multidisciplinary Journal of Tourism.*

Ringuet, C.A. (2007). Volunteers in sport motivations and commitment to volunteer roles. *International Journal of Event Management Research, 3*(1).

Pamungkas, R. (2012). *Discussion about management of Kickfest.* Bandung.

Pamungkas, R. (2013). *Discussion about event management & Kickfest.* Bandung.

Pamungkas, R. (2016a). *Interview about Kickfest and Volunteers (1).* Bandung.

Pamungkas, R. (2016b). *Interview about Kickfest and Volunteers (2).* Bandung.

Padi, T. (2016). *Interview about Volunteers and Kickfest.* Bandung.

Yen, L. (2015) *Successful strategies for recruiting, training, and utilizing volunteers.* U.S. Department of Health and Human Services. www.samhsa.gov.

Zulkifli, 2016. *Interview about Volunteers and Kickfest.* Bandung.

Ideas for 21st Century Education – Abdullah et al. (Eds)
© 2017 Taylor & Francis Group, London, ISBN 978-1-138-05343-4

Stress at work and well-being: Study of stress level at work to improve employee well-being on Pertamina's operators with standard 'Pertamina Way' in Bandung

M. Batubara
Universitas Padjadjaran, Bandung, Indonesia

ABSTRACT: Pertamina, as the only oil company in Indonesia, establishes a system of quality standards, namely 'Pertamina Way', to improve the quality of customer service. The operators not only needs to provide excellent service with smiles and greetings to customers with various characteristics, but also need to face complaints. On the other hand, their work can be considered quite dangerous because they are every day exposed to smoke from motor vehicles, and inhalation of vapors of diesel fuel. The working conditions have symptoms of stress. Therefore, this study aimed to get an overview of the level stress at work of operators. The design applied was non-experimental with descriptive method and administered to 34 operators in four gas stations in Bandung. Operators filled in a questionnaire about demographic data and measurement tools, that was adopted from a stress diagnostic survey from the theory of Ivancevich and Matteson (1980) and modified by the researcher. The results indicated the stress levels of operators as being low (18 respondents – 52.94%), moderate (15 respondents – 44.12%) and high (1 respondent – 2.94%). Most of the operators have work stress work at the low and moderate levels. It indicated that any external factor in a gas station after they gave services to their costumers is felt as being a psychological and/or physical demand that was appropriate for their job. Low and moderate levels of stress indicated that they have a positive impact stress, called Eustress. Eustress can be one of the indicators of well-being, especially psychological, that makes the operator feel happy with his/her work, builds a positive relationship with customers, and allows them to deal with the stress environment and also to regulate emotion.

1 INTRODUCTION

The existence of world-class oil and gas companies in Indonesia such as Shell, Petronas and Total means that Pertamina no longer monopolizes the fuel retail market in Indonesia, and it brings in new policies to maintain its presence and improve productivity. One effort at improvement is to set a standard quality system in accordance with the Bureau Veritas Certification Indonesia, the applicable standard 'Pertamina Way'. 'Pertamina Way' covers standards as follows: (1) The staff are well-trained and highly motivated. (2) Guarantee of the quality and quantity. (3) Knowledge of the product. (4) The physical format is consistent. (5) A process that maintains good maintenance, preventive maintenance, and maintenance details.

The standard 'Pertamina Way' is implemented optimally by the Operator. The operator is the operational officer on duty at the pump for filling the fuel. The operator can be considered as the spearhead of the company, being at a position of direct contact with the public, so that the form of services provided by the operator of gas stations is a series of public appraisal of Pertamina.

Based on the survey results, with the implementation of 'Pertamina Way', the operator feels his job harder. Operators should appear clean and tidy. Operators are obliged to wear a uniform, cap, ID card, pouch waist, and wear socks and shoes. Operators who do not use all the equipment will be reprimanded by the supervisor and have a reduced assessment of his/her work. In addition, the operator must have the ability to communicate, have etiquette and show good body language to the customer, maintain the physical appearance of gas stations and maintain the cleanliness of the toilets. They also must be patient in providing service to customers who cannot wait, or who even lodge a complaint.

Working time is around eight hours a day, with a fifteen-minute break a day, which implementation is based on agreement among employees. In the morning, afternoon and evening there are significant differences prevailing, due to differences in the weather and physical conditions. On the morning shift, the weather is fresh and is a normal time for work, so that the physical condition of the operator can be maximized. On the afternoon shift, the weather is getting hot and air pollution starts to increase, thus lowering the physical condition of the operator. On the night shift, the weather is cold and so becomes an obstacle in the performance of carriers and operators, which can be more harmful to health. In addition, drowsiness is also a constraint on the night shift because these hours would normally be used to rest. There are new regulations among other carriers that should not be ignored unless otherwise advised by doctors, such as tolerance to arriving late being given three times a month.

The operators are advised not to wear a mask covering the nose for the sake of service procedures that require them to keep smiling when providing service to the customers. This has given them more headaches, chest tightness, eye irritation, or more severe respiratory problems. Every day they are exposed to smoke from motor vehicles, and they inhale vapors of gasoline/diesel fuel (hydrocarbons), which contain a lot of dangerous organic substances that can cause hallucinations and disrupt the body's organs, especially the function of the brain, liver, kidneys, lungs, and potentially cause cancer if exposed to continuously in the long term. For pregnant officers, exposure can harm the fetus.

Based on this phenomenon, it is known that the working conditions for operators of gas stations caused symptoms of perceived physical stress, and psychological and behavioral effects. This means that the operators consider that the working conditions is one of the conditions that causes stress in themselves. According to the American Psychological Association (APA), the top stressors for people in the workplace, in order of importance, are low salaries (43 percent), heavy workloads (43 percent), lack of opportunity for growth and advancement (43 percent), unrealistic job expectations (40 percent), and job security (34 percent) (Stress in the workplace: Meeting the challenge, National Women's Health Resource Center, 2009). Some research has already studied stressors frequently reported as sources of distress (e.g., time pressure, role and work overload, excessive paperwork, unfair organizational practices, insecure relationships, or monotonous work that hinders personal development (Ivancevich et al., 1984) to be appraised as sources of distress (Kozusznik et al., 2015).

On the other hand, the gas station can produce a good performance with the retention of Pertamina Way. This indicates that the level of stress felt by operators is at moderate levels, so that the stress level can be considered as eustress. Quick & Quick (1979) identified eustress as 'healthy, positive, constructive results of stressful events and stress response'. Thus, eustress is considered as the result of the body's response to a stressor (Kupriyanov & Zhdanov, 2014).

According to those phenomena, the purposes of the present study are: 1) The examination of the level stress of the operators. 2) The examination of sub-dimension stressors which affect the formation of stress.

2 LITERATURE REVIEW

According to Ivancevich and Matteson (1980), stress work is an adaptive response, which is mediated by individual differences and/or a psychological process, which is a consequence of the action of the external (environmental) situations or events that make particular demands, which may be a result of psychological and physical tensions. External actions, events and situations are known as a source of stress (stressor).

Stress work internalized by the operator is mediated by psychological processes and individual differences. Each operator brings their individual differences such as the needs, goals, personality, gender, age, individual values espoused, and duration of work. Through the cognitive process, the operator will perceive and judge whether a stressor is perceived as something threatening, challenging or dangerous to himself/herself.

When an operator is better able to cope with stressors than other operators, then the operator can adjust his/her behavior to cope with stressors in the work environment. On the other hand, if an operator is unable to cope with the stressor and cannot customize the behavior of the stressor, then the operator may experience job stress.

Ivancevich and Matteson (1980) describes the distress and eustress to be able to explain the stress. Distress is defined as the negative side (which can cause pressure) because the stresses caused by it are bad. Distress is associated with the stress response that is both unsatisfactory and upsets the balance of the body. Eustress is interpreted as a positive side (the side of a pleasant stress caused by something good). Eustress is a response to stress that is satisfying and can generate the optimal functioning of both the body's physical and psychological functions.

According Ivancevich and Matteson (1980), a low stress level (underload stress) will provide low impact work (low performance) and consequently decrease motivation to work because they feel bored with work that does not challenge them to work better. Similarly, for high levels of job stress (overload stress), the work impact also becomes low (low performance) because the pressure is so heavy that the operators cannot work properly. Moderate levels of job stress (optimal stress) will yield high employment (high-performance). In this condition, individuals feel challenged and motivated by their work. Operators who feel happier will have well-being at work, so they can work well and give a high performance.

3 RESEARCH METHODS

This research method is applied using a non-experimental approach with descriptive method.

The sampling technique used is nonprobability-incidental sample type. The verification method is done to test the hypothesis by using statistical tests. In this research, data collection is conducted by using a questionnaire. The questionnaire includes demographic data (gender, age, tenure and marital status) and a modified Stress Diagnostic Survey (SDS).

3.1 Participant

Researchers used a non-probability technique sampling. The population of this research is the operators in a gas station in Bandung that have implemented 'Pertamina Way' for a minimum of two years. To determine a selected sample, operators were selected for a minimum term of six months, on the assumption that they had been exposed long enough to stressor. 34 operators that fitted these criteria were selected.

3.2 Instrumental

Data was collected using questionnaires about demographics, and 44 measurement tool items that were adopted from the SDS from the theory of Ivancevich and Matteson (1980). The SDS was modified by the researcher into two dimensions, individual stressor and organizational stressor. The individual stressor has one sub-dimension: work load. The organizational stressor has three dimensions: organizational climate, work condition and influence of leader.

Analysis of validity used Spearman rank where all items having a value above 0.3 indicates that the item is valid. Analysis of reliability measured by Alpha Cronbach shows the level of high reliability, with results of 0.805 for individual stressor and 0.967 for organizational stressor. Analysis of data gender and marital status used Mann Whitney. Analysis of data tenure and age used Chi-Square.

Data processing was to sum the total value of a questionnaire to determine the level of stress (frequency distribution). The answers were given using the Likert Verbal Frequency Scale, with a four-point scale ranging from 1 (do not feel anything), 2 (a little depressed), 3 (often depressed) and 4 (very depressed). The categorization level is outlined in Table 1.

Table 1. Total value and level of stress.

Stressor	Low	Moderate	High
Individual Stressor	44–87	88–132	133–176
Organizational Stressor	36–71	72–107	108–144

4 RESULTS AND DISCUSSION

4.1 Results

The results of 34 operators show that operators have stress work in low level (18 operators, 52.94%), moderate level (15 operators, 44.12%) and high level (1 operators, 2.94%). Most of the operators are in low and moderate level. It indicates that any external environment and situation at the gas station after applying customer service procedure 'Pertamina Way' felt as a psychological and/or physical demand that was appropriate with their job.

According to distress and eustress (Figure 2), it points to the link between stress level with the type of stress and performance. Operators at low and moderate levels are indicated as having eustress, while operators with high stress levels have indicated distress. Operators with eustress will produce a good performance.

Figure 3 shows a comparison between sub-dimensions. The percentage values are almost the same and were not significantly different. This means that there are four sub-dimensions into sources of stressors that make up the varied levels of stress on the operator.

Organizational climate and work conditions are stressors that have a significant impact on the incidence of stress on the operator. If they are associated with distress and eustress, then these stressors can also be distress and eustress.

Demographic data is not shown to have a significant impact in influencing the formation of operator stress. Through different tests, by calculating

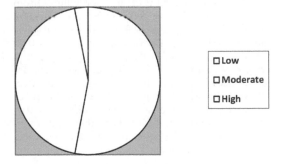

Figure 1. The level of stress on the operators.

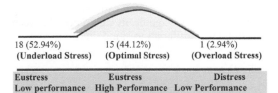

Figure 2. The level of stress on operators and associated with the types of stress and performance.

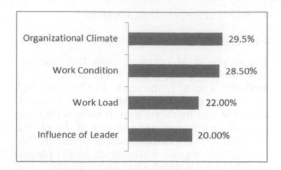

Figure 3. Comparison of four sub-dimensions of stressors.

the Mann Whitney, obtained the following data of gender = 0.902. From the calculation of the value obtained by Mann Whitney, $Z_{count} = -0123$ and the value of A Symp. Sig. = 0.902, using $\alpha = 0.05$ obtained $Z_{table} = -1.96$. Because $Z_{count} > Z_{table}$ and $\alpha sig > \alpha$, with a 5% risk alleged, then there was no difference in the average influence of gender on the job stress, meaning that there is no significant difference between the stress level of male and female operators as there is no significant difference between workers who are married and who are not. The following data for marital status also indicates the same thing with value = 0.914. It indicates that there is no difference in the average effect of marital status against work stress, meaning that there is no significant difference between the level of work stress for an unmarried operator and the level of job stress for a married operator.

The following data is the age and tenure. From the calculation using the Chi-Square, and the obtained value $\chi_{count} = 19.471$ and Asymp value. Sig. = 0.000 using the obtained value $\alpha = 0.05$ obtained $\chi_{table} = 9.21$. Because $\chi_{count} > \chi_{table}$ and $\alpha sig < \alpha$, with a 5% risk alleged, then there are differences in the average age range of the effect of work stress, meaning that there is a difference of work stress in the age range of different carriers, but it is not significant. The other data is tenure. From the calculation results obtained value $\chi_{count} = 0.765$ and Asymp value. Sig. = 0.682 using the obtained value $\alpha = 0.05$ obtained $\chi_{table} = 9.21$. Because $\chi_{count} < \chi_{table}$ and $\alpha sig > \alpha$, with a 5% risk alleged, then there was no difference in the average tenure effect on the level of work stress, meaning that there are no significant differences between job stress with different working times.

4.2 *Discussion*

Operators who have a moderate stress level (optimal stress) are operators that produce high performance at the gas station. Operators feel their position is a challenge and it gives them even more enthusiasm for work. They can smile and greet every customer without feeling any pressure. Meanwhile, operators who have a low level of job stress (under load stress) are operators who keep running standard Pertamina Way, but who are simply doing the work procedures with an absence of a desire to provide excellent service. Although not significant, the operator with a high level of job stress (overload stress) felt the implementation of Pertamina Way burdensome, and difficult to meet the given demands.

Referring to the comparison between sub-dimensions, the organizational climate and working conditions were assessed to play a significant role in influencing the level of stress on the operator. The existence of new regulations that operators should not be absent, should not arrive late more than three times in one month, that rest periods are given only 15 minutes, they had to use all the operator's equipment, and were prohibited to use masks that cover the nose, breathing air that can harm health, and their a reduction in the value of the work if neglected, were all perceived as a quite burdensome stressor. Workload that should be done by carriers such as giving a smile and a greeting, and resolving complaints from customers, is rated as a liability in carrying out the work and has been a challenge. The influence of leaders is a stressor with the smallest percentage, which indicates that their monitoring and regular assessment by the supervisor when the operator works, does not necessarily make operators feel threatened and burdened.

On the other hand, besides the stressor that can be burdensome (distress), the fourth sub-dimensions can also be a stressor that strengthens (eustress). From the interviews conducted, data showed that there are some facts that these stressors also act as support (eustress). Periodically, the operator is equipped with a special training and upgrading on a clear understanding of how to spearhead the front directly facing the consumer. They are given the opportunity to resolve complaints from consumers not satisfied with the services provided, while overseen by supervisors. This opportunity is perceived as a challenge but can be a pride of its own if successfully completed.

Supervisors, who evaluate the performance of the operator, are rated by the operator as someone who affects the performance and as a role model. The warning given for negligence by the operators' supervisors count as strikes and provides feedback that is positive in order to have better operators. Pertamina and gas stations also gave an award to the operators and supervisors who are getting the best value every month, which will be announced each period. Operators and supervisors who are selected will receive incentives and the title of Best Employee.

A specific policy for health insurance after retirement for operators of Pertamina does not exist and the policy is left entirely to the individual retail outlets. Gas stations each have a special doctor (personal), intended for all employees including filling station operators, as a form of appreciation to employees, and the service of these doctors is free of charge. The operator is also given milk regularly three times a week.

According to the results of interviews with supervisors, gas stations implement the kinship system and help each other in the relationship between employees and fellow employees. This is proven as a reminder for each other that if there are colleagues who do not use the full gear, help to bring receipt of purchase, share food and mutually encourage one another. Similarly for the supervisors, when gas stations enter a busy hour of consumers, supervisors do not hesitate to help with consumer convenience, as well as to ease the task of the operator.

According to the theory, low and moderate level of stress indicates that they have a positive impact stress called Eustress. Eustress can be one of the indicators of well-being, especially psychological, that makes operators feel happy with his/her work, builds a positive relationship with customers, enabling operators to deal with the stress environment and also to regulate emotion. Operators with such conditions are able to produce a good performance.

5 CONCLUSIONS

Based on the results of the study, it can be concluded that there are low and moderate levels of stress on the operator belonging to the eustress.

Their working conditions are potentially stressful, basically still being felt in accordance with the demands of the job. It is influenced by the presence of supporting activities that are positive. It is based on four stressors but are contrary because they act as support that strengthens (eustress). It can decrease stress conditions. Warm relationships among operators, supervisors who provide positive feedback, autonomy to solve problems, and the workload being considered by the company and appreciated, can be a source of eustress and also an indicator for well-being at work.

REFERENCES

Ivancevich, J.M., Matteson, M.T. & Smith, S.V. (1984). Relation of Type A behavior to performance and satisfaction among sales personnel. *Journal of Vocational Behavior, 25*(2), 203–214.

Ivancevich, J.M., Konopaske, R. & Matteson, M.T. (2011). *Organizational behavior and management* (9th ed.). New York, NY: McGraw-Hill Irwin.

Ivancevich, J.M. & Matteson, M.T. (1980). Stress and work. Scott, Foresman and Company.

Kozusznik, M.W., Lloret, S., Rodriguez, I. & Peiro, J.M. (2015). Hierarchy of eustress and distress: Rasch calibration of the Valencia eustress-distress appraisal scale. *Central European Journal of Management, 2*(1).

Kupriyanov, R. & Zhdanov, R. (2014). The eustress concept: Problems and outlooks. *World Journal of Medical Sciences, 11*(2), 179–185.

National Women's Health Resource Center. (2009). *Stress in the workplace: Meeting the challenge.* Pennsylvania: Health Advocate, Inc.

Quick, J.C. & Quick, J.D. (1979). Reducing stress through preventive management. *Human Resource Management, 18*(3), 15–22.

Course Management (CMT)

Ideas for 21st Century Education – Abdullah et al. (Eds)
© *2017 Taylor & Francis Group, London, ISBN 978-1-138-05343-4*

Preceptors' perceptions of preceptorship at Surgical Care Room General Hospital Haji Adam Malik Medan

R.E. Nurhidayah, Y. Aryani & C.T. Siregar
Faculty of Nursing, University of Sumatera Utara, Medan, Indonesia

ABSTRACT: Preceptorship is a method of clinical learning. Interest preceptorship is to build and develop the basic capabilities of students at the stage of professional education programs. The instructors are called preceptors and the learners are called preceptees. This study aims to determine the perception of the Preceptor of the implementation of preceptorship at the Surgical Care Room, General Hospital Haji Adam Malik Medan. Design research is qualitative. The data collected was through focus group discussions at surgical care room, general hospital Haji Adam Malik Medan. A Focus Group Discussion (FGD) consist of six preceptors of surgical care room, two preceptors of the nurse educational institution and four case managers of surgical care room and the deputy head of the surgical care room. Based on the results of FGD has found four themes. The first theme, which was the duration of time for students that needed guidance proved still be inadequate; the second theme is the decrease in ability of the students; the third theme is the cooperation of educational institutions and the preceptors still be improved; the fourth theme that was obtained was a moral burden of becoming a preceptor. Based on themes found it advisable that preceptors involvement at every stage of learning. A preceptor should have additional knowledge compared to nurses and the cooperation of educational institutions and the preceptors of clinical institutions that still needs to be improved.

Keywords: duration, ability, cooperation, moral burden

1 INTRODUCTION

Public hospital Haji Adam Malik Medan, is known to be a teaching hospital. The public hospital Haji Adam Malik Medan is also a referral hospital for the northern part of Sumatra Island. Almost all students of educational institutions of nursing, medicine, and obstetrics in northern Sumatera carry out nursing practices in Public hospital Haji Adam Malik Medan. As a referral hospital, the hospital of Adam Malik has many patients with very diverse causes.

According to Nurhidayah (2011), a competent professional nurse in Indonesia must pass through two stages: an academic education and a professional education stage. Phelps (2009) states that the primary purpose of nursing education is to make students competent nurses. While Nurachmah (2011) states that, the learning pattern for the education profession is internship and a pattern called preceptorship guidance, so that educators at the clinic can be called Preceptors and learners can be called perseptees.

The focus of learning during the stage of professional education is gradually delegating an authority. Maitland (2012) states that preceptorship is the period of support for newly qualified professionals that enables them to make the transition from student to become a registered practitioner. Hilli and Malender (2015) states that preceptorship has become an important part, which begins with a comprehensive orientation.

Preceptorship is symbolized as the heart of professional education programs. If the heart experiences problems then the professional education program is where the problem starts. In fact, a professional education program is a stage that must be passed to be a competent nurse. The emphasis of Preceptorship is mainly on critical thinking, reflection and focusing on ethics.

2 METHOD

This study aims to identify the perception of a preceptors of preceptorship at Surgical Care Units in the Haji Adam Malik Hospital. This study used a qualitative design. Participants are preceptors in the surgical treatment of Haji Adam Malik hospital. Data were collected through focus group discussions. A focus group discussion (FGD) consist consist of six preceptors of surgical care room, two preceptors of the nurse educational institution and four case managers of surgical care room and the deputy head of the surgical care room.

The collected data was analyzed by content analysis, which was inspired by Huberman (2010) description of the method. Step by step analysis of the data is as follow:

a. Data reduction, known as the electoral process, focuses on the simplification of the data in the form of a detailed and systematic description
b. Data Display, attempts to present the data by looking at the overall picture
c. Conclusions and Verification, looks for meanings of the collected data through looking at patterns, relationships, similarities that often arise.

3 RESULT

The outcome of the discussion was a slightly adjusted new list. After establishing a form of consensus, the original text was analyzed and coded again to ensure that the categories and subheadings in the final list covered all aspects of the participants' answers. An ethical aspect the study conducted and the Ethical Advisory Board at the University Obtained ethical approval.

Based on the results of FGD has found four themes. The first theme, which was the duration of time for students that needed guidance proved still be inadequate; the second theme is the decrease in ability of the students; the third theme is the cooperation of educational institutions and the preceptors still be improved; the fourth theme that was obtained was a moral burden of becoming a preceptor.

"I am happy to guide students, however, I feel the time to guide students is still very poor." (P2).

"Preceptors' off duty is always the longest, because they would take over for those with evening shifts" (P7).

"Preceptors must be sincere, our working duration is longer than nurses." (P10)

"The workload is a lot, assignments from campus are also extreme, we should know how to balance time." (P11).

"Preceptors does not only guide students, but they are still obliged to provide nursing care, this is mainly the reason for spending little time with students." (P10).

The opinion above is accordance with Muir et al research, (2013) that states that most of the preceptors perceive preceptorship programs and their role in the program positively, although the difficulty of preceptors to make appointments with perceptees is still a problem.

This research is supported by the opinions of Nurhidayah, Ariani and Siregar, (2016) that states all this while, the preceptors in most hospitals is someone who is given the assignment in accordance to the Decree (SK) manager. In addition to being a preceptor, they still have the responsibility to take care of patients. High workload leads to a lack of interaction between the preceptors with perseptees.

The preceptors reported that they experienced pride in the role of being a preceptor and described the challenges they faced, such as workload and the need to adapt to different learning styles Chang et al. (2013).

Many other opinions of research Indiarini, Rahayu, and Pindani (2015) states that Preceptors are expected to provide more time in the routine orientation rooms for the participants during the internship process so that new nurses can clearly understand the routine of the office room. Following, this is also directed to preceptors from educational institutions, the schedule to guide students in the hospital is still minimal, resulting in some guidance being implemented in educational institutions in the form of discussion.

The second theme that was acquired is the decrease in ability of students. It is based on some of the following opinions:

"Comparing to the past few years, the performance of students worsened." (P2).

"For theory, USU students are ahead, but to grade their practical skills, I'm sorry it's still very inadequate, perhaps it's because other institutions have begun since fifth and sixth semester." (P5).

"Private nursing educational institutions have started studying at the clinic since the third year of academic education, they indeed became more proficient." (P3).

"We feel that the guidance given is not optimal." (P8).

The third theme that was obtained is the cooperation of educational institutions and the preceptors of service institutions that still needs to be improved. It is based on some of the following opinions:

"Before students attend professional education, the preceptors of educational institutions and the preceptors of institutions should meet in accordance to make their views and targets the same." (P7).

"If there is a problem regarding students, communication is done over the phone because we are both busy." (P13).

"Students of USU have guide books, but we preceptors of other service institutions don't have it." (P2).

"Our prints are limited, we once gave guide books to the preceptors of service institutions." (P13).

"Ideally, preceptors of existing service institutions should be placed in the hospital." (P7).

Educational institution has made the plan several times, but the busy education agenda and a limited number of lecturers resulted in the plan to be delayed.

Several studies supported the third theme, which is the study of Burns, et al (2006) which states preceptors are urgently needed to prepare the next generation of clinicians and to provide access to Patients that is so important to clinical learning. In turn, preceptors Obtain satisfaction from meeting a professional obligation.

Information regarding rewards, preceptors was perceived as desirable also Obtained. Findings from this study can be used to develop Preceptor programs that increase of job satisfaction for preceptors and improve the learning environment for oriented, Stevenson et al (1995).

Through a review of available literature and the authors' experience as preceptors and faculty, it is clear that it is possible to implement a mutually beneficial preceptor experience even in today's productivity-based practice models. Preplanning and use of suggested strategies can make precepting an enjoyable and rewarding experience, Barker and Pittman (2010).

Three study correlations Reached statistical significance, suggesting that the commitment to the Preceptor role is positively associated with: (a) preceptors 'perception of benefits and rewards, (b) preceptors' perception of support, and (c) the number of Preceptor experiences. The results have implications for nursing administrators and nursing educators to Ensure that adequate benefits, rewards, and supports are available to preceptors.

The fourth theme that was found was a moral burden of becoming a preceptors.

It is based on some of the following opinions:

"A Preceptors is leaded by the head of the room, before becoming a case manager we are all perceivers." (P9).

"Preceptors must learn again to be able to guide students well." (P6).

"Preceptors used to have special incentives, but since the remuneration system stopped operating, the incentives no longer happened." (P7).

"Advantages of being a preceptor is that if there's training, we are prioritized." (P3).

"There are also nurses who are not willing to be preceptors, although it has been proposed, they still refuse." (P4).

There are many trials of becoming a preceptor—even a perfect preceptor comes with its own disadvantages. As revealed from the study of Widyastuti, Winarni and Imavike (2014) a clinical instructor is a moral burden. This theme forms its own two themes: the helplessness in providing

guidance and maximum responsibility towards the next generation of nurses. In contrary, Omansky (2013) stated that preceptors and mentors asked for recognition and support for a number of jobs involving in teaching students. This is expected to accentuate well-defined roles and responsibilities of a preceptor.

Bengtsson, M and Carlson (2015) adds that the preceptors is also expected to have the skills to be able to establish an effective learning environment and facilitate constructive clinical learning experiences for students and new employees. Meanwhile, according to Tang, Chou, and Chiang, a preceptor is the person that is responsible for ensuring students to learn, apply theory, gain experience, practice techniques, and develop into a skilled nurse, (2005).

4 CONCLUSION

Based on the results of this study, it has been concluded that the Preceptors' perceptions of a preceptorship in the Surgical Care Room at Adam Malik Hospital Medan has found four themes. The first theme, which was the duration of time for students that needed guidance proved to still be inadequate; the second theme is the decrease in ability of the students; the third theme is the cooperation of educational institutions and the preceptors still be improved; the fourth theme that was obtained was a moral burden of becoming a preceptor.

Based on themes found it advisable that preceptors involvement at every stage of learning. A preceptor should have additional knowledge compared to nurses and the cooperation of educational institutions and the preceptors of clinical institutions that still needs to be improved.

REFERENCES

Barker ER., Pittman O. Becoming a super preceptor: a practical guide to preceptorship in today's clinical climate. J Am Acad Nurse Pract. 2010 Mar; 22(3):144–9. doi: 10.1111/j.1745–7599.2009.00487.x.

Bengtsson, M., dan Carlson, E. 2015. *Knowledge and skills needed to improve as preceptor: development of a continuous professional development course, a qualitative study part I*, BioMedCentral (BMC) Nursing Journal. 14:51, p: 1–7.

Burns, C., Beauchesne, M., Ryan-Krause, P., Sawin, K., Mastering the Preceptor Role: Challenges of Clinical Teaching. Journal of Pediatric Health Car. Volume 20 Number 3 p:172–183, May/June 2006. doi:10.1016/j.pedhc.2005.10.012.

Chang A., Douglas M., Breen-Reid K., Gueorguieva V., Fleming-Carroll B. *Preceptors' perceptions of their role in a pediatric acute care setting*.

Hilli, Y. dan Melender, H.L. 2015. Developing *preceptorship* through action research: part 2. *Scandinavian Journal of Caring Sciences*: 2015 Sep; 29 (3):478–85.

Indriarini, MY., Rahayu, BM., dan Pindani B. 2015. Pengalaman Dukungan Preceptor Pada Perawat Baru Selama Proses Magang Di Rumah Sakit Santo Borromeus Bandung. Skripsi.

Maitland, A. 2012. A study to investigate newly-qualified nurses' experiences of *preceptorship* in an acute hospital in the south-east of England. Dissertation of Master of Science in Learning and Teaching Faculty.

Muir J., Ooms A., Tapping J., Marks-Maran D., Philips S, Burke L. Preceptors' perceptions of a *preceptorship* programme for newly qualified nurses. Nurse education Today, 2013 Jun;33(6):633–8.

Nurachmah, Elly. 2011. Kurikulum Pendidikan Profesi Ners di Indonesia. Hand Out. Tidak dipublikasikan.

Nurhidayah, R.E. 2011. *Pendidikan Keperawatan*. Pendekatan KBK. Medan: USU Press.

Nurhidayah, R.E., Ariyani, Y., Siregar, C.T. 2016. Persepsi Mahasiswa Stikes Swasta di Medan terhadap Implementasi Preceptorship Pasca Pendidikan Profesi Ners. Prosisding Seminar Nasional. Bandung: Unpad.

Phelps, L.L. 2009. Effective Characteristics of clinical Instructors. *A Research Paper Submitted to the Graduate School.*

Stevenson B., Doorley J., Moddeman G., Benson-Landau M. The preceptor experience: a qualitative study of perceptions of nurse preceptors regarding the preceptor role. J Nurs Staff Dev. 1995 May-Jun;11(3):160–5.

Tang, F., Chou, S., & Chiang, H. 2005. Students' perceptions of effective and ineffective clinical instructors. *Journal of Nursing Education*, 44(4), 187–192.

Widyastuti, M., Winarni, I., dan Imavike, F. 2014. Pengalaman Menjadi Pembimbing Klinik Mahasiswa Keperawatan Di Instalasi Gawat Darurat Rumkital Dr. Ramelan Surabaya. Skripsi.

Curriculum, Research and Development (CRD)

Ideas for 21st Century Education – Abdullah et al. (Eds)
© *2017 Taylor & Francis Group, London, ISBN 978-1-138-05343-4*

Improving the competences of vocational teachers: Graduate profile and learning outcomes of the agro-industry technology education program

M.N. Handayani
Universitas Pendidikan Indonesia, Bandung, Indonesia

ABSTRACT: The existence of the ASEAN Economic Community (AEC) requires educational institutions to improve the competences of their graduates as professional teachers. The Agro-Industry Technology Education Study Program at UPI (Universitas Pendidikan Indonesia) is a pioneer of agro-industry vocational teacher education institutions in Indonesia. Therefore, the program needs to establish a variety of strategies to create graduates who are ready to compete in the AEC era. This study aims to compile graduate profiles and learning outcomes of the Agro-Industry Technology Education Study Program in order to improve the competences of vocational teacher candidates. The data was collected through interviews, documentary studies, and focus group discussions. The data was analyzed using a qualitative approach. The results showed that (1) the agro-industry technology education graduate profile represented vocational teachers or professionals in technological processing of agricultural products, professional researchers in technological processing of agricultural products, and entrepreneurs in technological processing of agricultural products; (2) Learning outcomes comprised of the ability to apply science and technology, the mastery of theoretical concepts and practical technology education processing of agricultural products, the ability to take strategic decisions in problem-solving in the agricultural processing technology field, the responsibility in work, and good attitude to achieving organizational goals.

1 INTRODUCTION

The establishment of the ASEAN Economic Community (AEC) in 2015 has influenced many aspects. Indonesia as a member of ASEAN should be able to compete with other countries in the era of free trade that includes products, services, investment and skilled labor. Education is the key factor for improving the quality of human resources so it can deliver competitive products, the best services and skilled labor (The ASEAN Secretariat, 2015).

Education in Indonesia consists of academic education and professional education. Academic education is aimed at preparing students to develop academic potential at the level of high school, undergraduate, master and doctoral degrees. Whereas professional education is aimed at preparing learners to increase their potential of competence and appropriate expertise. Professional educational terms used in Indonesia include vocational education at the high school level named SMK (Sekolah Menengah Kejuruan) or vocational high school; vocational education at higher education level is such as polytechnics and diploma programs. Both academic education and professional education should be able to create graduates who are competitive in this AEC era.

The Agro-Industry Technology Education Study Program at UPI (Universitas Pendidikan Indonesia), as a pioneer of producing vocational teacher education graduates as prospective teachers in SMK agro-industry, must have a graduate profile and clear outcomes in the curriculum system. It is necessary for the educational process to run effectively and efficiently, and to be appropriate to the stakeholders' needs and requirements.

Indonesia has a great potential with more than 60% of its young population as a productive workforce. However, the facts show alarming conditions. Based on data from the Indonesian statistics bureau (Central Bureau of Statistics, 2015) the youth unemployment rate in 2015 increased by 31.12%. More specifically, 20% of the unemployed are graduates of vocational high school. This is partly because of the low competences of graduates, who are unable to compete in the AEC era. In order to produce qualified and competent graduates of vocational high school, graduate profiles and learning outcomes of a vocational study program need to be established.

Therefore, this research aims to compile graduate profiles and learning outcomes of the Agro-Industry Technology Education Study Program in order to improve the competences of vocational teachers who are able to compete in the AEC era.

2 RESEARCH METHODS

This research was conducted in three stages: (1) data collection; (2) data analysis; and (3) data analysis presentation. Data was collected through documentary studies, interview and Focus Group Discussions (FGD). Documentary studies were conducted on curriculum documents of the Agro-Industry Technology Education Study Program. Interviews were conducted with agro-industry vocational school teachers. FGD is a qualitative research method that was carried out by a focus group to discuss a specific problem, in an informal and relaxed situation.

Parties involved in FGD were all stakeholders of the Agro-industry Technology Education Study Program, consisting of principals and vice principals of the SMK agro-industry, teachers of the SMK agro-industry, and employees of the agricultural industry. FGD aimed to obtain suggestions and recommendations on the competencies of teachers and competencies of employees needed by agricultural industries. SMKs involved in the FGD were ten schools around West Java Province having an agricultural processing technology program, and four agricultural industries.

Data was analyzed using a qualitative descriptive approach. This study was a qualitative research and used a purposive sampling technique, where schools were selected as samples representing SMKs around West Java. The vocational schools were also selected based on the on-going partnerships with the UPI Agro-Industry Technology Education Study Program, such as field practice and employment of graduates.

3 RESULTS AND DISCUSSION

The Agro-Industry Technology Education Study Program is an institution that aims at producing teachers in vocational high school (SMK) and professionals in the field of agro-industry (Handayani & Mukhidin, 2014). Therefore, the education system and the curriculum development were designed through the steps described in Figure 1.

Graduate profiles and learning outcomes form the basis for the formulation of study materials that had been established in the curriculum structure. Meanwhile, the graduate profile is formulated from a strengths, weaknesses, opportunities, and threats (SWOT) analysis and needs assessment. Curriculum development should be in line with the socio-economic development, and the develop-

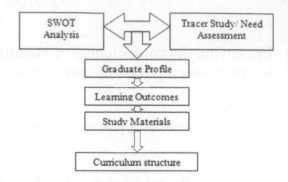

Figure 1. Curriculum development of agro-industry technology study program on education. (Handayani & Mukhidin, 2014).

ment of science and technology to fit the needs of users (Hung, 2013).

In this study, SWOT analysis and needs assessment was obtained through FGD and interviews with stakeholders.

On that basis, it is known that the presence of an agro-industrial technology education study program is needed for the provision of agro-industry SMK teachers. The respondents stated that currently many teachers were facing retirement, but the regeneration system was not yet established. This is an excellent opportunity for study program development. In addition, based on self-evaluation, an agro-industrial technology education study program supported by policy makers and legal university became its own power from the SWOT analysis aspect.

However, according to self-evaluation, interviews and FGD, it was revealed that the facilities and machinery laboratory equipment for learning support were insufficient. This was regarded as a weakness that needs to be resolved. According to the stakeholders from industry, the AEC era becomes a challenge for education because the industry will only use a ready graduate, so it is a threat and challenge to SMK, and also the agro-industrial technology education study program.

A tracer study/need assessment was based on interviews with stakeholders and an FGD (Focused Group Discussion). The following were identified as the required competencies:

– Industry requires a ready workforce, so that prospective vocational teachers are required to be more skillful than vocational school students in order to produce the more competitive SMK graduates. Therefore, the composition of subject matter about science and skills in the agricultural industry should be reviewed.
– Basic knowledge and competencies necessary, based on the first point, include: food processing technology and by-product, herbal beverage

52

food processing technology, physiology, chemistry, biochemistry, packaging technology and packaging design of agricultural products, quality testing of materials, agricultural products, machine tool processing of agricultural products, processing equipment, good manufacturing process, good agricultural practices, sanitation, statistics, food safety, HACCP and ISO system certification, and computer skills and IT applications. It is similar to Myers and Washburn (2008) who revealed that curriculum development in agricultural education should be strengthened with biology, chemistry and physics to improve the understanding and significance of learning.

- Agro-industry vocational schools requires teachers who have the competencies as follows: to understand innovational studies and models of teaching and learning technology of agricultural products so that the process is understood more easily, to have the skills in using computer applications and IT for the development of learning, to have skills in operating and maintaining the machine tool production, processing, and food quality control laboratory equipment, to master sufficient communication skills in foreign languages such as English, to have entrepreneur skills, to have good etiquette, leadership, communication skills, and also the willingness to learn and be able to work in teams. It is supported by Roberts et al. (2007), stating that agricultural vocational teachers are expected to act as facilitators, planners, and executors of learning so that students are actively involved in the whole processes.
- The stakeholders also gave recommendations to the agro-industrial technology education study program to equip students with organization skills, to be actively involved in scientific work involving innovative competition, and to deepen understanding of the national curriculum.

3.1 Agro-industry technology education graduate profile

In accordance with results of research through the study of documentation, interviews and FGD, the agro-industry technology education graduate profile can be formulated as follows:

- Professional teachers in SMK agro-industry, with expertise and competence in processing of agricultural products.
- Professional researchers in technology of agro-industrial processing of agricultural products.
- Entrepreneur in the field of agro-industrial processing of agricultural technology.

3.2 Learning outcomes of agro-industry technology education

Graduate profile is in need of adjustment to learning outcomes so that more clearly in the formulation of the study materials in the curriculum structure. Curriculum development in agricultural education programs should include general education, technical agricultural content and pedagogical knowledge, and also teaching skills (American Association for Agricultural Education, 2001). Consequently, the learning outcomes have to cover such issues as follows: apply science and technology, mastering theoretical concepts and practical technology education of agricultural product processing; able to take strategic decisions in problem-solving agricultural processing technology field and having a responsible attitude to work and achievement of organizational goals.

The following are the learning outcomes of the agro-industrial technology education study program:

- Being able to utilize science and technology in teaching, research, and entrepreneurial technological processing of agricultural products.
- Has the ability to adapt to difficult situations related to education, research, and entrepreneurship in technological processing of agricultural products.
- Mastering the principles of vocational education in particular areas of expertise in technological processing of agricultural products.
- Mastering the principles of vocational education in particular areas of expertise in agricultural product processing technology.
- Mastering the concepts of the pedagogy theory as psychological development of learners, curriculum, lesson planning, teaching models and learning evaluation system in agro-industry technology education.
- Mastering the basic concepts of the general theory that supports the theory of expertise technology of agricultural products such as mathematics, statistics, chemistry, biology, Indonesian, and English.
- Mastering the basic concepts of food processing technology such as: physicochemical and microbiological characteristics of foodstuffs; post-harvest handling technology; quality testing of materials and food products; physicochemical analysis of food products; technological processing of agricultural products (vegetable, animal, herbal, fisheries, plantation crops, forest products); good manufacturing process; food packaging technology; quality management systems, food safety, and food sanitation.
- Mastering the general principles of information communication technology in the implementation

of agro-industrial technology education, research, and entrepreneurship in agricultural products.
- Mastering the concepts and basic principles of research methods in educational research, and also in studies in the field of agro-industrial processing of agricultural products.
- Mastering the basic principles of entrepreneurship, the basic concepts in the analysis of agribusiness agricultural products, as well as marketing-product processing of agricultural products.
- Has the ability to apply principles of vocational education in carrying out education, especially teaching in the field of agro-industrial processing technology.
- Has the ability to apply theoretical concepts of science education and provide alternative solutions to the various problems that occur in the administration of education agro-industrial processing of agricultural products.
- Has the ability to apply principles and concepts of agribusiness and agricultural technology innovation in a study for the community, or in making a business opportunity in agribusiness.
- Able to use computer applications, information technology in agro-industry for teaching, research or entrepreneurship in agro-industrial processing of agricultural products.
- Has the ability to compose research papers, publications, and also dissemination of research results appropriately.
- Has the ability to make decisions and provide various alternative solutions to the problems that occur in the administration of agribusiness agricultural products.
- Responsible for the job, either as teachers, researchers, or as entrepreneurs in agribusiness.
- Able to demonstrate good ethics, professional performance, and a high integrity so as to be held accountable for the achievement of the work of vocational educational institutions, research institutions, or in business/industry.
- Able to work in a team and have a position of leadership in doing good professional work as a prolific teacher, researcher, or entrepreneur.
- Being able to communicate effectively with the various parties within the work undertaken both as a teacher, researcher, or entrepreneur.

4 CONCLUSIONS

The agro-industry technology education graduate profile comprises of: A vocational teacher professional in technological processing of agricultural products; a professional researcher in technological processing of agricultural products; an entrepreneur in technological processing of agricultural products.

Learning outcomes include: (1) ability to apply science and technology; (2) mastery of theoretical concepts and practical technology education processing of agricultural products; (3) ability to take strategic decisions in problem-solving in the agricultural processing technology field; and (4) having a responsible attitude to work and achievement of organizational goals.

REFERENCES

American Association for Agricultural Education. (2001). *National standards for teacher education in agriculture.* http://aaae.okstate.edu/files/ncatestds.pdf.
Central Bureau of Statistics. (2015). *Unemployment data.* Available online at https://www.bps.go.id/linkTabelStatis/view/id/972.
Handayani, M.N. & Mukhidin. (2014). *Development of agroindustry education technology curriculum based on national framework for Indonesia.* Proceeding Seminar of APTEKINDO VII.
Handayani, S., Handayani, M.N. & Cakrawati, D. (2014). *Role of study program of agroindustry technology education UPI in preparing vocational teachers agroindustry.* Proceeding International Conference of UPI UPSI.
Hung, H.X. (2013). *Comparative study on curricula for vocational teacher education in mechanical and electrical engineering.* Available on line at http//www.tvet-online.asia.
Myers, B.E. & S.G. Washburn. (2008). Integrating science in the agriculture curriculum: Agriculture teacher perceptions of the opportunities, barriers, and impact on student enrollment. *Journal of Agriculture Education, 49*(2), 27–37.
Roberts, T.G., Dooley, K.E., Harlin, J.F., & Murphey, T.P. (2007). Competencies and traits of successful agricultural science teachers. *Journal of Career and Technical Education, 22*(2).The ASEAN Secretariat. (2015). *ASEAN economic community blueprint 2025.* Available online at http://www.asean.org/storage/images/2015/November/aec-page/AEC-Blueprint-2025-FINAL.pdf.

Ideas for 21st Century Education – Abdullah et al. (Eds)
© 2017 Taylor & Francis Group, London, ISBN 978-1-138-05343-4

Authentic assessment analysis based on the KKNI curriculum in applied statistics learning

V. Yustitia
Universitas PGRI Adi Buana Surabaya, Surabaya, Indonesia

I.S. Wardani
Universitas Pendidikan Indonesia, Bandung, Indonesia

ABSTRACT: This research aims to describe lecturers' understanding and implementation of authentic assessment based on the Indonesian National Qualification Framework (KKNI) curriculum in Applied Statistics. This study was conducted using qualitative descriptive research. The subjects are Applied Statistics lecturers and students of Class C, year 2014, at Adi Buana PGRI University. The data collection techniques used were observation, interviews, and documentation. Data analysis was done through data reduction, data presentation, and conclusion. The validity of the data was tested by implementing data checking and triangulation. The results showed that: 1) the lecturer knows authentic assessment in the teaching of Applied Statistics; 2) the lecturer has implemented authentic assessments in the teaching of Applied Statistics, which include attitude, knowledge, and skill competency assessments. Attitude competency assessment has been carried out through observation, self-assessments, peer assessments, and assessments of the journals. Competency assessment has been carried out through technical knowledge, written tests, oral tests, and assignments. Skills competency assessment has been carried out through performance assessments, project assessments, product assessments, and portfolio assessments.

1 INTRODUCTION

Education is an integral part of the development and progress of a nation. The quality of education can be seen in the competencies of graduates from educational institutions, including university. Assessment is needed to measure the competencies of graduates. Assessment aspect is one of the keys determining the purpose of learning competencies. Hence, proper attention should definitely be given to assessment (Nurgiyantoro, 2008).

Learning assessment standards are used in the assessment system within the Indonesian National Qualification Framework (KKNI). The standards are defined in the Ministry of Education and Culture Regulation Number 49 (2014) Article (18) paragraph 1 as being minimum criteria of process and learning outcomes of students' assessment in the context of reaching graduates' learning achievement criteria. The assessment of students' process and study outcomes include: (1) the assessment's principles; (2) the assessment's techniques and instruments; (3) the mechanisms and procedures of assessment; (4) the execution of assessment; (5) assessment reporting; and (6) students' graduation.

Universities running good quality assurance systems from institution level to program study generally have implemented learning activity, based on learning achievements from the experiences of the curriculum development team for high education. The Indonesian directorate of high education has implemented training in the development of the curriculum in all Kopertis, a coordinator for private universities in Indonesia, raising some main problems. The problems are: that assessment activities focused on giving points to the students, instead of giving advice and guidance to release their potentials; and that the instruments of assessments tend to have characteristics of summative assessment, instead of formative assessment.

Applied Statistics is one of the subjects in the primary school teacher education department (PGSD) of Adi Buana PGRI University, Surabaya. Based on the researchers' own observations, currently the evaluation process in Applied Statistics subjects only focuses on the mastery of the topics by doing objective and subjective written tests. This is one indicator of less effort in teaching activities done by the lecturer in focusing on attitudes and skills development of the students.

Several problems which frequently appear in the learning and evaluation process are: (1) the lecturer only puts a score on the assessment, without giving any detailed explanation; (2) the lecturer often finds difficulty in assessing students and differentiating expected end criteria for learning outcomes; (3) the lecturer often has difficulty determining the method to conduct appropriate assessment (Kemenristek DIKTI, 2014). Hence, authentic assessment is regarded as appropriate in the teaching and learning process in university.

In fact, the observation found that many lecturers in Adi Buana PGRI University Surabaya had not fully understood what authentic assessment is and how to implement the authentic assessment in the teaching and learning processes. This condition was due to the short explanation of authentic assessment they received from the demo or training. Hence, they expected continuous training and workshops for lecturers related to the implementation of authentic assessment in the university based on the KKNI curriculum.

Based on the description above, the researchers were interested to know more about the implementation of authentic assessment based on KKNI in Applied Statistics subjects in the primary school teacher department of Adi Buana PGRI University, Surabaya. The purpose of this study was to describe the knowledge of lecturers and the chairman about authentic assessment and the implementation of it based on the KKNI curriculum in teaching and learning Applied Statistics.

2 LITERATURE REVIEW

2.1 The Framework of Indonesia National Qualification (KKNI)

The national standard of higher education, as stipulated in the Ministry of Higher Education of Research and Technology Number 44 (2015) Article 1, states that the curriculum is a set of plans and arrangements of learning achievement criteria, the material of the study, process, and assessment used as a guidance to the course of study. The curriculum for higher education institutions is always renewed according to the development of the needs, science, and technology as stated in the learning objectives. The university, as a body that produces educated human resources, needs to measure whether its graduates will have capability and skills equivalent to those that has been formulated based on the KKNI qualification (Kemenristek DIKTI, 2016).

KKNI is a guide to leveled competency qualifications for work recognition (Solikhah, 2015). KKNI was formulated by the Indonesia Ministry of Education and Culture in 2010 as the basis for curriculum development. KKNI was then enacted through presidential regulation number 8 (2012) about the framework of the Indonesian National

Qualification. Thus, every university must adopt this regulation.

2.2 Authentic assessment

Educational assessment is a formal attempt to determine the status of a student, in respect to educational variables of interest (Popham, 1995). According to Kemenristek DIKTI (2016), the assessment for learning is the assessment of process and outcomes of learning. The assessment of process and learning outcomes of the students includes: the assessment's principles; techniques and instruments of assessment; mechanisms and procedures of assessment; implementation of assessment; assessment reporting; and the students' graduation.

According to Budiyono (2015), an authentic assessment usually includes a series of tasks to be done by the students. Those tasks are measured with the help of assessment rubrics. Through rubrics, the students' tasks can be measured. Furthermore, an important aspect of providing students with an authentic context where learning can take place is to intrinsically link the content to be learned with the assessment in a meaningful way (Rourke & Coleman, 2011). Authentic assessment needs to support learning in general and be driven by the learner, to foster the attributes we expect of graduates and help learners prepare for a lifetime of learning (Falchikov & Thompson, 2008).

3 RESEARCH METHODS

This research was qualitative. It was conducted from March 7 until May 16, 2016. It was held in the PGSD of Adi Buana PGRI University, Surabaya. The subjects of this research were the lecturer of Applied Statistics subjects, and the students of Class C, year 2014. The research object was the knowledge of the lecturer of authentic assessment and its implementation based on the KKNI curriculum.

Data collection methods used in this research were observation, interviews, and documentation. Data analysis was done through data reduction, data presentation, and conclusion. Triangulation used in this research was triangulation sources and triangulation method. The triangulation source done in this research was to check degrees of trust obtained from the head of study program of PGSD, lecturers, and college students. The triangulation method is done by checking the degrees of trust discovery both in research results and data collection.

The instruments used were observation sheets, authentic assessment sheets, and interview guidelines. The observation sheet was used to monitor the implementation of an authentic assessment, while an authentic assessment sheet was used to explore the implementation of an authentic assessment. In addition, the interview guidelines were used to avoid questions arising which were not in accordance with research purposes.

4 RESULTS AND DISCUSSION

4.1 Lecturer's knowledge about the implementation of authentic assessment based on KKNI

Based on the interview conducted by the researchers with IS, the head of the primary school teacher education department, she defined authentic assessment as an assessment attached to the learning process which includes attitude, knowledge, and skills assessments. Those assessments were conducted with prepared instruments. IS could also explain very well about the methods that can be used to conduct an authentic assessment based on the KKNI curriculum.

The researchers already have knowledge of authentic assessment based on the KKNI curriculum. This knowledge was obtained through some training and demos, such as The United States Agency for International Development (USAID) and *Program Peningkatan Keterampilan Dasar Teknik Instruksional* (PEKERTI) training. Moreover, authentic assessment on the Applied Statistics subject has been conducted. Based on the results of the interview with one of the students, Rey, the lecturer has already used authentic assessment. The lecturer not only assessed the cognitive ability but also assessed attitudes and skills. The students were asked to demonstrate and apply knowledge and skills obtained in the class to the real life.

4.2 Implementation of authentic assessment based on KKNI

Attitude aspect learning through Applied Statistics aims to improve the quality of living in the society, in Indonesia as a country, and to reach advanced civilization based on Pancasila. To achieve those learning objectives, the lecturer incorporated observation, self-assessments, peer assessments, and journal assessment.

The researchers used observation to assess the attitude of the students in learning activities. The instrument was in the form of the rubric. The lecturers supervised and assessed the attitude of the students in accordance with the rubric's assessment. Self-assessments were conducted twice in a semester. The lecturer shared an instrument assessment themselves with the students and asked them to conduct self-assessment.

Peer assessments were conducted by the lecturer twice in the first semester. They were done by the lecturer through several steps. The lecturer shared assessment instruments with the students, and then the lecturer equalized perception about every indicator which was to be assessed. After that, the lecturer determined and asked them to perform peer assessments in the assessments sheets.

Journal assessments were conducted incidentally by the lecturer, either positively or negatively. The first one was by observing the behavior of the students. After that, the lecturer took notes about the behavior and attitudes for the assessment chronologically by giving time stamps for every student assessment. It was done to identify the strength and weakness of the students.

The aim of cognitive aspect learning through Applied Statistics was to be able to explain basic research and procedures that can solve learning problems in the primary school teacher's education sector. To reach this learning objective, the lecturer used written tests, oral tests, and assignments.

A written test was done and consisted of the daily quiz, mid-semester test (UTS) and final semester test (UAS). Based on the observation, the lecturer has provided written daily quizzes in the form of the same questions provided to discuss particular topics. The students would answer them on a piece of paper.

The oral test was done in each meeting. The lecturer conducted the oral test with some of the students one by one. The questions list was based on the learning objective of Self-Assessment Programs (SAP). The lecturer gave enough time for the students to think about the answer to the questions. The lecturer avoided repressive behavior in conducting this assessment.

Moreover, assignments were given in each meeting. Those were in the form of homework related to the learning objectives of each meeting. Assignment assessment was done by the lecturer in several steps. The first was giving the assignments to the students. After that, those must be submitted at the next meeting. The lecturer would assess the assignment with learning objective criteria provided and give feedback to the students.

The objective of skill competence in the Applied Statistics subject was to make the students able to apply conceptual knowledge in primary school, which is math, by learning in the statistics subject. To reach that objective, the lecturer used performance assessment, product assessment, and portfolio assessment.

Performance assessment was done by incorporating observation sheets for the students. The lecturer did the assessment and compared the results with the rubric. The lecturer would take notes of the assessment results and documented them.

The lecturer only conducted this kind of assessment once in the third observation. The project given to the students was to collect real data from the field, organize, analyze, and report the results back to the lecturer. The lecturer monitored every project and gave proper feedback. The students would compare the performance of other students by using the assessment rubric. The lecturer would take notes of them as the result of the assessment.

Product assessment was conducted by the lecturer in several steps. The steps were in accordance with the steps that must be done for product assessment. Those were: preparation phase, which was the assessment in preparation and development of ideas; and product development, which was the phase in which the students were assessed in selecting and using the materials, tools, and techniques. The next phase was the product assessment of the students based on particular criteria, such as the look, function, and the aesthetics of the products. The products were in the form of statistics learning media, such as the summary of particular learning topics in the form of a Statistic Encyclopedia.

Meanwhile, the portfolio was conducted by giving planned and structured assignments. Other than to assess the learning outcomes of the students, it can be used as a means to monitor the progress of students learning. It was supported by Mueller (2005) who states that the portfolio is a collection of students' works specially selected to show the state of the students in accordance specifically with learning progress. Portfolio assessment was done based on the learning plan of the lecturer. The students' assignments were collected as portfolios. However, there was less than expected because of the limited capability of the students themselves.

4.3 Discussion

KKNI is defined as quality and identity embodiment of the Indonesian nation related to the national education system, national job training system, and national equal learning outcomes system owned by Indonesia in order to get high quality and productive national human resources.

The idea of authentic and sustainable assessment is one that focuses on assessment tasks that have applicability to the world outside the classroom, and that foster autonomous learning. According to Bloxham and Boyd (2007), assessment has four purposes: certification, student learning, quality assurance, and lifelong learning capacity.

An authentic assessment is the proper assessment process for learning process in the university. Callison (Budiyono, 2015) explains that authentic assessment is an evaluation that involves multiple forms of performance measurement reflecting student's learning, achievement, motivation, and attitudes on instructional-relevant activities. Mueller (2005) also stressed that authentic assessment is a form of assessment in which students are asked to perform real-world tasks that demonstrate meaningful application of essential knowledge and skills.

The lecturer and the chairman of the PGSD in Adi Buana PGRI University, Surabaya, have already been able to explain authentic assessment based on the KKNI curriculum in the Applied Statistics subject. A professional education practitioner should be able to: (1) choose appropriate assessment procedure; (2) conduct the assessment, and (3) use the results of assessment to make learning decisions (Reynolds et al., 2010).

The lecturer has done the authentic assessment based on the KKNI curriculum in the Applied Statistics subject, covering the assessment for attitude, knowledge, and skills. The principles and techniques used were in accordance with the KKNI curriculum. According to Popham (1995), the attitude assessment objective focuses on the attitude itself, while the knowledge assessment objective focuses on the intellectual operations of the students, and the skill assessment objective focuses on the physical skills of the students.

Competence assessments were done by conducting written tests, oral tests, and assignments. Cognitive assessments would be developed if the lecturer explored cognitive learning activities properly. Supardi (2015) states that cognitive learning needs cognitive assessment supported by assessment instruments and rubrics.

Skill assessment was conducted in performance assessments, project assessments, product assessments, and portfolio assessments. According to Budiyono (2015), almost all physical activity involves cognitive processes.

In authentic assessment based on the KKNI curriculum practices in the Applied Statistics subject, the lecturer encountered some obstacles. This was due to the limitation of students' knowledge and skills, the less active students when doing authentic assessments, and the limitation of time for the learning activity, so that the assessment was not conducted optimally.

5 CONCLUSIONS

Based on the research done and discussion, it can be concluded that the lecturer and the chairman of the PGSD has already understood authentic assessment based on the KKNI curriculum. The lecturer conducted authentic assessment in the Applied Statistics subject in attitude, knowledge, and skill competencies assessments.

REFERENCES

Bloxham, S. & Boyd, P. (2007). *Developing effective assessment in higher education: A practical guide*. Berkshire: Open University Press McGraw-Hill Education.

Budiyono. (2015). *Pengantar penilaian hasil belajar*. Surakarta: UNS Press.

Falchikov, N. & Thompson, K. (2008). Assessment: What drives innovation? *Journal of University Teaching and Learning Practice*, 5(1), 49–60.

Kemenristek DIKTI. (2014). *Kurikulum pendidikan tinggi*. Jakarta: Direktorat Jenderal Pendidikan Tinggi Kementerian Pendidikan dan Kebudayaan.

Kemenristek DIKTI. (2016). *Kurikulum pendidikan tinggi*. Jakarta: Direktorat Jenderal Pendidikan Tinggi Kementerian Pendidikan dan Kebudayaan.

Mueller, J. (2005). Authentic assessment toolbox: Enhancing student learning through online faculty development. *Journal of Online Learning and Teaching*.

Nurgiyantoro, B. (2008). Penilaian otentik. *Jurnal Cakrawala Pendidikan*, XXVII(3), 101–108.

Popham. W.J. (1995). *Classroom assessment*. Boston: Allyn and Bacon.

Reynolds, C.R., Livingstone, R.B. & Wilson, V. (2010). *Measurement and assessment in education* (2nd ed.). London: Pearson Education, Inc.

Rourke, A.J., & Coleman, K.S. (2011). Authentic assessment in elearning: Reflective and Collaborative writing in the arts. *In ASCILITE-Australian Society for Computers in Learning in Tertiary Education Annual Conference* (Vol. 2011, No. 1, pp. 1089–1095).

Solikhah, I. (2015). KKNI dalam kurikulum berbasis learning outcomes. *Jurnal Lingua*, 12(1), 1–12.

Supardi. (2015). *Penilaian autentik*. Jakarta: PT Raja Grafindo Persada.

REFERENCES

Ideas for 21st Century Education – Abdullah et al. (Eds)
© *2017 Taylor & Francis Group, London, ISBN 978-1-138-05343-4*

The career competence profile of public elementary school students in Jakarta, Indonesia

A. Tjalla & H. Herdi
Universitas Negeri Jakarta, Jakarta, Indonesia

ABSTRACT: This research is motivated by results of theoretical studies, government policies and empirical facts about the importance of improving student's career competence at early ages. The purpose of this research was to get the student's career competence profile of public elementary school students in Jakarta, Indonesia. The study used a survey method. The research sample was 603 grade IV students, 30 headmasters, and 30 teachers of public elementary schools in Jakarta in the academic year of 2016/2017. The research sample was gained by using a multistage cluster random sampling technique. The data collection employed Need Assessment Inventory and Career Competence Scale for elementary school students. Data was analyzed by using percentage, mean, and standard deviation. Operational data analysis was performed by using software IBM SPSS version 20.0 for Windows. The results show that first, headmasters and teachers judge that all of the student's career competence profile is important. Another important thing is master the skills to investigate the world of work, master working strategy, understand the relationship between personal qualities, education and training. Second, the student's career competence profile shows that the condition of the fourth grade students in Jakarta has not reached an optimal level of career competency.

1 INTRODUCTION

The government mandates the specialization program for students in school from an early age, including elementary school, to obtain data and information related to the tendency of students in the direction of their specializations (Agency for Human Resources Development and PMP Kemendikbud, 2013a). One important component in service specialization/career is a specialization program services (guidance and career counseling) to develop an awareness of a career or career competence of learners. With an awareness of career (specialization clear direction) from an early age, students are expected to have completed elementary school career and have awareness of what they want to do after finishing elementary school.

The policy is urged to immediately implement massively with full planning and preparation. The argument is the empirical fact that the open unemployment rate (TPT) in Indonesia in February 2014 rose by 5.7% and unemployment of people graduating from college (open) in the productive age group in February 2014 amounted to 5.71%, and TPT elementary school graduates reached 3.69% (The Central Bureau of Statistics, 2014). According to the Agency for Human Resources Development and PMP Kemendikbud

(2013b) this condition is caused by the tendency of learners who graduated from specified levels (one of them elementary school) who continue their studies to a higher level yet do not possess adequate potentials to do so.

If this is analyzed using development theories, facts emerge about the unpreparedness of graduate schools, including elementary schools, competing in the world of work, and the increasing number of educated unemployment caused by the lack of knowledge about the potential of self (self) and information about the world of work (world of work). Experts in career counseling (Super, 1957; Super in Sharf, 1992; Crites, 1986; Zunker, 1990; Osipow & Fitzgerald, 1983; Herr & Crammer, 1996; Patton & Lokan, 2001; Hassan, 2006; Versnel et al., 2011) state that if people have adequate information about themselves and the world of work, they will achieve career competence – in elementary school: it is called career awareness competence.

The other problem is that the government has not appointed special counselors, or guidance and counseling teachers, in elementary school to support career counseling and guidance services. In fact, the government, through Permendikbud No. 111 of 2014 on the Implementation of Guidance and Counseling in Elementary and Secondary Education Line, issued a policy on mandatory implementation of career guidance and counseling programs and specialization in elementary school by counselors, or guidance and counseling teachers. To solve the problem, the classroom teachers hold the role of career guidance and counseling, and run specialization programs in elementary school. This condition can have an impact on the implementation of the career guidance and counseling program and specialization in elementary school, making it less than optimal, since it was organized by classroom teachers who do not have sufficient competence in the field of career guidance and counseling.

Based on the theoretical study and empirical facts, it is necessary to have systems support to improve the career guidance service, and the counseling program and specialization in elementary school. One solution is through the development of a career guidance and counseling module as a guide for classroom teachers to improve the career competence of elementary school students. Modules have some advantages, one of them being that they 'can be studied independently by participants' (Regulation of State Administration Institution No. 5 Year 2009 on Guidelines for Education and Training Writing module) so that classroom teachers can learn the material and procedure guidance, and the career counseling to be done. On the basis of this rationale, the research focused on the development of career guidance and counseling modules for classroom teachers to improve the career competence of students.

2 LITERATURE REVIEW

2.1 Career development stages of elementary school

Super (1957) (Brown & Lent, 2005; Herr & Cramer, 1996; Gothard et al., 2001; McMahon & Patton, 2006; Osipow, 1983; Sharf, 1992) explains that the career development of students in the schools is at the phase of growth which starts from birth through to about 15 years old. In this phase, students have to achieve a development task which is called career awareness or career competence. In this phase, a child develops various potentials, attitudes, interests, and needs, which are incorporated in the structure of self-concept. The self-concept evolves through the process of identifying key figures in the family and the school. The development phase of the fourth-grade students is in a fantasy sub-phase (age ≤ 10 years), which is characterized by the dominance of aspects of need for curiosity towards self and career.

In this growth phase, elementary students are expected to develop appropriate career competencies. If they succeed to achieve career competency, it is predicted that they will achieve career success at the next career phase. Conversely, if they fail,

it is predicted that they will experience failure and disappointment in their careers at a later stage.

2.2 Career competencies of elementary school students

According to the theory of career development (ASCA, 2005; Cobia & Henderson, 2007; Gysbers & Henderson, 2006) career competencies that must be mastered by elementary school students are as follows.

Standard A: master the skills to investigate the world of work in relation to the manufacture of self-knowledge and career information. Standard B: master working strategy for achieving successful and happy career goals in the future. Competence C: understand the relationship between personal qualities, education and training within the world of work.

3 RESEARCH METHODS

This research uses survey descriptive to describe a phenomenon, an event that occurs in the present (McMillan & Schumacher, 2001). It describes data of the career competence profile of students at elementary schools in Jakarta.

The population is all principals, class teachers, and fourth-grade students of elementary schools in Jakarta in academic year 2016/2017. The sample was selected using a multistage cluster random sampling technique.

The data was collected using an instrument of need and career competence for elementary school students. Research instruments were developed based on constructs of career competence raised by Super (1957), Cobia & Henderson (2007), Gysbers & Henderson (2006) and ASCA (2005). Data analysis used statistic descriptive, namely mean and percentage. Operationally, data analysis used IBM SPSS software version 20.0 for Windows.

4 RESULTS AND DISCUSSION

4.1 Result of validation of construct career competence of elementary school students

The construct of career competence of students is based on the ASCA National Standards for students (ASCA, 2005). In order to construct the career competency in accordance with the needs and context of Indonesia, the principals and classroom teachers of elementary schools in Jakarta were validated.

The results showed that the principals and teachers of public elementary schools in Jakarta assess the construct of career competence of elementary school students as important, with a mean score of 4.2 on a scale of 1–5.

4.2 Profile of career competence of elementary school students

The results showed career competence in fourth-grade elementary school students in Jakarta in the academic year 2016/2017 in the category of less competent (45.2%), fairly competent (41.2%) and competent (13.6%).

Based on the results of the study and after review of the published literature, they showed that the fourth-grade students in Jakarta have not reached an optimal career competency. It is characterized by the following conditions. First, the inability of students in developing skills to put, evaluate, and interpret career information. In this indicator, the students have not been able to identify a source for information about further education after elementary school.

Second, the inability of students to learn a variety of traditional and non-traditional occupations. In this indicator, the students have not been able to determine/choose five kinds of traditional and modern works, and are not familiar with the five duties/responsibilities of the traditional and modern main job.

Third, the inability of elementary school students in developing awareness of their abilities, skills, interests and personal motivations. In this indicator, the students do not know the work they are interested in the future, continuing education that must be taken after completing elementary school in accordance with its ideals, and efforts have been made to achieve the interest in the job.

Fourth, the inability of students in learning how to interact and work together in a group. In this indicator, the students have not been able to identify the stages of completing the task group, or manage time while completing the task group.

Fifth, the inability of students in learning how to make career decisions. In this indicator, the students have not been able to establish how to make a choice from some school extracurricular activities, as a matter of fact, they have to be able to choose their school after graduating from elementary school.

Sixth, the inability of elementary students in learning how to set career goals. In this indicator, the students have not been able to establish continuing education that will be selected after elementary school and identify what will be done to successfully enter desired further education. Seventh, the inability of elementary students in understanding the importance of making career plans.

Eighth, the inability of students in developing hobbies and interests in specific jobs. On this indicator, students have not been able to choose the activities done in their spare time, followed by extracurricular activities at school, extracurricular activities followed outside the school, and tutoring that can be followed outside the school.

Ninth, the inability of students in balancing between work and leisure time. In this indicator, the students have not been able to determine the number of hours of study each day, the number of hours following extracurricular activities per week, and the number of hours used to fill free time every day.

Overall characteristics of learners who have not qualified is certainly contrary to some opinions of experts such as the concept of career awareness (Super in Sharf, 1992), and career competency (ASCA in Cobia & Henderson, 2007; Gysbers & Henderson, 2006). The research result is contrary to constructing the ideal career competence based on the results of validation done by principals and teachers of the fourth-grade students of elementary school in Jakarta (Tjalla et al., 2015). The same research result was found by Nazli (2014) that elementary school students in Turkey know their personal characteristics and can associate them with their future selected career. However, they know less about the implications towards their lives, choice of career, and the way to manage tasks of the selected career.

Based on theoretical constructs and the results of these empirical studies, ideally fourth-grade students have career awareness, such as: (1) develop skills to localize, evaluate, and interpret career information; (2) study various traditional and non-traditional occupations; (3) develop awareness of their abilities, skills, interests, and personal motivation; (4) learn how to interact and work together in teams; (5) study the decision-making; (6) understand the importance of career planning; (7) develop competence in the area of interest; (8) develop hobbies and vocational interests; and (9) the balance between work and free time.

The theoretical and empirical studies reinforce and strengthen the importance of career development of learners at an early age. One of the efforts to do so is to set up a program of career guidance and counseling/specialization in elementary schools. Career guidance and counseling program/specialization in elementary schools is not intended to prematurely make elementary school students take career decisions. Career guidance and counseling/specialization in elementary school is focused on the development of career awareness competence, both in the choice process, how to plan and anticipate choice, and the relationship between personal characteristics, planning, and career choice. Elementary school students need to know that they have the opportunity to choose and have the competence career awareness. Elementary school students need to have self-awareness, awareness of change and personal development, and how to use the school experience to explore and prepare for the future (Herr & Cramer, 1996).

5 CONCLUSIONS

This research has generated constructs for career competence of elementary school students adapted from ASCA National Standards for Students (ASCA, 2005) and is validated by principals and teachers in the elementary schools in Jakarta. The results of the validation showed that constructs career competency, good standard of competence, basic competence, and that the indicator is an important category with a mean score of 4.2 on a scale of 1–5.

This research shows that headmasters and teachers judge that all of the student's career competence profile is important, the other is master the skills to investigate the world of work, master working strategy, and understand the relationship between personal qualities, education and training. Furthermore, we found career competency of elementary students in the category of less competent (45.2%), fairly competent (41.2%), and the remaining competency (13.6%) show that the condition of the fourth-grade students in Jakarta has not reached an optimal career competency.

The researchers can next feed the results of this study into a pilot project to examine the other variables associated with competency development and career development efforts done to develop career competence of elementary school students.

REFERENCES

American School Counselor Association. (2005). *ASCA national standards for students*. Alexandria, VA: Author.

Badan, Pengembangan SDM dan PMP Kemendikbud RI. (2013a). *Modul asesmen peminatan peserta didik*. Jakarta: Kemendikbud.

Badan Pengembangan SDM dan PMP Kemendikbud RI. (2013b). *Panduan khusus bimbingan dan konseling: Pelayanan arah peminatan peserta didik*. Jakarta: Kemendikbud.

Brown, S.D. & Lent, R.W. (Eds.). (2005). *Career development and counseling: Putting theory and research to work*. New Jersey: John Wiley & Sons, Inc.

Cobia, D.C., & Henderson, D.H. (2007). *Developing an Effective and Accountable School Counseling Program (2e)*. Cincinnati: Merrill Publish-Eng.

Coogan, T.A. (2016). Supporting school counseling in Belize: Establishing a middle school career development program. *International Electronic Journal of Elementary Education*, 8(3), 379–390.

Crites, J.O. (1986). *Career counseling: Models, methods, and materials*. New York: McGraw-Hill Book Company.

Gothard, B., Mignot, P., Offer, M. & Ruff, M. (2001). *Career guidance in contexts*. London: Sage Publications.

Gottfredson, L.S. & Lapan, R.T. (1997). Assessing gender-based circumscription of occupational aspirations. *Journal of Counseling Psychology*, 28(6), 545–579.

Gysbers, N.C. (2005). Comprehensive school guidance programs in the United States: A career profile. *International Journal for Education and Vocational Guidance*, 5, 203–215.

Gysbers, N.C. & Henderson, P. (2006). *Developing & managing: Your school guidance and counseling program (4th ed.)*. USA: American Counseling Association.

Hassan, B. (2006). Career maturity of Indian adolescents as a function of self-concept, vocational aspiration and gender. *Journal of the Indian Academy of Applied Psychology*, 32(2), 127–134.

Herr, E.L. & Crammer, S.H. (1996). *Career guidance and counseling through the life span*. Toronto: Little, Brown & Company.

Knight, J.L. (2015). Preparing elementary school counselors to promote career development: Recommendations for school counselor education programs. *Journal of Career Development*, 42(2), 78–85.

McMahon, M. & Patton, W. (2006). *Career counseling: Constructivist approaches*. London: Routledge.

McMillan, J.H. & Schumacher, C. (2001). *Research in education: A conceptual introduction*. New York: Addison Wesley Longman, Inc.

Nazli, S. (2007). Career development in primary school children. *Career Development International*, 12(5), 446–462.

Nazli, S. (2014). Career development of upper primary school students in Turkey. *Australian Journal of Guidance and Counseling*, 24(1), 49–61.

Osipow, S.H., & Fitzgerald, L.F. (1983). *Theories of career development (p. 322)*. Englewood Cliffs, NJ: Prentice-Hall.

Patton, W. & Lokan, J. (201). Perspectives on Donald Super's construct of career maturity. *International Journal for Educational and Vocational Guidance*, 1, 31–48.

Sharf, R.S. (1992). *Applying career development theory to counseling*. California: Brooks/Cole Publishing Company.

Super, D.E. (1957). *Scientific careers and vocational development theory: A review, a critique and some recommendations*. Oxford, England: Columbia University.

Tjalla, A., Herdi, & Kustandi, C. (2015). Model online career counseling untuk meningkatkan kematangan karir peserta didik Sekolah Menengah Kejuruan (SMK). *Laporan Penelitian Hibah Bersaing DIKTI: Tahun Ketiga*. Jakarta: Tidak diterbitkan.

Versnel, J. DeLuca, C., Hutchinson, N.L., Hill, A., & Chin, P. (2011). International and national factors affecting school-to-work transition for at-risk youth in Canada: An integrative review. *The Canadian Journal of Career Development/Revue [canadienne de'veloppement de carri'ere]*, 10(1).

Zunker, V.G. (1990). *Career counseling: Applied concepts of life planning (2nd ed.)*. Monterey, California: Brooks/Cole Publishing Company.

Educational Foundation (EDF)

Ideas for 21st Century Education – Abdullah et al. (Eds)
© *2017 Taylor & Francis Group, London, ISBN 978-1-138-05343-4*

Promoting undergraduate students' critical thinking skills in zoology vertebrate courses

S. Sa'adah, F. Sudargo & T. Hidayat
Universitas Pendidikan Indonesia, Bandung, Indonesia

ABSTRACT: The goal of science education is to develop critical thinking skills. Critical thinking skill is important in education as well as in everyday life. In this study to promote students' critical thinking skills we used a team-based learning instructional model. The purpose of this study was to evaluate the potential effects of a Team-Based Learning (TBL) instructional model on undergraduate students' critical thinking skills. A quasi-experimental pre-test/post-test with control group design was used to determine critical thinking gains in a TBL and a non-TBL group. The instrument to collect data was open-ended questions that represented by critical thinking elements. The result showed that TBL can improve undergraduate students' critical thinking skills and also showed that the TBL group had significantly quantitative differences compared to the non-team-based learning group in improving students' critical thinking skills. Through the results of this study, it is hoped that the faculties who value both research and critical thinking will consider using the TBL instructional model.

1 INTRODUCTION

21st-century competencies that must be possessed by students are critical thinking and problem-solving skill (Trilling & Fadel 2009; Ledward & Hirata, 2010). This is also in line with the goal of science education, namely to develop critical thinking skills (Bailin, 2002), which means avoiding memorizing terms, but instead building relationships between concepts, applying the appropriate framework for solving the problem, and drawing conclusions critically (Bransford et al., 2000). The importance of critical thinking has been proven since the time of Socrates (Quitadamo et al., 2008). Critical thinking skills are regarded as one of the essential skills that directly influence academic and professional success (Bassham et al., 2010).

Critical thinking skills have a strategic role in the field of education. Many college faculties consider critical thinking to be one of the most important indicators of student learning quality (Quitadamo & Kurtz, 2007). This is in line with the demands of the working world that higher education stakeholders seek individuals who are able to think critically and communicate effectively are to join their company (Benjamin et al., 2013). Therefore, higher education faculties need to make practical instructional changes, in order for students to better compete on the international stage (Quitadamo et al., 2008). The learning paradigm in higher education

should change from conventional learning that emphasizes the low-level thinking skills towards learning that emphasizes learning higher-order thinking skills or where less emphasis is placed on content-specific knowledge and more is placed on critical thinking skills, such as analytic and quantitative reasoning, and problem-solving (Benjamin et al. 2013). According to the AAAS (1989) and NRC (1996), science is learned and taught as is done in real life. Rutherford and Ahlgren (1990) state that the education of science should help students develop an understanding and habits of thought to face their future life. Learning that is not an emphasis on the development of higher level thinking skills (critical thinking skills) tends to condition students into rote learning. Students very easily forget previously learned material. Bassham et al. (2011) reported that the learning of most schools tends to emphasize lower-level thinking skills. Therefore, the purpose of this study was to improve students' critical thinking skills through the team-based learning strategy.

2 LITERATURE REVIEW

Critical thinking is a general term given to a variety of cognitive skills or cognitive activity and intellectual dispositions associated with the use of the mind (Cotrell, 2005; Basham et al., 2011). Various definitions of critical thinking have been offered by the researchers. Norris and Ennis (1989) give the definition of critical thinking as to think reasonably and reflectively with emphasis on making decisions about what to believe or do. Inch et al. (2006) state that critical thinking is a process in which a person tries to answer rational questions that cannot easily be answered and where no relevant information is available. Critical thinking requires consideration. Halpern (2014) states that critical thinking is the use of cognitive skills or strategies that increase the likelihood of desired results. It is used to describe the thinking as being purposeful, reasoned, and goal-directed.

Cotrell (2005) suggested that good critical thinking skills bring many benefits such as increased attention and observation, more focused reading, improved the ability to identify important and less important, and analysis skills that can be applied in different situations. Basham et al. (2011) suggested that the benefits of students' critical thinking are that students can understand, critically evaluate, and build arguments. Further, Basham et al. (2011) state that critical thinking is beneficial to the world of work and in everyday life. For example, critical thinking skills can help in making decisions carefully, clearly and logically, so as to reduce the possibility of making a fatal mistake when making decisions.

Critical thinking skills are supposed to be part of the curriculum in schools. Students must be given meaningful experiences for learning in order to develop their critical thinking skills. Thus, teachers as educators are obliged to condition the learning so that students are able to develop intelligence and critical thinking skills. To meet the expectations of the above, it is necessary to develop a learning program that allows students the opportunity to practice using critical thinking skills, because critical thinking skills is an ability that can be learned (Halpern, 2014).

One of the strategies that are expected to improve critical thinking skills is Team-Based Learning (TBL). TBL is a pedagogical strategy that uses groups of students working together in teams to learn course material. The main learning objective in TBL is to provide students the opportunity to practice course concepts during class time (Clair & Cihara, 2012).TBL is one form of collaborative learning which consists of three phases: (1) During the first phase, learners read and study material independently outside class. (2) Learners complete an Individual Readiness Assurance Test (IRAT), After the IRAT, pre-assigned teams of 5–7 learners re-take the same test (Group Readiness Assurance Test (GRAT)). (3) Application and integration of information that has been obtained in phases 1 & 2 (Michaelsen & Sweet, 2008).

3 RESEARCH METHODS

The study took place at a Program Studi Pendidikan Biologi, UIN SGD, Bandung. All participants were undergraduate students on a zoology vertebrate course. The participants consisted of two classes – the control and experimental class. The research was quasi-experiment research with randomized pre-test/post-test control group design (Fraenkel & Wallen, 1990). The instrument to collect data was open-ended questions that were representative by critical thinking skill elements (Noris & Ennis, 1989). A feasibility instrument test had been tested with reliability 0.8 and was categorized as high (Arikunto, 2005). Analysis of the differences critical thinking skill enhancement was conducted with parametric statistical tests by using t-tests on N-gain critical thinking skill, which have previously been tested for normality and the homogeneity of F distribution.

4 RESULTS AND DISCUSSION

Comparison of the critical thinking skill between experiment classes and control classes as a whole can be seen in Table 1. Based on Table 1, the means of experiment class pre-test, post-test and N-gain are higher than the control class, and statistically using t-test showed that for TBL there were significantly quantitative differences compared to the non-TBL group (sig. 0.00, $\alpha = 0.05$). So, it can be concluded that a TBL strategy can promote critical thinking skill more significantly than can a traditional strategy.

Comparison of the percentage of category N-gain critical thinking skills of students between the experimental and control classes is shown in Figure 1.

Based on Figure 1, the percentage of high category in the experimental class was more than the control class. Enhancement of critical thinking skill for each indicator/element can be seen in Table 2.

Based on Table 2, undergraduate students' critical thinking skills improved in all indicators of critical thinking skill. Students in class experiments had higher critical thinking skill enhancement than

Table 1. Recapitulation of the pre-test, post-test, and N-gain critical thinking.

Sample	Mean Pre-Test	Mean Post-Test	Mean N-Gain $<g>$
Experiment Group	41	80	0.7
Control Group	42	67	0.4

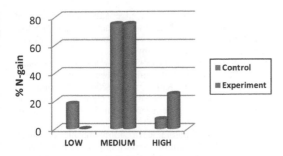

Figure 1. Comparison of the percentage of category N-gain critical thinking skills of students between the experimental and control classes.

Table 2. Enhancement of critical thinking skill for each indicator.

Indicator critical thinking skill	Pre-Test	Post-Test	N-Gain	Category
Decide on an action.	53	81	0.6	Medium
Suggests an attitude orally or in writing.	59	89	0.7	Medium
Making the induction and consider induction.	46	79	0.6	Medium
Analyzing argument.	33	79	0.7	Medium
Ask and answer questions.	25	74	0.7	Medium
Consider the credibility of the sources.	21	70	0.6	Medium
Focusing questions.	38	81	0.7	Medium
Identifying assumptions.	33	59	0.4	Medium
Identifying similarities and differences.	46	87	0.8	High

the control class. This may be influenced by the students themselves and the learning processes factor. The factor of the student had a very large impact on student learning outcomes; students with high motivation and attention will be able to achieve optimal learning (Makmun, 2008; Sudjana, 2009).

We found that TBL was an effective teaching strategy that focused on students' practice in applying concepts in class. By assessing students on both individual work and team work, students were motivated to come to class and engage in the group activities. The students were motivated to prepare for class every day, because they did not want to let their teammates down. Students learned good habits such as reading the material before coming to class. Generally, students enjoyed the TBL strategy for a variety of reasons, such as it not being boring, it was fun, and they gained a better understanding of learning materials, but this learning process requires thorough preparation and good time management.

Previous studies have shown that the use of a variety of learning strategies can improve students' critical thinking skills (Addy & Stevenson, 2014; Carson, 2015; Caruso et al. 2016; Aebli & Hutchison, 2016).

5 CONCLUSIONS

Based on the results and discussion, it can be concluded that the TBL strategy may increase undergraduate students' critical thinking skills. The results also indicated that the TBL strategy significantly facilitated critical thinking skill enhancement compared to a non-TBL strategy. Through the results of this study, it is hoped that faculties who value both research and critical thinking will consider using the TBL strategy.

REFERENCES

Addy, M.T. & Stevenson, M.O. (2014). Evaluating biological claims to enhance critical thinking through position statements. *Journal of Microbiology & Biology Education, 15*(1), 49–50.

Aebli, K. & Hutchison, E. (2016). Classroom activities to engage students and promote critical thinking about genetic regulation of bacterial quorum sensing. *Journal of Microbiology & Biology Education, 17*(2), 284–285.

American Association for the Advancement of Science. (1989). *Science for all Americans*. A Project 2061 Report on Literacy Goals in Science, Mathematics, and Technology, Washington, DC.

Arikunto, S. (2005). *Dasar-dasar evaluasi pendidikan.* Bumi Aksara: Bandung.

Bailin, S. (2002). Critical thinking and science education. *Science & Education, 11*, 361–375.

Bassham, G., Irwin, W., Nardone, H. & Wallace, J.M. (2010). *Critical thinking A student introduction* (4th ed.). New York: McGraw Hill Company.

Benjamin, R., Klein, S., Steedle, J., Zahner, D., Elliot, S. & Patterson, J. (2013). *The case for critical-thinking skills and performance assessment.* Council for Aid to Education.

Bransford, J.D., Brown, A.L. & Cocking, R.R. (2000). *How people learn: Brain, mind, experience, and school (Expanded Ed.).* Washington, DC: National Academy Press.

Carson, S. (2015). Targeting critical thinking skills in a first-year undergraduate research course. *Journal of Microbiology & Biology Education, 16*(2), 148–156.

Caruso, J.P., Israel, N., Rowland, K., Lovelace, M.J. & Saunders, M.J. (2016). Citizen science: The small world initiative improved lecture grades and California critical thinking skills test scores of nonscience major students at Florida Atlantic University. *Journal of Microbiology & Biology Education, 17*(1), 156–162.

Clair, K. & Cihara, L. (2012). Team-based learning in a statistical literacy class. *Journal of Statistics Education*, 20(1).

Cotrell, S. (2005). *Critical thinking skills developing effective analysis and argument*. New York: Palgrave Macmillan.

Fraenkel, J.R. & Wallen, N.R. (1990). *How to design and evaluate research in education*. McGraw-Hill, Inc.: Washington.

Halpern, D. (2014). *Thought and knowledge: An introduction to critical thinking* (5th ed.). New York: Psychology Press.

Inch, E.S., Warnick, B. & Enders, D. (2006). *Critical thinking and communication: The use of reason in argument*. Boston: Pearson Education.

Ledward, B.C. & Hirata, D. (2010). *An overview of 21st century skills*. Pacific Policy Research Center.

Makmun, A.S. (2008). *Psikologi pendidikan*. Bandung: Rosda.

Michaelsen, L.K. & Sweet, M. (2008). The essential elements of team-based learning. *New Directions for Teaching and Learning,* 116.

National Committee of Science Education Standard and Assessment (National Research Council)/NRC. (1996). *National science educational standard*. Washington: National Academy Press.

Norris, S.P. & Ennis, R.H. (1989). *Evaluating critical thinking*. Pacific Grove California: Midwest Publication.

Quitadamo, I.J. & Kurtz, M.J. (2007). Learning to improve: Using writing to increase critical thinking performance in general education biology. *CBE-Life Science Education, 6*, 140–154.

Quitadamo, I.J., Faiola, C.L., Johnson, J.E. & Kurtz, M.J. (2008). Community-based inquiry improves critical thinking in general education biology. *CBE-Life Science Education, 7*, 327–337.

Rugierro, V.R. (2012). *Beyond feelings: A guide to critical thought* (9th ed.). New York: McGraw Hill Company.

Rutherford, F.J. & Ahlgren, A. (1990). *Science for all Americans*. Oxford: Oxford University Press.

Sudjana, N. (2009) *Dasar-dasar proses belajar mengajar*. Bandung: Sinar Baru Algesindo.

Trilling, B. & Fadel, V. (2009). *21st century skills: Learning for life in our times*. San Francisco: Jossey-Bass A Wiley Imprint.

Ideas for 21st Century Education – Abdullah et al. (Eds)
© *2017 Taylor & Francis Group, London, ISBN 978-1-138-05343-4*

Information processing capability in the concept of biodiversity

S. Rini, A. Rahmat & T. Hidayat
Universitas Pendidikan Indonesia, Bandung, Indonesia

M. Gemilawati & D. Firgiawan
SMA Negeri 8 Bandung, Bandung, Indonesia

ABSTRACT: The purpose of this study is to investigate and to compare the ability to process information on biodiversity issues in Indonesia using video and text. Information processing capability was measured using an essay adapted from Marzano which includes four components: identification of information, interpretation of information, relate the information, and use the information. This test was given to 30 students at 10th grade in one of the *Sekolah Menengah Atas* (SMA) at Kota Bandung. The results showed that processing the information using video was better than by using text. The average obtained for processing information using video was 3.41 and processing information using text was 3.38 on a scale of 4. It can be concluded that information given using video was more easily understood by students in information processing than that given using text. Using video as a media source of information can help students to imagine the information clearly and convert the information correctly into meaningful information.

1 INTRODUCTION

1.1 *Background*

With the fast growth of scientific development, students are able to get any information and knowledge from many sources. The information may not be appropriate for the concept that has been learned by the students. Before the students connect the various concepts, they need to process the information.

The information shorting process, which generally means information summarizing, needs experience and understanding about the issues. Students with less experience and low understanding will find it difficult to connect various information. The impact is that students have a high cognitive load because they receive a lot of information, but they have strong logical reasoning to help them organize their knowledge (Rahmat et al., 2015). Common issues in society involve a lot of information and they can make knowledge increase.

Environment and biodiversity becomes a common issue that is often discussed, especially environmental damage or loss of the source of biodiversity. The damage to the environment is the actions that cause the changes directly or indirectly against physical characteristics and or declared that resulted in the environment not being able to function again in supporting sustainable development (Yudhistira et al., 2011). The damage to the environment or biodiversity due to human activities, such as construction or environmental changes into a new habitat, can cause a negative impact on the environment.

However, having many well-known issues and information does not make them meaningful to the students. Moreover, the information is considered as general information and causes misconceptions if it is used. The error in processing the information becomes a reason to reveal how the image of information is processed and to see students' social sensitivity, especially towards the biodiversity issue in Indonesia. This research aims to know the ability of high school students in processing the information on the issue of biodiversity in Indonesia.

1.2 *Literature*

Brain activity is an important process and we cannot ignore it (Iskandar, 2011). The process that occurs in the information processing includes metacognition, which is related with use and strategy to complete the cognitive tasks and the ability to manage their own thinking process (Kandarakis et al., 2008). To more understand the things to be accepted by our senses, people will try to organize their past experiences. The experience of organizing and separating important and unimportant ideas is becoming crucial. Moreover, people will try to connect one idea with other ideas to create new ideas in the thinking system.

Knowledge or new ideas that are owned by each person are the result of processing information. Knowledge in every person is not merely cognitive but is the result of the process of cognitive interaction and experience that already exists, or the conjunction of the various kinds of knowledge that have been acquired by the brain (Caldwell, 2012). Processing information that occurs in the human brain is seen from the newly received information that can be combined with the existing information by matching the concept that exists in the brain to prevent wrong retention in brain memory.

There are several components that must be passed in processing information, developed by Marzano et al. (1993). In the framework, there are four supporting components and each component is related to one another. The four components are:

1. Identification of the information
2. Interpretation of the information
3. Connecting the information
4. Using the information

Processing information starts with information selected that is considered important and effective. The shorting of the information that is considered important in an issue or learning materials becomes the main thing that should be noted in order to avoid misconception. The next process is the interpretation of the information where the information that has been selected and previously owned is translated and given an appropriate meaning, using the experience that is obtained in the acceptance of the information. The result of the interpretation of this information is the retention of information in the form of memory that will ease up the application of the information that has been obtained in the end.

The third process is connecting the information. This process, that will make the information that is accepted by each individual, will be different, depending on the thinking process that is done by the individual. Each individual must integrate knowledge, ideas and information for new information; they are required to think critically and logically against the information in front of them. It may be said that if the information-connecting process becomes difficult, it is because it requires the ability to think critically and logically. The final process is to use the information in a different situation. This is the application of the information, that has been linked with various ideas to be the more meaningful information.

2 RESEARCH METHODS

2.1 Method

This research was a descriptive research that aimed at investigating the ability of high school students in processing the information on the issue of biodiversity in Indonesia. Processing information was assessed using worksheets and observation sheet during the learning activities. The questions were given in the form of the information presented during the lesson from video or text and according to standard information processing (Marzano et al., 1993).

2.2 Subject

The subject in this research was 30 students at 10th grade in one of the senior high schools at Kota Bandung. This research used purposive sampling because they usually use many media as a source of information.

2.3 Analyze data

This research was carried out in three days to investigate the development in the processing information capability. Every day, students received information from video and text about biodiversity. The data was analyzed using observation sheets to find the ability to process information. Observation sheets helped to describe the information they had received, and asked them to make a real illustration if needed. After that, worksheets were also used to measure the progress in every component at processing information. Results from observation sheets and worksheet were then analyzed using rubrics on processing information.

3 RESULTS AND DISCUSSION

3.1 Results

Processing information was carried out via two sources, video and text. The students' abilities were then measured in terms of four components, namely identification of information, the interpretation of information, connecting information, and the using of information. The components were related to each other so that if one of the components was not processed properly, other related components were affected. The result from processing information using video and text is given in Figure 1.

Based on Figure 1, the score of processing information using the text and video was 3.40 out of 4.00. This may indicate that the students were able

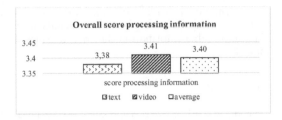

Figure 1. The results of the entire processing information.

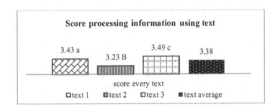

Figure 2. Result of the processing of information using the text.

to process the information obtained from identifying it until they finally come up with the correct information. The information that may not be completely received through the text makes the information processing scores slightly lower. The text shorting on the whole was actually influenced by the results of the information processing on each the text.

Based on Figure 2, text 2 resulted in the lowest information processing scores. Text 2 discussed about the benefits of mangrove forest conservation for the community. The given text was actually easily understood but there were a number of items of new information about the benefits of mangrove forest that were considered difficult to comprehend.

The use of video as a media enabled students to visualize the information. The result obtained at least gave an overview of the students' ability to respond to the video as a learning resource. The results of the information processing average score on the use of video as a learning resource are shown in Figure 3.

The use of video can help students in processing the information. Based on Figure 3, video 1 gets the lowest score for information processing. In video 1, students cannot receive the information clearly.

3.2 Discussion

The average obtained for processing information using video was 3.41 and processing information

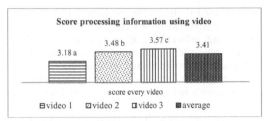

Figure 3. Result of the processing of information using video.

using text was 3.38, out of a scale of 4. Different media had different influences to stimulate each person. The media has a different speed of communication to stimulate the development of students (Skrzypek et al., 2011). Using text as a source of information is one way to facilitate the students who have a visual learning. In addition, media texts make it easy for students to use the learning resources repeatedly until they think they no longer need them. Easy access to the text provides the opportunity for students to equate perception of the information provided.

The disadvantage of using only text as media for sharing information is the difficulties in visualizing technical information. At this stage, an error occurred because each student can create a different concept/image about the information. When the students read the text, and were required to make simple visualization about information in their minds, some alternative meanings/errors occurred. Errors when describing the concept may create errors in processing the information or misconceptions in the students' minds.

To minimize the error in describing the situation clearly, the use of video can help to illustrate the situation on biodiversity in Indonesia correctly. Using video can help those students who learn not only using visual aid, but also using audio aid. The information is obtained through the screenings of videos that will directly be processed by the brain. The brain has a special sensitivity to change about movement, contrast and color that occurs in the video, so that it is easy to understand (Jensen, 2011). In addition, the use of color can provide more long memories (Jensen, 2011).

Students gave extra focus to watching video because video has interesting pictures, and detailed explanation will provide knowledge (Jun & Holland, 2012). Processing information using video actually has a great impact in the dissemination of information. Less sighting information processing is obtained through the videos that are not used yet. According to research by Jun & Holland (2012), the processing of information will be increased if the picture presented is interesting and accompanied by a strong explanation, while processing information does not increase if the picture presented is just normal, even though the explanations provided are very detailed.

When new information is received by students, it will be connected to *long-term memory* for processing in the brain (Yahaya, 2009). Information processing begins with the processes of the identification, interpretation, connecting and applying the information. The process of identification in the acceptance of information is the beginning of the process in the brain when receiving the information. This process has a high effect because if there is a mistake in identifying information and the information is accepted, there can be misconceptions. Identifying the information becomes unique because the brain has a different strategy to make meaningful information (Jensen, 2011).

If students are able to perform the identification information components but have not yet acquired a conclusion, then the identification process will be unsuccessful. The identification of the information will be successful if the information received can be united with the right concept or information. The merger of new and old information has its own difficulty because the students needed the understanding and broader perspective in order to combine the various information to form the information that has meaning.

4 CONCLUSIONS

The average obtained for processing information using videos was 3.41 while processing information using the text was 3.38. This indicated that the use of video was more easily understood by students in processing information compared with the use of the text of the discourse. Not all of the information that was in the discourse and video was used in the hard skill information, which indicated that the information provided was not entirely meaningful for students. Therefore, the students were able to use other information that was considered important.

REFERENCES

Caldwell, R. (2012). Systems thinking, organizational change and agency: A practice theory critique of Senge's learning organization. *Journal of Change Management*, 1–20.

Iskandar, Tb.Z. (2011). *Psikologi lingkungan teori dan konsep.* Bandung: PT. Refika Aditama.

Jensen, E. (2011). *Pembelajaran berbasis otak paradigma pengajaran baru edisi kedua.* Jakarta: PT. Indeks.

Jun, S.H. & Holland, S. (2012). Information-processing strategies: A focus on pictorial information roles. *Journal of Travel Research*, *51*, 205–218.

Kandarakis, A.G. & Poulos, M.S. (2008). Teaching implications of information processing theory and evaluation approach of learning strategies using LVQ neural network. *WSEAS Transactions on Advances in Engineering Education*, 111–119.

Marzano, R.J., Pickering, D. & McTighe, J. (1993). *Assessing student outcomes: Performance assessment using the dimensions of learning model.* Virginia: Association for Supervision and Curriculum Development.

Rahmat, A., Nuraeni, E., Soesilawaty, S.A., Hernita, Alawiyah, D., Garnasih, T. & Noorwahidah. (2015). Beban kognitif & kemampuan penalaran siswa SMA, MA dan SMA berbasis pesantren pada pembelajaran Biologi. *Prosiding Seminar Nasional Sains & Entrepreneurship* II, 240–245.

Skrzypek, W., Potyrala, K. & Walosik, A. (2011). Formation of biology student social skills with use of ICT tools. *Western Anatolia Journal of Education Science*, 219–224.

Yahaya, A. (2009). *Information processing and its implications to teaching and learning.* Kuala Lumpur: Faculty of Education, University Technology Malaysia.

Yudhistira, Y., Hidayat, W.K. & Hadiyarto, A. (2011). Kajian dampak kerusakan lingkungan akibat kegiatan penambangan pasir di Desa Keningar Daerah Kawasan Gunung Merapi. *Jurnal Ilmu Lingkungan*, 1–9.

The contribution of creative thinking skills to students' creativity on enzyme kinetics practical projects using local materials

D.K. Sari
Doctoral Student in Science Education, Universitas Pendidikan Indonesia, Bandung, Indonesia
Universitas Sriwijaya, Palembang, Indonesia

A. Permanasari & F.M.T. Supriyanti
Universitas Pendidikan Indonesia, Bandung, Indonesia

ABSTRACT: The purpose of this study is to learn the contribution of creative thinking skills to students' creativity on the enzyme kinetics topic in the Biochemistry Experiment course. The subject of the project-based practical work is 40 students in the Chemistry Program. In the class, the students were asked to do a creative thinking skills test (before and after class) as well as they could, so that their creative products could be assessed. The instruments used were a creative thinking skills test and a creative product rubric. Correlation analysis was performed using SPSS 22.0. The results showed that there is a significant increase in the students' creativity on the project-based enzyme kinetics experiment using local materials. The increase of the students creative thinking skills shown by the <g> value was 0.46 (intermediate category). The Spearman correlation analysis shows a strong correlation between creative thinking skills and the students' creativity. It can be concluded that the project-based learning on enzyme kinetics laboratory work can enhance creative thinking skills. Furthermore, creative thinking skills give a high contribution to the students' creativity.

1 INTRODUCTION

Experimental activities should make students do minds-on as well as hands-on activities. The shifting of laboratory activities from the traditional with laboratory recipes into inquiry-based laboratory will make a positive impact on the experience, understanding, and creativity of the students alongside learning (Bellin et al., 2010; Coleman, 2010; Witherow & Carson, 2011; Di Trapani & Clarke, 2012). Student activities in the practical laboratory work will build the creative thinking skills of students (Olson & Loucks-Horsley, 2000; Minstrell & van Zee, 2000; Chin & Chia, 2005; Hong & Kang, 2009). Moreover, creative thinking skills will lead to creativity as is mentioned by Paul et al. (in Al-Suleiman, 2009). It is very possible to investigate how creative thinking skills contribute to the creativity. According to Paul and Elder, and Williams and Torrance (Al-Suleiman, 2009), some indicators of creative thinking skills to be used related to this study include **fluency** (the ability to generate ideas), **flexibility** (the ability to express a variety of solutions or approaches to the problems), **originality** (the ability to spark ideas in ways that are original), and **elaboration** (the ability to describe in detail).

Essentially, creative thinking skills will lead to creativity. The learning conditions that allow a student to create the meaningful creative product are the personal conditions and environmental conditions, namely the extent to which both encourage people to involve themselves in the creative process (activity, activity). The product itself is very diverse, ranging from the discovery of the mechanism and new chemical processes, to the new solutions or new statements about an issue in mathematics and science.

Amabile (1983) argued that a product or a person's response is stated to be creative if the experts or observers who have the authority in the field say that it was creative. Thus, creativity is the quality of a product or a creative response, assessed by an expert or observer. Amabile also revealed, that a product is a creative thing if (a) it is new, unique, useful, correct, or valued in terms of specific needs, and (b) more heuristic, which displays the methods that have never or rarely been applied by others before. Meanwhile, according to Bassemer and Treffirger (in Bassemer, 2005), creative product is classified into three categories, known as the Creative Product Analysis Matrix (CPAM), such as novelty, solving, and detail. The first category is the **novelty**; this category assesses the extent to which the product is new, in the number and area of new processes, new techniques, new materials, new concepts, and creative products in the future. The products were original which is very rare among products made by people with the same experience and training, also provoked surprise (surprising) and germ (may lead to the idea of the original product, etc.). The second category is **solving** (resolution); this category concerns the degree of an extent to which it meets the needs to address the problem. There are three criteria in this dimension: product must be meaningful, logic, and useful (can be applied in practice). The last category is **detail** (elaboration); this dimension refers to the degree of an extent to which the product is not the same merger to finally become sophisticated and coherent.

In this study, researchers tried to develop the creative thinking skills in enzyme kinetics laboratory work. The project-based learning model using local materials was used to facilitate the development of creative thinking skills as well as building creativity to find a creative product. Hsieh et al. (2013) revealed that the lecture-based project provides the opportunity for students to do teamwork and stimulate creativity and design ideas, as well as in solving problems. Furthermore, Bell (2010) stated that project-based learning is student-centered and driven by the need to create an end product. As some references stated the close connection between creative thinking skills and creativity, it is very interesting to know how creative thinking skills contribute to the creative product in enzyme kinetics lab-based projects with local materials.

2 RESEARCH METHODS

The pre-experiment study, followed by correlation study, was used to elaborate on what is the enhancement of the creative thinking skills, and what is the contribution of creative thinking skills to the creativity of student. The design used in this study is *pretest-posttest control group design* (Fraenkel & Wallen, 1993). The subject of the study was 40 students who attended lectures in biochemistry laboratory at the University of Sriwijaya (UNSRI) in February 2016, and who were divided into six groups.

2.1 Procedure study

All groups of four students had to make a procedure of enzyme kinetics experiment as a creative product that would be assessed. They had to learn the basic concept involved before doing the project. At the beginning and the end of the class, students had a creative thinking skills test and their creative product was then assessed.

2.2 Instrument of data collecting

Quantitative and qualitative data was collected simultaneously in this study. Creative thinking skills tests were used as quantitative data. The test consisted of seven question descriptions which measured the creative thinking skills of the student laboratory activities. The questions represented all indicators of creative thinking skills (fluency, flexibility, originality, and elaboration (Torrance, 1984)). Analysis of creative products was also used as quantitative data based on indicators, such as originality, reflecting problem-solving and detail. The qualitative data was obtained through analysis of student response to the implementation of the project-based learning.

2.3 Data analysis

Data was analyzed using SPSS 22.0 to see the normality, the value of $<g>$ formula Hake (Meltzer, 2002; Archambault, et al., 2008), and the average/ the mean of the pretest and posttest. A correlation test was also done to see if creative thinking skills contribute to the creative product produced. Qualitative data (questionnaires) were analyzed descriptively.

3 RESULTS AND DISCUSSION

Previously, the biochemistry experiment for the enzyme kinetics topic used commercial enzymes. Students did the experiment without any challenge to build/train the higher-order thinking such as creative thinking. In this study they were asked to seek, to choose, and to think of the alternative enzyme which can be used. They had to seek from local materials around them. They also had to try the procedure out, to ensure that the procedures that they created could be used properly. Each group has to propose a title, destination, local materials and procedures independently. The local materials were obtained from sourced enzyme extract *polyphenol oxidase*, which was then optimized to the factors that affect enzyme kinetics. Summary results of the student activities during practical work of enzyme kinetics are shown in Table 1.

The implementation of lecture can be assumed will increase the creative thinking skills, and the increase can be assessed by using *pretest-posttest control group design* (Fraenkel & Wallen, 1993). The creativity was assessed through the product (procedure for enzymes kinetics experiments) through the section analysis. Pre-posttest mean scores and the score for creativity (Product) can be seen in Table 2.

The $<g>$ value 0.46 in Table 2 shows the increase of creative thinking skills of students in project-based enzyme kinetics laboratory using local materials.

Table 1. Summary of results of enzyme kinetics laboratory activities.

Local materials	Potatoes	Apples	Pears	Banana (Ambon)
Source	Bulbs	Fruit	Fruit	Fruit
Substrates	Catechol	Catechol	Catechol	Resorcinol
[S] (drops)	25	20	25	25
[E] (drops)	15	15	5	1
Temperature (°C)	4	4	4	50
pH	11	11	11	11
Inhibitor	EDTA	EDTA	EDTA	EDTA

Table 2. The mean score pretest-posttest, products and value $<g>$.

	Score		
	Pretest	Posttest	Products
Average	41.92	68.80	82.08
Sd	7.63	13.24	2.44
$<g>$	0.46 (medium)		

Table 3. Normality test.

	Kolmogorov-Smirnov			Shapiro-Wilk		
	Statistic	Df	Sig	Statistic	Df	Sig
Creative product	0.209	40	0.000	0.829	40	0.000
Creative thinking skills tests	0.076	40	0.200	0.964	40	0.236

This is in line with Robinson (2013), that laboratory activity can enhance creative thinking skills if it is well planned. Furthermore, Ashraf (2013) concluded that creative thinking skills of students will develop when the student laboratory work is associated with everyday life. On this laboratory project, students used local materials as enzymes, such as apples, pears, potatoes, and bananas. The procedure used during the practical work varied according to local materials chosen by the students. Laboratory procedures were analyzed using the rubric of creative products. From Table 2 it is also seen that the average score for creative product based on analysis section is 82.08. By using SPSS 22.0, the normality test results are shown in Table 3.

The normality test results in Table 3 show that the p-value calculated by the Shapiro-Wilk test for creative thinking skills is $p > 0.05$, and meanwhile for the creative product is $p < 0.05$. This means that the data was not normally distributed so that the

Table 4. Correlation test for creative thinking skills tests with the creative products.

			Creative products	Creative thinking skills tests
Spearman's Rho	Creative products	Correlation coefficient	1.000	0.710*
		Sig (2-tailed)		0.000
		N	40	40
	Creative thinking skills tests	Correlation coefficient	0.710*	1.000
		Sig (2-tailed)		0.000
		N	40	40

non-parametric statistics (Spearman correlation test) was used. The next hypothesis test of correlation is at the level of 5% (α 0.05). If the significance value is less than 0.05, then Ho is accepted that there is a significant correlation between creative thinking skills with creative products produced. If the significance value is more than 0.05, then Ho is rejected, and it means there is no significant correlation between the two variables. Correlation test results are shown in Table 4.

From Table 4, it can be seen that the value of Sig. (2-tailed) test showed a correlation of 0.000 with a confidence level of 5% (α 0.05). This means that Ho is accepted, which means there is a significant correlation between creative thinking skills tests with creative products produced. This is in line with research conducted by Lou et al. (2012) that creativity in project-based learning can produce creative products on using three indicators, such as originality, acceptance, and usability. Furthermore, this study reveals that the creative products result from project-based learning should be in accordance with the topics required, have value, and apply new concepts for designing a unique product.

4 CONCLUSIONS

From the research, the project-based learning on enzyme kinetics laboratory work enhanced creative thinking skills as well as creativity. The enhancing of creative thinking skills highly contributed to the creativity of students in high category.

ACKNOWLEDGMENT

The authors would like to thank the Ministry of Research, Technology and Higher Education of Indonesia for the scholarship of BPPDN and the Post-Graduate education in Indonesia University of Education.

REFERENCES

Al-Suleiman, N. (2009). Cross-cultural studies and creative thinking abilities. *Journal of Educational and Psychologic Science, 1*(1), 42–92.

Amabile, T.M. (1983). *The social psychology of creativity.* New York: Springer-Vedag.

Archambault, J. Burch, T., Crofton, M., & McClure, A. (2008). *The effect of developing kinematics concepts graphically prior to introducing algebraic problem solving techniques.* Action Research required for the Master of Natural Science Degree with Concentration in Physics. Arizona State University.

Ashraf, S.S. (2013). Raising environmental awareness through applied biochemistry laboratory experiments. *The International Union of Biochemistry and Molecular Biology, 41*(5), 341–347.

Bassemer, S.P. (2005). Be creative using creative product analysis in gifted education. *Creative Learning Today, 13*(4), 1–4.

Bell, S. (2010). Project-based learning of the 21st century: Skills for the future. *The Clearing House, 83*, 39–43.

Bellin, R.M., Bruno, M.K. & Farrow, M.A. (2010). Special section: Innovative laboratory exercises purification and characterization of taq polymerase: A 9-week biochemistry laboratory project for undergraduate students. *The International Union of Biochemistry and Molecular Biology, 38*(1), 11–16.

Chin, C. & Chia, L. (2005). Problem-based learning: Using ill-structured problems in biology project work. *Science Education, 90*(1), 44–67.

Coleman, A.B. (2010). New ideas for an old enzyme: A short, question-based laboratory project for the purification and identification of an unknown LDH isozyme*. *The International Union of Biochemistry and Molecular Biology, 38*(4), 253–260.

Di Trapani, G. & Clarke, F. (2012). Biotechniques laboratory: An enabling course in the biological sciences. *The International Union of Biochemistry and Molecular Biology, 40*(1), 29–36.

Fraenkel, J.R. & Wallen, N.E. (1993). *How to design and evaluate research in education.* Singapore: McGraw-Hill.

Hong, M. & Kang, N.H. (2009). South Korean and the US secondary school science teachers' conceptions of creativity and teaching for creativity. *International Journal of Science and Mathematics Education, 8*(1), 821–843.

Hsieh, H.Y., Lou, S.J. & Shih, R.C. (2013). Applying blended learning with creative project-based learning: A case study of wrapping design course for vocational high school students. *Journal of Science and Technology, 3*(2), 18–27.

Lou, S.J., Chung, C.C., Dzan, W.Y. & Shih, R.C. (2012). Construction of a creative instructional design model using blended, project-based learning for college students. *Creative Education, 3*(7), 1281–1290.

Meltzer, D.E. (2002). The relationship between mathematics preparation and conceptual learning gains in physics: A possible "hidden variable" in diagnostic pretest scores. *American Journal of Physics, 70*(7).

Minstrell, J. & van Zee, E.H. (2000). *Inquiry into inquiry learning and teaching in science.* Washington, DC: American Association for the Advancement of Science.

Olson, S. & Loucks-Horsley, S. (2000). *Inquiry and the national science education standards: A guide for teaching and learning.* Washington, DC: National Research Council.

Robinson, J.K. (2013). Project-based learning: Improving student engagement and performance in the laboratory. *Anal Bioanal Chem- Springer, 405*, 7–13.

Torrance, E.P. (1984). *Torrance test of creative thinking streamlined (revised) manual including norm and direction for administering and scoring figural A and B.* Bensville, IL: Scholastic Testing Service.

Witherow, D.S. & Carson, S. (2011). A laboratory-intensive course on the experimental study of protein—protein interactions. *The International Union of Biochemistry and Molecular Biology, 39*(4), 300–308.

Ideas for 21st Century Education – Abdullah et al. (Eds)
© *2017 Taylor & Francis Group, London, ISBN 978-1-138-05343-4*

The effect of 'Everyone is a teacher here' strategy on students' results in geography

M. Meilia, G.N. Nindya & Z.K. Habibah
Universitas Pendidikan Indonesia, Bandung, Indonesia

ABSTRACT: The learning strategy 'Everyone is a teacher here' gives an opportunity for the student to act as a teacher for the other students. Therefore, this study aims at investigating the impact of the implementation of the 'Everyone is a teacher here' strategy on the result study of students in the Geography subject. The participants of this study were the students of eighth grade at SMP Negeri 1 Belitang III East OKU. The research was conducted through quasi-experiment with a non-equivalent pretest-posttest control group design. The data was collected by an instruments test. The obtained data was analyzed statistically using an independent t-test. The results show that the students' results at an experiment class were higher than the students' results at a control class. This means that the implementation of the 'Everyone here is a teacher' strategy gave positive impact in enhancing the students' learning outcome. The t-test analysis confirms the significance of the students' results improvement at an experiment class compared to the students at a control class.

1 INTRODUCTION

Based on the observations of pre-researchers who conducted observations in SMP Negeri 1 Belitang III East OKU, it is known that the learning process is still conventional, meaning that the teaching is still using lectures, notes and assignments, so that the learning process does not encourage the students to be more active, but makes the students passive. Another phenomenon shows that an average yield minimum obtained completeness criteria (KKM) of 75. For the class VII SMP Negeri 1 Belitang III east OKU, the subject of integrated social studies (Geography) still remains low, with only 45–50% of students who pass.

In the article entitled "The Science", teachers need to choose the right strategy that can drive the students to learn actively, to have sense of curiosity, and to enhance meaningful knowledge. (Handelsman et al., 2004). The process of learning is student-centered. There are several stages which the active students will pass through: having a new idea, planning, the implementation, evaluation, and development (Gottlieb, 2004). The teacher as the spearhead of the learning should not just settle the case. The teacher should engage the students in the learning process, give them an active role to express their opinions, provide communication training, and get them to collaborate in solving problems (Abdurrahman, 2010).

Based on the last analysis that has been done (Nasrawati, 2013), the results of the observation show that the strategy 'Everyone is a teacher here' is very effective to the result of students' learning in Class VIII SMP Negeri 2 Parepare at the subject matter of mean material reaction of oxidation reduction. The results of the observation show that the strategy 'Everyone is a teacher here' is able to increase the results of student learning in Sociology lessons in class XI IPS 1 SMA Negeri 1 Pejagoan (Khanifah, 2014).

As it is stated above that one of the learning strategies is to increase the quality of students' learning outcomes. 'Everyone is a teacher here', which is one of the learning strategies that are designed to obtain the participation of the class as a whole and individually. It also gives a change for the students, providing opportunities for the students to take on the role as a teacher to his/her friends. With this strategy, students will engage in active learning and develop their creative thinking (Zaini, 2008). On the other side, the purpose of the active learning is in helping to develop the studying environment by engaging the students, to develop their willingness to participate in active learning, to create the norms of the positive classrooms, and to help to renew team building and take the students' interest in the subject (Silbermen, 2012: 40). Based on the related opinion with the learning strategy 'Everyone is a teacher here', so the strategy can become a solution to increase the students' learning outcome on integrated social studies (Geography), especially the forms of material of environmental damage and causes.

In the specification to answer this research, the researchers formulated the problem 'Is there the Effect of Learning Strategy "Everyone is a teacher here" for the Geography learning outcomes in eighth grade students at SMP Negeri 1 Belitang III East OKU?'. The aim of this research is to assess the effect of the learning strategy 'Everyone is a teacher here' for Geography learning outcome for the students' class eight SMP Negeri 1 Belitang III southern Sumatera, East OKU.

2 LITERATURE REVIEW

2.1 Learning strategy 'Everyone is a teacher here'

The learning strategy 'Everyone is a teacher here' is designed to obtain the participation of the class as a whole or individually. This strategy provides an opportunity for students to act as a teacher to his or her friends. With this strategy, the students who before do not want to be actively involved in learning, will be involved in active learning (Djamarah, 2010: 44). The active learning includes a variety of ways to make students active from the beginning through the activities that build teamwork, within a short time: to make them think about the lesson, be responsible, lead the study for the whole class or for the small groups, to stimulate the discussion and debate, to practice the skills, to encourage their questions, and even for each student to teach one another or peer tutor (Silbermen, 2012: 22).

By placing the student as the object in the learning process strategy 'Everyone is a teacher here', it provides a great way to involve students directly and allow participation for the whole class or individually. On the other hand, this strategy gives an opportunity for the student to act as tutor to his friends or peer tutor (Suprijono, A, 2011: 110; Silberman, 2012: 183). The strategy 'Everyone is a teacher here' has the benefit of being able to make the student gain the opportunity individually or in groups to present and answer the questions, and to train the student to be responsible and have courageous opinion. The teacher is able to measure the students' understanding for the material (Silberman, 2012).

2.2 Learning outcomes

The results of the study show the students' achievement, while the learning achievement is the indicator of the degree for the change in the students' behavior (Hamalik, 2012: 159). Based on the theory of Bloom's Taxonomy (Suprijono, 2011: 6) the learning outcome is achieved through three domain categories: cognitive, affective, and psychomotor. In this research, the result of the study is used in the cognitive only; where students do not only hear but are able to look for the problem herself or himself. The learning result is a changing pattern of the values, attitudes or someone's actions. The

skills are shown by one of the results of the study that are implicit in the purpose of learning effectively and efficiently (Suprijono, 2011: 5; Djamarah, 2010: 87), to study from the learning result that the student obtained is reported as an illustration of the success during the learning process from the academic result, skills and attitudes (Gelmon et al., 2001). The results of study include three cognitive areas, namely cognitive (intellectual mastery), the field of affective (attitude and value), and an a psychomotor (ability) (Suprijono, 2011: 5–6).

The Council of Higher Education Accreditation stated that the results of the correct students' learning has the definition in knowledge, skill, and abilities that the student has reached in the end or as a result of the involvement in the learning process (Svanstrom et al. 2008). The result of learning that is obtained by someone has a direct benefit and there are rewards and encouragement to keep trying to get better. Besides that, the results of learning becomes an accreditation measurement that indicates the success of reading comprehension. The other side the result of learning is changing someone from not being able to do something to being able to do so, as a result of their learning experiences (Watson, 2002: 208).

3 RESEARCH METHODS

This kind of research is a quasi-experimental design with non-equivalent control group pre-test-posttest, where the students' learning outcomes are compared with the learning strategy 'Everyone is a teacher here' and with the lecture method. This research is done in the class eight SMP Negeri 1 Belitang III Sumsel East OKU, in the first semester of the academic year 2013/2014. This research population was all students in grade eight SMP Negeri 1 Belitang III East OKU with a total of 212 students. The sample technique using random sampling was taken randomly. The samples were VIII.1 class with a total of 34 students as the experiment class using learning strategy 'Everyone is a teacher here', and in VIII.3 class as the control class with a total of 34 students by using the lecture method. Here is a draft of the research planning:

Figure 1. Research design.
Note:
X_1 = Learning strategies 'Everyone is a teacher here'
X_2 = Conventional (lecture method)
$O_{1,3}$ = Pre-test
$O_{2,4}$ = Post-test.

Before the treatment is given, both of the classes were given the pre-test, then given the treatment, and then ended with the post-test.

The item test that was used in this observation was 25-items test in multiple choice with the tacsonomi kognitive indicator Anderson and Krathwhol kognitive Class VIII.3, using a formula:

Validity (Arikunto, 2006).

$$r_{XY} = \frac{N\sum xy - \left(\sum x\right)\left(\sum y\right)}{\sqrt{\left\{N\sum x^2 - \left(\sum x\right)^2\right\}\left\{N\sum y^2 - \left(\sum y\right)^2\right\}}}$$

Reliability (Arikunto, 2006).

$$r_{11} = \frac{2x r_{\frac{1}{2}\frac{1}{2}}}{\left(1 + r_{\frac{1}{2}\frac{1}{2}}\right)}$$

From the 25 items test were derived the validity result and readability 20 items test, so 5 items tests were discarded. Instruments that were used were 20 items test in multiple choice questions that had been tested on the validity and reliability according to the indicator of cognitive taxonomy Anderson and cognitive Krathwhol by remembering, understanding, applying, analyzing, evaluating and creating. The statistical analysis used the correlation of product moment to examine the correlation of independent variables to dependent variable, with the determination of value (r) not more than the value $(-1 \leq r \leq +1)$ in research of value is t_{table} 0.339. To know the differentiate of hypothesis testing, parametric t-test is used using statistical parametric t-test or test the different experiment that given treatment with the control class that was not given treatment by using a learning strategy 'Everyone is a teacher here', with the testing criteria if $t_{count} \geq$ from t_{table}, so Ha is accepted, and if $t_{count} \leq$ from t_{table} Ha is rejected.

4 RESULTS AND DISCUSSION

The process steps of data analysis that the researcher did was to collect the test data of the student learning outcome, obtaining the average value of learning outcome after application of the learning strategy 'Everyone is a teacher here' in VIII.1 class is 81.76, while learning outcomes that do not use learning strategy 'Everyone is a teacher here' in the Class VIII.7 was 70.14. This can be more clearly seen in the following Table:

Table 1. Comparison of the results of experiment class learning and control classroom.

Class	Average	Criteria
Eksperimen	81.76	Good
Kontrol	70.14	Enough
Comparison	11.62	

Based on the comparison of the average student's learning outcome after being given the treatment by using the learning strategy 'Everyone is a teacher here' is better, and so the different response to that of the control classroom that only used the lecture, is 11.62. Furthermore, normality test data obtained for the experiment classes, in the Class VIII.1 gained 0.03, whereas for data normality test which did not use learning strategy 'Everyone is a teacher here' in grades VIII.7 obtained 0.76. It can be concluded that the test data with applied learning strategy 'Everyone is a teacher here' was not normally distributed.

The hypothesis testing in this research by using t-test, with the following criteria: Ha is received if $T_{count} \geq T_{table}$ and rejected Ha if $T_{count} \geq T_{table}$. Based on the data analysis about the students' learning outcome through t-test technique, so it can be derived that $t_{count} = 6.18$. While $t_{table} = 1.98$ from the result of counting can derive that $t_{count} \geq$ from t_{table} or $6.18 \geq 1.98$. Ha is accepted it means it is significant and the hypothesis test is accepted, there is the implementation of learning strategy 'Everyone is a teacher here' for the student learning outcome in social studies integrated (Geography) basic competence of forms for the damage environment and cause factor.

The results of the above statistical hypothesis test proves that the learning strategy 'Everyone is a teacher here' has an influence on student learning outcome, where the result of student learning outcomes is higher than the control class, where the students tend to be more active in the learning process. This result is supported by the opinions of Hamruni (2011: 163) that strategy 'Everyone is a teacher here' is developed to motivate the student to be more active in the learning process, as well as having the ability to think, the skills to ask and to express an opinion.

5 CONCLUSIONS

Based on the results and discussion obtained, the average value of student Class VIII.1 is 81.76 by using the learning strategy 'Everyone is a teacher here', which is greater than Class VIII.7 70.14 with the lecture. It means that the learning strategy

'Everyone is a teacher here' can increase learning outcomes for Integrated Social Science (Geography) Class VIII students of SMP Negeri 1 Belitang III East OKU. Thus, learning strategy 'Everyone is a teacher here' tends to be more effective in increasing the students' learning, where the students are positioned as objects that are directly involved in the learning process, playing an active role in learning. This is different in the control classroom that uses lectures during the learning process, and it can be seen that the result of learning that is obtained is lower.

The advice given in this research is that the teacher should pay attention to the basic competencies that will be achieved, and also customize the character and the students' material, not usually asking the student to hear but instead involve the student as the object that can take care of their own problems, and also actively participates in the learning process and the teacher only acts as the facilitator.

REFERENCES

Abdurrahman, K. (2010). Learning-centered micro teaching in teacher education. *International Journal of Instruction, 3*(1).

Arikunto, S. (2006). *Prosedur penelitian suatu pendekatan praktik*. Jakarta: Rineka Cipta.

Djamarah, S.B. (2010). *Aswan Strategi belajar mengajar*. Jakarta: Rineka Cipta.

Gelmon, S., Holland, B., Driscoll, A., Spring, A. & Kerrigan, S. (2001). *Menilai KKN dan sipil keterlibatan: Prinsip dan teknik*. Providence, RI: Kampus Compact.

Gottlieb, S. (2004). *Innovative assessment in competency based student centered learning*. Sakarya, Esentepe Campus of Sakarya University.

Hamalik, O. (2012). *Kurikulum dan pembelajaran*. Jakarta: Bumi Aksara.

Hamruni. (2011). *Strategy pembelajaran*. Yogyakarta: Insan Madani.

Handelsman, J., Ebert-May, D., Beichner, R., Bruns, P., Chang, A., DeHaan, R., Gentile, J., Lauffer, S., Stewart, J., Tilghman, S.M. & Wood, W.B. (2004). Scientific teaching. *Science, 304*(5670), 521–522.

Khanifah, M.N. (2014). Penerapan model pembelajaran aktif tipe "everyone is a teacher here" untuk meningkatkan hasil belajar siswa pada mata pelajaran sosiologi di kelas. XI IPS 1 SMA N 1 Pejagoan Tahun 2013/2014. *Jurnal: FKIP UNS*.

Nasrawati. (2013). Keefektifan strategi everyone is a teacher here terhadap hasil belajar siswa kelas X SMA Negeri 2 Parepare. *Jurnal Jurusan Kimia, FMIPA UNM*.

Silberman, M. (2012). *Active learning*. Bandung: Nuansa.

Suprijono, A. (2011). *Cooperative learning teori & aplikasi paikem*. Yogyakarta: Pustaka Belajar.

Svanstrom, M., Gracia, F.J.L. & Rowe, D. (2008). Learning outcomes for sustainable development in higher education. *International Journal of Sustainability in Higher Education, 9*(3).

Watson, P. (2002). The role and integration of learning outcomes into the educational process. *Active Learning in Higher Education, 3*(3), 205–219.

Zaini, H. (2008). *Strategy pembelajaran aktif*. Yogyakarta: Pustaka insan.

Ideas for 21st Century Education – Abdullah et al. (Eds)
© *2017 Taylor & Francis Group, London, ISBN 978-1-138-05343-4*

Students' misconceptions on titration

H.R. Widarti
Doctoral Student in Science Education, Universitas Pendidikan Indonesia, Bandung, Indonesia

A. Permanasari & S. Mulyani
Universitas Pendidikan Indonesia, Bandung, Indonesia

ABSTRACT: The potency of misconceptions has been investigated in the analytical chemistry subject, especially on titration concepts. A descriptive method was used in this study, involving 66 students who were taking the Analytical Chemistry course. The seven items of multiple-choice tests with reasons were used as an instrument and the results were then analyzed using a modified Certainty of Response Index technique. The results showed that misconceptions mainly occurred in macroscopic and symbolic level of representation on choosing measuring equipment for titration, using titration equipment, and calculating titration.

1 INTRODUCTION

Misconception may occur due to the understanding of concepts not corresponding to the actual concepts (Suparno, 2005; Berg & Brouwer, 1991). Luoga et al. (2013) also stated that misconception is inconsistency between students' and experts' views. Misconceptions will affect students' understanding level if not resolved immediately. Pinarbasi et al. (2009) reported misconception on the colligative properties subject, and advised that the strategies of learning should be applied using substantial review. Moreover, Yarroch (in Nakhleh, 1992) has carried out the research on misconceptions associated with chemical reaction. He reported that all students managed to equalize the equation, but most of them could not draw molecular diagrams correctly to explain the similarities sub-microscopically. Regarding the size of atoms, Eymur et al. (2013) reported that the misconceptions of students who are prospective teachers and high school students were almost similar. Misconceptions are highly dependent on many factors, such as experience, creativity, perception, and textbooks.

Pinarbasi (2007) reported that Turkish students have a number of common misconceptions on the topic of acids and bases. The research about misconceptions on acids and bases has also been performed by Demircioglu et al. (2005), Rahayu et al. (2011), Damanhuri et al. (2016) and Tumay (2016). Pan & Henriques (2015) proposed at least six types of misconceptions on acids and bases concepts. The type of misconceptions that mostly occurred

was the notion that the endpoint or equivalence point of a titration is always at pH 7.

Titration concept is a basic concept for learning analytical chemistry. The previous research by Sheppard (2006) on acid-base titration reported that junior high school students have major difficulty with the acid-base concept. They also are not able to accurately describe the concept of acid-base, such as pH, neutralization, acid-base strength, and acid-base theory. Several factors contribute to these difficulties: the basics of chemistry on acid-base are too visceral, with more emphasis on mathematical calculations for learning, and are textbooks-dependent. The density of the acid-base concept, the confusing acid-base terminology and lack of consensus on what material should be included in the curriculum of chemistry were also identified as problems. Widarti et al. (2016) reported that they found misconceptions on redox titration, where misconceptions occurred on concepts involving concentration measurements with chemical equation, species existing in solution from titration process, and the characteristics of redox titration involving potential calculation.

A previous research on titration was carried out involving 38 students who were taking the Basic Analytical Chemistry (BAC) course. The study showed that although the students' ability in BAC was considered as fair in category, three times tests indicated significant deterioration (the scores respectively: 72.79, 66.00, and 59.00). Based on the analysis on students' answers, they are generally weak in mathematics literacy and the ability to provide relevant analysis of how to determine the choice of indicators in a titration (Widarti et al., 2016). It is highly associated with a student's low ability in explaining phenomenon in a submicroscopic and symbolic way. Moreover, the characteristics of the misconceptions are durable, being firmly entrenched in one's mind (Louga et al., 2013). Therefore, the aim of current research was to investigate students' misconceptions on the concept of titration.

Relating to personal belief, the use of Certainty of Response Index (CRI) techniques is relevant to reveal student misconceptions. The modified CRI, as proposed by Hakim et al. (2012), was used to accommodate bias from the Indonesian students' culture. Mostly, students understand the concept but are not confident enough with their answers. Therefore, showing the additional reasoning in every student's answer will help to analyze the data. According to Kurbanoglu and Akin (2010), students who mastered the concept well also showed high confidence. The modification of CRI scaling was also performed as recommended by Potgieter et al. (2005), by using four scales (1–4): 1 = guessing, 2 = not certain, 3 = certain, 4 = very certain.

2 RESEARCH METHODS

A descriptive method was used in this research. Subjects of the research were 66 students in the 3rd semester in the department of Chemistry Education of a university in East Java. Misconceptions in the study are revealed by the merger of two instruments that have been developed, namely the multiple-choice test of seven questions on the concept of titration with open reason and the CRI technique by Hasan et al. (1999) which has been modified. The technique can distinguish among knowing the concept well, not knowing the concept and having misconceptions. The reasoning is needed as a reflection of their thinking and understanding of the proposed concept. From the answers and the reasoning, the congruence of the students' understanding with the scientific concept can be discovered.

Based on the answers, reasoning and CRI, there are some possibilities happening. If a student's answer is correct with correct reasoning and with CRI > 2, then the student is classified as understanding the concept well. If the student's answer is correct with correct reasoning but with CRI < 2, then the student is still classified as understanding the concept well. If the student's answer is wrong with correct or wrong reasoning and with CRI > 2, then the student is classified as having misconception. If the student's

answer is wrong with correct or wrong reasoning and with CRI < 2, then the student is classified as not understanding the concept. If the student's answer is correct with wrong reasoning and with CRI < 2, then the student is classified as not understanding the concept.

3 RESULTS AND DISCUSSION

A data test on volumetric analysis in the BAC course reflected the students' skills level. Each student has a different level of comprehension and understanding of concepts although they study the same material with the same lecturer. The different ability may cause the different comprehension and misconceptions of the concepts learned. As was stated before, titration concepts have already been taught since senior high school. In reality, the BAC course could not reduce all of the misconceptions. The total number and percentage of the students' answers on titration are shown in Table 1.

The students' answers to concepts 4, 5, and 7 (washing and rinsing the erlenmeyer, rinsing technique along titration, and stoichiometric titration, respectively) lead to misconceptions, of which more than 50% of students had misconceptions (Table 1). In contrast, there were many fewer students having misconceptions on concepts 2 and 3 (titration processing, and washing and rinsing burette, respectively). The number of students who had misconceptions on concepts 1 and 6 (selection of equipment, and selection of titration method, respectively) was comparative to the number of students who had understood the concepts. Moreover, there were only a few students who did not understand the concepts, around 13.3% of all students.

The misconception which most commonly occurred was in the answer to question number 4 (64%). The question asked was 'what has to be done when provided erlenmeyer flasks has already rinsed off but not dried and it will be used immediately for titrating HCl with NaOH solution?'. Most of the students answered with misconceptions potential. Students explained that the erlenmeyer has to be rinsed by HCl solution instead of aquadest. Some of them said that the erlenmeyer should be dried, and some others answered to use the erlenmeyer immediately. The erlenmeyer as a container for sample or titrant (HCl) has to be clean and dry. If not, it will affect the quantitative result. Wet apparatus can be used but it must be rinsed off with aquadest. However, if the apparatus is already dry and clean, there is no need to rinse it off. When dried with tissue, erlenmeyer will become dirty because the wet tissue will stick to the erlenmeyer.

Table 1. Total number and percentage of students' misconceptions on titration.

No	Concept	Misconception (M)		Not Understand (N)		Understand (U)	
		Total	Percentage	Total	Percentage	Total	Percentage
1	Selection of equipment	29	44	4	6	33	50
2	Titration processing	10	15	3	5	53	80
3	Washing and rinsing burette	5	8	2	3	59	89
4	Washing and rinsing Erlenmeyer	42	64	3	4	21	32
5	Rinsing technique along titration	41	62	3	4	22	34
6	Selection of titration method	1	1	31	47	34	52
7	Stoichiometric titration	34	52	16	24	16	24
Average			35.1		13.3		51.6

Total and Percentage (%) of student having misconeption

Number of students: 66.

The same case also occurred on question number 5 regarding the process of titration. The students were asked to explain whether it is allowed to rinse the erlenmeyer wall with aquadest along titration. As many as 62% of students showed misconceptions by answering that it is not allowed because the amount of substance will change. However, rinsing or adding aquadest to erlenmeyer flask does not affect the amount of substance, because the number of moles of solute (HCl) will not change, regardless of the addition of solvent volume. The students must understand that the amount of dissolved substance or the number of moles of a solute (HCl) do not change by solvent addition. The change will only occur in the concentration and the molarity of a solution. There is a relationship between the concentrations and the amount of solute. Concentration is the number of moles of solute in a volume of solution.

Moreover, misconceptions also occur in concept 7 (the calculation of moles of NaOH needed for titration with 10 mL 0.1 M oxalic acid solution). In students' perception, the moles of oxalic acid are equal to moles of NaOH at the endpoint of titration, without considering the chemical equation. It is also predicted that students possibly do not know the chemical formula of oxalic acid. Actually, they have to take into account the equivalence of H^+ and OH^- ions in the chemical equation. Oxalic acid produces two H^+ ions. The reaction between oxalic acid and NaOH is as follows:

$$2H^+ + C_2O_4^{2-} + 2Na^+ + 2OH^- \rightarrow 2H_2O(l) + C_2O_4^{2-} + 2Na^+$$

One mole of oxalic acid (produces two moles H^+) is equivalent to two moles of NaOH (produces two moles).

Test number 1 is about the equipment needed to measure the volume of sample solution for titration. To measure the volume of solution sample, as many as 44% of students chose "measuring pipette" instead of "volumetric pipette". The students still assumed that the measuring pipette has the highest accuracy in measuring the volume of a solution, instead of choosing the volumetric pipette.

Most of students did not understand the concepts that should be answered in question number 6. The concept is about titration application, especially in the selection of an appropriate method that can be used to determine the content of a sample containing caustic soda. Commonly, they chose the other techniques instead of acid-base titration. They proposed some reasons. Some of them thought that caustic soda can be precipitated using $AgNO_3$ to produce silver hydroxide. Some others argued that the hydroxide ion can be oxidized into oxygen gas.

Most of students have a good understanding of the concepts on what to do to the washed but not dried burette when it will soon be used to titrate the HCl sample with NaOH solution. Most of the students gave the argument that they have to rinse the burette with NaOH solution that will be used as titrant.

Based on research, there were still found some potential misconceptions on the titration subject. It should not be the case because actually this concept has been studied both during senior high school as well as in the basic chemistry course at graduate level. Mostly, the misconceptions occurred on macro and symbolic representation. Students were confused when choosing the appropriate apparatus for measuring volume of sample. Students were also ambiguous in titration calculations. These phenomena occur in less meaningful learning as stated by Berg (1991), Cross et al. (1986), and Pinarbasi (2007). Some researchers recommend that learning, especially in chemistry, has to be planned by involving more active students that can accommodate multiple representatives. (Pinarbasi, 2007, 2009; Luoga et al., 2013; Pan & Henriques, 2015). The important thing is that the students have to have a high motivation on learning. Some strategies are recommended by some researchers, such as cognitive dissonance (Linenberger & Bretz, 2012). The cognitive dissonance strategy has not been used in chemistry learning yet, so it is interesting to combine the multiple representation and cognitive dissonance strategies, as a challenge interactive learning.

4 CONCLUSIONS

Misconceptions will affect the level of students' understanding, if it is not resolved immediately. The descriptive research involving 66 students of the BAC course showed that 51.6% of students understand the concepts of titration well; 13.3% of students do not understand the concept, and 35.1% have misconceptions. The misconceptions mainly occurred in macro and symbolic skills on choosing measuring equipment for titration, using titration equipment, and calculating titration.

Sustainable misconceptions are fatal in students' understanding and learning outcomes. The learning reform is recommended to minimize misconceptions. In the case of learning chemistry, especially in the BAC course, the use of a multi-representative strategy combined with cognitive dissonance can be considered. Multiple representations are expected to make learning more meaningful while cognitive dissonance can lead to student learning motivation.

ACKNOWLEDGMENT

The writers are grateful for the BPPS-doctor scholarship from the Directorate of Higher Education, Republic of Indonesia. The writers also would like to convey gratitude for their college (Universitas Negeri Malang) who have been extremely helpful along the study.

REFERENCES

Berg, T., & Brouwer, W. (1991). Teacher awareness of student alternate conceptions about rotational motion and gravity. *Journal of Research in science teaching, 28*(1), 3–18.

Cross, G.A., Fox, J.A., Duszenko, M., Ferguson, M.A., & Low, M.G. (1986). Purification and characterization of a novel glycan-phosphatidylinositol-specific phospholipase C from Trypanosoma brucei. *Journal of Biological Chemistry, 261*(33), 15767–15771.

Damanhuri, M.I.B., Treagust, F.D., Won, M. & Chandrasegaran, L.A. (2016). High school students' understanding of acid-base concepts: An ongoing challenge for teachers. *International Journal of Environmental & Science Education, 11*(1), 9–27.

Demircioglu, G., Ayas, A. & Demircioglu, H. (2005). Conceptual change achieved through a new teaching program on acids and bases. *Chemistry Education Research and Practice, 6*(1): 36–51.

Eymur, G., Cetin, P. & Geban, O. (2013). Analysis of the alternative conceptions of preservice teachers and high school students concerning atomic size. *Journal of Chemical Education, 90*, 976–980.

Hakim, A., Liliasari & Kadarohman, A. (2012). Student understanding of natural product concept of primary and secondary metabolites using CRI modified. *International Online Journal of Educational Sciences, 4*(3), 544–553.

Hasan, S., Bagayoko, D. & Kelley, E.L. (1999). Misconceptions and the certainty of response index (CRI). *Physics Education, 34*(5), 294–299.

Kurbanoglu, N.I. & Akin, A. (2010). The relationships between university students' chemistry laboratory anxiety, attitudes and self-efficacy beliefs. *Australian Journal of Teacher Education, 35*(18), 48–59.

Linenberger, J.K. & Bretz, L.S. (2012). Generating cognitive dissonance in student interviews through multiple representations. *Chemical Education Research & Practice, 1*, 172–178.

Luoga, N.E., Ndunguru, P.A. & Mkoma, S.L. (2013). High school students' misconceptions about colligative properties in chemistry. *Tanzania Journal of Natural & Applied Sciences, 4*, 575–581.

Nakhleh, B.M. (1992). Why some students don't learn chemistry: Chemical misconceptions. *Journal of Chemical Education, 69*(3), 191–196.

Pan, H. & Henriques, L. (2015). Students' alternate conceptions on acids and bases. *School Science and Mathematics* 115(5): 237–244.

Pinarbasi, T. 2007. Turkish undergraduate student's misconceptions on acids and bases. *Journal of Baltic Science Education, 16*(1), 23–34.

Pinarbasi, T., Sozbilir, M. & Canpolat, N. (2009). Prospective chemistry teachers' misconceptions about colligative properties: boiling point elevation and freezing point depression. *Chemistry Education Research and Practice, 10*, 273–280.

Potgieter, M., Rogan, M.J. & Howie, S. (2005). Chemical concepts inventory of grade 12 learners and UP foundation year students. *African Journal of Research in SMT Education, 9*(2), 121–134.

Rahayu, S., Chandrasegaran, L.A., Treagust, F.D., Kita, M. & Ibnu, S. (2011). Understanding acid-base concepts: Evaluating the efficacy of a senior high school student-centered instructional program in Indonesia. *International Journal of Science and Mathematics Education, 9*, 1439–1458.

Sheppard, K. (2006). High school students' understanding of titrations and related acid-base phenomena. *Chemistry Education Research and Practice, 7*(1), 32–45.

Skoog, A.D., West, M.D., Holler, J.F. & Crouch, R.S. (2004). *Fundamentals of analytical chemistry* (8th ed.). Southbank, Australia: Thomson Learning.

Suparno, P. (2005). *Miskonsepsi & perubahan konsep pendidikan fisika*. Jakarta: Grasindo.

Tumay, H. (2016). Emergence, learning difficulties, and misconceptions in chemistry undergraduate students'

conceptualizations of acid strength. *Science and Education, 25*, 21–46.

Van den Berg, E. (1991). *Miskonsepsi fisika dan remediasinya*. Salatiga: Universitas Kristen Satya Wacana.

Widarti, R.H., Permanasari, A. & Mulyani, S. (2014). *Analisis kesulitan mahasiswa pada perkuliahan dasar-dasar kimia analitik di Universitas Negeri Malang*. Malang: Seminar SNKP.

Widarti, R.H., Permanasari, A. & Mulyani, S. (2016). Student misconception on redox titration: A challenge on the course implementation through cognitive dissonance based on the multiple representations. *Journal Pendidikan IPA Indonesia, 5*(1), 56–62.

Ideas for 21st Century Education – Abdullah et al. (Eds)
© 2017 Taylor & Francis Group, London, ISBN 978-1-138-05343-4

Parent-adolescent conflict: Is there a difference of main sources between intergeneration?

T.H. Dahlan, I.H. Misbach & D.Z. Wyandini
Universitas Pendidikan Indonesia, Bandung, Indonesia

ABSTRACT: The aim of this study was to identify a difference of conflict sources between parents and adolescents. This research used cross-sectional survey design to compare conflict sources between parents and adolescents within family structures. Samples were selected using accidental sampling technique which consisted of 729 Indonesian adolescents (11–21 years old), 448 mothers, and 395 fathers, who lived in Bandung. The majority of respondents were Sundanese and Islamic. The result showed that conflict sources perceived by adolescents were different from those of parents. Most adolescents (37%) perceived family relationships as being the main conflict sources with parents whereas most mothers (39%) and fathers (41%) perceived values and attitudes were the main conflict sources with adolescents. It can be concluded from this difference in perspective that most adolescents think their parents cannot understand their feelings and emotional needs. On the other hand, parents think that adolescents should obey the set of values and attitudes from the parents' standard. This finding will underlie the design of a parent-adolescent conflict resolution model as a part of a family support program.

1 INTRODUCTION

In family relationships, parents often do not realize that adolescents are undergoing a change or transformation into adulthood (Grotevant & Cooper, 1986; Steinberg, 1993; Allison & Schultz, 2004). Based on the adolescent development tasks, adolescents demand to get greater autonomy and responsibility, which is often confusing to many parents. The adolescents seek recognition as being independent individuals; on the other hand, they still have high dependence on their parents (Collins & Laursen, 2004). These problems lead to different perspectives between them that cause an unresolved conflict, but this conflict is not related to the fundamental values (Steinberg, 1993).

The intensity of the conflict is higher in early adolescence and declines in the late adolescence period (Allison & Schultz, 2004). Three domains of conflict resources are usually derived from personal character, disruptive behavior, and academic performance in school (Shek & Ma, 2001).

The influence of ethnic background also plays a role, influencing a parent-adolescence conflict. This assumption was examined among Croatian early adolescents by Brkovich et al. (2014), who explored changes in parent-adolescent school-related conflict rate and academic achievement over a 5–year period. The data was collected three times, one year apart, from 851 adolescents and 1,288 parents. The results showed that parent-adolescent school-related conflict rate is increasing and adolescents' academic achievement is decreasing.

Another study, that measured the role of the ethnic background to parent-adolescent conflict, looked at autonomy and its relation to adolescent-parent disagreements regarding ethnic and developmental factors (Phinney et al., 2005). Findings showed that the adolescents with American ethnic backgrounds aged 14 to 22 years expressed autonomy and its relation to their responses to disagreements with parents, such as their projected actions (compliance, negotiation, self-assertion) and reasons for their actions in response to disagreements and family interdependence. The adolescents from non-European ethnic backgrounds complied more with parents, compared with adolescents of European ethnicity, but they were not different with regards to autonomy. The older European American adolescents were more conventionally family-oriented than younger ones, while older Armenian and Mexican American adolescents were more assertive than younger ones. Family interdependence mediated ethnic differences in compliance and predicted self-assertion.

Compared to American ethnic backgrounds, a cross-cultural study of Dutch, Moroccan, Turkish, and Surinamese adolescents, living in the Netherlands, revealed ethnic similarities in the mean levels of support, disclosure, self-esteem, delinquent behavior, authoritative control, and positive quality of the parent-adolescent relationship. A negative quality of the relationship showed significant associations with each adolescent outcome in all ethnic groups. For both aggressive behavior and self-esteem, the same model was applicable to all ethnic groups. The linkage between the quality of the parent-adolescent relationship and delinquent behavior was also similar, but the relationship between parenting behavior and delinquent behavior differed across groups (Liddle et al., 2017).

Many research findings related to parent-adolescent conflict rate mostly involve white and middle-class adolescents. Compared to one of Indonesian ethnic backgrounds, the study from Christiyanti (2010) in Semarang, Indonesia, shows that sources of parent-adolescent conflict derived from poor communication quality. This often leads to juvenile delinquency outside the home. Another study shows that parent-adolescent conflict comes from a lack of parental knowledge about the effect of puberty, which is naturally experienced by adolescents (ages 12–18). The onset of puberty is associated with it being a trigger that activates hormones that can trigger an increase in adolescent emotions (Allison, 2000). Parents are often not prepared to see that previously their children were well behaved and complied with the parents' standards, but that the children then changed drastically to reject their parents. Most adolescents resolve conflicts by giving a victory to their parents to avoid conflict. On the other hand, parents usually win the arguments. (Santrok, 2007, 2011).

However, there is almost no research finding about parent-adolescent conflict sources in the Sundanese ethnic community in Indonesia. The purpose of the study was to address this significant gap in the Western literature by examining whether in Eastern culture, particularly in the Sundanese ethnic group, there is either a similar or different pattern of conflict sources between parent and adolescent, like in many other Western and Asian cultures.

The conflict sources in this study refer to the factors that lead to the debate because of differences in thoughts, feelings, and behaviors between parents and adolescents. These factors include: the family relationships (closeness, respect, personal opinion, affection and attention); values and attitudes (beliefs, education, and employment); personal style (hairstyles, clothing options, and the use of leisure time).

2 RESEARCH METHODS

The research used a cross-sectional survey design to compare parents with adolescents in terms of their conflict sources. The researcher designed pilot instruments to identify conflict sources perceived between parents and adolescents. Following the explanation of Creswell (2012), in designing the survey instruments, the researcher implements the following processes in order: identifying began with identifying the purpose of the instrument, reviewing

the literature, writing closed- and open-ended questions, using clear language that was applicable to all participants, inviting two expert judges to complete qualitative evaluation, and revising items based on the judges' feedback. Afterwards, the pilot instrument was administered to a small number of six mothers, six fathers, and six adolescents. Then a repeat revision was made based on their feedbacks.

The data was collected from mailed questionnaires and web-based questionnaires (Creswell, 2009), which consisted of 22 items classified into three domains, namely family relationships, attitudes and values, and personal style. The respondents were permitted to choose more than one domain. The instrument also identified the ethnicity, religion, and age of respondents.

The samples were selected using accidental sampling technique consisting of 729 Indonesian adolescents (11–21 years old), 448 mothers, and 395 fathers, who lived in Bandung, Jawa Barat.

3 RESULTS AND DISCUSSION

Findings about the conflict sources between parents and adolescents, as perceived by adolescents, mothers, and fathers, is presented in Table 1.

Table 1 shows that the highest percentage of conflict source for adolescents is different from the highest percentage for the parents. Adolescents feel that the domain of family relationship (37%) is the main conflict source they experienced with their parents, whereas the percentage for family relationship is the lowest as the perceived conflict source for mother (24%) and father (25%) with the adolescent.

On the other hand, both mothers (39%) and fathers (41%) perceive that the domain of values and attitudes is the main conflict source they experienced with adolescents. The second highest percentage of conflict source for parents is the personal taste of the adolescent (mother (36%) and father (34%)), whereas the percentage of personal taste is the lowest conflict source perceived for adolescents (29%).

Table 1. Sources of parent-adolescent conflict perceived by adolescents, mothers, and fathers.

Domain	Perceived by adolescents	Perceived by mothers	Perceived by fathers
Family relationship	37.33%	24.23%	25.39%
Values and attitudes	34.08%	39.37%	40.87%
Personal tastes	28.59%	36.40%	33.74%
Total	100%	100%	100%

90

These different perspectives show that the adolescents' evaluation of the family relationships domain (relationship closeness, respect, personal opinion, affection and attention), was influenced because of an unequal relationship so that adolescents tend to feel that most of their parents cannot understand their feelings and emotional needs. The existence of an unequal relationship between them causes adolescents to perceive that parents still play a major role in holding the unilateral authority to control adolescents. They assessed that parents cannot shift power to give them greater autonomy and responsibility, especially in terms of the decision-making process. On the contrary, parents think that adolescents should obey their set of values and attitude standard. Parents tend to feel that the domain of values and attitudes (in terms of belief, education, and employment), with the domain of personal styles (such as fashion and musical style), and the domain of leisure time activities, are all more important issues than the family relationship domain.

Findings based on the ethnicity and religions of respondents are presented in Table 2.

Based on ethnic backgrounds, most adolescents are Sundanese (76%), mother Sundanese (52%), and father Sundanese (46%). The findings indicate that most of them are influenced by the internalization of the traditional values of the Sundanese ethnic group, those values that uphold manners and respect for older people (Ekadjati, 1995). In line with religious background (which are 96% Moslem adolescents and 97% Moslem mothers and

Table 2a.   Ethnicities of respondents.

| Ethnic | Frequency | | |
	Adolescents	Mothers	Fathers
Bali	1	1	–
Batak	11	–	4
Bengkulu	1	–	–
Jawa	25	27	44
Karo	1	–	–
Minang	4	2	1
Papua	1	–	1
Sunda	551	234	181
Toraja	1	1	–
Banjar	–	1	–
Betawi	–	1	2
Bugis	–	1	–
Menado	–	1	–
Maluku	–	–	1
Melayu	–	–	1
Papua	–	–	1
Not filled out	133	179	160
Total	729	448	395

Table 2b.   Religions of respondents.

| Religion | Frequency | | | Percentage | | |
	A	M	F	A	M	F
Muslim	703	435	385	96%	97%	97%
Christian	25	12	10	3.4%	2.7%	2.5%
Hindu	1	1	–	0.1%	0.2%	–
Total	729	448	395	100%	100%	100%

*Note: A = Adolescent, M = Mother, F = Father.

fathers), the findings indicate that the Sundanese families tie strongly on the patriarchy hierarchy where fathers act as leaders in the family structure and the role of the Islam religion greatly influences one's position in the family structure to determine the total of respect to elders in all aspects of Sundanese family life (Rosidi, 1984; Alwasilah, 2006).

The study also identified the respondents' age, which is presented in Table 3.

The different viewpoints between what adolescents and parents perceived of conflict sources can also be caused by the gap between them of the development stage. Based on the demographic ages, the mothers' ages ranged between 36–50 years (77%) and fathers' ages ranged between 36–55 years (88%), which is entering the *middle adulthood stage* (Hurlock, 1999), while adolescent ages ranged between 12–17 years (98%), which is going through late childhood into the adolescence stage (Papalia et al., 2007).

Both of their development stages have similar transitions or changes. The parent respondents are undergoing their middle adulthood which has a transition period of several roles that begin at the age of approximately 35 to 45 years old, up to the age of six decades. They are experiencing a transition period to expand the involvement of personal responsibility and social role inside the family, to take care of their children as their next generation to be mature people. Meanwhile, outside the family, they are required to have a social role to achieve and maintain their own career satisfaction (Hurlock, 1999).

The adolescent respondents are also experiencing a period of transition from early childhood to early adulthood. In this development stage, the attainment of independence and self-identity are very prominent as they spend more time with peers outside the family (Hurlock, 1999).

The difference changes during adolescence indicate growing well through the puberty phase. Conversely, changes during middle age are degenerative. Both of these development stages have similar awkward behaviors and feelings, because parents and adolescents are in similar transitions but in different contexts, in terms of physique,

Table 3. Ages of respondents.

	Age	Frequency	Percentage
Adolescents	11–12	159	21.8%
	13–14	280	38.4%
	15–16	198	27.2%
	17–18	89	12.2%
	Not filled out	3	0.4%
	Total	729	100%
Mothers	<30	2	0.4%
	30–35	49	10.9%
	36–40	119	26.6%
	41–45	129	28.8%
	46–50	96	21.4%
	51–55	35	7.8%
	>55	12	2.7%
	Not filled out	6	1.3%
	Total	448	100%
Fathers	<36	9	2.3%
	36–40	60	15.2%
	41–45	117	29.6%
	46–50	111	28.1%
	51–55	61	15.4%
	50–60	23	5.8%
	>60	9	2.3%
	Not filled out	5	1.3%
	Total	395	100%

emotion, values, lifestyle and choice of social interaction (Santrock, 2011).

4 CONCLUSIONS

In general, there is a perspective difference in parent-adolescent conflict sources. Adolescents feel that the domain of family relationships is a major source of conflict when they negotiate with parents because of an unequal relationship. On the contrary, parents perceive that the domain of values and attitudes along with personal style are conflict sources. The big difference in the conflict sources shows a different pattern of thought, feeling, and behavior between parents and adolescents. In the context of the Eastern culture such as with the Sundanese ethnic group, the natural relationship between parents and children have caused the formation of the value internalization of adherence that makes children leave their decisions to their parents as a form of moral obligation.

Understanding about the perspective differences between what adolescents and parents are experiencing as being the conflict sources, will underlie the design of the parent-adolescent conflict resolution model that focuses on the solution-talk method.

REFERENCES

Allison, B.N. & Schultz, J.B. (2004). Parent-adolescent conflict in early adolescence. *Adolescence*, *39*(153), 101–119.

Allison, B.N. (2000). Parent-adolescent conflict in early adolescence: Precursor to adolescent adjustment and behavior problems. *Journal of Family and Consumer Sciences 9*(5), 53–55.

Alwasilah, A.C. (2006). *Pokoknya sunda*. Bandung: Kiblat.

Brkovich, I., Keresteš, J. & Puklek, M. (2014). Trajectories of change and relationship between parent-adolescent school-related conflict and academic achievement in early adolescence. *The Journal Early Adolescent*, *34*(6), 792–815.

Christiyanti, D. (2010). *Memahami komunikasi antar pribadi orang tua anak yang terlibat dalam kenakalan remaja*. (Doctoral dissertation). Diponegoro University.

Collins, W.A. & Laursen, B. (2004). Changing relationship, changing youth: Interpersonal contexts of adolescent development. *Journal of Early Adolescence*, *24*(1), 55–62.

Creswell, J.W. (2009). *Research design: Qualitative, quantitative, and mixed methods approaches* (3rd ed.). California: SAGE Publications.

Creswell, J.W. (2012). *Research design: Planning, conducting, and evaluating quantitative and qualitative research* (4th ed.). Boston: Pearson Education.

Ekadjati, E.S. (1995). *Kebudayaan sunda, suatu pendekatan sejarah*. Jakarta: Pustaka Jaya.

Grotevant, H. D, & Cooper, C.R. (1986). Individuation in family relationships: A perspective on individual differences in the development of identity and role-taking skill in adolescence. *Human Development, 29*, 82–100.

Hurlock, E.B. (1999). *Psikologi perkembangan: Suatu pendekatan sepanjang rentang kehidupan*. Jakarta: Erlangga.

Liddle, J., Wishink, A., Springfield, L., Gustafsson, L., Ireland, D., & Silburn, P. (2017). Can smartphones measure momentary quality of life and participation? A proof of concept using experience sampling surveys with university students. *Australian Occupational Therapy Journal*.

Papalia, D.E., Olds, S.W. & Feldman, R.D. (2007). *Human development* (10th ed.). New York: McGraw-Hill Companies.

Phinney, S.J., Kim-Jo, T., Osorio, S. & Vilhjalmsdottir, P. (2005). Autonomy and relatedness in adolescent-parent disagreements: Ethnic and developmental factors. *Journal of Adolescent Research*, *20*(1) 8–39.

Rosidi, A. (1984). *Manusa sunda*. Jakarta: Giri Mukti Pusaka.

Santrock, J.W. (2007). *Remaja Jilid 2 (edisi 11)*. Jakarta: Penerbit Erlangga.

Santrock, J.W. (2011). *Life-span development, perkembangan masa hidup Jilid 1 (edisi ketigabelas)*. Jakarta: Penerbit Erlangga.

Shek, D.T. & Ma, H.K. (2001). Parent-adolescent conflict and adolescent antisocial and prosocial (2010). Parent-adolescent conflict and adolescent antisocial and prosocial behavior: A longitudinal study in a Chinese context.

Steinberg, L. (1993). *Adolescence* (3rd ed.). New York: McGraw-Hill.

Ideas for 21st Century Education – Abdullah et al. (Eds)
© 2017 Taylor & Francis Group, London, ISBN 978-1-138-05343-4

Students' mental model profile of microorganism after the implementation of mental model-based microbiology course

Y. Hamdiyati, F. Sudargo, S. Redjeki & A. Fitriani
Universitas Pendidikan Indonesia, Bandung, Indonesia

ABSTRACT: The mental model indicates what someone understands about a concept at a certain point of time, including their knowledge and belief about the concept. The construction of the mental model is the core of meaningful learning. This descriptive study identifies biology students' mental model after the implementation of a mental model-based microbiology course. Respondents were 32 biology students enrolled in sixth semester at the Department of Biology Education of Universitas Pendidikan Indonesia. Data was collected using a writing-drawing test about microorganism structure. Respondents were asked to draw from their imagination about a microorganism structure and then they were asked to explain the microorganism structure in writing through open-ended questions. Their responses represented the biology students' mental model after the implementation of the mental model-based microbiology course. The biology students' mental models were classified into five levels (level 1–5): 'there is no drawing/writing' (D1/W1), 'wrong or irrelevant drawing/writing' (D2/W2), 'partially correct drawing/writing' (D3/W3), 'the drawing/writing that has some deficiencies' (D4/W4), and 'completely correct and complete drawing/writing' (D5/W5). Results suggested that drawing and writing illustrated various mental model levels, but overall the biology students' mental models were improved after the implementation of the mental model-based microbiology course. The highest mental model level, D5/W5, was found in almost all microorganism concepts, but for archaea, the highest mental model level was D4/W4. The lowest mental model, D1/W1, was not found for all four microorganism concepts after the implementation of the mental model-based microbiology course. Thus, it can be concluded that the mental model-based microbiology course can improve or complete the biology students' mental model about microorganisms.

1 INTRODUCTION

The mental model is someone's depiction, personal ideas, or internal representation about a phenomenon, a set of ideas, or concepts. The construction of a mental model is the core of meaningful learning. As stated by Ausubel (1984, in Byrne, 2011), meaningful learning requires structuring or restructuring of specific concepts related to concepts in the learner's cognitive structures. Therefore, study about the mental model in science education has grown rapidly as a way to understand the learning process, particularly in understanding basic knowledge representation (Greca & Moreira, 2000). The same goes for the microbiology course.

The study of student mental models about the microorganism concept of pre-service biology teachers in 2013/2014 showed that most students believe that microorganisms are small creatures that can only be seen with a microscope. Most students described microorganisms as being bacteria and viruses. Fungi, microalgae, and protozoa were rarely depicted. How microorganisms proliferate was rarely fully described. Most said that microorganisms have beneficial or detrimental effect but

the explanation about the effect was not further described (Hamdiyati et al., 2014). A group of non-experts (students and teachers) had a complex mental model but it was not scientifically correct compared to the expert group (Jee et al., 2013). Pre-service biology teachers in Turkey already have knowledge about some microorganism concepts, but there are still concepts that they have not mastered yet and misconceptions still occur (Kurt, 2013). The same thing is also found in the case of the virus concept, in which describing virus definitions was prevalent and misconceptions about virus concepts still occur (Kurt & Ekici, 2013). The Hamdiyati et al. (2016) study showed that the students' mental model level before taking the microbiology course, most commonly found on each concept from the drawing-writing test (D/W), was bacteria (D2/W2), Archaea (D1/W1 and D2/W2), virus (D3/W3), and fungi (D2/W1). These indicates the students' low conceptual mastery. Therefore, it is necessary to study the students' mental model after the implementation of the mental model-based microbiology course.

The role of microorganisms in life is increasingly essential. The conference on microbiology organized by The American Academy of Microbiology in South Carolina in 2003 discussed important issues about the role of microorganisms for life on earth, research directions for the 21st century, and the need to educate society about microbiology (Schaechter et al., 2004). Therefore, the detection of conception and interconnection between microbiology concepts is important in microbiology courses, either before or after the mental model-based microbiology course is implemented, as a way to revise or supplement student mental models. The purpose of this study was to observe the students' mental model profile on the concept of microorganisms' structures (bacteria, archaea, viruses, and fungi) after the implementation of the mental model-based microbiology course.

2 LITERATURE REVIEW

Johnson-Laird (1983 in Byrne, 2011) stated that the mental model is a dynamic representation that is never complete, continuously growing and developing along with information accretion. Its dynamic and continuous nature means that the mental model will undergo modification when new information is added. The mental model indicates what someone understands about a concept at a certain point or time, including their knowledge and belief about the concept. The formation of a mental model happens through several intermediate mental models (intermediate/M1, M2, ... Mn) before it finally reaches the targeted mental model: the consensus mental model according to experts (Rea-Ramirez et al., 2008). The mental model is generative which means it can lead students to new information and use it to predict and give an explanation (Wiji, 2014).

3 RESEARCH METHODS

This research used descriptive method. Respondents were 32 biology students enrolled in sixth

Table 1. Mental model categories for drawing-writing test results.

Level	Statement	Drawing	Writing
Level 1	There is no drawing/writing	D1	W1
Level 2	Wrong or irrelevant drawing/ writing of question	D2	W2
Level 3	Partially correct drawing/ writing	D3	W3
Level 4	The drawing/writing that has some deficiencies	D4	W4
Level 5	Completely correct and complete drawing/writing	D5	W5

(Adapted from Yayla and Eyceyurt, 2011).

semester at the Department of Biology Education of Universitas Pendidikan Indonesia. Data was retrieved after students enrolled in the mental model-based microbiology course in the sixth semester. Microorganisms material discussed were structure of bacteria, archaea, viruses, and fungi. The instrument to probe mental models was a drawing-writing test. Students were asked to describe their imagination about the structure of microorganisms, and they were subsequently asked to explain the structure of microorganisms in writing through open-ended questions. The students' answers, after the implementation of the mental model-based microbiology course, were then compared with scientists' mental models as the targeted mental model. The students' mental models are categorized into five levels (Table 1). The mental model level for each concept (the structure of bacteria, archaea, virus, and fungi) is described in detail for each level of drawing and writing (Hamdiyati et al., 2016).

4 RESULTS AND DISCUSSION

The students' mental model of microorganism structure after the mental model-based microbiology course is illustrated in Tables 2–5 and Figures 1–2. Figures 1–2 show the example of students' mental models about microorganisms' structure

Table 2. Students' mental model about bacteria structure.

	Bacteria				
	W1	W2	W3	W4	W5
D1	–	–	–	–	–
D2	–	3	1	–	–
D3	–	5	11	–	–
D4	–	1	4	4	1
D5	–	–	–	1	1

D = Drawing, W = Writing.

Table 3. Students' mental model about archaea structure.

	Archaea				
	W1	W2	W3	W4	W5
D1	–	–	–	–	–
D2	–	9	3	–	–
D3	–	5	10	2	–
D4	–	–	2	2	–
D5	–	–	–	–	–

D = Drawing, W = Writing.

Table 4. Students' mental model about virus structure.

	Virus				
	W1	W2	W3	W4	W5
D1	–	–	–	–	–
D2	–	–	2	–	–
D3	–	6	12	2	–
D4	–	1	4	3	-
D5	–	–	1	–	1

D = Drawing, W = Writing.

Table 5. Students mental model about fungi structure.

	Fungi				
	W1	W2	W3	W4	W5
D1	–	–	–	–	–
D2	–	1	3	1	–
D3	–	1	10	6	–
D4	–	–	3	3	–
D5	–	–	–	1	3

D = Drawing, W = Writing.

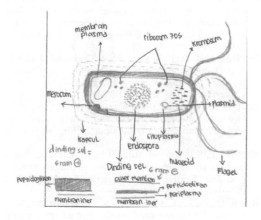

Figure 1. Student's mental model about bacteria structure (D5).

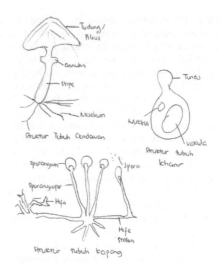

Figure 2. Student's mental model about fungi structure (D5).

through a drawing-writing test. The Tables suggest that the mental model commonly found through the drawing-writing test was at the level of D3/W3. Therefore, the mental model-based-microbiology course improved or enhanced image representation and verbal representation. The students' mental model through the drawing test shows a higher level than writing (W), except in the structure of a virus concept in which the level 1 mental model was still found for the drawing-writing test on all the microorganism's concepts (Hamdiyati et al., 2016).

After the implementation of the mental model-based microbiology course, the lowest mental model (level 1) was not found in the drawing or writing test. The number of students having level D2/W2 had significantly decreased compared to before the model-based microbiology course was implemented. Although, there were still nine students having D2/W2 level for archaea structure.

Table 2 depicts the students' mental model about a bacteria's structure. The lowest mental model level from the drawing-writing test was D2/W2 and the highest was D5/W5.

Table 3 depicts the students' mental model about archaea's structure. The lowest mental model level from the drawing-writing test was D2/W2 and the highest was D4/W4.

Table 4 depicts the students' mental model about a virus's structure. The lowest mental model level from drawing-writing test was D2/W3 and the highest was D5/W5.

Table 5 depicts students' mental model about fungi's structure. The lowest mental model level from drawing-writing test was D2/W2 and the highest was D5/W5.

Figure 1 depicts a student's mental model example about a bacteria's structure (level 5/D5). Students drew the bacteria's structure in accordance with the targeted mental model. Students drew the bacteria's cell components as well as its cell walls completely.

Figure 2 shows a student's mental model example about the structure of fungi (level 5/D5). Students drew three structure forms of fungi (yeast, mold, and mushrooms) and also provided explanations for each part according to the targeted mental model.

The students' initial mental model about microorganisms is strongly influenced by the experience in adding new information into the existing schemas. Information about the initial mental model from these study results is a result of model evolution in which prior knowledge is combined with new information to form a new mental model. This mental model building process can be through model reinforcement, model-elaboration and revision, or model-rejection (Buckley & Boulter, 2000).

The mental model-based microbiology course implemented in this study is based on a constructivism framework where the learner's ideas about a concept, phenomenon, or event are usually obtained from previous experience, both in everyday life as well as in formal occasions (Hamdiyati et al., 2016). The mental model-based microbiology course applied was based on constructivism learning steps according to Duit et al. (2007) with the framework of model-based-learning paths according to Rea-Ramirez et al. (2008).

The improvement of the mental model level about microorganisms in the mental model-based microbiology course was due to exploration of the students' preconceptions, by lecturer, to find out the students' knowledge about the concepts of microorganisms. Concept restructuring was then carried out. At this stage, the students were led to achieve a higher-level mental model. The students were then asked to apply the new concept. The mental model-based microbiology course then ended with a mental model-based review and the evaluation of new ideas that do not correspond to the targeted mental model.

New ideas about the structure of microorganisms were assimilated into existing schema or accommodated by creating a new scheme. This may cause cognitive dissonance, for example, if new ideas are scientifically accurate but are not consistent with the scheme that the students possess. The rejection of the new ideas could happen. When new ideas are accepted, and connected with the existing ideas, then the mental model changes through improving or completing the mental model.

5 CONCLUSIONS

Results suggest that the level of the students' mental model through the drawing-writing test for microorganism concepts were varied. The improvement in mental model level was observed when the mental model-based microbiology course was applied. The highest mental model level, drawing-writing (D5/W5) level, was found in almost all microorganism concepts, but for archaea, the highest mental model level was D4/W4. The lowest mental model (D1/W1), was not found in the microorganism concepts after the implementation of the mental model-based- microbiology course. Thus, from this study it can be concluded that the mental model-based microbiology course can

improve or complete the biology students' mental model about microorganisms.

REFERENCES

Buckley, B.C. & Boulter, C.J. (2000). Investigating the role of representations and expressed models in building mental models. In J.K. Gilbert & C.J. Boulter (Eds.), *Developing Models in Science Education* (pp. 119–135). Dordrecht: Kluwer Academic Publishers.

Byrne, J. (2011). Models of micro-organisms: Children's knowledge and understanding of micro-organisms from 7 to 14 years old. *International Journal of Science Education*, 1–35.

Duit, R., Widodo, A. & Wodzinski, C.T. (2007). Conceptual change ideas: Teachers' views and their instructional practice. In S. Vosniadou, A. Baltas & X. Vamvakoussi (Eds.), *Reframing the Conceptual Change Approach in Learning and Instruction*. Netherlands: Elsevier.

Greca, I.M. & Moreira, M.A. (2000). Mental, physical, and mathematical models in the teaching and learning of physics. *Science Education*, 86, 106–121.

Hamdiyati, Y. (2014). *Profil perkuliahan mikrobiologi dan model mental mahasiswa tentang mikroorganisme*. Field Study Paper for Science Development Program Course for Doctoral Degree. SPs UPI Bandung.

Hamdiyati, Y. Sudargo, F., Redjeki, S., & Fitriani, A. (2016). *Biology students' initial mental model about microorganism*. The International Seminar on Mathematics, Science, and Computer Science Education (MSCEIS) FPMIPA UPI.

Jee, B.J. Uttal, D.H., Spiegel, A., & Diamond, J. (2013). *Expert–novice differences in mental models of viruses, vaccines, and the causes of infectious disease*. Public Understanding of Science (PUS): 1–16.

Kurt, H. & Ekici, G. (2013). What is a virus? Prospective biology teachers' cognitive structure on the concept of virus. *International Online Journal of Educational Sciences*, 5(3), 736–756.

Kurt, H. (2013). Turkish student biology teachers' conceptual structures and semantic attitudes towards microbes. *Journal of Baltic Science Education*, 12(5), 085–093.

Rea-Ramirez, M.A., Clement, J. & Nunez-Oveido, M.C. (2008). An instructional model derived from model construction and criticism theory. In J.J. Clement and M.A. Rea-Ramirez (Eds.), *Model Based Learning and Instruction in Science* (pp. 23–43). London: Springer.

Schaechter, M., Kolter, R. & Buckley, M. (2004). *Microbiology in the 21st century: Where are we and where are we going?* Charleston, South Carolina: American Academy of Microbiology Charleston.

Wiji. (2014). *Pengembangan desain perkuliahan kimia sekolah berbasis model mental untuk meningkatkan pemahaman materi subyek mahasiswa calon guru kimia*. Doctoral Thesis. Sekolah Pasca Sarjana UPI Bandung. Unpublished.

Yayla, R.G. & Eyceyurt, G. (2011). Mental models of pre-service science teachers about basic concepts in chemistry. *Western Anatolia Journal of Educational Science (WAJES)*. Selected Papers Presented at WCNTSE. Dokuz Eylul University Institute, Izmir, Turkey ISSN 1308-8971: 185–294.

Building meaningful learning through coherence learning among mathematics, language and science lessons

A. Permanasari, T. Turmudi & V. Vismaia
Universitas Pendidikan Indonesia, Bandung, Indonesia

B. Rubini
Universitas Pakuan Bogor, Bogor, Indonesia

ABSTRACT: Building coherence between science and other subjects such as mathematics and languages leads to meaningful learning. This study was done to investigate how far learning mathematics and language (Indonesian language) on using themes about science has an impact on the science literacy of secondary schools' students. The study was done using quasi experiment and descriptive methods, in a collaboration between math, science, and language teachers and university researchers. The teaching and learning processes of all subjects were done by each teacher, while the researchers enrolled as observers. Classes of mathematics and languages took place before classes of science. The results show that learning mathematics and language using themes about science close to their daily life made learning more meaningful. This is indicated by an increase in math and language literacy, which appears to be better than the class that does not promote theme about science. Moreover, the treatment had the positive impact of meaningful learning in science, because they got an initial understanding of science. This led to the enhancement of science literacy, compared to the control class. The important thing was that teachers felt the positive impact of the collaboration between teachers such as a sense of togetherness, sharing experiences and expertise, and finally having a nurturing effect mainly in the teaching practice of meaningful learning.

1 INTRODUCTION

Literacy is closely related to reading, science, and math. Reading literacy includes reading and writing skills, as well as the skill to use language properly, effectively and critically. Alwasilah (2012) and Miller (1983) stated that science literacy can be defined as skills to read and write about science and technology. Similar with Miller, Shamos (1995) stated that understanding science needs language to read and criticize science properly. This is why the one who has literacy in science has to have reading literacy as well.

The science skills are strongly affected by how systematic, logical, and rational one's thinking is. All those ways of thinking can be trained through mathematics. Mathematics can also be used when someone solves the problems in a science context. Logical and rational thinking are some aspects of math literacy. Ewwel (2001) wrote that reading literacy is closely correlated with math literacy. This is in line with Schoenfeld (2001) thinking, that reading skills are closely affected by one's skills in logical, systematical, and critical thinking. All the statements above make science, math and reading literacy very important things to enhance. This is why the three are internationally recognized as parameters for education quality. *The Program for International Student Assessment (PISA)* arranged by the Organization for Economic Cooperation and Development (OECD) assesses continually the three literacies of students in secondary school levels in so many countries in the world. In Indonesia's context, low literacy level of reading, science, and math is still a problem that need to be solved immediately. First, we need to map all of the dimensions, so it can be used as a basis for finding the solution.

2 METHODOLOGY

The research was done during 2014–2015 and focused on mapping all of literacy's aspects, as well as the relationship among the three variables, being mathematics, languages, and science. The research methods used were an experimental quantitative approach with multivariate correlation design (Fraenkel, Wallen & Hyun, 1993, Creswell & Clark 2007). Cluster random sampling was used to get a representative sample. The sample consisted of students from secondary schools from higher, moderate, and lower categories of schools in the Bandung city regions.

Every category was represented by three schools, chosen randomly. The instruments used were science, math and reading literacy tests from PISA instruments tests after translation and adaptation for the Indonesian context. Interviews were also conducted in order to include teachers' opinions. A student questionnaire was also prepared to know the students' perspectives on learning conditions.

Data analysis was done initially through the change of raw score to the final score. Furthermore, a correlation test of the three variables was done using multivariate correlational statistics with multiple linear regression. a descriptive analysis and data triangulation were done to find the weaknesses, strengths, and opportunities for mutual learning to be implemented.

3 RESULTS AND DISCUSSION

The tests were administered using validity and reliability tests. All of the tests were constructed in themes. Every theme consisted of at least five items. The type of test could be multiple choice, simple choice, as well as an open-ended question. All of the tests reflected competency, knowledge, and attitude aspects. Scopes of the science test included health, natural resources, environment, science and technology. The concepts to be assessed were pollution, acid-base, biodiversity, photosynthesis, body health, chemical reactions, biotechnology, ecosystem, and density. The scopes in the math literacy test included the use of technology/IT, financial transactions, room planning, pollution, health, flora and fauna, economy/socio-economy, geometry, exchange rates, arithmetic, and knowledge of graphics. Meanwhile, reading tests used some scopes, such as earth and space, health, social life, phenomena around us, and saving energy. The skills to be assessed was narrative interpretation, writing appreciation, and reading.

The results show that science, math, and reading literacies of students in secondary schools are generally in the medium and some even in the low levels.

Table 1 shows a trend of the science score, from the higher to the lower school category. All achievements are in the moderate tending to low score. However, there were several students who reached the same or the higher score of other student s in the higher school level. Based on questionnaires, it was revealed that the atmosphere academic in learning process was not meet with curriculum expectation. The existing teaching and learning process focuses merely on the subject fulfillment without considering students' needs. Holbrook, Laius, & Rannikmäe (2003) stated that science learning is too difficult for students so they do not like it. Piirto (2011), Permanasari (2010), and Chang (2008) agreed that science literacy can be achieved by students if learning was meaningful and in a creative way.

Moreover, the research reveals a similar case for math and reading literacy as it is shown in Table 2. Students stated that learning mathematics is very boring, because so many formulas have to be memorized. Compared to science and math, a better achievement is shown in reading literacy. Students stated that language learning, particularly reading, is easier than math and science, but still tedious and less challenging.

To overcome the problem, the contextual teaching-learning was tried out in the same schools with different students in the following year. The themes were used in all subject matters promoting the Contextual Teaching and Learning (CTL) approach. The achievements in science, math and reading literacy after the treatment are shown in the right side of Table 2. A learning process that accommodates the needs of students, as well as bringing in the context of everyday life, fostered curiosity and encouraged students to learn. This finding is in line with some previous research, such as Nentwig et al. (2002) and Mayer & Moreno (2003), that learning would be relevant if it is packed in the context found around students' daily life.

Table 1. Students' science literacy profile in three categories of school (max score = 100).

Category	School/class	Average score	Lowest score	Highest score	Average/category	Lowest/category	Highest/category
	JHS A	58	29.6	75.6			
	JHS B	49.3	34.3	69.2	52.8 ± 4.61	29.6	75.6
Higher	JHS C	51	31.3	72.7			
	JHS D	46.7	25.3	68.7			
	JHS E	44.4	28.2	73.6	43.8 ± 2.42	15.3	73.6
Moderate	JHS F	40.2	15.3	64.9			
	JHS G	39.7	8.3	62.9			
	JHS H	36.3	8.9	62.3	36.5 ± 3.12	7.2	62.9
Lower	JHS I	33.6	7.2	57.3			

Table 2. The average science, math and reading literacy scores based on school's category after two different treatments.

School category	Average score after conventional treatment in			Average score after inovative treatment in		
	Science	Math	Reading	Science	Math	Reading
Higher	52.8 ± 4.6	50.3 ± 3.5	61.3 ± 4.7	84.0 ± 5.2	77.2 ± 3.4	79.2 ± 4.0
Moderate	43.8 ± 2.4	40.5 ± 3.1	57.5 ± 3.8	84.1 ± 6.5	66.3 ± 2.9	73.6 ± 4.9
Lower	36.5 ± 3.1	30.4 ± 3.2	56.3 ± 2.9	83.1 ± 6.1	75.5 ± 3.5	83.4 ± 6.7

Table 3. Correlation among math, science, and language literacies.

	Science	Math	Reading
Science		+ 0.71	+ 0.31
Math	+ 0.71		+ 0.43
Reading	+ 0.30	+ 0.43	

Based on the data in Table 2, the relationship among the achievement of the three literacies was studied, and the result are shown by Table 3. A strong relation is shown between math and science achievements. A moderate relation is shown between reading and science, as well as reading and math.

Using the same context in all three subject matters provides literacy enhancement, and leads to the enhancement of student literacy in science, math as well as reading. Cockroft (1982) stated that learning math on using themes about sciences would lead to increasing math literacy. The important thing from the research was a nurturing effect caused by this collaboration. Students felt that learning science is fun and easy because they have the initial knowledge from the math and language learning conducted before. They were also very happy learning math and language because they were all brought into the real life contexts. Teachers of all subjects felt the collaboration benefitted them. Togetherness, shared experience, good academic atmosphere were some efforts they needed to continue. They also thought that this project would lead to the enhancement of PISA results as was reported.

4 CONCLUSIONS

The use of contexts that are close to the daily life of students led to meaningful learning. Using the same context (science context) in learning mathematics and language learning, particularly reading skills, enhanced the literacy of all subjects. Learning math and language (within a science context) prior to science had a positive impact of a meaningful learning of science, because the students have had an initial understanding of science. The important thing was that teachers felt the positive impact of collaboration between teachers such as building togetherness, sharing experiences and expertise, and finally having a nurturing effect mainly in the teaching practice of meaningful learning.

ACKNOWLEDGEMENT

Many thanks to School of Post graduated Study of Universitas Pendidikan Indonesia (SPs UPI) for the funding contribution to this project. Thank you to all of student in SPs who have assisted in the retrieval data. Thank you for all of Schools that have participated in the project.

REFERENCES

Alwasilah, A.C. (2012). *Pokoknya rekayasa literasi*. Bandung: Kiblat.

ChanLin, L.J. (2008). Technology integration applied to project-based learning in Science. *Innovations in Education and Teaching International. 45*(1) 55–65.

Cockroft, W.H. (1982). *Mathematics counts*. Report of the Committee of Inquiry into the Teaching of Mathematics in Schools. London: Her Majesty's Stationery Office.

Creswell, J.W. & Clark, V.L. (2007). *Designing and conducting mixed methods research*. Oxford, UK: Oxford University Press.

Ewell, Peter T. (2001). Numeracy, mathematics, and general education. "In Lynn Arthur Steen (Ed.). *Mathematics and Democracy: The Case for Quantitative Literacy* (pp. 37–48). Princeton, NJ: National Council on Education and the Disciplines.

Fraenkel, J.R., Wallen, N.E., & Hyun, H.H. (1993). *How to design and evaluate research in education* (Vol. 7). New York: McGraw-Hill.

Holbrook, J., Laius, A., & Rannikmäe, M. (2003). *The Influence of Social Issue-Based Science Teaching Materials On Students' Creativity*. University of Tartu, Estonian Ministery of Education.

Mayer, R.E. & Moreno, R. (2003). Nine ways to reduce cognitive load in multimedia learning. *Educational Psychologist, 38*(1): 43–52.

Miller, J.D. (1983). Scientific literacy: A conceptual and empirical review. *Journal of the American Academy of Arts and Sciences, 112*(2): 29–48.

Nentwig, P., Parchmann, I., Demuth, R., Graesel, C., & Ralle, B. (2002). Chemie im Kontext, from situated learning in relevant contexts to systematic development of chemical concepts. In *second IPN_YSEG Symposium on context-based curricula* (pp. 10–13).

Permanasari, A., Mudzakir, A., & M. Mahiyudin. (2010). The influence of social issue-based chemistry teaching in acid base topic on high school student's scientific literacy. *Proceedings of the First International Seminar of Science Education, Science Education Program Graduate School, Indonesia University of Education (UPI)*.

Piirto, J. (2011). *Creativity for 21st century skills*. Rotterdam: Sense Publishers.

Schoenfeld, Alan H. (2001). Reflections on an impoverished education. In Lynn Arthur Steen (Ed.), *Mathematics and Democracy: The Case for Quantitative Literacy,* (pp. 49–54). Princeton, NJ: National Council on Education and the Disciplines.

Shamos, M.H. (1995). *The myth of scientific literacy*. New Brunswick, NJ: Rutgers University Press.

Ideas for 21st Century Education – Abdullah et al. (Eds)

The analysis of junior high schools' educational facilities, infrastructure needs and location determination based on a social demand approach and geographical information system

T.C. Kurniatun, E. Rosalin & L. Somantri
Universitas Pendidikan Indonesia, Bandung, Indonesia

A. Setiyoko
National Institute of Aeronautics and Space, Indonesia

ABSTRACT: This study aimed to identify the infrastructure needs of junior high schools' (or equivalent grades) as well as location determination using a social demand approach and Geographical Information Systems (GISs) in the coastal areas of the West Java province. This study focused on limited educational facilities and infrastructure needs, particularly related to the number of new schools and new classrooms. The method used in this research was the explorative method followed by descriptive method. The data analysis techniques used in this study were ratio analysis, infrastructure needs projection analysis and GIS. The results show that the number of junior high schools in Sukabumi and Cianjur is still not significant in comparison with the number of elementary schools. The difference average was > 80%. The study concludes that up to the year 2020, both in Sukabumi and Cianjur district, there will be a need to build new school building(s) in accordance with the SPM (minimum service standard). The locations will then need to be based on the population distribution of identified locations, which later on can then be proposed as the appropriate locations of junior high schools.

1 INTRODUCTION

Conceptually, educational planning approaches, particularly ones that are related to the government's compliance, need to meet a social demand approach. In general, the social demand approach is usually carried out in countries that have newly gained independence from colonialism and ones that are classified as developing countries (Combs, 1970; Babalola, 2003; Adekoya & Gbenu, 2008). It is the government's obligation to meet the needs of social services including education. The implication is that, if this approach is used, educational planners must estimate the needs of education in the future by analyzing demographics, enrollment, current students and desires of the community about the types of education (Adekoya & Gbenu 2008; Udin & Ma'mun, 2005; Timan., 2014; Gbenu, 2012).

In an effort to plan education using a social demand approach, the role of Geographical Information Systems (GISs) is very important, especially to determine the location of the schools as the educational facilities. The GIS data and spatial analysis provide input for policy decision-making (Fleming, 2014; Burrough et al., 2015; Agrawal and Gupta, 2016). This study aimed to identify the needs of secondary school education facilities and infrastructures in the framework of the mandatory time of study in 9 consecutive years in West Java as well as the determination of the exact location in the southern coastal areas of West Java using the approach of social demand and GIS.

2 RESEARCH METHODS

The locations of observation of this study were the Agrabinta Subdistrict of Cianjur and Cisolok Subdistrict of Sukabumi. These two locations represent the characteristics of the coast in the province of West Java.

This study used an explorative approach using analytical techniques as follows:

2.1 *Ratio analysis*

Compares the existing with the ideal conditions for later analysis and quality gaps.

2.2 *Projection method*

In line with the scope of work, among others, this research is projecting facilities and education infrastructure at junior level within the academic years 2015–2020, by conducting a certain method, namely the specific projection level enrollment method. This method is described as follows.

$$EG_t = K_t + E_t + \sum_{j=1}^{8} G_{ji}$$

$$SG_t = S_t + \sum_{j=9}^{12} G_{ji}$$

where
- i = Subscript denoting age
- j = Subscript denoting grade
- t = Subscript denoting time
- K_t = Enrollment at the nursery and kindergarten level
- G_{jt} = Enrollment in grade j
- G_{1t} = Enrollment in grade 1
- E_t = Enrollment in elementary special and ungraded programs
- S_t = Enrollment in secondary special and ungraded programs
- P_{it} = Population age i
- RK_t = Enrollment rate for nursery and kindergarten
- RG_{1t} = Enrollment rate for grade 1
- RE_t = Enrollment rate for elementary special and ungraded programs
- RS_t = Enrollment rate for secondary special and ungraded programs
- RPG_t = Enrollment rate for postgraduate programs
- EG_t = Total enrollment in elementary grades (K–8)
- SG_t = Total enrollment in secondary grades (9–12)
- R_{jt} = Progression rate for grade j: the proportion that enrollment in grade j in year t is of enrollment in grade j–1 in year t–1.

Source: Shryock & Siegel (1976)

2.3 GIS technique analysis

This analysis is used to determine the condition of existing schools with the flow as shown in Figure 1.

The source of the data used in this research was satellite image data, in addition to secondary data from the Central Bureau of Statistics. In this study, the satellite imagery used was the image of the Pleiades, WorldView 2, and GeoEye. The national images were provided by the National Institute of Aeronautics and Space (LAPAN) through the Center for Technology and Remote Sensing Data (Pustekdata). The character of this type of satellite imagery is shown in Table 1.

The three types of satellite imagery used were included in the same category, namely the very high-resolution satellite imagery with a spatial resolution or pixel size of approximately 0.5 m. Pleiades satellites which began their operations in performing data acquisition in 2011 have a spatial resolution of 2 m for the band multispectral (MS) and 0.5 m for the band panchromatic (Pan) with a width greater than the sensor sweep of WorldView 2 and Geo-Eye 1. WorldView 2 and GeoEye 1 satellites had the same specification spatial resolution, but differ in the number of bands. WorldView 2 which has been operating since 2009 has 8 bands, thus the spectral resolution is better than the Pleiades and GeoEye1. The differences were not very significant as the spatial object information can be interpreted in three different types of satellite images.

3 RESULT AND DISCUSSION

The results of this study in the form of data and information about the needs of educational facili-

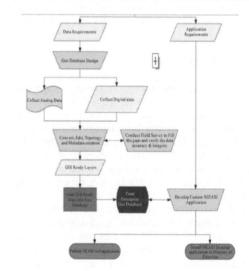

Figure 1. Workflow. Source: Rasheed dan Gamily (2103).

Table 1. Satellite imagery specification used.

| No | Type | Operated since | Band total | Spatial resolution | | Swath Width (km) |
				MS (m)	Pan (m)	
1	Pleiades	2011	4	2	0,5	20
2	WorldView 2	2009	8	1,84	0,46	16,4
3	GeoEye1	2008	4	1,84	0,46	15,2

Source: http://inderaja-catalog.lapan.go.id. (2016).

ties and determining the location of junior high schools in the coastal areas of West Java, include:

3.1 Cisolok subdistrict, Sukabumi

For the needs of the educational facilities in the Cisolok Subdistrict, Sukabumi, in the academic years 2017–2019, the following need to be prepared:

- Renovation of classrooms in need of major repairs: 34 rooms
- Construction of a latrine/toilet: 45 units
- Construction of libraries: 11 units
- Construction of Laboratory: 10 units

Furthermore, new classrooms needs will be associated with the analysis of the current school within the settlements, which will be discussed in the study in 2017.

To determine the approximate locations of the junior high schools, or equivalent, distribution in Cisolok subdistrict, Sukabumi, based on Pleides 2014 image interpretation of the distribution settlements in Cisolok subdistrict, there are many settlements in Cimaja, Cisolok, Karangpapak, Cikahuripan, Pasirbaru, and Caringin villages. Further north, the settlements were increasingly rare; such as in the Cileungsing, Sirnarasa, Sirnaresmi, Cicadas, Gunung kramat, and Cikelat villages, even in the Sirnaresmi, Sirnarasa, and Cileungsing village. Settlements were very rare in the third village bordering the Mist Mountain Salak National Park. Settlement in the Cisolok Subdistrict concentrated in the southern regions that form the lowlands, while getting to the northern settlements they become increasingly rare because of the topography of the high hills. The settlements are mostly scattered around in the south area, since the area is more easily accessible in comparison to the north one. Settlements in the south were scattered among rice fields. Further north, ridge settlement follows the contours of the hills. The use of land in the hills consists of shrubs and mixed farms. Most population is commonly in the southern region bordering the sea.

There are eight ideal establishment locations for junior school buildings based on the density of settlements. Densely populated areas are the Cisolok, Karangpapak, and Pasirbaru villages; which made these villages ideal in establishing junior high school building. As for reaching the hilly areas, taking into account the density of settlement in the villages of Caringin, Cikakak, and Mount Tanjung, the establishment of a junior high school building was considered to be accessible. To reach mountain areas, junior high school students need to go across the Cicadas and Sirnarasa border, due to the long distance.

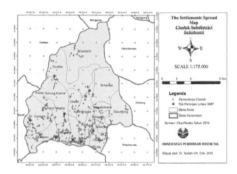

Figure 2.

As a result, the possible junior high school locations for Cisolok Subdistrict Sukabumi are as shown in Figure 2.

3.2 Agrabinta subdistrict, Cianjur

For the educational facilities needs in the Agrabinta Subdistrict, Cianjur, in the academic years 2017–2019, the following need to be prepared:

- Construction of new classrooms: 11 new classrooms
- Renovation of classrooms in need of major repairs: 8 rooms
- Construction of latrines/toilets: 193 units
- Construction of a library: 1 unit
- Construction of Laboratory: 2 units
- In 2019, should it need to build one new school, then it will require another laboratory, library and a teachers' room

To determine the approximate distributed locations of the junior high schools, or equivalent, in Agrabinta Subdistrict, Cianjur, then based on the WorldView and GeoEye 2014 image interpretation, the distribution of settlements in Agrabinta Subdistrict was widely spread in Puncak Wangi, Walahir, Sirnasari, Sukasirna, and Sukamulya villages. The settlements were rather numerous in Pusakasari, Neglasari, Sirnalaut, and Bojongkaso villages. Many settlements are far from the sea due to difficult access to the roads. Plantations are mostly rubber and chocolate. Settlements in Bojongkaso village follow along the river channel of Cibuni. The topography of Agrabinta Subdistrict is undulating to hilly, with narrow plains in the south bordering the sea so that the concentration of the population is mostly in the countryside. The settlements are common in South Trans, which connects West Java Cianjur and Sukabumi in southern coastal areas.

There are five ideal locations for the potential establishment of a school building in the Agrabinta Subdistrict, the first location is at Walahir village. This is due to the dense settlement that

103

Figure 3.

includes Puncak Wangi Walahir village. This would be accessible to children at junior high school ages in both villages. The fact that Neglasari and Sukamulya are far from the center of the city proves that it is not easy to build new schools there. Other villages in need of having schools are Sukamanah, Wanasari, and Karangsari. Furthermore, other schools are needed in Sirnasari and Bojongkaso villages as both are far in the south part of West Java. A junior high school which would be established in Bojongkaso village can also be reached by the children of Sinarlaut village; which is bordering the sea and has a sparse population.

As a result, the possible junior high school locations for Agrabinta Subdistrict, Cianjur are as shown in Figure 3.

Based on the above results, it should be clarified that the results of this study are relevant to the conceptual approach of planning based data (Campbell, 2002; Gbadamosi, 2005; Hite, 2008) and the spatial approach (Al-Hanbali et al., 2005). The results of the spatial analysis, presented above, can be a reference in determining the locations of schools in accordance with the population distribution. This is consistent with the role of GIS information as a basis for determining the infrastructure needs of schools (Agrawal and Gupta, 2016; Attfield et al., 2002). In addition, spatial information also provides tools and techniques for analyzing educational needs from a geographical perspective (Mendelsohn, 1996). Some studies (Fabunmi, 2007; Hite, 2008; Galabawa et al., 2002 Odhiambo and Imwati, 2014) have also deliberated about the concept and the use of GIS in school mapping.

4 CONCLUSION

Currently the number of junior high schools in Sukabumi and Cianjur is still much smaller than the number of elementary schools. The average difference is more than 80%. This illustrates that in order to achieve the target of 100% compulsory education program, educational facilities and infrastructure are in strong need of improvement. There should be more junior high schools to anticipate the high number of elementary school graduates. Up to 2020 both in Sukabumi and Cianjur will still requisite to build new school buildings in accordance with Education Minimum Service Standards. The exact location used as the location of new schools is an area close to dense residential population, however, need to analyze other land use aspects.

REFERENCES

Adekoya, S.O.A. & Gbenu, J.P. (2008): Fundamentals of educational planning (revised and enlarged). Lagos: Micodex Nig. Ltd.

Agrawal, S., & Gupta, R.D. (2016). School Mapping and Geospatial Analysis of the Schools in Jasra Development Block of India. *International Archives of the Photogrammetry, Remote Sensing and Spatial Information Sciences*, 145–150.

Al-hanbali, N., Al-kharouf, R. and Alzoubi, M.B. (2005). Integration of geo imagery and vector data into school mapping GIS data-model for educational decision support system in Jordan. *SYSTEM*, 2(2), 1–6.

Babalola, J.B. (2003). *Basic text in educational planning*. Ibadan: The Department of Educational Management.

Burrough, P.A., McDonnell, R., McDonnell, R.A., & Lloyd, C.D. (2015). *Principles of geographical information systems*. Oxford University Press.

Campbell, O.O. (2002). *Educational planning, management and school organization*. Lagos: Babs Olatunji Publishers.

Fabunmi, M. (2007). *Perspectives in educational planning*. Ibadan: Odun Prints.

Flemming, C (ed). (2014). *The GIS guide for elected official*. California: Esri Press.

Galabawa, J.C., Agu, A.O. and Miyazawa, I. (2002). The impact of school mapping in the development of education in Tanzania: an assessment of the experiences of six districts. *Evaluation and Program Planning*, 25(1): 23–33.

Gbadamosi, L. (2005). *Basics of educational planning*. Ogun State: Triumph Publishers.

Gbenu, J.P. (2012). The adoption of the principles of social demand approach (SDA) as a strategy towards ensuring the success of the UBE programme in Nigeria. *Knowledge review*, 24(1), 1–5.

Hite, S.J. (2008). *School mapping and GIS in education microplanning*. Paris: UNESCO Publishing IIEP.

Mendelsohn, J.M. (1996). *Education planning and management and the use of geographic information system*. Paris: UNESCO Publishing IIEP.

Odhiambo, O.G. & Imwati, A.T. (2014). Use of geoinformation systems for educational services provision and planning in Asal areas: A case study of Garissa County Kenya. *International Journal of Science and Research*, 3(9): 2432–2446.

Timan, A. (2004). *Perencanaan pendidikan. Study Book*. Malang: Jurusan AP.

Ideas for 21st Century Education – Abdullah et al. (Eds)
© 2017 Taylor & Francis Group, London, ISBN 978-1-138-05343-4

Debriefing teachers' competence based on reflective teaching to facilitate creative thinking skills of elementary school students

R. Witarsa, A. Permanasari & U.S. Saud
Graduate School of Universitas Pendidikan Indonesia, Bandung, West Java, Indonesia

ABSTRACT: The objective of this study was to strengthen the teachers' capabilities related to creative thinking skills. The research methodology used in this research was research and development. The results show that the use of a guidance book about debriefing teachers' competence based on reflective teaching (DTCRT) by teachers of elementary schools improved their creative thinking skills by 51.53%. It also improved the teachers' skills in preparing lesson plans based on creative thinking skills and in making assessments based on creative thinking skills by 53% and 63%, respectively. The use of the DTCRT guidance book also improved the skills of the teachers in demonstrating learning activity based on creative thinking skills by 51%. This implies that the use of the DTCRT should be continued and can be applied at educational institutions and for the professional education of personnel and teachers.

1 INTRODUCTION

1.1 *Background*

The paradigm of the 21st Century National Education stated that one of the strategies of educational attainment in the future is the application of the creative learning method in Elementary Schools (ESs). This method is focused on the principle that each individual is unique and has their individual talent, so the learning methodology must pay attention to the diversity of learning styles of each individual. A learning model that emphasizes the distinctive features and diversity of individuals should be developed. An example of such a learning model is the Problem Based Learning (PBL). In addition, the cooperative learning model should be emphasized among individuals to improve the interpersonal competence in their social life, such as: Cooperative Learning and Collaborative Learning.

Unfortunately, the learning process in primary schools is still far from expectations. For example, in science, learning in the classroom is often boring for the students so that it becomes meaningless. Learners simply sit still and silent, listening to the teacher, the teacher is stood in front of the class to explain the subject matter, and teachers tend to lecture. The described conditions might be due to a lack of knowledge and a low ability of the teacher in creative thinking. There are also concerns about subject knowledge, pedagogical competence and academic skills of elementary school teachers.

It can be seen that the ability of teachers is categorized as low in science subjects. In a test done by the teachers prior to the research, it is shown that the average score was 46.5, with a minimum score of 80.0 and a maximum score of 100. The lowest score was 15.56 and the highest score was 82.22. Meanwhile, the results of the tests of the creative thinking skills of the teachers showed that the lowest score was 41.34 while the highest score was 55.82 (a maximum score is 100). The average score of teachers in creative thinking skills was 33.22 out of 100 (Witarsa 2011).

In the field, the teachers rarely used fun methods in teaching science, for example, by direct observation, experiment, or simulation. As a result, science is regarded as subjects based on memorization. Science teaching and learning should enable students to improve their motivation, innovation, and creative thinking skills so that they will be able to face the challenges in the future, particularly in relation to science.

A low ability of teachers in teaching science has an impact on learners' achievements. The results of the Programme for International Student Assessment (PISA) in 2012 showed that Indonesia's position in science was ranked 64th out of 65 participating countries. The average score of science is 382 out of 528. In PISA, an Indonesian once reached the second level out of the six levels available. Level six of the PISA is the ability to synthesize a variety of knowledge and information that is stated explicitly to solve complex problems or make decisions. It shown that 24.7% of Indonesian learners have not even reached its lowest requirements (The World Bank Report 2011, PISA 2012).

Results of assessing teachers' ability and the results of the PISA is indicators that learning science in Indonesia is stagnant and has even tended to decline from 2006 (Science 393 PISA results), in 2009 (Science 383 PISA results).

The learning outcomes of learners in science are strongly influenced by the ability of teachers in giving instruction during learning activities in the classroom. If teachers have the ability to give better instruction in teaching and learning science, it is expected that the results of the learning will also improve. Elementary school teachers should be able to create fun learning situations during teaching science i.e. learning science hands-on (Lee 2006, Pine 2006, Foulds 1996).

The results of observation in several trainings that have been conducted in West Bandung Region (WBR) show that trainings that are supposed to focus on developing creative thinking skills for primary teachers promoted merely conceptual information and theories. The instructor/facilitator rarely gives a concrete example of a learning model after the presentation of the concepts or theories, thus, the thing that is retained by the teachers in training activities is only limited to theoretical information.

Based on the analysis above, from both direct observation during training and through document analysis of the training that has been conducted by WBR, we conclude that it is necessary to repair the implementation of creative thinking skills training for primary school teachers. The material of the training should be focused on developing creative thinking in creating LP of science and implementing creative-based science learning. The focus of the training should pay attention to: training objectives, training materials based on a needs analysis, use of learning resources from the environment, extracting scientific concepts learned through teacher interaction. Moreover, science learning also needs creativity elements such as sensitivity, fluency, flexibility, originality, detailing, and evaluating. (Witarsa 2011).

Debriefing teachers' competence of creative thinking through a reflective teaching activity is expected to impact their creative thinking. For instance, teachers can finally create science learning that is not monotonous. (Langer et al. 2003, York-Barr et al. 2001).

Based on the explanation above, it is necessary to develop a program for debriefing teachers' competence based on reflective teaching to facilitate the creative thinking skills of elementary school children.

1.2 Problem formulation research

Based on the description that has been given as the background of the research, the formulation of the research problem is stated as follows: "How to debrief teachers' competence based on reflective teaching to facilitate the creative thinking skills of elementary school children?".

1.3 Research questions

Problems in this study have been specifically formulated as research questions as follows:

1. What is the DTCRT program?
2. How is the teacher's understanding of developing science learning in elementary schools based on creative thinking skills?
3. How to improve the skills of elementary school teachers in developing lesson plans for science subjects based on creative thinking skills following the DTCRT program?
4. How to improve the skills of primary school teachers in making assessments of science subjects based on creative thinking skills as the impact of the implementation of DTCRT program?
5. How to improve the skills of primary school teachers in teaching science based on creative thinking skills as the impact of the implementation of DTCRT program?

1.4 Research objectives

The purpose of this study was based on the formulation of the problem which has been described previously. The general purpose of this research is to create such product as debriefing teacher's competence of creative thinking skills through reflective teaching for elementary school students. In particular, the purposes of this study are as follows:

1. Describing the profile of the DTCRT program for elementary school teachers.
2. Describing teacher understanding in developing ES-based science learning in creative thinking skills.
3. Mastering the skill development of the teachers in preparing lesson plans in science subjects based on creative thinking skills as the impact of the implementation of DTCRT program.
4. Mastering the skill development of the teachers in making assessments of science subjects based on creative thinking skills as the impact of the implementation of DTCRT program.
5. Mastering the skill development of the teachers in teaching science based on creative thinking

skills as the impact of the implementation of DTCRT program.

1.5 Benefits of the research

The results of this study are expected to provide benefits for theory development and contribute practical benefits as follows:

1. Theoretical Benefits

The DTCRT program is expected to be an innovative and effective training program to achieve learning objectives. The DTCRT program will increase the number and the variety of training programs, in line with the government's commitment to improve teacher competence. Thus, this DTCRT program has great benefits that can be adopted by education and training institutions (*Diklat*), and can be adapted by the Institute Producing Personnel (LPTK) which has the authority to improve the professionalism of teachers through Teacher Educational Program.

2. Practical benefits

The practical benefits of implementing the DTCRT program are: (1) providing direct experience for the teachers who are involved in the training program to improve their creative thinking skills, and teaching science to elementary school based on creative thinking skills; (2) providing input to the Principal for improvement in elementary school science lessons in their school to improve the skills of creative thinking; (3) providing contributions to officials in the Education Department or District to improve the ability of creative thinking and learning science in elementary schools based on creative thinking skills through training programs; (4) as a reference for developers and trainers to develop innovative, effective, and efficient training programs.

2 METHODS

This study aims to produce a product that is a set of teacher training programs that can facilitate the creative thinking abilities of teachers. In accordance with the research objectives, this study used a strategy of mixed methods (quantitative and qualitative) as well as research and development or R&D. The chosen strategy of research and development is a powerful strategy for improving a practice model. It is a process used to develop and validate educational products. This supports the reason the research method selected is in accordance with the purpose of the current research. (Borg & Gall 1989).

In research and development, its phases are a cycle that includes the discovery of various field research findings related to the product that will be developed. There are ten steps to be taken in the implementation of research and development method. These are as follows: (1) research and information gathering; (2) planning; (3) development of a preliminary product form; (4) pre-trials; (5) a revision of the primary products; (6) the main trial based on preliminary test results; (7) the revised operational products; (8) operational trials; (9) the revision of the final products; (10) dissemination and implementation (Borg and Gall, 1989).

The ten steps are simplified into the following three steps: (1) The preliminary study consists of a literature review and field studies; (2) the development of a draft of the training program, which includes the drafting of initial, limited testing, and comprehensive testing; (3) validation of training programs implemented in the form of experiments (Sukmadinata 2006).

3 RESULTS AND DISCUSSION

3.1 Profile Preparatory Program Competency Based Teacher Reflective Teaching (DTCRT)

Debriefing Program Profile Reflective Teaching Competency Based Teacher (DTCRT) developed in this study refers to the stages of research and development expressed by Borg & Gall (1989), in which the stages are as follows:

1. Research and information gathering.
2. Planning.
3. Development of a preliminary product form.
4. Test the introduction.
5. Revisions to the main product.
6. The main test that is based on preliminary test results.
7. Revision of operational products.
8. Test operational.
9. Revision of the final product.
10. Dissemination and implementation.

DTCRT profiles that have been researched and developed are illustrated and were made into a book that essentially comprised the following:

1. Rationale.
2. Definition of the DTCRT Program.
3. Interesting factors of DTCRT Program.
4. Curriculum.
5. Curriculum Structure.
6. DTCRT Learning Systems Program.
7. Stages of Implementation Workshop.
8. Duration Implementation Workshop.
9. Practice Learning.
10. Evaluation.
11. References.

Table 1.   The profile of teachers in elementary school (SD) in Group Padalarang in the district of West Bandung.

Nama Sekolah Binaan	Jumlah Guru			Guru Penjaskes		Gugus
	PNS	NON PNS	Jumlah	PNS	Non PNS	
SDN 1 Cimerang	7	3	10	1	–	1
SDN 2 Cimerang	4	5	9	1	–	1
SDN 3 Cimerang	5	4	9	–	1	1
SDN Ciampel	5	6	11	–	1	1
SDN Tipar	5	3	8	1	–	1
SDN Sindangsari	4	9	13	1	–	1
SDS Damian School	–	6	6	–	1	1
BAIS	–	22	22	–	–	1
Jumlah Guru	30	8	88	4	3	
SDN 1 Margalaksana	5	4	9	1	–	2
SDN 2 Margalaksana	3	5	8	1	–	2
SDN 3 Margalaksana	6	5	11	–	1	2
SDN Cibacang	6	3	9	1	–	2
SDN 1 Cipeundeuy	7	4	11	1	–	2
SDN 1 Cibacang	4	4	8	–	1	2
SDN 2 Cipeundeuy	9	3	12	1	–	2
SDN 2 Cibacang	4	5	9	–	1	2
Jumlah Guru	44	33	77	5	3	
SDN 1 Krida Utama	5	8	13	1	–	3
SDN 2 Krida Utama	6	5	11	1	–	3
SDN 1 Curug Agung	9	4	13	1	–	3
SDN 2 Curug Agung	9	1	10	1	–	3
SDN 1 Kerta Jaya	7	4	11	1	–	3
SDN 3 Kerta Jaya	8	4	12	1	–	3
SDS Cahaya Bangsa	–	33	33	–	1	3
SDS Al Irsyad	–	34	34	–	1	3
Jumlah Guru	44	93	137	6	–	
SDN 1 Padalarang	6	5	11	–	1	4
SDN 2 Padalarang	7	4	11	1	–	4
SDN 3 Padalarang	6	3	9	1	–	4
SDN 4 Padalarang	5	3	8	1	–	4
SDN 5 Padalarang	8	3	11	1	–	4
SDN 1 Kertamulya	10	1	11	–	1	4
SDN 2 Kertamulya	6	9	15	1	–	4
Jumlah Guru	48	28	76	5	2	
SDN 1 Sudimampir	7	6	13	1	–	5
SDN 2 Sudimampir	10	14	24	1	–	5
SDN Babakan Loa	8	8	16	–	1	5
SDN Bina Bakti	5	7	12	1	–	5
SDN Budi Asih	8	3	11	–	1	5
SDN Sukamaju	7	3	10	1	–	5
SDN 1 Cipadangmanah	7	1	8	1	–	5
Jumlah Guru	52	42	94	5	2	
SDN 2 Purabaya	8	2	10	1	–	6
SDN 3 Purabaya	13	9	22	–	1	6
SDN 4 Purabaya	7	7	14	1	–	6
SDN Kertasari	6	5	11	1	–	6
SDN Karya Bakti	5	4	9	1	–	6
Jumlah Guru	39	27	66	4	1	
SDN Jayamekar	5	9	14	1	–	7
SDN Cipondoh	5	7	12	–	1	7
SDN Darma Bakti	4	4	8	–	1	7
SDN 1 Gunung Bentang	4	8	12	–	1	7

(*Continued*)

Table 1. (*Continued*)

SDN 2 Gunung Bentang	5	7	12	1	–	7
Jumlah Guru	23	35	58	2	3	
SDN Sunan Giri	8	4	12	1	–	8
SDN Pamucatan	9	3	12	1	–	8
SDN 1 Ciburuy	13	15	28	1	–	8
SDN 2 Ciburuy	8	3	11	1	–	8
SDN 1 Kamulyan	8	4	12	1	–	8
SDN 2 Kamulyan	6	3	9	–	1	8
Jumlah Guru	52	32	84	5	1	
SDN 1 Tagogapu	8	6	14	1	–	9
SDN 2 Tagogapu	8	6	14	1	–	9
SDN 1 Parigi	5	4	9	–	1	9
SDN 2 Parigi	7	4	11	–	1	9
SDN Neglajaya	8	3	11	1	–	9
Jumlah Guru	36	23	59	3	2	
SDN Sadang	5	5	10	1	–	10
SDN Margarahayu	10	5	15	1	–	10
SDN Cadas Mulya	4	9	13	1	–	10
SDN Mekar Jaya	3	4	7	1	–	10
SDN 1 Medal Sirna	7	2	9	–	1	10
SDN 2 Medal Sirna	4	6	10	1	–	10
Jumlah Guru	33	31	64	5	1	
Jumlah Guru Total	401	402	803	44	18	

Table 2. List name elementary school stage trial introduction (The first stage).

Nama Sekolah Binaan	Jumlah Guru			Guru Penjaskes		Wilayah Sekolah
	PNS	NON PNS	Jumlah	PNS	Non PNS	
SDN 1 Tagogapu	8	6	14	1	–	Kota
SDN Neglajaya	8	3	11	1	–	Semi Kota
SDN Mekar Jaya	3	4	7	1	–	Pinggiran
Jumlah Guru Total	19	13	32	3	–	

12. Appendix Creative Thinking Skills.
13. Appendix Creative Thinking Skills Assessment.
14. Appendix RPP-Based Creative Thinking Skills.

DTCRT's validity has been tested by five experts. Two of them are experts in science, one is an expert in pedagogic to ES, one is an expert of basic education, and the last is an expert of teacher training. Based on the validation test by five experts, and after making improvements a couple of times, the handbook of DTCRT program can be directly applied in the field.

Books of the PKGRT program were discussed in the teacher forum at Padalarang, West Bandung. Broadly speaking, the profile of teachers in elementary school (SD) in the districts was as follows:

Based on Table 1 above, the chosen three elementary schools to be a place of research in the first phase of this I say as a preliminary test. ES selected third is the result of discussions between the promoters, kopromotor, the Primary Schools in the District and Chairman PGRI Padalarang

sub district. The third elementary spread in the City District, the semi City, and suburban areas. The following three elementary profile that became a test phase of preliminary studies (the first stage).

Based on the program planned, DTCRT was implemented on August, 4–6, 2016 at Padalarang. The DTCRT's workshop first phase, was attended by three elementary schools that are also the research sites that attended the workshop, and was also attended by other ES who were in the group IX (SDN 2 Parigi, SDN 1 Parigi, SDN 2 Tagogapu), the watchdog in the District Padalarang, and was opened by the Head of Unit District of Padalarang. The total number of participants in the first phase of activities reached 78 people.

4 CONCLUSION

The results showed that the use of the guidance book about debriefing teachers' competence based on reflective teaching (DTCRT) by teachers of elementary schools increased their creative thinking skills by 51.53%. It also increased the teachers' skills in preparing lesson plans based on creative thinking skills and in making assessments based on creative thinking skills by 53% and 63%, respectively. The use of the DTCRT guidance book also improved the skills of the teachers in demonstrating learning activity based on creative thinking skills by 51%. It implies that the DTCRT should be continued and can be applied in educational institutions and for the professional education of personnel and teachers.

REFERENCES

Badan Standar Nasional Pendidikan. (2010). *Paradigma Pendidikan Nasional Abad XXI*. Jakarta: Badan Standar Nasional Pendidikan.

Borg, W.R. & Gall, M.D. (1989). Educational research: An introduction, Fifth Edition. New York: Longman.

Foulds, W. (1996). The enhancement of science process skill in primary teacher education students. *Australian Journal of Teacher Education, 1* (12): 16–23.

Hasbi, H. (2007). *Tanggapan Guru terhadap Profesi*. Banda Aceh: Ar-Raniry Press.

Langer, G.M., Colton, A.B., & Goff, L. (2003). Collaborative analysis of student work: Improving teaching and learning. Alexandria, VA: ASCD (Association for Supervision and Curriculum Development).

Lee, O. (2006). Science inquiry and student diversity: Enhanced abilities and continuing difficulties after an instructional intervention. *Journal of Research in Science Teaching, 10* (4): 607–636.

Pine, J. (2006). Fifth graders science inquiry abilities: A comparative study of students in hands-on and textbook curricula. *Journal of Research in Science Teaching, 43* (5): 467–484.

PISA, O. (2012). *Results in Focus: What 15-year-olds know and what they can do with what they know*. Retrieved from http:////www.oecd.org/pisa,/keyfindings,/pisa-2012-results-overview,pdf.

Sukmadinata, N.S. (2006). *Metode Penelitian Pendidikan*. Remaja Rosdakarya: Bandung.

Witarsa, R. (2011). Analisis Kemapuan Inkuiri Guru yang Sudah Tersertifikasi dan Belum Tersertifikasi dalam Pembelajaran Sains SD. (Tesis). Sekolah Pascasarjana, Universitas Pendidikan Indonesia, Bandung.

World Bank, Kementerian Pendidikan Nasional, Kingdom of the Netherlands. (2011). Mentransformasi Tenaga Pendidikan Indonesia, Volume I: Ringkasan Eksekutif. Kantor Bank Dunia: Jakarta.

World Bank, Kementerian Pendidikan Nasional, Kingdom of the Netherlands. (2011). Mentransformasi Tenaga Pendidikan Indonesia, Volume II: Dari Pendidikan Prajabatan hingga ke Masa Purnabakti: Membangun dan Mempertahankan Angkatan Kerja yang Berkualitas Tinggi, Efisien, dan Termotivasi. Kantor Bank Dunia: Jakarta.

York-Barr, J., Sommers, W.A., Ghere, G.S., & Montie, J. (2001). Reflective practice to improve schools: An action guide for educators. Thousand Oaks, CA: Corwin Press.

Global Issues in Education and Research (GER)

Ideas for 21st Century Education – Abdullah et al. (Eds)
© *2017 Taylor & Francis Group, London, ISBN 978-1-138-05343-4*

The awareness of risk prevention level among urban elementary school students

R. Effendi
Universitas Pendidikan Indonesia, Bandung, Indonesia

ABSTRACT: Improvement of the risk prevention awareness for children of elementary school age group is becoming more necessary due to massive technology advancement in the future. Therefore, there is the need for awareness to prevent risks before an adverse event occurs. This study aims to clarify the extent of awareness level to reduce the risks among urban elementary school students. The research investigated the awareness levels of students in identifying and declining the purchase of risky products, in refusing the use of risky products, in identifying the content of the product, and in having awareness of the emerging risks of the associated products. The method used in this research was survey, while the research location was determined purposively, namely two elementary schools located in Jakarta, Bandung, Surabaya, Yogyakarta, Medan, Makassar, and Denpasar. The results of the study for four components of risk prevention awareness were at 'apathy' level. It means that students have apathetic attitudes towards the danger of risks which take place around them.

1 INTRODUCTION

The emergence of risk in various scopes of life in Indonesia shows weak risk awareness in society in Indonesia. Risk occurs in various dimensions of life, both at home and outside of the home. Risks take place both within the scale and scope of the lives of individuals and community groups, both nationally and transnationally. If this condition is not closely paid attention to, it will bring a greater risk in which people will be even less concerned with their safety and security.

Similarly, in a group of elementary school students, the phenomenon regarding the lack of risk awareness is also found. It can be seen through the use of information and communication tools which ignores the emerging risks. Using toys or games, in fact, have negative impacts. In addition, the students' behavior is prone to endanger themselves, other people, and the environment in which they live. Furthermore, they also consume food or drinks that are assumed to be very dangerous to their health, due to the contamination of excessive artificial colorings or preservatives.

Students' awareness and understanding of risks has not been sufficient. Their understanding of risky products, such as food, toys, and communication tool products, still needs to be advanced. Similarly, in his study, Effendi (2012: 277) argues that, 'students' awareness of the risk is almost entirely at the level of apathy, apathy or apathetic level shows indifference to the problem of risk in the vicinity'.

The challenges at present as well as in the future had also been highlighted by Ulrich Beck, a German sociologist around 1986, in his *Risikogesellschaft*. Beck (1992: 97) came with the opinion that modern society is encountered with the conditions of risk. According to Beck (1992: 97), in industrialized societies, it was initially found that distribution of goods (an economic) has now moved towards the distribution of risks and hazards, referred to as community risk (risk society). Modern life has created social productions which were so-called wealth distribution or social production of wealth (Beck, 2008: 19).

Giddens (1999) argues that risks contain a boomerang effect. As it is explained, '... that widespread risks contain a "boomerang effect", in that individuals producing risks also will be exposed to them'. In addition, people who have been benefited by wealth created greater risks. 'This argument that wealthy individuals suggests whose capital is largely responsible for creating pollution also will have to suffer when, for example, the contaminants seep into the water supply' (Beck, 2008: 20–23).

Based on the elaboration above, the researcher noticed that the growth of the dangers and risks of changes in society, as a result of industrialization and modernization, must be followed by the students' awareness in an attempt to prevent risks. The real conditions on the ground have comprehensively not been found regarding the kind of awareness that can protect elementary school students in urban areas from the risks.

This paper is viewed as an important and interesting work because it aims at elaborating the levels of students' awareness in risk prevention. The above objectives can be divided into several issues covering: in what levels students are considered as having awareness in order to cancel the purchase of risky products, to cancel the use of risky products, to become aware of the content of the products, and to become aware of the emerging risks.

To get all information about the above objectives, the writer used survey method. This is because the writer wanted to gather a large amount of data and cover seven cities in Indonesia.

2 LITERATURE REVIEW

The concept of awareness of (public) risks (risk society) was initially proposed by Ulrich Beck (2010). He was a German sociologist as well as the author of *Risikogesellschaft* (1986). His work had been translated into English under the title of *Risk society – Towards a new modernity* (1992). Beck (2012) expressed this idea in relation to the development or change in local communities as well as the consequences of those changes.

The concept of risk (risk) is often associated with something possible and impossible to happen, calculable and incalculable, and predictable and unpredictable. Basically, there are some concepts of risk which are in the circle of science and technology, and there are also some concepts which do not belong.

Meanwhile, risk society as defined by Ulrich Beck (1992) came to the concepts as '... a systematic way of dealing with hazards and insecurities induced and introduced by modernization itself' (1992: 21). Giddens (1999: 3), as quoted by Wikipedia (en.wikipedia.org/wiki/risk_society), defined risk society as 'a society increasingly preoccupied with the future (and also with safety), which generate the notion of risk'. In fact, these definitions had been taken as the reference by Beck (2010) and Giddens (1999) in discussing risk society.

Risk society is a society in which almost all aspects of life are built on risk awareness. However, it does not mean that all parts of people's lives are at risk. In fact, risk awareness and how to respond to risks influence all on their own social processes.

With respect to the concept of risk, Beck (1992) and Giddens (1999) have similar as well as different perspectives. On the one hand, the similarity can be seen in the emergence of risks as the effect of the advancement of science and technology utilized to prepare human beings' lives in the future. They assume that this term emerges when human beings had been in touch with modernization, particularly when science and technology dominate human beings' lives. When people think of a better welfare, they try to find a way out through science and technology. However, as a result of the development of science and technology, a boomerang effect automatically comes up, specifically the risks that have impacts on human beings.

Beck (2012) argued that risks contain boomerang effects, as he explains, '... that widespread risks contain a "boomerang effect", in that individuals producing risks also will be exposed to them'. In addition, people benefited by wealth, create greater risks. This argument suggest that 'wealthy whose capital is largely responsible for creating pollution also will have to suffer when, for example, the contaminants seep into the water supply' (Beck, 1992). However, another opinion is also expressed by Giddens. He sees from the positive side and he also suggests that there 'can be no question of merely taking a negative attitude towards risks. Risks need to be disciplined, but active risk-taking is a core element of a dynamic economic and an innovative society' (Giddens, 1999: 10).

According to Beck (2010), the life of risky society is commonly associated with how to prevent, minimize, and distribute risks. Beck (2010: 48) states that the public response towards risks can be divided into three stages covering: (1) denial, (2) apathy, and (3) transformation. First, denial responses occur in some communities in modern cultures. Second, apathy or apathetic responses occur in most postmodern nihilism. Third, transformation occurs in partly cosmopolitan society. Denial is a response that is characterized by a rejection of understanding or thinking, and feeling, which in fact exist, and it accommodates different opinions. However, apathy is a response which is characterized by ignorance and indifference. Then, transformation is a response which is characterized by the existence of better transformation, improvement, or characterized functions in order to initiate a transformation.

3 RESEARCH METHODS

The research survey of the risk prevention awareness was measured by applying Likert scale that has been modified into four alternative answers. Likert scale modification removed the middle answer categories based on a number of reasons, such as avoidance of multiple-meaning answers (multi-interpretable). It was in fact influenced by social and cultural conditions, in which for Indonesian people, they were assumed to have a tendency to take a stand in the middle.

The research location was determined purposively and consisted of two elementary schools located in Jakarta, Bandung, Surabaya, Yogyakarta, Medan, Makassar, and Denpasar. This selection was based on the reason that the locations selected as the research sites were some schools which have more concerns in science and technology as being compulsory in the curriculum. Meanwhile, the awareness of risky society as the topic of this study was associated with human beings' lives in the process of science and technology advancement (Beck, 2010: 34).

The participants involved in this study were taken from 4, 5, and 6 grade students respectively in two schools in each chosen city. It was estimated that there were 840 students spread across seven cities. The data objects under investigation would focus primarily on the attitudes of the students regarding the risk awareness in relation to the risk prevention. An instrument was developed according to Beck's concept (2010), where he states that risky society is commonly associated with how to prevent, minimize, and distribute risks (Beck, 2010: 48). The writer develops a concept of 'prevent' in four categories, namely cancelling the purchase of risky products, cancelling the use of risky products, becoming aware of the emerging risks, and understanding the content of the product.

The instrument used is validated. The result is output from SPSS which provides Cronbach's Alpha value for the whole scale of measurement of 0.873. Cronbach's Alpha value is clearly above the minimum limit of 0.70 so that it can be concluded that the attitude of measurement scale of public awareness of the risk has a good reliability. Thus, this instrument can be used as a data collector in connection with these studies. The obtained data was classified into several types of quantitative data. Then, it was classified into groups of denial, apathy, and transformation.

4 RESULTS AND DISCUSSION

4.1 Results

The results of the study on students' awareness in risk prevention taken from students in grade 4, 5, and 6 of the seven cities, covering Jakarta, Bandung, Yogyakarta, Surabaya, Medan, Makassar and Denpasar are summarized in Table 1. Based on the Table, it is clear that the highest number of students' awareness in risk prevention in the entire cities was found in Medan with a total score around 74.9, while

the lowest one was found in Surabaya with a total score around 72.8. This awareness in risk prevention is elaborated into the cancellation of the purchase of risky products, cancellation of the use of risky products, becoming aware of the emerging risks, and becoming aware of the importance of product contents.

Based on the types of awareness above, it also can be seen that the highest score of awareness was obviously found in the cancellation of the purchase of risky products in Denpasar with the score total around 80.1, while, as a comparison, the lowest one was found in Surabaya with a score of 73.1. Secondly, the highest score of awareness in the cancellation of the use of risky products was undeniably found in Yogyakarta with the score total around 70.6, while the lowest one was found in Jakarta with the score around 64.0. Thirdly, the highest score of awareness in the awareness of the emerging risks was found in Denpasar with the score around 77.7, while the lowest one was found in Makassar with the score around 73.4. Fourthly, the highest score of awareness in terms of the importance of product content was found in Denpasar with the score around 87.9, while the lowest one was clearly found in Jakarta with the score around 77.9.

Meanwhile, the recapitulation results of the students' awareness in risk prevention in Jakarta, Bandung, Yogyakarta, Surabaya, Medan, Makassar and Denpasar can be seen in Table 2. Based on the Table, it can be seen that the students' awareness in risk prevention, namely the cancellation of the purchase of risky products, was 76.6, cancellation of the use of risky products was 68.0, awareness of the emerging risks was 76.1, and awareness of the importance of product contents was 84.4. Based on Figure 1 below, it is obvious that the score total obtained was around 74. This number can be easily found in the apathy column (Figure 1). It means that in general if it is viewed from all students in grade 4, 5, and 6 SD in Jakarta, Bandung, Yogyakarta, Surabaya, Medan,

Table 1. Students' awareness of risk prevention from seven cities.

| Awareness of risk prevention. | Score | Cities | | | | | | |
		Jakarta	Bandung	Yogyakarta	Surabaya	Medan	Makasar	Denpasar
Cancellation of the purchase of risky products.	F	1,113	1,081	1,064	1,053	1,116	1,134	1,154
	%	77.2	75.0	73.8	73.1	77.5	78.7	80.1
Cancellation of the use of risky products.	F	1,299	1,286	1,357	1,315	1,330	1,312	1,247
	%	64.0	66.9	70.6	68.4	69.2	68.3	64.9
Awareness of the emerging risks.	F	1,073	1,118	1,099	1,077	1,109	1,075	1,120
	%	74.5	77.6	76.3	74.7	77.0	73.4	77.7
Awareness of the importance of product contents.	F	374	421	415	401	401	402	422
	%	77.9	87.7	86.4	83.5	83.5	83.7	87.9
Score total.	F	3,859	3,906	3,935	3,846	3,956	3,923	3,943
	%	73.0	73.9	74.5	72.8	74.9	74.2	74.6

Table 2. Recapitulation of students' awareness in risk prevention.

Awareness of risk prevention.	Score	Class			
		IV	V	VI	Total
Cancellation of the purchase of risky products.	F	2,622	2,570	2,523	7,724
	%	78.0	76.4	75.0	76.6
Cancellation of the use of risky products.	F	3,061	3,112	2,972	9,145
	%	68.3	69.4	66.3	68.0
Awareness of the emerging risks.	F	2,557	2,538	2,576	7,671
	%	76.1	75.5	76.6	76.1
Awareness of the importance of product contents.	F	909	988	939	2,836
	%	81.1	88.2	83.8	84.4
Score total.	F	9,149	9,208	9,010	27,376
	%	74.2	74.7	73.1	74.0
	27,376				
	74				

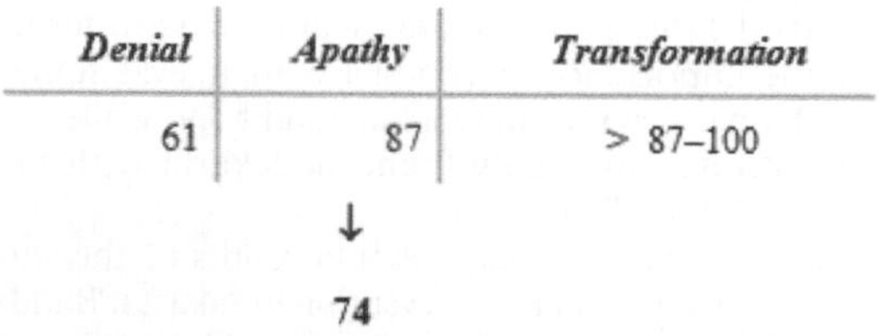

Denial	Apathy	Transformation
61	87	> 87–100

↓

74

Figure 1. The categories of students' awareness in risk prevention.

Makassar and Denpasar, the levels of awareness of risk prevention is found at the level of apathy.

4.2 Discussion

The awareness of risk prevention among elementary school students in urban areas is found at the level of apathy (score 74). Apathetic levels show indifference to the efforts in risk prevention which means that the students tend to disregard the risks that may take place in the future. Apathy in risk prevention is one of risk awareness attitudes. In addition, apathy is one of the individual or community response attitudes towards risks.

All scores (four components) from seven cities are in the range 68.0–84.4, and are all located in the apathy level. Apathy is a response characterized by ignorance and indifference. Apathy or apathetic attitude indicates a lack of interest, indifference or attention, especially in several aspects related to the public interest. In this case, the ignorance and apathy towards the problem of risk undeniably takes place around them.

Beck (2012) said that the public response towards risks can be divided into three stages covering,

denial, apathy, or transformation. The characteristics of these three levels of awareness in risk prevention in fact can be identified. These characteristics are the results of field identification and reviewing related literature, as shown in the following Table 2.

Risk awareness is one of the characteristics of the late modern society (late modernity). Society of late modernity is characterized by two aspects: 1) Reflexive modernity, and 2) Risk awareness. First, the late modern society is 'reflexive' and the word reflexive in this case is associated with "knowing" or knowledge, as stated by Giddens (1999). Second, society of late modernity is characterized by the emergence of people who are aware of the risks (risk society). Public awareness of risk is characterized by: 1) Prevention of the risk awareness, 2) Awareness of minimizing risks, and 3) Awareness of distributing risks.

5 CONCLUSIONS

Students' awareness in risk prevention takes place at the level of apathy as shown in Tables 1 and 2. They show that the students have less concern in terms of: cancelling the purchase of products, reducing product usage, realizing the importance of product content, and realizing the risks that will probably take place in the future.

The implications of apathy level of awareness in preventing the risk show that if it is left without any solution, the phenomenon of ignoring risk will continuously intensify. As a consequence, it will indirectly threaten human beings' lives in which the development of technology automatically will result in danger or risk.

REFERENCES

Beck, U. (1992). From individual society to risk society. *Theory, Culture & Society*, 9(1), 97–123.
Beck, U. (2008). *Risk society – Towards a new modernity*. Translated by Mark Ritter, London: SAGE Publication.
Beck, U. (2010). *World at risk*. UK: Cambridge Polity Press.
Beck, U. (2012). *Living in and coping with world risk society*. The Cosmopolitan Turn, Lecture in Moscow.
Effendi, R. (2012). *Isi dan modus pembelajaran pendidikan kewarganegaraan sebagai sarana peningkatan kesadaran masyarakat risiko (Risk Society)*. Sekolah Pascasarjana Universitas Pendidikan Indonesia, Bandung.
Giddens, A. (1999). Runway world: How globalization is reshaping our lives. BBC Reith Lecture.
Wikipedia (en.wikipedia.org/wiki/risk_society).

Ideas for 21st Century Education – Abdullah et al. (Eds)
© *2017 Taylor & Francis Group, London, ISBN 978-1-138-05343-4*

The role of academic self-management in improving students' academic achievement

A.L. Kadiyono & H. Hafiar
Universitas Padjadjaran, Bandung, Indonesia

ABSTRACT: This research was based on the academic performance phenomenon of the student during the learning process that was based on their Grade Point Average (GPA). In this regard, it was affected by the capability of students to control the factors which influence the learning process, called Academic Self-Management. A successful student is regarded as the one who is able to control the factors which influence their learning process. This research was conducted to verify the role of academic self-management in improving students' academic achievement. The quantitative methodology with a simple random sampling technique was employed. The total samples were 105 students of Padjadjaran University. The results showed that 78% of the subjects possessed high academic self-management, meaning that the majority of subjects used academic self-management to control factors which influence the learning process. The strategy category mostly used was Motivational Strategies; meanwhile the Learning and Study Strategies was rarely used. Additionally, the Taking Exam dimension was mostly used, but Learning From Textbook was rarely used. The final model of academic self-management can be used in order to empower students to improve their academic achievement, so that they have capital to move forward and build their future.

1 INTRODUCTION

Students are expected to be able to adjust their academic life in order to develop their potential optimally. Successful students are not simply individuals who know more than others. They also have more effective and efficient learning strategies for accessing and using their knowledge, can motivate themselves, and can monitor and change their behaviors when learning does not occur.

University education should be about more than just obtaining a paper qualification. Education has to be relevant to the social, political and economic environments and should, therefore, focus on the holistic development of the student (Harris, 2001; Johnson et al., 2000). It has been well documented that school leavers need to possess certain core competencies in order to help them cope with the demands of life after school (Lindhard & Dhlamini, 1990) but, according to recent research (Wood & Olivier, 2004), schools are not adequately equipping the learners with the necessary skills to do this.

The causes of students being underprepared in South Africa have been extensively discussed in literature (Wood & Olivier, 2004; Lethoko, 2002; Yoon, 2002; Johnson et al., 2000) and result in the students arriving at tertiary education lacking in the skills needed to cope with the academic and social demands of the tertiary environment.

Students have to adjust their academic field to per sue their optimal achievement, which can be seen in their GPA or the time needed to finish their study. The key to reaching academic success is in practicing the learning strategies taught so that the students will learn automatically. As they practice, students will be able to learn more material in less time than previously using these new strategies, in the process of developing the necessary expertise to meet the academic demands of college learning. Much of the same self-discipline and self-motivation will apply in these areas of expertise to pursue their academic excellence.

The students' beliefs about learning and motivation influence their behaviors. The following beliefs can impact achievement. If students believe they are less capable than others, they may spend considerable time using failure-avoiding strategies in the classroom (e.g., trying not to be called on, copying material from friends, and appearing to be trying hard when they really are not). Other students who believe they can achieve are more likely to spend their time using effective learning and study strategies, and tend to persist longer on difficult tasks.

The achievement of student learning performance can be reflected in the GPA that students achieved. It can be measured as an internal indicator showing the level of productivity of a certain education institution on whether the curriculum they implement is relevant to the needs of the society or not.

According to Zimmerman and Martinez-Pons (1988:284), people who possess effective self-management skills know how to set goals, are effective problem-solvers, think positively when faced with academic demands and challenges, use available resources, structure their environment to suit their goals, and are able to reflect on the reasons for failure and reset goals for future improvement.

Previous study about holistic student development, found that academic and life skills of students should be programmed in institutions to ensure students can perform at an acceptable level of learning (Wood et al., 2004). How to reach this student learning performance can be identified by learning how to become more successful learners, using appropriate strategies to manage their motivation, behavior, and learning. It is important to identify specific behaviors that influence the level of academic success and use a process to self-manage students' academic behavior.

The purpose of this study is to examine academic self-management students.

Academic self-management is frequently cited as one of several necessary skills that lead students with disabilities towards being more self-determined youngsters who can appropriately and proactively take control of aspects of their lives, in and out of school settings. But, there is still less research about academic self-management in Indonesia and what dimension in academic self-management should be developed.

2 LITERATURE REVIEW

Self-management is defined as the capacity to work effectively toward meaningful goals, and to be flexible in the face of setbacks. Note that the first half of this definition reflects planning and goal-striving behavior, and the second half reflects resilience behavior, but both components are intertwined with the process of flexibly working towards meaningful personal outcomes. Students with higher self-reported behavioral self-management have higher self-reported well-being, resilience, academic performance, and capacity to adapt to change (Agolla & Ongori, 2009). The evidence-based activities listed in this manual are associated with higher academic performance and well-being, and are based on research in goal-setting, motivation, time management, mindfulness, procrastination, psychological flexibility, and positive psychology. They are grouped around four categories, which reflect the proactive motivational sequence including the consideration of consequences of changing circumstances and varying outcomes:

- Defining meaningful goals
- Working toward meaningful goals
- Staying flexible
- Facing setbacks.

These categories are based on both the motivation and resilience literature.

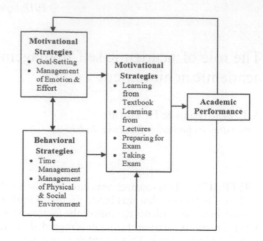

Figure 1.   Academic self-management (Dembo, 2004).

Dembo (2004) identifies that there are six components that students need to control to become successful learners. These are motivation, methods of learning, time management, physical and social environment, and performance. These components serve as the basis for organizing and integrating the content throughout the text. This focus allows for the integration of both motivation and learning strategies. As students learn new learning strategies, they must develop the motivation to use them.

To achieve certain academic performance, students implement strategies in academic self-management (Figure 1). These are motivational strategy, behavioral strategy, and learning & study strategy. Motivational strategy consists of goal-setting, and management of emotion & effort. Behavioral strategy consists of time management, and management of physical & social environment. Learning and study strategy consists of learning from textbook, learning from lectures, preparation for exam, and taking exam.

3 RESEARCH METHODS

This research approach was a non-experimental research approach. The method used was a descriptive and verification research method. According to Sugiyono (2012), the descriptive method is a method that is used to describe or analyze research results but is not used to make broader conclusions. While the verification method, according to Mashuri (2009), checks the validity of a test whether they improve when implemented in several places and contexts.

To determine the sample in the study, researchers used non-probability sampling. The sampling technique used specifically was convenience sampling. Convenience sampling is a technique for determining the sample based on coincidence, namely, anyone who by chance meets with investigators can be used as a sample; if the person accidentally suits the data source needed. Convenience sampling is a type of non-probability sampling, where samples are taken because of the easy access to geographic proximity, availability, time, and the willingness of participants (Etikan et al., 2015).

3.1 Participants

Total samples were as many as 105 students. The verification method was chosen to test the hypothesis by using a statistical test, which is simple linear regression correlation analysis. In this study, data collection was conducted by using as a measuring instrument a main questionnaire, and supporting data in the form of open-ended questions to broaden the discussion of research. The Academic Self-Concept Scale by Dembo (2004) was used to identify the students.

3.2 Instrument

The Academic Self-Concept Scale (ASCS) was developed by Dembo (2004). It has 30 items. Each item's value is coded as 1 = Not Sure; 2 = Least Sure; 3 = Sure; 4 = Very Sure. The instrument demonstrated high internal consistency with Cronbach alpha of 0.935 for this study.

4 RESULTS AND DISCUSSION

The results showed that 78% of the subjects possessed high academic self-management (Figure 2). This shows that students used motivational strategies, behavioral strategies, and learning and study strategies in achieving their academic achievement, although these were in varying degrees.

The students had 'moderate high' academic self-management. It could be argued that the use of academic self-management is often done to control the factors that affect the learning process. The major strategy that had been used to achieve

their academic achievement is using motivational strategy. Motivational strategy was used to control motivation and emotional factors, as well as efforts to address the problem of academic achievement.

It means that the majority of the subjects often used academic self-management to control factors influencing the learning process. The strategy category most used was the motivational strategies, and the learning and study strategies was rarely used.

There is an internal process that provides the power and direction for students to strive and to achieve the target that has been determined. This indicates that the targets need to be set in directing the academic achievement of students. The mapping of strategies that students used can be seen in Figure 3.

Motivational strategy is the ability to direct the internal processes that give power and direction for students to strive and to achieve self-determined targets. This is reinforced by supporting data, that students have parents who have high expectations of them, so that they have confidence that is strong enough to be able to achieve their targets, and a strong enough commitment to strive and to achieve the set targets. They in average also have a GPA above 3.00. This is in accordance with the theory of motivation which states that motivation is internal factors which include goals, beliefs, perceptions, and expectations that will provide direction and power on a behavior (Dembo, 2004).

To be a successful learner in college, students must be able to concentrate and deal with the many potential personal and environmental distractions that may interfere with learning and studying. Students use many different processes to control aspects of their behaviors. Dealing with distracting factors in learning is an important aspect of self-management, because it helps protect one's commitment to learn.

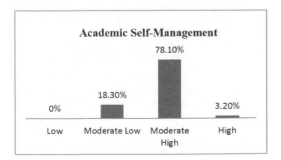

Figure 2. Academic self-management students.

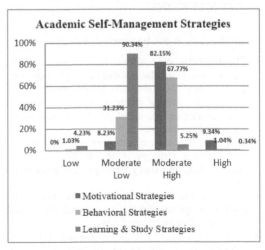

Figure 3. Academic self-management strategies.

A number of important motivational self-management techniques can be used to develop and maintain these important beliefs. The first is goal-setting. Educational research indicates that high achievers report using goal-setting more frequently and more consistently than low achievers (Zimmerman & Martinez-Pons, 1988). When individuals establish and attempt to attain personal goals, they are more attentive to instruction, expend greater effort, and increase their confidence when they see themselves making progress. It is difficult to be motivated to achieve without having specific goals.

In summary, to control students' motivation, the students need to set goals, develop positive beliefs about their ability to perform academic tasks, and maintain these beliefs while faced with the many disturbances, distractions, occasional failure experiences, and periodic interpersonal conflicts in their life. Someone will have difficulty managing his/her behavior if they do not have confidence in their ability to succeed. So, the students must develop confidence in their ability by learning how to use different learning and study strategies that lead to academic success.

In behavioral strategy, students will manage time, their physical environment, and social environment, to control the factors that affect learning. Meanwhile, on learning strategies, students will use their time to learn, practicing, preparing and doing the test. From the academic self-management dimension, they often use learn from lectures, preparation for the exam, and taking the exam. However, learning from textbook was rarely used.

The final model of academic self-management can be used in order to empower students to improve their academic achievement, so that they have capital to move forward and build their future.

5 CONCLUSIONS

From the data, it can be concluded that:

- The majority of students have academic self-management in the moderate high category. This shows that students using motivational strategy, behavioral strategy, and learning and study strategies to control the factors that affect the learning process.
- Based on the strategy category, the category of motivational strategy is the highest category and learning and study strategies category is the lowest.
- Based on the frequency, the dimensions of the highest usage are taking exam, and the lowest use is learning from textbook.
- There are still many students who are in the low category at the dimensions of management of

physical and social environment, learning from the lecturer and on exam preparation.
- Students need to scale back the use of dimensional learning from textbook, considering that the students are required to master the textbook as a main source of lecture material.
- Management of physical and social environment, learning from lecturer, and preparation for exam, also need to be managed to improve the students' academic achievement.

REFERENCES

Agolla, J.E. & Ongori, H. (2009). *An assessment of academic stress among undergraduate students: The case of University of Botswana*. Botswana: University of Botswana.

Dembo, M.H. (2004). *Motivation and learning strategies for college success: A self-management approach*. New Jersey: Lawrence Erlbaum Associates, Inc.

Etikan, I., Musa, S.A. & Alkassim, R.S. (2016). Comparison of convenience sampling and purposive sampling. *American Journal of Theoretical and Applied Statistics*, 5(1), 1–4.

Harris, B. (2001). Facing the challenges of education reform in Hong Kong: An experiential approach to teacher development. *Pastoral Care*, 21(31).

Johnson, S., Monk, M. & Hodges, M. (2000). Teacher development and change in South Africa: A critique of the appropriateness of transfer of Northern/Western practice. *Compare: A Journal of Comparative Education*, 30, 179–183.

Kerlinger, F.N. (2004). *Asas-asas penelitian behavioral*. Edisi Keempat. Yogyakarta: Gajah Mada University Press.

Lethoko, M. (2002). The role of teachers in the culture of learning and teaching. In L. Calitz, O.L. Fuglestad & S. Sillejord, *Leadership in Education*. Sandown: Heinemann.

Lindhard, N. & Dhlamini, N. (1990). *Lifeskills in the classroom*. Cape Town: Maskew Miller Longman.

Mashuri. (2009). *Penelitian verifikatif*. Yogyakarta: Penerbit Andi.

Nazir, M. (2005). *Metode penelitian*. Bogor: Ghalia Indonesia.

Sugiyono. (2012). *Metode penelitian kuantitatif kualitatif dan R&D*. Bandung: Penerbit Alfabeta.

Warsito, H. (2009). Hubungan antara self-efficacy dengan penyesuaian akademik dan prestasi belajar akademik. Universitas Negeri Padang: *Jurnal Ilmiah Ilmu Pendidikan* IX(1).

Wood, L.A. & Olivier, M.A. (2004). A self-efficacy approach to holistic student development. *South African Journal of Education*, 24(4), 289–294.

Yoon, J.S. (2002). Teacher characteristics as predictors of teacher-student relationships: Stress, negative affect and self-efficacy. *Social Behavior & Personality: An International Journal*, 30, 485–494.

Zimmerman, B.J. & Martinez-Pons, M. (1988). Construct validation of a strategy model of student self-regulated learning. *Journal of Educational Psychology*, 80(3), 284–290.

Ideas for 21st Century Education – Abdullah et al. (Eds)
© 2017 Taylor & Francis Group, London, ISBN 978-1-138-05343-4

Identifying research supporting factors: What should institutions provide?

M.C. Sondari, C. Rejito & L. Layyinaturrobaniyah
Universitas Padjadjaran, Bandung, Indonesia

ABSTRACT: This paper aims to identify factors that should be provided by institutions to support research activities of faculty members to increase research productivity in higher education institutions. This research has two objectives. First, this research aims to identify factors that are considered as research supporting factors. Second, to validate the items found at the exploration stage. Exploratory research was carried out, using a focus group discussions technique. Four informants, representing life sciences, health sciences and social sciences, were involved in focus group discussion. The result of the focus group discussion was analyzed using thematic analysis and converted into the form of questionnaires. A survey of 63 respondents was applied to collect data for the validity and reliability of the instruments. The result of the focus group discussion revealed that there are nine factors that should be provided by institutions, namely: time for conducting research; facilities and access; financial support and funding; staff assistance; constructive working environment; research productivity-based rewards; fair career system; capacity development; and good research policy. However, one factor representing the dimension of 'time' is not valid, based on the item validation result.

1 INTRODUCTION

The strategic plan of the Ministry of Research, Technology and Higher Education in Indonesia, showed that higher education is required to become an agent of education and research. As an agent of education, Higher Education Institutions (HEI) give a major contribution in the process of educating the public.

Meanwhile, as research agents, HEIs provide a major contribution towards solving the basic problems faced by the community and can provide a solution that is applicable. Those contributions can be achieved through publications, patents, and citations. However, in term of publications, HEIs in Indonesia are still lag-behind other major HEIs within the region. Even within Malaysia, for example, a head to head comparison between the most productive HEI from each country shows that the most productive HEI in Indonesia (Bandung Institute of Technology) only has one tenth of the total documents that belong to the most productive HEI in Malaysia (University of Malaya).

To be able to compete with other universities within the region, let alone to achieve the vision of becoming a world class university, HEIs in Indonesia should strive to increase the number of publications being driven by high research achievements. Therefore, the institution should provide factors that are believed to be related to the productivity of research.

Previous studies have been searching for factors that are considered to correlate with research productivity. However, most of the early studies put too much focus on the demographic of individuals. Those studies were also conducted mostly in the United States. Thus, it may not be appropriate to be applied in the context of developing countries such as Indonesia. Therefore, this paper has two objectives: first, to identify the factors that should be provided by institutions to support research activities of faculty members in order to increase research productivity in the higher education institution; second, to validate the instruments which consist of factors that have been identified as a result of the process in point one which has been discussed previously.

2 LITERATURE REVIEW

2.1 Research productivity

The term research here refers to the academic research which is defined as research conducted in the higher education sector (Vincent-Lancrin, 2006). Research productivity can be defined as the total output of the study compared with the specific input (Lertputtarak, 2008). Research results are usually published to strengthen the utilization; thus, some other researchers use the term 'publication productivity'. The input of the research process refers to the time used for doing research or the production period (Ab. Aziz et al., 2013).

2.2 Research supporting factors

Research supporting factors are associated with organizational resources and policies that allow individuals to do the research and produce publication. In the context of research productivity study, Gregorutti (2010) revealed that respondents who were professors in Mexico or the US expressed concerns relating to the timing for teaching and administrative roles, supportive work environment, and collaborative and training needs. Byrne and Keefe (2002) examined the role of guidance in improving the productivity of research and found that faculty members who have access to mentoring have higher productivity. Balakrishnan (2013) explores a lot of practice that can be provided by HEIs to increase research productivity, such as by giving sponsorship for the association membership, grant support, investing in training, providing guidance, encouraging collaboration and providing editing services in English. Some factors related to facilities and funding also were found to significantly influence research productivity of faculty members, such as library facilities (Noh, 2012; Shariatmadari & Mahdi, 2012), research funding (Webber, 2012; Sulo et al., 2012); and funding for attending conference (McGill & Settle, 2012).

Other factors are the type of institution, and whether the department is a non-research institution or research institution which is also a doctorate-granted institution (Jung, 2012). It is related to the supply of staff support that is believed can support the research activities to be productive (McGill & Settle, 2012). Research institutions which are also doctorate-granted institutions would be able to provide doctoral students, who are needed to help the faculty in conducting research (Valle & Schultz, 2011) and the faculties whose supervised postgraduate students were reported to have high research productivity (Alghanim & Alhamali, 2011).

The most frequently found as significant organizational factor that influences research productivity was the proportion of time for conducting research compared to other functions, such as teaching and administrative tasks. Institutions play an important role in allowing faculty members to have less time to conduct teaching so that those faculty members can conduct research (White et al., 2012; Webber, 2012), and it is significantly related to research productivity (Sulo et al., 2012; White et al., 2012; McGill & Settle, 2012; Webber, 2012; Shariatmadari & Mahdi, 2012; Jung, 2012). It is found that faculty members who reported involvement in administrative activities (Alghanim & Alhamali, 2011) and having heavy teaching loads (Hesli & Lee, 2011) were less likely to publish.

The absence of resources, facilities and other organizational factors perceived by faculty members, were found to become barriers or to have a negative impact in producing research. Studies in developing countries start to focus on the barrier factors that might influence the low rate of research productivity in those countries. Okiki (2013) investigated the barriers perceived by teaching faculty members in Nigerian federal universities. The study found that the most perceived barriers included low internet bandwidth and financial constraints. A study from Alghanim and Alhamali (2011) found that the most frequently found barriers, perceived by faculty members, included lack of time, lack of research assistants, lack of funds for research, and being busy with teaching load.

Extrinsic reward also becomes one of the institutional factors that is frequently being discussed (Chen et al., 2010; Monroe & Kumar, 2011; Shariatmadari & Mahdi, 2012). It is provided by institutions to motivate faculty members to do research. Therefore, many studies relate this factor with the motivational aspect or the satisfaction of faculty members (Tafreshi et al., 2013). One of extrinsic rewards that is significantly found related to research productivity is the promotion system (Monroe & Kumar, 2011).

In this paper, research supporting factors is conceptualized as institutional resources and policies perceived by faculty members are available to facilitate him/her to do the research. Institutional resources and policies are believed to be a key to the success of academic research in higher education (Research Universities Futures Consortium, 2012).

3 RESEARCH METHODS

In this paper, exploratory research was carried out to identify research supporting factors. To answer the first objective of this research, the exploration study employed a focus group discussions technique to explore a contributing factor for the research of the faculty.

The four informants, who are faculty members in one of the big universities in Indonesia, were involved in the focus group discussion. They had been purposely selected, based on criteria that those persons must be a productive researcher and represent various disciplines, includes life sciences, health sciences and social sciences. The results were obtained from the focus group discussion and then analyzed using a thematic analysis technique. The output of the focus group discussion was then converted into the form of questionnaires, that is also supported by theory and previous research about research productivity in institutions.

Working on the second objective of this research, a survey towards 63 respondents was applied to collect data for the validity and reliability of the instruments. The respondents were lecturers in one of the top ranked universities as the institution that became the object of this research. The respondents were faculty members representing the disciplines of life science, health science, physical science, and social science. The results of all the collected questionnaires was calculated using SPSS v23 to determine the validity and reliability of each question. Cronbach's alpha was used to measure the reliability of the instrument. The instrument validity was evaluated using the correlation value.

4 RESULTS AND DISCUSSION

4.1 *Results*

As explained above, this research began by conducting a Focus Group Discussion (FGD). During the FGD, respondents were given a lot of opinions regarding the research in the institution. Some of the informants' opinions are summarized in Table 1.

As can be seen in Table 1, there are some examples taken from discussion with informant regarding research productivity. As illustrated in Table 1, authors summarized each dimension in one sentence discussed by informant. Then after the discussion was held, the verbatim result became the indicator that was used in the questionnaire.

The output of thematic analysis towards the verbatim of focus group discussion transcript, was converted into 38 items of indicators for representing the research supporting factors construct, as illustrated in Table 2.

After completing the survey process, the author then calculated reliability and validity using the SPSS v23 statistics application. The author summarized all the questionnaire results, and then input the results into the statistics application, and the output is the result. The result of the reliability statistics using Cronbach's alpha shows that the construct is reliable with the value of 0.955. The result of item validity is illustrated in Table 2. The item is considered valid if the value of correlation is more than 0.209. The amount of 0.209 is taken from r Table.

From Table 3 we can see that for items 1 to 3, it is considered not valid. It left the construct with 35 remaining items.

4.2 *Discussion*

It is interesting that based on survey result, all three items related to dimension 'Time' are considered not valid. This dimension seemed to be important for all informants in the focus group discussion session. For example, informant A explained that there are a lot of young doctoral lecturers trapped in a comfort zone as structural position because of its high compensation, but with the consequence of work load, so that it is difficult to focus on research activities. Informant B believed that 'Time' is the most important factor related to research productivity. In detail, informant B explained that lecturers must divide their time between teaching, research, community service, and serving a role as a structural officer. However, there is usually little proportion they have for research. In addition, informant C said that in some cases several lecturers are more active outside the institution (for teaching, or are borrowed for structural positions) because they are more appreciated than in their own institutions. Thus, it will also reduce the time for research. Informant D believed that research should be regarded as a full-time job which should be given high compensation. Thus, there should be a correlation between the time they should spend

Table 1. Summary of informant's opinions.

Dimension	Opinion	Informant
Time	Lectures' time are taken to teach, structural and support.	Informant B
Facilities and access	The lack of laboratory facilities to support the research.	Informant A, B
Financial support and funding	Research funding system faces problem in how institution manages financial system.	Informant B, C, D
Staff assistance	Students could help informant while doing the research.	Informant B
Working environment	Informant stated its agreement that there need to be a research culture creates by regulatory/policy.	Informant C, D
Rewards provided	There are no differences between those doing publicity and not doing the publicity. There is no penalty for those who do not do publicity.	Informant D
Career system	In this institution, publication here is just for promotion.	Informant A, B
Capacity development	Informant network with other researchers can also help improve research productivity.	Informant D
Research policy	Systems created by institution are too complicated, then making it difficult for researcher.	Informant A, B, C

Table 2. The items of indicators.

Dimension	Indicators	Item
Time	Time for teaching (–)	RF1
	Time for administrative role (–)	RF2
	Time for other activities in institution (–)	RF3
Facilities and access	Library / journal – quantity	RF4
	Library / journal – quality	RF5
	Library / journal access – 24 hours access	RF6
	Internet access – quality	RF7
	Internet access – 24 hours	RF8
	Laboratory – quantity	RF9
	Laboratory – quality	RF10
	Laboratory access – 24 hours	RF11
Financial support and funding	Ease of research funding	RF12
	Financial support for attendance at conference	RF13
	Working capital for initial research	RF14
	Autonomy in managing funding	RF15
	Amount of fund	RF16
	Support in pursuing external funding	RF17
	Financial support for publication fees	RF18
	Financial support for pre-publishing service (proofreading)	RF19
	Research supplies purchasing scheme	RF20
Staff assistance	Research assistance from graduate students	RF21
Working environment	Conducive research environment	RF22
	Research culture	RF23
	Research group	RF24
Rewards provided	Incentive for output	RF25
	Credit acknowledgment for the process of research that has been conducted by faculty member	RF26
	Fairness between rewards for research and other activities	RF27
	Research merit acknowledgment	RF28
Career system	Promotion based on research achievement	RF29
	Fairness of credit weight based on the quality of publication	RF30
	Impersonal career policy	RF31
	Clarity in career policy	RF32
Capacity development	Research network	RF33
	Research skill-upgrading	RF34
	Research mentoring	RF35
Research policy	Research's terms & conditions	RF36
	Institution's research roadmap	RF37
	Decision maker's competence	RF38

Table 3. Item validity result.

	Corrected item-total correlation	Cronbach's alpha if item deleted
RO1	–0.295	0.959
RO2	–0.405	0.960
RO3	–0.176	0.958
RO4	0.617	0.954
RO5	0.657	0.953
RO6	0.634	0.953
RO7	0.590	0.954
RO8	0.540	0.954
RO9	0.675	0.953
RO10	0.733	0.953
RO11	0.533	0.954
RO12	0.687	0.953
RO13	0.651	0.953
RO14	0.543	0.954
RO15	0.551	0.954
RO16	0.633	0.953
RO17	0.658	0.953
RO18	0.645	0.953
RO19	0.643	0.953
RO20	0.747	0.953
RO21	0.630	0.953
RO22	0.799	0.952
RO23	0.780	0.953
RO24	0.516	0.954
RO25	0.671	0.953
RO26	0.660	0.953
RO27	0.803	0.952
RO28	0.815	0.952
RO29	0.659	0.953
RO30	0.576	0.954
RO31	0.534	0.954
RO32	0.576	0.954
RO33	0.811	0.952
RO34	0.791	0.953
RO35	0.742	0.953
RO36	0.801	0.952
RO37	0.751	0.953
RO38	0.600	0.954

for research and the compensation they will get. Otherwise, they will tend to do jobs other than research.

So, how can the survey lead to invalid results for those time-related items? Based on post-survey evaluation and some clarification interviews to the respondents, it is found that the respondents' perception towards the 'time' are various. Furthermore, the question listed in the questionnaires breaks the dimension into three different 'time' indicators, namely time for teaching, time for structural position, and time for other activities, which leads to the wider variation of answers.

Maybe it is better to put the 'time' indicators as an objective item such as to put the option of the amount of their teaching load or to put the option as to whether they serve as structural officers or not, and then proceed with cross-tabulation or correlation for those items with the productivity item.

5 CONCLUSIONS

This paper has identified the factors that should be provided by institutes to increase research productivity. There are eight factors or dimensions, namely: facilities and access; financial support and funding; staff assistance; conducive working environment; research productivity-based rewards; fair career system; capacity development; and good research policy.

One factor that seemed to be important during FGD sessions, turns out to be not valid based on the statistics result. The authors believed that it is difficult to have consensus related to the meaning of the 'time' dimension, since it is relative to different persons. Thus, the authors suggest putting the item of 'time' in another part of the questionnaire, that illustrates the more objective item of indicator, such as amount of teaching load, or whether or not the respondent serves as a structural officer.

REFERENCES

Ab Aziz, K., Harris, H., Zahid, S.M. & Ab Aziz, N. (2013). Commercialisation of university research: An investigation of researchers' behaviour. *Communications of the IBIMA,* Article ID 120942, DOI: 10.5171/2013. 120942.

Alghanim, S.A. & Alhamali, R.M. (2011). Research productivity among faculty members at medical and health schools in Saudi Arabia. Prevalence, obstacles, and associated factors. *Saudi Medical Journal, 32*(12), 1297–1303.

Balakrishnan, M.S. (2013). Methods to increase research output: some tips looking at the MENA region. *International Journal of Emerging Markets, 8*(3), 215–239.

Byrne, M.W. & Keefe, M.R. (2002). Building research competence in nursing through mentoring. *Journal of Nursing Scholarship, 34*(4), 391–396.

Chen, Y., Nixon, M.R., Gupta, A. & Hoshower, L. (2010). Research productivity of accounting faculty: an exploratory study. *American Journal of Business Education (AJBE), 3*(2).

Gregorutti, G., (2010). Moving from a predominantly teaching oriented culture to a research productivity mission: The case of Mexico and the United States. *Excellence in Higher Education, University of Montemorelos, Mexico, 1*(1&2). 69–83.

Hesli, V.L. & Lee, J.M. (2011). Faculty research productivity: Why do some of our colleagues publish more than others? *P.S.: Political Science & Politics, 44*(02), 393–408.

Jung, I. (2012). Asian learners' perception of quality in distance education and gender differences. *The International Review of Research in Open and Distributed Learning, 13*(2), 1–25.

Kaufman, R.R. (2009). Careers factors help predict productivity in scholarship among faculty members in physical therapist education programs. *Physical Therapy, 89*(3), 204–216.

Lertputtarak, S. (2008). *An investigation of factors related to research productivity in a public university in Thailand: a case study.* [Online] Available at: http://vuir.vu.edu.au/id/eprint/1459.

McGill, M. & Settle, A. (2012). Identifying effects of institutional resources and support on computing faculty research productivity, tenure, and promotion. *International Journal of Doctoral Studies, 7,* 167–198.

Monroe, S. & Kumar, R. (2011). Faculties motivations and incentives for academic research: A basis for improvement in publication productivity. *International Journal of Management and Strategy, 2*(3), 1–31.

Noh, Y. (2012). A study measuring the performance of electronic resources in academic libraries. *ASLIB Proceedings, 64*(2), 134–153.

Okiki, O.C. (2013). Research productivity of teaching faculty members in Nigerian Federal Universities: An investigative study. *Chinese librarianship: An International Electronic Journal, 36,* 99–118.

Research Universities Futures Consortium. (2012). s.l.: The current health and future well-being of the American Research University.

Shariatmadari, M. & Mahdi, S. (2012). Barriers to research productivity in Islamic Azad University: Exploring faculty members perception. *Indian Journal of Science and Technology, 5*(5), 2765–2769.

Sulo, T., Kendagor, R., Kosgei, D., Tuitoek, D. & Chelangat, S. (2012). Factors affecting research productivity in public universities of Kenya: The case of Moi University, Eldoret. *Journal of Emerging Trends in Economics and Management Sciences, 3*(5), 475.

Tafreshi, G.H., Imani, M.N. & Ghashlag, P.M. (2013). Designing a model for research productivity evaluation of faculty of district 2 of Islamic Azad University of Iran. *World Applied Sciences Journal, 21*(12), 1708–1720.

Valle, M. & Schultz, K. (2011). The etiology of top-tier publications in management: A status attainment perspective on academic career success. *Career Development International, 16*(3), 220–237.

Vincent-Lancrin, S. (2006). What is changing in academic research? Trends and future scenarios. *European Journal of Education, 41,* 169–202.

Webber, K.L. (2012). Research productivity of foreign-and US-born faculty: differences by time on task. *Higher Education, 64*(5), 709–729.

White, C.S., James, K., Burke, L.A. & Allen, R.S. (2012). What makes a "research star"? Factors influencing the research productivity of business faculty. *International Journal of Productivity and Performance Management, 61*(6), 584–602.

Ideas for 21st Century Education – Abdullah et al. (Eds)
© *2017 Taylor & Francis Group, London, ISBN 978-1-138-05343-4*

Science, technology, engineering, and mathematics literacy skills: Profiles and comparison amongst prospective science teachers

C. Rochman, D. Nasrudin & H.Y. Suhendi
UIN Sunan Gunung Djati Bandung, Bandung, Indonesia

ABSTRACT: In the era of globalization and modernization, as well as in the midst of an energy crisis, Science, Technology, Engineering, and Mathematics (STEM) literacy skills on energy and renewable energy (NRE) sources theme is one of the key competencies that must be possessed by science students. In this study, the literacy skills profiles of students who were prospective science educators were measured and compared. The results showed that STEM literacy skills profiles on renewable energy theme were in a good category for all students. The variation in the profile of STEM literacy skills, including concept, process, context, and attitude, among Physics, Biology, and Chemistry students was found on energy theme. Also the difference was found in the profile category of answers in renewable energy STEM literacy among students.

1 INTRODUCTION

Renewable energy is a form of energy that can produce power for a long time without any fuel input. The sources of this type of energy can be from wind, sunlight, sea wave, and geothermal. On the other hand, new energy is a type of energy that uses nuclear and biomass in place of fossil fuel. The studies and uses of new and renewable energy (NRE) are interconnected with Science, Technology, Engineering, and Mathematics (STEM).

NRE can be used as alternative electric sources. Although this type of energy generator cannot be produced massively, but the potential to use this form of energy sources in Indonesia is still wide open and well spread. Commonly, this type of energy generator will only produce small scale electric energy. In addition, it cannot fulfil growing demand for electricity. At last but not least, by understanding NRE, Indonesian people can be cognizance, and having affirmative approach towards development programs of NRE for life.

NRE has become urgently considered since Earth Summit in Rio de Janeiro in 1992. One of the results of the summit is that NRE will be developed furthermore as alternative energy sources for the reason that it will not emit carbon dioxide which creates greenhouse effect on earth. As for ratio of actual power generated versus maximum possible until 2013 from various NRE is shown as follow *nuclear* (90,9%), *geothermal* (67,2%), *biomass* (67,1%), *coal-fired* (58,9%), *natural gas, combined cycle* (50,3%), *hydro* (40,5%), *wind* (32,3%), and *solar* (24,4%) (Allman & Daoutidis 2016).

In Indonesia, until March 2011, the development of electricity from renewable energy touched 8.772,50 MW. This power is produced from several sources namely geothermal energy (1.189 MW), waste energy (13,5 MW), wind energy (1,96 MW), hydropower (5.711,29 MW), hydro energy (229 MW), and biomass (1.628 MW). The power produced is still far less than the electricity demand for all Indonesia. This condition drives the government to optimize NRE management (PDTI ESDM 2015).

NRE management for electricity plants using wind, sunlight, and nuclear power face many problems, including storing the power, capacity factor, area, supportive energy sources, economy (Santosa, Azrifirwan, Tp, & Eng 2012), and community support (Peranginangin 2014). The support from community for development of NRE is still low. The reason is that waste of NRE power plant still emits methane gas (CH_4) and carbon dioxide (CO_2) (O et al., 2014). Same response is given to nuclear power plant of which community still think more on the bad side effect and risk (Eheazu 2014). Whereas nuclear power plant is very potential to produce high power electricity, needs small land to build, and uses small fuel input (PDTI ESDM 2015). Generally, most of community are still illiterate on this NRE benefit for their life (Eheazu 2014; Allman & Daoutidis 2016). Energy literacy education program could help community to gain sufficient information about NRE and its benefit for their life (Rusli 2016; Rochman 2015; Hobson 2003).

2 RESEARCH METHODS

The study was conducted in an open question test design on a sample of 72 students from department of physics, chemistry, and biology education of teacher training institute (UIN SGD Bandung). The improvement of STEM literacy skills was gained through the test questions (Holden, 2010). The research instruments used were test comprehension of the three types of renewable energy: waste to energy, nuclear and geothermal energy. Students answered four questions about the content, process, context and attitudes towards such energy. Assessment was done by using a rubric of which the score for each were 4,3,2,1, and 0. Analysis of the data include: STEM literacy skills profiles of three types of renewable energy (waste to energy, nuclear and geothermal energy), and profile category of answers in STEM literacy and third groups of students.

3 RESULTS AND DISCUSSION

3.1 The profile of STEM literacy skills of the student on renewable energy theme

Based on data analysis of STEM literacy skills, a profile graph of physics undergraduate students' data analysis is shown in the Figure 1. It can be seen that the highest STEM skills score of Physics Students was on nuclear topic.

For Chemistry student group, STEM skills score profile is shown in the Figure 2. The highest STEM skills score of Chemistry Students was on geothermal topic.

For Biology student group, STEM skills score profile is shown in the Figure 3. The highest STEM skills score of Biology Students is on waste topic.

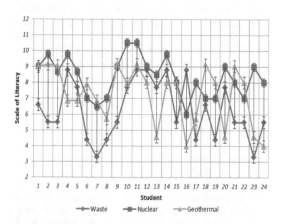

Figure 1. Physics student profile of STEM literacy skills for renewable energy.

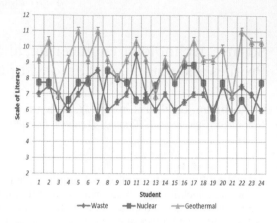

Figure 2. Chemistry student profile of STEM literacy skills for renewable energy.

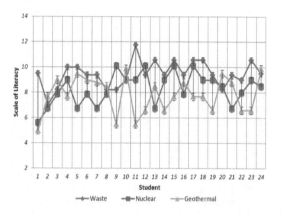

Figure 3. Biology student profile of STEM literacy skills for renewable energy.

According to Figures 1, 2, and 3, the STEM literacy skills profile of Physics, Chemistry, and Biology students have different characteristic (Akengin & Sirin 2013). The distinctive characteristic of the profile of three groups' students was presented on the results of students' worksheet about NRE (waste, nuclear, and geothermal energy) (Devick-Fry & LeSage 2010; Nwosu & Ibe 2014). In addition, each group had different strength in STEM. Physics students were eminent in nuclear topic; Chemistry students were eminent in geothermal topic; while Biology students were eminent in waste topic.

Curricula given for science students (Physics, Chemistry, and Biology) in the first year is similar for all. They get fundamental physics, general chemistry, and general biology. Supposedly it is assumed that all students have same basic knowledge. However probably there are other factors that contribute to students' strength diversity in those three different types of renewable energy topic. Physics students were prominent in nuclear energy topic. This is understandable, because physics students gain more information on physics concept under-lining nuclear energy, for instance fusion and fis-sion reaction concept, also radioactivity in physics, modern physics, or any other supportive courses. As for Biology students, they were more promi-nent in explaining about waste energy production (O et al., 2014). In waste energy production topic, more biology concepts are discussed such as micro-biology, ecosystem, plague, and so on (Agamuthu et al., 2015). Those concepts related to biology make Biology students more literate in waste energy production. As for Chemistry students, they were prominent in geothermal topic. This is acceptable since Chemistry students gain more information on chemical energy concept, thermochemistry, heat, thermal properties of materials, and so on. Those concepts are contributing in geothermal energy topic. Profile of The STEM literacy skills on renew-able energy among Physics, Biology and Chemistry students is shown in the Figure 4.

Figure 4 shows that the highest average score on NRE literacy was achieved by Biology Students on waste topic (9,47), Chemistry Students on geother-mal topic (9,31), Physics Students on Nuclear topic (8,36). STEM literacy skill of biology student stands out on waste topic. In nuclear topic, physics students were more dominant. On the other hand, STEM literacy skill for geothermal topic was mas-tered mostly by chemistry students.

3.2 Student Profile of STEM literacy on renewable energy

The score of STEM literacy skills (concepts, proc-esses, contexts, and attitude) of Physics, Chemistry,

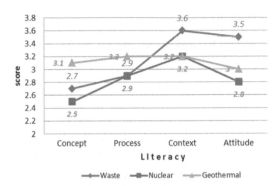

Figure 5. Student profile of STEM literacy for renew-able energy answer.

and Biology students is shown in the Figure 5. The Figure shows that the highest score at concept aspect, processes, contexts, and attitude (on the scale 0–4) were 3.1 for geothermal energy, 3.2 for geothermal energy, 3.6 for waste energy, and 3.5 for waste energy, respectively. From information given, students were still weak in grasping NRE concept, although their attitude toward those three NRE was better. Moreover, students were able in understanding value and impact (context) of these three NRE.

The variance of students' literacy component (concepts, processes, contexts, and attitude) indi-cated information gap on NRE in fundamental physics courses or any other courses. Further-more, fundamental concepts that display correla-tion among science, technology, engineering, and math were not communicated in a lecture in any course yet. On the other hand, information on value, impact, and encouragement to act positive toward NRE were presented very well in courses. The correlation amongst electrical energy, solar energy, wind energy, and so on had not been deliv-ered yet in a learning process of the course. This information intertwine and can affect to the energy demand from human (Tan et al., 2013).

4 CONCLUSIONS

It can be concluded that the profiles of STEM literacy skills of renewable energy which consists of nuclear energy, geothermal and waste tend to be good for all students of physics, chemistry, and biology. There were differences in the STEM lit-eracy skills (concept, process, context and attitude) of renewable energy among students of Physics, Biology, and Chemistry. The profiles of the stu-dent answers of STEM literacy skills on renewable energy were also distinct.

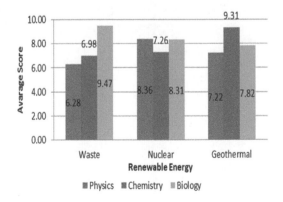

Figure 4. Student profile of STEM literacy skills for renewable energy.

REFERENCES

Agamuthu P., Milow P., Nurul A.M.N., Nurhawa A.R. & F.S.H. 2015. Impact of flood on waste generation and composition in Kelantan. *Malaysian Journal of Science* 34(2): 130–140.

Akengin, H. & Sirin, A. 2013. A comparative study upon determination of scientific literacy level of teacher candidates. *Educational Research and Reviews* 8(19): 1882–1886.

Allman, A. & Daoutidis, P. 2016. Optimal design of synergistic distributed renewable fuel and power systems. *Renewable Energy* 100: 78–89.

Chaerul Rochman. 2015. Analisis dan kontribusi kemampuan konsep dasar fisika, literasi kurikulum pembelajaran dan psikologi pembelajaran terhadap kemampuan penyusunan Lembar Kegiatan Peserta Didik (LKPD) mahasiswa pendidikan fisika. *Prosiding Symposium Nasional Inovasi Dan Pembelajaran Sains*: 1–5.

Devick-Fry, J. & LeSage, T. 2010. Science literacy circles: Big ideas about science. *Science Activities: Classroom Projects and Curriculum Ideas* 47(2): 35–40.

Eheazu, C.L. 2014. Acquisition of environmental literacy by Nigerian University students: An empirical study. *Journal of Education and Practice* 5(11): 20–27.

Hobson, A. 2003. Physics literacy, energy and the environment. *Physics Education* 38(2): 109–114.

Holden, I.I. 2010. Science literacy and lifelong learning in the classroom: A measure of attitudes among university students. *Journal of Library Administration* 50(3): 265–282.

Nwosu, A.A. & Ibe, E. 2014. Gender and scientific literacy levels: Implications for sustainable Science and Technology Education (STE) for the 21st century jobs. *Journal of Education and Practice* 5(8): 113–118.

O, Suagwu, C, Hiemeriwo, G, Odday, Ariatamby. 2014. Bio-hydrogen production from food waste through anaerobic fermentation (Pengeluaran bio hidrogen daripada sisa makanan melalui fermentasi anaerobik). *Sains Malaysiana* 43(12): 1927–1936.

PDTI ESDM. 2015. *Perkembangan penyediaan dan pemanfaatan migas batubara energi baru terbarukan dan listrik*. Jakarta: Kementerian ESDM.

Peranginangin, L.S.U. 2014. Partisipasi Masyarakat dalam Pengelolaan Kawasan Konservasi Lily Sri Ulina Peranginangin Penganalisa Bahan dan Pemanfaatan pada Balai KSDA Sumatera Barat. Jurnal Kebijakan Lily Sri Ulina Peranginangin & Administrasi Publik.

Rusli, A. 2016. Science awareness and science literacy through the basic physics course: Physics with a bit of metaphysics? *Journal of Physics: Conference Series* 739: 12012.

Santosa, I., Azrifirwan, M.P., Tp, S. & Eng, M. 2012. Hasil penelitian Jurnal Energi Alternatif (2012) Studi Tekno Ekonomi Pembuatan Biogas Di Pt. Shgw (Stichting Het Groene Woudt) Bio Tea Indonesia Oleh: Bindari Rahmadian Kenaikan harga bahan bakar turut memberikan efek yang besar dalam bidang industr. *Jurnal Energi Alternatif*.

Tan, C.S., Maragatham, K. & Leong, Y.P. 2013. Electricity energy outlook in Malaysia. *IOP Conference Series: Earth and Environmental Science* 16: 12126.

Ideas for 21st Century Education – Abdullah et al. (Eds)
© *2017 Taylor & Francis Group, London, ISBN 978-1-138-05343-4*

Developing community-based media on environmental education to conserve mangrove and coral ecosystem in Kepulauan Seribu

D. Vivanti, M. Miarsyah, R. Komala & A. Suryanda
Universitas Negeri Jakarta, Jakarta, Indonesia

ABSTRACT: Mangrove and coral ecosystem damage has been a serious problem that must be solved, including that in the Kepulauan Seribu area. An integrated program between various society elements addressing this damage will therefore urgently be required. This research aimed at developing community-based media on environment education to overcome mangrove ecosystem and coral reef damage. Methods used were descriptive study with survey followed by Research and Development (R&D). Medias developed were the Environmental Education Community Network (EECN), namely http://www. eecn.or.id, and three guidebooks on environment education. The survey was done by an NGO (Non-Governmental Organization) conservationist, while the media were assessed by experts. The media was assessed on stakeholders, such as educators and community groups of mangrove and coral reef conservationists, and also on groups of colleges. Data was analyzed qualitatively. Based on the surveys it was found that conservation on mangrove and coral reefs in Kepulauan Seribu did not involve many people. The local people understand the importance of conservation and are interested to work on it. Yet, it has not been coordinated and needs other community groups. Expert testing of results showed that the book and the website developed needed improvements on some items, but overall they were fair. It can be concluded that the media developed were good criteria and addressed effectively the building up of cooperation between many sectors.

1 INTRODUCTION

Indonesia is an archipelago, comprising of islands and surrounded by seas. Indonesia has 17,508 islands in the archipelago, with 62% of the entire territory area being sea, with a great maritime potential. The surface area is 1,922,570 km² and the sea area reaches 3,257,483 km². The Indonesian coastline is also one of the four longest coastlines in the world, at 95,181 km. Indonesia has coastal areas, with areas of land adjacent to the sea that are exposed to the tide, giving a very large potential of natural resources.

Two of the ecosystems in coastal areas are mangrove and coral reef ecosystems. Mangrove forests grow in almost all coastal areas of Indonesia, among others: Sumatra, Kalimantan, Java, Sulawesi, Bali, West Nusa Tenggara, East Nusa Tenggara, Maluku and Papua (Pramudji, 2003). Based on data from the Page et al. (2002), 9.2 million hectares of mangrove forest area, today more than half, has been damaged. The existence of mangrove forest is under pressure from the increasing number of people in Indonesia. According to Saparinto (2007), community needs for agriculture and fisheries are increasing and the public perception is that the mangrove forest is an alternative land source.

Some coastal areas of Indonesia previously had mangrove forests, which have now almost entirely disappeared. For example, Whitten et al. (1999) mention Cilegon and Indramayu, which have lost almost all mangrove forest, due to over exploitation for a range of human interests. Saparinto (2007) asserts that timber harvesting for commercial purposes as well as land conversion are activities that provide the greatest contribution to the destruction of mangrove in Indonesia. In general, a large number of converted mangrove forest areas have been designated for residential areas, aquaculture, agriculture and industry. Conversion of mangrove forest areas that are not controlled is the starting point of the complex problems in coastal areas.

Handayani (2017) noted that the general condition of 31% of coral reefs in Indonesia are in a bad state, with 37% fairly well, 27% good and only 5% very good (http://www.oseanografi.lipi.go.id). Praseno et al. (2000) concluded that the reefs rarely in Payung Island were quite good with hard coral cover of 67%, whereas for Pari Island it was moderate (36%), and the Lancang Island was classified as poor (15%). This study also showed that there was an increase in hard coral cover at a farther distance from the mainland of the island of Java.

Research of Santoso in 2011 noted the loss of coral reefs in Karang Congkak Island at

Kepualuan Seribu only reached 27.41% (classified as low-medium). Coral reef damage occurs due to increased temperatures, bathymetry, tidal wave in the waters nearby, as well as human activities such as fishing with explosives and toxic materials, digging for limestone reef, and beach tourism activities.

Mangrove ecosystems and coral reefs in Kepulauan Seribu have decreased and been damaged. One contributing factor was the result of eco-tourism activities in the Kepulauan Seribu, which has grown substantially in the last five years. According to data from Setyawan (2014), the number of domestic tourists increased from 226,234 people in 2010 to 1,482 million in 2013. The positive side of tourism is the income for local people. However, the increase in tourism was in line with the rate of environmental degradation. Mangrove destruction has a direct impact on the city due to tidal flooding, rising sea level, sea water abrasion, sea water intrusion, degradation of natural resources potential, and a decreased quality of life of coastal communities. Similarly, this applies for the destruction of coral reefs in Kepulauan Seribu. These include the type of 'Fringing Reef' (Reef Edge) covering an area of 4,750 hectares, 2,375 hectares of which are currently in good condition, with the remaining 2,375 hectares now damaged.

Some effort has been made by the community to address the damage to mangrove ecosystems and coral reefs, but they have not been able to resolve the problem of mangrove ecosystem and coral reef damage. Therefore, we need an integrated program between the various elements of society to work together to overcome the destruction of mangrove and coral reef ecosystems. This research was done to develop the community-based media through a website and books to overcome the damage to mangrove ecosystems and coral reefs. The community, which consists of students, university students, NGO conservationists, businesses and the public, were integrated into the network system of cooperation globally connected (locally integrated, globally connected).

Damage to mangroves and coral reefs in Kepulauan Seribu is very serious, so it is a problem that must be addressed in an integrated and sustainable exchange between communities and the government. If mangrove and coral reef damage is not immediately handled, the impact of environmental degradation on human life will be more severe. The media generated from this study can also be applied to cope with a variety of other environmental damage in Indonesia, and nationally will form a national salvation movement of mangrove ecosystems and coral reefs.

Examining the literature relating to damage to mangrove forests and coral reefs, as well as a variety of factors, the proposed research has characteristics of its own. No previous studies have been made that specifically integrate the concept of environmental education to overcome damage to mangrove forests and coral reefs through the concept of building a network (network) between community groups, local government, business, and education.

The research problems are: 'How to develop community-based media on environmental education to overcome the damage of mangrove ecosystems and coral reefs in Kepulauan Seribu?'

The general objective of this research was to develop community-based media on environmental education to overcome the damage of mangrove ecosystems and coral reefs in the Kepulauan Seribu. More specifically, the purpose of this study was to: (1) incorporate an element of education in dealing with damage to mangrove ecosystems and coral reefs in the Thousand Islands, (2) involve all elements of society to play a role in the rescue of mangrove ecosystems and coral reefs, (3) inculcate awareness of the dangers of damage to mangroves and coral reefs for life, and (4) establish a network of cooperation among communities to overcome the damage of mangrove ecosystems and coral reefs. The specific objective of the research was to produce community-based medias in environmental education to overcome the damage of mangrove ecosystems and coral reefs in Kepulauan Seribu.

1.1 Research methods

This research was conducted using a survey method followed by research and development. The target to be achieved is to produce community-based medias on environmental education, to be obtained through the following steps:

1.2 Introduction research and analysis needs

The study begins with a descriptive quantitative study (survey) done of conservationists in Kepulauan Seribu in: (1) Tidung, (2) Harapan Island, (3) Kelapa Island, (4) Pari Island, and (5) Pramuka Island, to dig up information about the damage to mangrove and coral reef ecosystems and the efforts that have been made by the community for the preservation of mangroves and coral reefs. The data that will be explored are as follows: (1) the conditions and the level of damage to mangroves and coral reefs, (2) public knowledge about mangrove ecosystems and coral reefs, (3) the behavior of society for the preservation of mangroves and coral reefs, (4) the values of local wisdom that developed, (5) the activities of community groups in conservation of mangroves and coral reefs, and (6) the cooperation between local governments and community group conservationists in the Thousand Islands. Data obtained from the survey was analyzed to get a database on development of community-based media on environmental education to overcome the damage of mangrove ecosystems and coral reefs.

1.3 Media development

The draft media was developed based on the results of preliminary research and analysis needs. The results of the study were based on a number of literature sources, and by taking into account the results of the mapping of: (1) the conditions and levels of damage to mangroves and coral reefs, (2) public knowledge about mangrove ecosystems and coral reefs, (3) the behavior of society for the preservation of mangroves and coral reefs, (4) the values of local wisdom that develops, (5) the activities of community groups in conservation of mangroves and coral reefs, and (6) the cooperation between local governments and community group conservationists in the Kepulauan Seribu.

Draft media to be developed include: (1) The Environmental Education Community Network (EECN) website, and the educational media EECN in a book-form guide to environmental education for the community, guidebooks of mangroves and coral reefs conservation, and guidebooks on cooperation between communities for the conservation of mangroves and coral reefs. The website will be packaged by applying the principles of learning to people who are multilevel and multicultural, through distance learning (distance learning), by applying the principles of self-learning (independent learning), and by the means of communication between community group mangrove and coral reef conservationists. (2) A draft model of organizational structure and schedule of activities of community group conservationists of mangrove and coral reefs. (3) A draft network model of cooperation between local governments, community groups, businesses, and education to preserve mangroves and coral reefs.

1.4 Focus group discussion

Media depth were assessed through Focus Group Discussion (FGD) involving stakeholders, such as educators and community groups of mangrove and coral reef conservationists, and college groups.

1.5 The media validation

Media has been given feedback from the FGD, and then validated by experts comprising of: (1) education expert, (2) nautical/marine expert, (3) environmentalists, and (4) Information and Communication Technology (ICT) expert.

The validation process was done by asking the experts to examine the media draft and then fill the evaluation instrument that contains a number of statements (evaluative), seen from their respective areas of expertise. Based on the results of validation and input from experts, the draft was revised, especially those parts that were still considered weak or poor.

1.6 Small group test

The media has been revised, based on the evaluation results that was tested on a limited target group. Participants were invited and asked to participate in utilizing the media. At the end of the program they were asked to evaluate and give feedback on what has been exploited in relation to the damage. This small group consists of five lecturers, five local conservationists, and 20 environmental group students.

1.7 Test the pitch (field evaluation trials)

Field trials were conducted to look at the effectiveness of the media. The participants consisted of 30 college students, 30 high school students, and 20 Pramuka island visitors.

1.8 Drafting model of Environmental Education Community Network

The guidebooks and the website will be the basis in the development of the draft model of the EECN. A network of community groups in conservation of mangrove and coral reefs will be interconnected in one's community of mangrove and coral reef conservationist. With this network of business and community activities in the preservation of mangroves and coral reefs, there will be more focus and there will be more people who are directly involved in environmental conservation.

2 RESULTS AND DISCUSSION

The results of the research were gathered by first conducting a needs analysis, with the following results:

1. Conservation of mangroves and coral reefs in the Kepulauan Seribu were in Harapan Island and Kelapa Island. Pari Island stated that not many people are involved directly in the activities of mangrove and coral reef conservation.
2. Residents of the islands who are interested in working directly in the preservation of mangroves and coral reefs are limited to only a few residents.
3. People understand the importance of preserving the mangrove and coral reef.
4. Conservationists of mangroves and coral reefs have not been well-coordinated.

5. The conservationist groups do not yet have a network of cooperation with other community groups such as NGOs, universities, local governments, community groups, and education.

The qualitative data mentioned above becomes the steps in the preparation of media. The guidebook developed in this study included three books, namely:

1. Guidebook 1: Profile of Thousand Islands, in geographical, social, cultural, and economic terms.
2. Guidebook 2: Mangroves and coral reefs are reviewed theoretically, with introduction, types, benefits and ways of cultivation.
3. Guidebook 3: Mangroves and coral reefs of the islands in the Kepulauan Seribu (Harapan Island, Kelapa Island, Pari Island, Tidung Island, and Pramuka Island).

The next stage after the preparation of the book was the expert testing that includes testing materials experts, linguists, and media expert test. The results of the expert test are summarized as improvements in product development as follows:

1. Some improvements in books covering content, terminology used, consistency of the terms used, and accuracy of species name.
2. The quality of the letters, pictures or photos that are not clear, and the naming and also image sources to be true.
3. Lay out the book that originally shaped revamped portrait to landscape, the use of color, the layout of the narrative and images that must be confirmed.
4. The website content to be completely revised.

The trial consisted of a small group of local conservationists of mangroves and coral reefs in the Kepulauan Seribu. Input results from this small group trial generally includes several terms for the mangroves and coral reefs that are poorly understood by local conservationists.

Improvements were carried out with reference to the input and suggestions of the expert test, and the test results of the small groups.

A large group trial was conducted with several community groups including students, teachers and students and lecturers, groups of environmentalists, and tourists visiting the Kepulauan Seribu. The test results provide a record of a large group that is mainly in a legible term that is sometimes poorly understood primarily in terms of mangroves and coral reefs. The whole large group of trial participants welcomed the guidebooks and EECN website, and declared that they are good books and a prospective website. Participants entirely plan to participate in the conservation of mangroves and coral reefs.

3 CONCLUSIONS

The development of media to address environmental damage was a good criteria and effectively addressed the building up of cooperation between many sectors. Furthermore, the development of the website was very effective in mobilizing environmentalists and conservationists on mangrove forests and coral reefs.

ACKNOWLEDGMENT

We greatly appreciate and give thanks to the Ministry of Research Technology and Higher Education for funding this research.

REFERENCES

Handayani, T. (2017). The Potency of Macroalgae in the Reef Flat of Lampung Bay. *Oseanologi dan Limnologi di Indonesia, 2*(1), 55–67.

Page, S.E., Siegert, F., Rieley, J.O., Boehm, H.D.V., Jaya, A., & Limin, S. (2002). The amount of carbon released from peat and forest fires in Indonesia during 1997. *Nature, 420*(6911), 61–65.

Pramudji, P. (2000). Hutan mangrove di Indonesia: Peranan permasalahan dan pengelolaannya. *Oseana, 25*(1), 1–15.

Pramudji, P. (2003). *Hutan mangrove di Indonesia dan permasalahannya.* Pusat Penelitian Oseanografi-Lembaga Ilmu Pengetahuan Indonesia, Jakarta. Orasi Ilmiah Pengukuhan Ahli Peneliti Utama Bidang Botani Laut Dengan Spesialisasi Mangrove: 1–6.

Pramudji, P. (2007). Mangrove in coastal zone of Lampung bay province of Lampung: A preliminary study. *Marine Research Indonesia, 32*(2), 179–183.

Praseno, D.P., Wanda, S.A., Imam, S., Ruyitno, R. & Bambang, S.S. (2000). *Komunitas fauna echinodermata di pulau-pulau seribu bagian utara.* Pesisir dan Pantai Indonesia IV. LIPI. Jakarta.

Santoso, A.D. (2011). Kondisi terumbu karang di pulau karang congkak kepulauan seribu. *Jurnal Hidrosfir Indonesia, 5*(2).

Saparinto, C. (2007). *Pendayagunaan ekosistem mangrove.* Daharaprize, Semarang: xii + 232.

Setyawan, D. (2014). The impacts of the domestic fuel increases on prices of the Indonesian economic sectors. *Energy Procedia, 47*, 47–55.

Whitten, T., Soeriaatmadja, R.E., Afiff, S.A. & Widyantoro, A. (1999). *Ekologi Jawa dan Bali.* Prenhallindo Press, Jakarta: xxii + 972.

Ideas for 21st Century Education – Abdullah et al. (Eds)
© 2017 Taylor & Francis Group, London, ISBN 978-1-138-05343-4

Social class and access to higher education in the secondary schools: Supporting the preparation of lessons and access for national exam

A. Konaah, A.L. Sugiarti, A.A. Lukman, S. Nurbayani & A.G. Abdullah
Universitas Pendidikan Indonesia, Bandung, Indonesia

ABSTRACT: The important role of parents in preparing their children to succeed in national examination and enrollment exam of higher education is by involving children in after-school activities in order to support the achievement of success and access to higher education. The purpose of this study is to determine how much the role of parents helps their children to be successful in higher levels of education, and the effects of after-school activities on the preparation of students in national exams. The method used in this research is the combination method in which the data search used was the technique of quantitative data, and the elaboration of the results used the techniques of qualitative data. For the data analysis, percentage analysis was used. The technique used is purposive sampling and the samples taken are learners who follow the after-school activities, as many as 181 students, using a krejcie table with a significance level of 0.05%. The results showed that the role of parents in supporting their children in the national exam and college enrollment test is so great. It can be seen from the parents who provide supportive learning facilities and advice to the children in choosing a tutor, although the children still choose their own tutoring.

1 INTRODUCTION

The participation of high school students in the public universities enrollment exam is supported by the efforts of parents on their preparation such as by joining in the after-school activities (Halsey, 1993). Parents who have higher income and job status tend to choose tutoring as a preparatory effort (Thomas et al., 2014). Differences in income and job status of parents are the main reasons in the selection of tutoring (Egerton & Halsey, 1993).

Activities that support the student's preparation for the college enrollment exam are formed by social and cultural capital (habit) of the student's family (Ball et al., 2002). Socio-economic status is the background which influences the selection of their studies and becomes the reason for making the decision on their educational paths (Sianou-Kyrgiou, 2010). Students who have parents with high social class have a greater opportunity to enter college (Connor, 2001).

Families, especially parents who have a strong influence, play an important role when making decisions about which choices are 'suitable' for their children to enter college (Brooks, 2003b). Due to the different and low parental backgrounds, the children are required to follow the activities presented by school. Generally, it tends to be difficult to get into college (Egerton & Halsey, 1993). It is said that the difference in social class is the key to understanding the decisions taken by students in higher education (Ball et al., 2002).

2 LITERATURE REVIEW

Parental involvement in determining the education for their children is something familiar, it has even become a habit, and is constituted by education and social background, or the notion of family 'institutions' and 'habits' (David et al., 2003). Determining the future for children who are going to take the national exam and college enrollment exam at the age of 16–18 years, requires openness in planning their futures (Brooks, 2003a).

In determining the successful access for college enrollment, it is inseparable from the involvement of families who have a strong influence on the children's conceptualization. From this sector, friends and peers also play an important role in informing the decision of what is 'suitable' to be selected (Brooks, 2003b). The social background of parents also becomes one of the reasons for the children to follow the national exam and college enrollment exam preparation, as well as the after-school activities (Connor, 2001).

In addition to the role of parents and friends, and habits and family background which support the preparation for college, the residence's geographical location and the location of the college also becomes one of the reasons for after-school activities and college choice (Johnston et al., 1999). Thus, there is a relationship between the college, national exams and access to college with stratification, in this case the social class of the parents (Johnston et al., 1999).

3 RESEARCH METHODS

This study was conducted in a public high school in Karawang with the position of the first cluster. The students are 347 students from nine classes: seven science classes and two social classes. The criteria of the school is being located in an urban area with differences in family background and jobs. The representative sample is the number of students who take part in after-school activity in the form of tutoring, both private and institutional. The sampling uses purposive sampling technique, in which the object of the research is the students who follow the activities outside of school. Purposive sampling technique is a sampling technique by choosing some particular samples which were assessed in accordance with the purposes and research problems in a population. Samples taken are learners who participate in after-school activity, with as many as 181 students in krejcie table with a significance level of 0.05%. This study used a closed questionnaire, where the answer had been

Table 1. Questions in questionnaire.

No	Statement
1	Parents support after school activities
2	Parents give advice of tutoring places
3	Place of tutoring is determined by parents
4	Parents provide facilities to support after school activities
5	Parents give more attention regarding after school activities
6	Background for entering tutoring is a personal desire
7	There is a coercion from parents to join tutoring
8	Diligently attending the tutoring
9	The schedule of tutoring is flexible
10	The main objective of attending tutoring is to pass college enrollment and national exam
11	The tutoring place has decent facilities to support the access to enroll in the university
12	Learning material is according to curriculum
13	Learning materials level with the national exam and college enrollment exam
14	Tutoring agency guarantee access to particular college(s)
15	Tutoring agency guarantee passing the national exam

provided by the researchers so the respondents only needed to choose the appropriate or suitable answers to respondents' preferences. The questionnaire consists of 15 questions. The researcher used a Likert scale of measurement, which is used to measure the opinion of someone to an event or social phenomena.

A validity test was used to test the items in the questionnaire which were tested on 20 non-sample students. In this study, the researchers only used quantitative data and findings that come out of the statistical analysis. The purpose of these questions is to gather information and to find out what is the role of parents and after-school activity in affecting students' preparation for national exam.

The questions are how big is the role of parents in preparing students for national exams, how prepared the students are in following after-school activities for national exams, and how much are the activities outside the school affecting students' access to higher education.

4 RESULTS AND DISCUSSION

This study has three formulas, each of which has been answered by some of the questions related to it. The first question is intended to determine the role of parents in supporting their children in preparing for college enrollment exam by following the tutoring. The results obtained from the first question about the role of parents are: more than half answered that the parents support after-school activity (59.12%). Less than half agreed that the parents give advice of tutoring place (48.67%). Half of the students feel that the tutoring place is not determined by the parents (49.72%), but they are free to choose the tutoring place. Less than half of the students said that their parents provide facilities to support after-school activities (46.41%). Less than half of the students agreed that their parents actually pay more attention regarding after-school activities (42.54%).

The second question is intended to determine the readiness of students. The results of this question are: less than half of the students stated that their background in entering the tutoring is a personal desire (47.51%). Less than half considered that there is no coercion from parents in attending the tutoring (44.21%). More than half admitted that they regularly attended tutoring (61.33%). Less than half said that the schedule of tutoring is flexible (37.02%). Less than half said that the main purpose of attending the tutoring is to succeed in the college enrollment exam and national exam (48.08%).

The third question is intended to investigate the effect of after-school activity. Less than half said

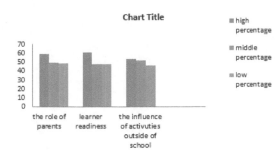

Figure 1.

that the tutoring facilities are decent and able to support the access to the university enrollment (46.41%). More than half said that the learning materials in tutoring are in accordance with materials in school (53.59%). More than half stated that the learning materials compare with the national exam and college enrollment exam materials (51.93%). Less than half stated that they are unsure if the tutoring agency ensures access into certain colleges (39.23%). Less than half stated that tutoring agencies ensure access to pass the national exam (34.81%).

This data can explain that the role of parents has a high percentage rate in terms of supporting after-school activities, the second percentage represents the voice that the tutoring place is not determined by the parents, and the third percentage explains that parents give advice to choose the tutoring place. The readiness of learners in terms of attending the learning activities has the most votes, in addition to the reason they participated in the learning activities is to get into college, and the last percentage that can be seen on the chart tells that they get tutoring activities based on their own willingness. The influence of after-school activities explains that the materials are in accordance with school materials, and also that the problems given are at the same level as the college enrollment exam, and the last percentage describes that the learning facilities owned by tutoring institutions are capable of supporting access to get into college.

5 CONCLUSIONS

The conclusion of this study is that the role of parents in supporting their children to go to college is so great. It can be seen from the parents who provide support for learning facilities and advice to the children in choosing a tutoring place, although the children still choose a tutoring place to be attended by themselves. The second fact obtained in addition to the role of parents is about the readiness of learners in attending the tutoring without coercion from parents, and their main goal of attending the tutoring is to enter the college. Besides investigating the role of parents and the readiness of learners, this study also takes aim at after-school activities attended by the student. As it turned out, although they participated in the tutoring, they doubt that tutoring could guarantee them to get into college. This is contrary to their purpose in following the tutoring activities. Materials provided by tutoring is according to their national exam and college enrollment exam, but still no more than half believe that tutoring could guarantee them to pass the national exam and college enrollment exam.

REFERENCES

Ball, S.J., Reay, D. & David, M. (2002). Ethnic choosing: Minority ethnic students, social class and higher education choice. *Race Ethnicity and Education, 5*(4), 333–357.

Brooks, R. (2003a). Discussing higher education choices: Differences and difficulties. *Research Papers in Education, 18*(3), 237–258.

Brooks, R. (2003b). Young people's higher education choices: The role of family and friends. *British Journal of Sociology of Education, 24*, 283–297.

Connor, H. (2001). Deciding for or against participation in higher education: The views of young people from lower social class backgrounds. *Higher Education Quarterly, 55*(2), 204–224.

David, M.E., Ball, S.J., Davies, J. & Reay, D. (2003). Gender issues in parental involvement in student choices of higher education. *Gender and Education, 15*(1), 21–36.

Egerton, M. & Halsey, A.H. (1993). Trends by social class and gender in access to higher education in Britain [1]. *Oxford Review of Education, 19*(2), 183–196.

Halsey, A.H. (1993). Trends in access and equity in higher education: Britain in international perspective. *Oxford Review of Education, 19*(2), 129–140.

Johnston, V., Raab, G. & Abdalla, I. (1999). Participation in higher education in Scotland: A geographic and social analysis. *Higher Education Quarterly, 53*(4), 369–394.

Sianou-Kyrgiou, E. (2010). Stratification in higher education, choice and social inequalities in Greece. *Higher Education Quarterly, 64*(1), 22–40.

Thomas, G.E., Alexander, K.L. & Eckland, B.K. (2014). *Higher Education: of. 87*(2): 133–156.

Ideas for 21st Century Education – Abdullah et al. (Eds)
© *2017 Taylor & Francis Group, London, ISBN 978-1-138-05343-4*

The gap of the economic background of the parents towards student achievement

L. Sirait, M.I. Triawan, T. Sulastri, B. Maftuh & A.G. Abdullah
Universitas Pendidikan Indonesia, Bandung, Indonesia

ABSTRACT: The gap of education can be affected by several factors, and one of them is the economic background of the parents. It cannot be denied that the economic background of the parents more or less affects student achievement. If the economic background of the parents is good, then the nurturing given to their children would also be good, and a good educational infrastructure provided for the children will automatically affect their achievements. The purpose of this research is to investigate the influence of the economic background of the parents towards student achievement. This research used a quantitative approach and involved 76 samples of senior high school students in Bandung. Data was collected through a questionnaire distributed to students. The results showed that there is the influence of the economic background of the parents towards student achievement. The influence of the economic background of the parents towards student achievement was 14.9%.

1 INTRODUCTION

The presence of varied family backgrounds has led to a social inequality educational impact on students. Research that examines these issues has appeared in journals of sociology, economics, and psychology. In the field of social capital, the neglected problem is how students acquire social capital from a variety of contexts in their favor to form their educational plans which consequently affect their future level of aspiration (Shahidul et al., 2015). Early childhood education is culturally sensitive to the needs of children and controlled by the community to improve and close the education gap (Nguyen, 2011). The gap in achievement is a function of a number of factors, such as income inequality and wealth, access to child care and preschool programs, nutrition, physical and emotional health, environmental factors, the structure of society and the family, differences in the quality of instruction and school achievement, and education (Lynch & Oakford, 2014). Educators, mentors, parents and the community are all responsible for achieving the milestone when considering closing the achievement gap for students of various cultural, social, and economic backgrounds in their quest for knowledge (Faitar, 2011). The existence of social inequality in education is caused by several factors that affect the decline in student achievement, such as family factors, a number of government fundings which aim to increase primary and secondary education, the social gap between poor students and rich, the human factor, and school infrastructures that are inadequate. The local governments in Indonesia

have a greater autonomy since the beginning of the reform movement in Indonesia. Most educational management was delegated to the regional government. Because of this, the level of education varies widely throughout Indonesia (Azzizah, 2015). An important component to student motivation is the ability of teachers to improve the relationship between parents/guardians (Ankrum, 2016). But there are some other factors, such as quality improvement, and the development of educational infrastructure is not comparable with the interests of the students that have an impact on student achievement.

2 LITERATURE REVIEW

Socio-economic status is a measure of economic and sociological conditions of a person's work experience, and of the economic and social position of individuals or families relative to each other, based on income, education, and employment. When analyzing the socio-economic gap, a family, household income, education seeker and the job are checked, as well as the combined revenue, compared with individuals, when their own attributes are assessed. A family's socio-economic status is based on family income, parental education, parental occupation, and social status in the community (such as community relations, group associations, and public perceptions of the family). Families with high socio-economic status often have more success in preparing their children for school because they usually have greater access to the resources to promote, explore and support young children's mental and physical development. Parents have more resources to focus on the growing needs of children, with their mental and physical care, access to better books, and educational toys (edutainment concept) that help in the formation of character. Because families with socio-economic status are better at doing most activities together, their togetherness at home also helps in developing better characteristics. These opportunities help parents understand the emotional, mental, social, physical, psychological, and most of all, cognitive growth or development. Socio-economic status is higher itself in building the trust of individuals to face the challenges of life as compared with the poverty-stricken people who are desperate to meet the goals in their lives, especially the challenges faced by children in schools.

Families with low socio-economic status are not only lacking in financial support, financial and social support, and education of their siblings, peers or society as a whole, but they have also lost the support of the community around them at a very important time in their lives. This is a very important factor that promotes and supports child development and school readiness. Parents with low socio-economic status find themselves struggling to raise financial resources, and lack time to inculcate values, good habits, and manners in their children, which may even end up in ignorance of immunization or basic nutrition for their children.

Poverty is not a curse for adults only, but even more so for children. It is children who become easy victims of hunger, poor growth, illness, physical and mental disability, abuse, early marriage, child trafficking, and others. This is a major environmental factor that contributes to children living in poverty, making them four times more likely to have learning disabilities than students who are not in poverty (Apple & Zenk, 1996). According to Casanova et al. (2005), it is a combination of environmental factors and the influence of the family that contributes to the academic success of students. If a student does not eat for days and has clothes that do not fit, how could he be expected to maintain focus in the classroom? Children who come from poverty are not provided with the same tools as the rich, and they had been left behind by those who do not live in the same conditions. The research shows that the problems start with the parents, and lack of education and understanding of the needs of children.

The nature of the relationship between Socio-Economic Status (SES) and student achievement has been debated for decades, with the most influential argument appearing in the unequal opportunity of education of the United States, and a number of questions that were asked in Australia. A person's education is closely related to their life opportunities, income and welfare. Therefore, it is important to have a clear understanding of what benefits or obstacles there are to the attainment of one's education.

Socio-economic background actually sets a roadmap to achievement. Therefore, since it is the determining factor for academic achievement, we have experienced it in our lives every day as well. There are a variety of topics that are closely related to academic achievement. These include the talents of the students, their approach to academics, the school environment, peer pressure, and their relationship with the mentor.

3 RESEARCH METHODS

This study used quantitative research approach. The objects of this study were students from Senior High School BPI 1 Bandung 1, located at Jl. Burangrang No. 8 Bandung. The population of this research is all the class XI students of Senior High School BPI 1 Bandung, numbering 320 students. Samples of this study were 76 students. The sampling in this study used probability sampling. Probability sampling is a sampling technique that gives the same opportunities to every member of the population to be sampled. Collecting data in this study used a questionnaire type of inquiry which is closed.

The questionnaire in this study consisted of ten questions with two variables: the independent variable is the economic background of the parents, and the dependent variable is the learning achievement. There are two alternative answers using a nominal scale that is 'yes' and 'no', which aims to determine whether there is influence of social class inequalities, especially later behind the economic status of parents on learning achievement. The list of questions in the questionnaire can be seen in Table 1 below.

Table 1. List of questions.

Questions
1. Do your parents work or not?
2. Are your parents income below/above UMR Bandung?
3. Do you have a complete school equipment?
4. Do you follow the guidance of learning outside school hours?
5. Is your achievement exceeds/below your friend?
6. Are you including 10 students who excel in the classroom?
7. Are you given the allowance/allowance of more than IDR 20,000?
8. Did you get the motivation to learn from your parents?
9. Is the task given by the teacher always collected on time?
10. Do you have obstacles in completing homework?

Source: Processed researcher, 2016.

Analysis of the data in this study used simple linear regression analysis. There are two stages in the data analysis of this study. The first is a simple linear regression to determine whether there is influence between the variable economic background of parents with variable student achievement. The second is the coefficient of determination (R square) to determine the contribution of variables which influence the economic background of parents with the variable student achievement.

4 RESULTS AND DISCUSSION

4.1 Results

The data processing and data analysis in this study used SPSS. The calculations show equation $Y = 2.208 + 0.484X$. The constant 2.208 states that if there is no value of the variable X (background of parent) then the value of the variable Y (student achievement) is 2.208. The regression coefficient of 0.438 X states that for each additional 1 value of the variable X (the economic background of parents), then the value of the variable Y (student achievement) is 0.438 larger. The results of this study are presented in Table 2 below.

Based on calculations, the value significantly smaller than 0.05 is 0.001. Then the t value of 3.605 while t table of this research 1.992 (t tables of 74). Because $0.001 < 0.05$ and t is greater from t table ($3.605 > 1.992$), it can be concluded that the variable economic background of parents (X) has a significant effect on student learning (Y).

The contribution of its influence can be seen from R square multiplied by 100 ($0.149 \times 100\%$), in order to obtain contributions from the influence of parents' economic background on student performance of 14.9%, while 85.1% is precisely the other variables not examined in this research.

4.2 Discussion

Research has found that socio-economic status, parental involvement and family size are very important factors in a student's academic performance. Family background is key to a student's life and in and out of school and is the most important

Table 2. The results of calculations SPSS.

Variable	R Square	T Count	Significance
Economic background of parents and student achievement	0.149	3.605	0.001

Source: Processed researcher, 2016.

influence on student learning. The environment at home is the primary socializing agent and influences the child's interest in school and aspirations for the future. The socio-economic status of the students is determined by combining the parents' education level, employment status and income level. Studies have repeatedly found that socio-economic status affects student achievement. Education will be meaningless if it is associated with socio-economic status of parents in the conservative and traditional sense, particularly parental education. The study set forth above and below is intended to test the impact of socio-economic status of parents on children's education. There is the influence of parents' economic background on student performance. The economic background of parents accounted for 14.9% influence on student achievement.

5 CONCLUSIONS

Based on the research that has been done, it can be concluded that there is significant influence of economic background of parents on student achievement. Economic background of parents accounted for 14.9% influence on student achievement.

Although the results show there is influence between economic backgrounds of parents on student achievement, student achievement is not absolutely influenced by the economic background of parents; there are many other factors that could affect student achievement. Therefore, students who come from families with low economic background need not despair and be discouraged because a lot of students who come from ordinary families can excel and beat the students who come from wealthy families. Differences in economic background of parents would give rise to social inequalities in education.

REFERENCES

Ankrum, R.J. (2016). Socioeconomic status and its effect on teacher/parental communication in schools. *Journal of Education and Learning*, 5(1): 167–175.

Apple, M. & Zenk, C. (1996). American realities: Poverty, economy, and education. *Cultural Politics and Education*, 68–90.

Azzizah, Y. (2015). Socio-economic factors on Indonesia education disparity. *International Education Studies*, 8(12), 218.

Casanova, F.P., Garcia-Linares, M.C., Torre, M.J. & Carpio, M.V. (2005). Influence of family and socio-demographic variables on students with low academic achievement. *Educational Psychology*, 25(4), 423–435.

Faitar, G.M. (2011). Socioeconomic status, ethnicity and the context of achievement in minority education. *Journal of Instructional Pedagogies*, 1–8.

Lynch, R., Center for A.P. & Oakford, P. (2014). The economic benefits of closing educational achievement gaps. Center for American Progress, (November). Retrieved from https://interactives.americanprogress.org/projects/2015/achievement-gap/.

Nguyen, M. (2011). Closing the education gap: A case for Aboriginal early childhood education in Canada, A look at the aboriginal Headstart program. *Canadian Journal of Education*, *34*(3), 229–248.

Shahidul, S.M., Karim, A.H.M.Z. & Mustari, S. (2015). Social capital and educational aspiration of students: Does family social capital affect more compared to school social capital? *International Education Studie*s, *8*(12), 255.

Ideas for 21st Century Education – Abdullah et al. (Eds)
© *2017 Taylor & Francis Group, London, ISBN 978-1-138-05343-4*

Perception of students towards campus internationalization

P.E. Arinda, R. Apriliandi, R. Pranacita & A.G. Abdullah
Universitas Pendidikan Indonesia, Bandung, Indonesia

ABSTRACT: The development of science requires each country to be able to compete in a global world. Internationalization has also been attempted in the world of education. This study aimed to describe the students' perceptions of internationalization, and the causes for them to enter the international study program. The respondents were 58 students of one of the international study programs at a university in Bandung, West Java. The data collection was done by using a questionnaire. Based on the research, it is known that the students' perceptions towards internationalization are at the level of medium category. The students' perceptions of internationalization are dominated by factors originating from outside of the student, such as the elderly, information obtained, the hope of future work, and student exchange programs and neighborhoods.

1 INTRODUCTION

Internationalization has been a key theme since 1990 in both policies of higher education and research of higher education. Internationalization challenges the dominance of the nation state as the main determinant of the character of the universities and colleges, student experience, graduates, and the people who work in it (Enders, 2004). Universities expect researchers, policymakers and experts in the wider field to produce graduates who have good international skills, as citizens of the world as well as professional workers, in facing the challenges of education in the era of globalization (Yemini et al., 2014; Scott, 2015).

In fact, not all countries are interested in internationalization. America, for example, is already quite confident in the ability of the country. However, global demand is eventually forcing them to be able to compensate. This starts with the increasing number in international education (Albers-Miller et al., 2015).

Internationalization is an ongoing process that involves learning and teaching, the development of the quality of teachers, curriculum development, student experience, development of school services, and so on (Warwick & Moogan, 2013). Internationalization is considered as very influential in changing the way of thinking in the world of education (Amborski et al., 2008). The benefit felt by students of internationalization is that students can face the reality of the challenges in working and socializing in diverse social groups (Trahar & Hyland, 2011). A fair international learning and teaching environment can be created if there is a cooperation between teachers, students,

policymakers and administrators of educational institutions in the world, such as those in English-language higher education in Australia and Canada (Daniels, 2012).

2 LITERATURE REVIEW

Internationalization is an ongoing process that involves learning and teaching, the development of the quality of teachers, curriculum development, student experience, development of school services, and so on (Warwick & Moogan, 2013). One of the goals of internationalization of education is to encourage students to understand, appreciate and articulate the reality of interdependence among nations (Montgomery, 2009).

The success of internationalization is characterized by performance indicators such as graduation rates or the gain of the Nobel Prize in faculty (depending on the institution). The number of overseas education programs offered or the proportion of students involved in study abroad becomes an indicator of success in public institutions. In an ideal world, competition between agencies is not a top priority. The main priority is how to increase the quality, so as to achieve the purpose (Green, 2012).

Motivation of internationalization is grouped into five areas, namely political, economic, social, cultural, and academic. Internationalization has developed into a mass phenomenon that is synonymous with globalization and commercialization of higher education (Berry & Taylor, 2014). What is expected of the students with their involvement in internationalization is to enhance the overall experience of their life, the opportunity to live in another culture, and to improve their work. In addition, they believe that by involving themselves in matters of an international nature, it will enhance their competitiveness (Shinn, 2010).

In addition, some factors that affect the decision making of students to continue their education internationally are based on knowledge, information, and destination of areas to be selected. In addition, student perception is also influenced by the recognition of institution, the support of parents or family, finances, as well as the distance of residence (Chou et al., 2012).

3 RESEARCH METHODS

This study uses quantitative method. The experiment was conducted at one of the universities located in Bandung, which in this case was held on an international study program. An international study program has been selected because the students were considered to have the knowledge, understanding and fairly good experience about internationalization. This study aims to illustrate the students' perceptions of internationalization.

The population in this study was all students of a postgraduate study program of international totaling 58 people. The samples in this study were all students of the population. Respondents came from three classes, each consisting of 20, 21 and 17 people. Researchers chose the entire population, as a sample for a limited number of respondents as well as to maximize the accuracy of the results. The independent variable in this study was 'internationalization' and the dependent variable was 'student perceptions'.

The data collection was done by using questionnaire. Researchers did not do any treatment to the respondents, but only provided a questionnaire containing a list of statements about students' perceptions of the international study program

Table 1. General overview of research questions.

No	Question
1	What is your parents' last education?
2	What is your reason entering study program of International?
3	Do you have the skill to speak good English?
4	Where did you obtain information about this study program of International?
5	Are you certain that you will get a good job after graduating from this study program?
6	Are you interested in student exchange program?
7	Are the people in your environment having good budget to study program of International?

to be answered later on by the respondents. Some questions given to the respondents can be seen in Table 1.

Researchers chose to use the questionnaire because they expect answers to be diverse and can describe the factors underlying students' perception of internationalization of education more clearly.

4 RESULTS AND DISCUSSION

4.1 Results

In this section we present and discuss the results of research on students' perceptions of internationalization. We describe internationalization activities that took place in the university and the extent of involvement of the internationalization from the student's perspective.

4.1.1 Characterization of the study population

In this section, we describe some descriptive information on the international study program based on the questionnaires distributed.

Based on the Table above, it can be seen that the percentage of male students amounted to 29.3% and the percentage of female students entering the international study program was 70.7%. Based on the Table above, the percentage calculation shows that female students who entered the international study program were more than male students by a margin of 41.4%. Then we look at the educational background of the parents' education based on the last education took. In the following Table the researchers distinguished between levels of education using three categories: (1) *Not Passing Senior High School (SMA)* to group parent education ranging from those who did not go to school, completed primary school, completed junior high

Table 2. Percentage of students entering international study program.

Sex	Percentage (%)
Male	29.3
Female	70.7
Total	100%

Table 3. Levels of education of parents.

Last education	Percentage (%)
Not Passing SMA	19.0
Passed SMA	22.4
Passed Higher Education (PT)	58.6
Total	100%

school, up to high school dropout, (2) *Passed SMA* for parents whose last education was high school, and (3) *Passed Higher Education (PT)* to group parent education up to their last education from the program of D3, S1, S2, to S3. Based on the research that has been done on the education levels of parents, Table 3 shows the percentages obtained.

The Table above shows the levels of education of parents, from the students who entered the international study program. Based on the Table above, it is known that the percentage of education of the parents who 'Passed Higher Education' was 58.6%, then 'passed SMA' was 22.4%, and not passing SMA' was 19.0%. The results of the percentage showed that more than half of the total number of students who chose the international study program were supported by the parents with a relatively good education background, where they completed their education at higher education, either D3, S1, S2, to S3, with the percentage of 58.6%.

4.1.2 Students' perception towards internationalization

In this research, the students' perception was described. Based on the results of statistical calculation by using the SPSS 23 application, it is known that the students' perception amounted to 41 people (70.7%) in category medium, 14 students (24.1%) in category high, whereas 3 students (5.2%) were in category low, as shown in Figure 1.

Then, from the answers of the respondents about internationalization and based on the results of statistical calculation, it is known that internationalization mostly amounted to 38 students (65.5%) who were in category medium, and 12 students (20.7%) who were in category high, whereas 8 students (13.8%) were in category low (as shown in Figure 2).

The next step in this research is to reveal the influence of students' perceptions of internationalization. Based on data analysis, it appears that the influence of students' perceptions of internation-

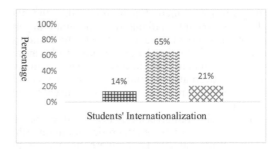

Figure 2. Percentage category of students' internationalization.

alization is the direction (positive), which means that the increase in students' perceptions will result with an increase of the internationalization.

Based on the statistical calculation, the value of determination coefficient of 0.135 or 13.5% was obtained. This means that 13.5% of internationalization is caused by the perception of students, while the remaining 86.5% can be caused by other factors such as family background, information and knowledge, and surroundings.

4.2 Discussion

Internationalization is now regarded as an important activity in higher education (Berry & Taylor, 2014). Internationalization has been identified as one of cultures and learning approach, and has been perceived to benefit through international higher education and international curriculum in order to improve the ability to socialize in groups and improve the competitiveness of work in the era of globalization (Trahar & Hyland, 2011).

This is in line with Warwick and Moogan (2013) that internationalization helps students to take part in becoming global citizens in the global labor market. Thus, the desire to internationalize the students should be further improved and the universities should help students to develop teaching and learning strategies appropriate to a diverse student population.

Perceptions of students in the international study program at a university in Bandung showed positive results that student perception affects internalization enhancement. Parental education, gender, information, future work, foreign language skills, student exchange programs, and the surrounding environment also have a positive impact on students' perceptions of internationalization.

From the aspect of parental education, the majority of students selecting the international study program have parents with high educational backgrounds. Then, from the aspect of gender, the majority of students in the international study

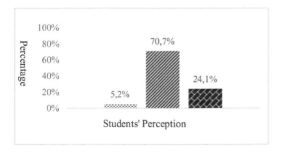

Figure 1. Percentage category of students' perception.

program are women. Furthermore, the students also chose the international study program for job competitiveness in the era of globalization, and to enhance their foreign language skills to be more qualified. Students also hoped that with the entry into the international study program they could have overseas experience through student exchange programs.

In line with Yemini et al. (2014), the education of parents, overseas experience, proficiency in English, and institutional efforts to internationalize were found to be positively influential on students' perceptions to campus internationalization initiatives and characteristics. This means that in addition to the perception of the individual students themselves, students' perceptions of internationalization are also influenced by external factors of students.

5 CONCLUSIONS

Based on the data processed by using the SPSS Statistic 23 application, it is known that students' perceptions affect international enhancement in the medium category. Factors that influence this perception are dominated by factors external to the students, such as the elderly, information obtained, the hope of future work, student exchange programs and the surrounding environment. Gender and foreign language skills also have a positive impact on students' perceptions to internationalization. Students who take the study program of internationalization is dominated by women, as well as the educational background of parents with recent education of university graduates, either D3, S1, S2, and S3.

REFERENCES

Albers-Miller, N.D., Prenshaw, P.J. & Straughan, R.D. (2015). Student perceptions of study abroad programs: A survey of US colleges and universities. *Marketing Education Review*, *9*(1), 29–36.

Amborski, D., Boyle, R., Fubini, A., Oc, T., Heath, T., Watson, V., Frank, A. & Peel, D. (2008). The internationalisation of planning education: issues, perceptions and priorities for action. *The Town Planning Review*, *79*(1), 87–123.

Berry, C. & Taylor, J. (2014). Internationalization in higher education in Latin America: policies and practice in Colombia and Mexico. *Higher Education*, *67*(5), 585–601.

Chou, C.P., Roberts, A. & Ching, G.S. (2012). A study on the international students' perception and norms in Taiwan. *International Journal of Research Studies in Education*, *1*(2), 71–84.

Daniels, J. (2012). Internationalization, higher education and educators' perceptions of their practices. *Teaching in Higher Education*, *18*(3), 236–248.

Enders, J. (2004). Higher education, internationalisation, and the nation-state : Recent developments and challenges to governance theory. *Higher Education*, *47*(3), 361–382.

Green, M.F. (2012). Measuring and assessing internationalization. *NAFSA e-Publications*, 1–21.

Knight, J. (1997). A shared vision? Stakeholders' perspectives on the internationalization of higher education in Canada. *Journal of Studies in International Education* 1(1), 27.

Montgomery, C. (2009). A decade of internationalization: Has it influenced students' views of cross-cultural group work at university? *Journal of Studies in International Education*, *13*(2), 256–270.

Scott, P. (2015). *Globalisation and higher education: Challenges for the 21st century*. The University of Iowa Libraries.

Trahar, S. & Hyland, F. (2011). Experiences and perceptions of internationalisation in higher education in the UK. *Higher Education Research & Development*, *30*(5): 623–633.

Warwick, P. & Moogan, Y.J. (2013). Compare : A Journal of Comparative and International Education A comparative study of perceptions of internationalisation strategies in UK universities. pp. 37–41.

Yemini, M., Holzmann, V., Fadilla, D., Naturd, N. & Stavanse, A. (2014). Israeli college students' perceptions of internationalization. *International Studies in Sociology of Education*, *24*(3): 304–323.

Ideas for 21st Century Education – Abdullah et al. (Eds)
© *2017 Taylor & Francis Group, London, ISBN 978-1-138-05343-4*

The influence of gender differences in mathematics achievement of high school students

A. Riyanti, R. Anggraini, S. Nurohim, S. Komariah & A.G. Abdullah
Universitas Pendidikan Indonesia, Bandung, Indonesia

ABSTRACT: Gender differences in mathematics achievement have received enough scrutiny in the fields of sociology, economics and psychology. This study seeks to explain the influence of gender differences on the achievement of students in mathematics. The data set in this study was drawn from primary data of a questionnaire which was administered to 162 grade 9 students in high school. This study used linear regression to describe the influence of gender differences on the outcomes of learning mathematics. The results found no substantial difference in scores statistically on the outcomes of learning mathematics between girls and boys, but girls have diverse scores and more at the top of the distribution by gender. Analysis on mathematics achievement based on the calculation of the National Examination finds that male students had higher scores than female students.

1 INTRODUCTION

Gender differences in school test scores overall, and particularly in mathematics achievement, have attracted attention in the last few years and a lot of research has been done from various perspectives. Studies that explore these issues have appeared in the Journal of Sociology and Psychology (Felson & Trudeau, 1991; Helwig et al., 2016; Hall & Hoff, 1988; Li, 1999; Muller, 2014; Niederle & Vesterlund, 2010; Preckel et al., 2008; Quinn & Cooc, 2015; Robinson, 2015; Spanias, 2015; Wilson & Boldizar, 2013).

This interest can be explained by several factors. Teachers have different beliefs about girls and boys; they tend to have stereotypes that boys are more dominant in their ability to learn mathematics. The differences in motivation and interest between girls and boys towards mathematics becomes one of the motivations to do this research. Some previous studies also showed that girls experienced more anxiety in facing mathematics tests than boys.

There are several other factors that can be an influence. Gender, the personality of students, class attendance, parental involvement, neighborhood, ethnicity, and age of the students can be predictors of academic achievement.

Mathematics achievement in schools can affect career choices, for college electoral, and future income.

The study of gender differences in mathematics achievement of high school students can explain the extent to which it relates to gender inequality on the outcomes of learning mathematics.

Mathematics is one of subjects taught in schools, aiming to help students prepare themselves to be able to face changing circumstances, in life and in the world, that has been always evolving, through the practice of acting on the basis of thinking logically, rationally and critically. Math lessons taught in school also aim to prepare students to use mathematics and develop a mindset of mathematics in everyday life and in studying science.

Many factors must be considered in the studying of mathematics, among others the willingness, ability and specific intelligence, teacher readiness, student readiness, curriculum, and methods which will be presented. An equally important factor is the gender factor; not only gender but the character, nature and intelligence. In general, a large-scale study on the influence of gender differences in mathematics achievement based on secondary data was obtained from several concerned institutions. This study used primary data from questionnaires distributed to the grade 9 students in high school, majoring in science in one school in Bandung, with a population of 272 students. To determine the number of samples, the formula from Slovin was used, and then the results gained as many as 162 students to be the respondents.

2 LITERATURE REVIEW

In this study, the authors present several other studies that have been conducted by previous researchers, which are relevant to the influence of gender differences on the outcomes of learning mathematics in a high school in Bandung. This study attempts to answer whether there is influence of gender differences between girls and boys on the outcomes of learning mathematics. Felson and Trudeau (1991) examined the standard socialization explanation of gender differences in mathematics performance by using data from students in grades 5 to 12. It has been found that significant gender differences are in the overall value (F(1) = 25.2; p < 0.001), reflecting the fact that girls get better grades than boys. These results showed that girls' performance is better than boys' in mathematics, as girls do in their other academic subjects.

Hall and Hoff (1988) examined gender differences in children tested at grade levels two, four and six, over an eleven year period. Total Mathematics, Mathematical Computation, and Mathematical Concept Normal Curve Equivalency (NCE) scores were collected from the past eleven years of administrations of Science Research Associates (SRA) Achievement Series tests. No significant gender differences or gender-by-grade-level differences were found.

Li (2006) investigated the beliefs in gender differences in mathematics learning and gender differences of teachers. Ratings show that these two aspects have not been dealt with in a fairly big way in the research literature. Meanwhile, Preckel et al. (2008) investigated gender differences in 181 gifted and 181 average-ability 6th graders in achievement, academic self-concept, interest, and motivation in mathematics. Results support the assumption that gender differences in self-concept, interest, and motivation in mathematics are more prevalent in gifted than in average-ability students.

Middleton and Spanias (1999) examined the recent research on motivation in mathematics education and discussed findings from research perspectives in this domain. The findings in theoretical orientation show that students' perception of success in mathematics is very influential in shaping the attitudes of their national motivation. One final criticism is directed against the use of theory in the study of motivation.

In their study, Wilson and Boldizar (2013) attempted to answer two questions about the gender distribution of college curricula: (1) To what extent do aggregate mathematics achievement level, income potential, and aggregate high school aspirations account for the gender segregation of bachelor's degrees?, and (2) How did gender segregation change between 1973 and 1983? They found that mathematics achievement level and income potential of college curricula exert a powerful influence on gender segregation of bachelor's degrees, but that practically all the influences work through gender-differentiated aspirations in high school. The findings of this study regarding of the important influence of the high school aspirations on gender segregation in higher education show that many obstacles to parity in major college selection were in place before students graduated from high school. However, there are many changes in the direction of increasing parity, in some previously dominated by boys, such as higher potential income, and curriculum encouragement.

Using longitudinal study, Muller (2014) attempted to reveal the impact of parental involvement in adolescents' mathematics achievement from grades 8 to 12. The study found that gender differences in achieving test scores on mathematics tests were small but consistent among senior high schools. Gender differences in the test scores of grade 8, and the advantage of grades 8 to 10, were found only when parental involvement is controlled.

Robinson and Lubienski (2011) found that there is no math gender gap in kindergarten. But, based on this analysis, it is revealed that boys are higher in achieving mathematics scores than girls, while girls are superior in reading. Niederle and Vesterlund's study (2010) also showed that boys understand mathematics better than girls. At the core, male students are not competitive in the mathematics subject while female students are more competitive, so that their scores test is higher than male students, although boys have superior understanding in mathematics than girls.

Quinn and Cooc (2015) examined that science achievement disparities by gender and race/ethnicity often neglects the beginning of the journey in the early grades. This study addresses this limitation using nationally representative data following students from grades 3 to 8. The study finds that the Black–White science test score gap (–1.07 Strongly Disagree (SD) in grade 3) remains stable over these years, the Hispanic–White gap narrows (–0.85 to –0.65 SD), and the Asian–White grade 3 gap (–0.31 SD) closes by grade 8. The female–male grade 3 gap (–0.23 SD) may narrow slightly by 8th grade.

3 RESEARCH METHODS

This research is a correlation research. The method used is descriptive method with quantitative approach. This study attempts to know the influence of gender differences in mathematics achievement. In this research there is a main research question, which is *what is the influence* of gender differences in mathematics achievement of students in high school. The object of this research is gender and mathematics scores based on the calculation of the National Examination with gender as the independent variable, which is developed into three sub-variables: character, nature, and intelligence, while the dependent variable in this study is the mathematics scores of junior high school students from the exam.

The subject of this research is grade 9 students majoring in Mathematics and Science from one of the senior high schools in Bandung, with the population amounting to 272 students, with the samples obtained by using the formula from Slovin, gaining as many as 162 students to be the respondents. The reason why the sample was taken from students who majored in Mathematics and Science is because in high school, Mathematics and Science students have more hours to learn mathematics than the students who majored in Social Sciences and Languages. Mathematics and Science students have two schedules to learn mathematics. Firstly, they must take a mathematics lesson for three hours, and secondly, they must take a mathematics specialization for two hours, so it was decided to make Mathematics and Science students as the sample in this study. Sampling was determined by using purposive sampling technique, which means that sampling is used as the needs of researchers. Data collection techniques used questionnaires and documentation technique in the form of documents such as list of mathematics score of junior high school student from National Examination, and list of students by gender. Table 1 presents several statements that become the core of research related to the influence of gender differences in mathematics achievement of high school students in Bandung.

Table 1. Research statements of the influence of gender in mathematics achievement.

No	Statements
1	I like mathematics subject
2	I think mathematics is difficult subject
3	I think mathematics is boring subject
4	Mathematics is a subject that requires a high concentration
5	I always study mathematics that will be taught next day at home
6	I note mathematics that are considered important
7	I never skip during hours of mathematics lesson
8	I would ask the teacher or a friend if there is a mathematical matter I do not understand
9	I get more easily to understand mathematics subject compared to other subjects
10	I always get high scores during the mathematics subject test
11	I always do the mathematics task properly and on time
12	I always ready if there is a mathematics test given by the teacher

The influence of gender on mathematics achievement is measured by using Likert Scale from 1 to 4. Ranging from Strongly Agree (SA), Agree (A), Disagree (D), and Strongly Disagree (SD). Learning outcomes in this study were measured by using mathematics scores of junior high school students from National Examination. Apart from mathematics scores of junior high school students from National Examination, data collection was done by spreading questionnaires that identify gender variable for the character, nature and intelligence between boys and girls. Measurements for gender differences in math scores used a linear regression analysis approach to calculate the extent to which there is a causal relationship between gender differences and mathematics achievement.

4 RESULTS AND DISCUSSION

4.1 Results

4.1.1 Male gender
This section explains about male student's perceptions towards mathematics subject where the indicators are character, nature, and intelligence. Based on statistical calculation, it is known that the minimum score is 32 and a maximum score is 70, so it can be obtained that interval categorization of male gender variables to mathematics is 32–44 categorized as low category, 45–57 categorized as medium category, and 58–70 categorized as high category. Furthermore, regarding male respondent's answer in mathematics subject classified by frequency and percentage calculation, it is known that the male gender to the subjects of mathematics largely or as many as 66 students (81.5%) is in the medium category, then as many as 15 students (18.5%) are in the low category, and none of the students is in the high category (see Figure 1).

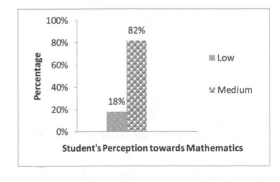

Figure 1. Male student's perception towards mathematics.

4.1.2 Female gender

This section explains about female student's perception towards mathematics subject with indicators of the character and nature and intelligence. Based on statistical calculation, it is known that the minimum value is 38, and the maximum value is 63, so it can be obtained that interval categorization of male gender variables to mathematics is 38–46 categorized as low category, 47–55 categorized as medium category, and 56–64 categorized as high category. Furthermore, from female respondent's answer in mathematics classified by frequency and percentage calculations, it is known that the female gender of the subjects of mathematics largely or as many as 70 students (86.4%) were in the medium category, as many as 11 students (13.6%) were in the low category, and none of the students was in the high category (see Figure 2).

4.1.3 Description of gender and national examination

This section explains about the gender differences in the mathematics learning achievement by using indicators of National Examination score in the mathematics subject. The results of the analysis find that there is no significant difference in the achievements of boys and girls in mathematics, but the boys are superior to girls because they are at the top of the distribution.

The results of statistical calculations about the difference between gender and National Examination indicators above show that males with UN 3–6 are 15 students and those with UN 7–10 are 66 students. Meanwhile, female with UN 3–6 are 19 students and those with UN 7–10 are 62 students.

4.1.4 Linear regression analysis

This study used linear regression to measure the influence of gender differences in mathematics achievement. The results of linear regression analysis by using SPSS was t count is 4.564 and t table with degree of freedom (DF) = 162 – 2 – 1 = 159 is 1.654. Because t count > t table or 4.564 > 1.654, it can be

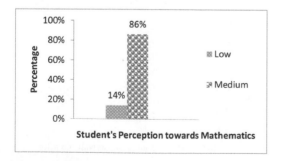

Figure 2. Female student's perception towards mathematics.

seen from the significance value which is 0.000 or less than 0.05 (0.000 <0.05) so that H0 is rejected, which means that gender differences have an influence on the outcomes of learning mathematics.

Based on the calculations, the value of determination coefficient is 0.115 or equal to (0.115 × 100%) = 11.5%. This means that 11.5% of mathematics learning outcomes is caused by gender differences, while the remaining 88.5% can be caused by other factors which are not examined in this study.

4.2 Discussion

This research resulted some conclusions. Firstly, gender differences in mathematics achievement of high school students in Bandung by using character, nature and intelligence as the indicators, are not too different. The value of the standard deviation for male is 6.725 and the value of the standard deviation for female is 5.184. For the analysis of the gender based on statistical calculations it is known that the minimum score of mathematics learning outcomes is 32 and the maximum score is 70, the obtained interval categorization variables male gender to mathematics is 32–44 categorized as low category, 45–57 categorized as medium category, and 58–70 categorized as high category.

Male student's answer (perception) to mathematics subjects known as many as 66 students (81.5%) is in the medium category, then as many 15 students (18.5%) is in the low category, and none of the students is in the high category. The minimum value is 38 and the maximum value of 63, so it can be obtained interval categorization of male gender variables to mathematics is 38–46 categorized as low category, 47–55 categorized as medium category, and 56–64 categorized as high category.

Furthermore, the respondents' answers to female in mathematics are classified by frequency and percentage calculations. It is known that the female gender of the subjects of mathematics largely or as many as 70 students (86.4%) is in the medium category, as many as 11 students (13.6%) is in the low category, and none of the students was in the high category. The results of our analysis on gender differences in mathematics achievement in Bandung city, by using the indicators are nature, character, and intelligence denies the popular myth that states that on average male students are more talented and accomplished in the field of mathematics compared to female students.

Secondly, it contrasts with the results of using nature, character and intelligence as the indicators. The analysis of gender differences in mathematics achievement using indicators of the score of National Examination shows that male students are at the top of the distribution and superior compared to girls, but this difference is not too significant.

The analysis using linear regression shows that the influence of gender perception to the mathematics subject is in the same direction (positive), which means that the increase in gender perception will make an increase in the outcomes of mathematics learning. It can be shown from the regression coefficient, which is 0.023. The meaning of the regression equation is that each increase of one gender perception will be followed by an increase to the outcomes of mathematics learning as 0.023. The other way around, if the gender perception has decreased by 1, the outcomes of mathematics learning will have decreased also by as many as 0.023.

This study uses linear regression to measure the influence of gender differences in mathematics achievement. Results of linear regression analysis by using SPSS show that acquired t count is 4.564 and t table with degree of freedom (DF) = 162 − 2 − 1 = 159 is 1.654. Because t count > t table or 4.564 > 1.654, it can be seen from the significance value which is 0.000 or less than 0.05 (0.000 < 0.05), so that H0 is rejected, which means that gender differences have an influence on the outcomes of learning mathematics.

Based on the calculations, the value of determination coefficient is 0.115 or equal to (0.115 × 100%) = 11.5%. This means that 11.5% of mathematics learning outcomes is caused by gender differences, while the remaining 88.5% can be caused by other factors which are not examined in this study.

Overall, the results in this study are the same as in previous studies which were conducted by Hall and Hoff (1988), Robinson and Lubienski (2011), Niederle and Vesterlund (2010), and Quinn and Cooc (2015), which show that boys and girls are almost the same or very close in their mathematics achievement. Although there are gaps in the distribution, however on the other side, the gaps are not too large. Achievement differences between boys and girls are mostly caused by factors other than gender differences. Those factors, which are not examined in this study, are those such as motivation, interests, cognitive, teachers' skills, and socio-economic conditions.

Note that this study only took 162 students who majored in Mathematics and Science as the sample from its population is 272 students, the calculation using the formula from Slovin with a significance level of 5%. In this research please also note that the gender differences are not the only indicators that are factors in the differences in mathematics achievement between boys and girls, so that other studies are needed to show what factors are being the cause of differences in mathematics achievement which makes the gender gap in mathematics achievement.

5 CONCLUSIONS

This research attempts to reveal to what extent there are gender differences in mathematics achievement for high school students in Bandung. Methodologically, this study contributes to literature by analyzing the distribution of National Examination scores in mathematics subject, and by showing the effects of gender characteristics, such as character, nature and intelligence. The data used in this study is a list of mathematics scores of junior high school students from the National Examination.

The results revealed that male and female students are almost the same or very close in mathematics achievement. Although there are gaps in the distribution, but on the other side, the gaps are not too large. Achievement differences between boys and girls are mostly caused by other factors other than gender differences. Factors which are not examined in this study are those such as motivation, interests, cognitive, teachers' skills, and socio-economic conditions.

REFERENCES

Felson, R.B. & Trudeau, L. (1991). Gender differences in mathematics performance. *Social Psychology Quarterly, 54*(2), 113–126.

Helwig, R., Anderson, L., & Tindal, G. (2001). Influence of elementary student gender on teachers' perceptions of mathematics achievement. *The Journal of Educational Research, 95*(2), 93–102.

Hall, C.W. & Hoff, C. (1988). Gender differences in mathematical performance. *Educational Studies in Mathematics, 19*(3), 395–401.

Li, Q. (1999). Teachers' beliefs and gender differences in mathematics: A review. *Educational Research, 41*(1), 63–76.

Middleton, J.A., & Spanias, P.A. (1999). Motivation for achievement in mathematics: Findings, generalizations, and criticisms of the research. Journal for research in Mathematics Education, 65–88.

Muller, C. (2014). Gender differences in parental involvement and adolescents. *Mathematics Achievement, 71*(4), 336–356.

Niederle, M., & Vesterlund, L. (2010). Explaining the gender gap in math test scores: The role of competition. *The Journal of Economic Perspectives, 24*(2), 129–144.

Preckel, F., Goetz, T., Pekrun, R., Kleine, M., Goetz, T. & Kleine, M. (2008). Gifted child quarterly. http://doi.org/10.1177/0016986208315834.

Quinn, D.M. & Cooc, N. (2015). Science achievement gaps by gender and race/ethnicity in elementary and middle school: Trends and predictors. *Educational Researcher, 44*(6).

Robinson, J.P. (2015). Joseph Paul Robinson Sarah Theule Lubienski University of Illinois at Urbana-Champaign, *48*(2), 268–302.

Robinson, J.P., & Lubienski, S.T. (2011). The development of gender achievement gaps in mathematics and reading during elementary and middle school: Examining direct cognitive assessments and teacher ratings. *American Educational Research Journal, 48*(2), 268–302.

Spanias, P.A. (2015). Motivation for achievement in mathematics: Findings, generalizations, and criticisms of the research. *30*(1), 65–88.

Wilson, K.L., & Boldizar, J.P. (2013). Gender segregation in higher education: Effects of aspirations, mathematics achievement, and income. *Sociology of education*, 62–74.

Ideas for 21st Century Education – Abdullah et al. (Eds)
© *2017 Taylor & Francis Group, London, ISBN 978-1-138-05343-4*

Student participation in the tutoring program (comparative study between socio-economic schools high and low)

L. Utami, M. Azis, H.M. Yusuf, R. Kartika, W. Wilodati & A.G. Abdullah
Universitas Pendidikan Indonesia, Bandung, Indonesia

ABSTRACT: Tutoring becomes one of the non-formal education institutions that are believed to assist in improving student achievement in formal schooling. This article explains how tutoring depicts the level of participation between students from high socio-economic status and those who come from the low one. The sample in this study is 130 students from high and low public high schools, as well as private schools. The analysis of data used linear regression. The result of this study indicates that there is a significant difference in the intensity of the participation of the student in tutoring in both public and private schools with high and low socio-economic status, which creates a strong social stratification climate.

1 INTRODUCTION

The tutoring program is now becoming a trend among students as a way to improve performance in school. This is also supported by their parents. As well as in Indonesia, the tutoring system is conducted in some countries to make up for the lack of formal education systems regarding the equality of educational opportunities (Dawson, 2010). Tutoring has become a major phenomenon in East Asia, including Japan, South Korea, and Taiwan. Factors underlying the growth of private lessons vary, but in all cases they have implications for achievement and livelihoods (Bray, 2006). In addition, in the United States, tutoring has the support of the family as a complement to formal school education (Ireson, 2004). Studies in Korea show that parental involvement in children's education becomes an economic burden, but this is done to find out the academic progress of their children in tutoring (Park et al., 2011).

The ability of students to participate in learning programs is strongly influenced by the socio-economic background of the family. This happened in the territory of China: Shanghai, Taiwan, Hong Kong and Macau. Students who come from families whose social status is higher have more opportunities to follow lessons than students who come from the low socio-economic family status (Zhou & Wang, 2014).

This research aims to look at the extent to which there is a difference between participating school tutoring programs. In addition, this study investigates the quality of various socio-economic statutes in every school. The hypotheses in this study are:

H1: The influence of the participation of cluster tutoring high-intensity tutoring
H2: The influence of the participation of cluster tutoring low-intensity tutoring.

2 LITERATURE REVIEW

Private tutoring is defined as education outside the formal school system. It provides additional instruction for children in the lessons they learned in the educational system. It provides financial benefits for the tutors as well as providing learning opportunities to better serve students (Dang & Rogers, 2008). The development has become a strategy increasing various markets and corporates. (de Castro & de Guzman, 2012).

The influential factors of private tutoring are parental income, participation, and student achievement, as well as the location of the school (Aurini & Davies, 2013; Chung & Choe, 2001; Šťastný, 2016). In the capital city of Seoul, families of middle and high school students spend up to 30% percent of their household income on private tutoring. This is greater than the cost of their monthly food consumption, as much as 16% (Yi, 2013). Active participation and high motivation are very influential to student achievement after attending private lessons (Jung & Lee, 2010). In Korea and Taiwan, an intense student in private lessons will have good learning outcomes in mathematics and English (Kim & Park, 2010; Park et al., 2011; Zhang, 2013). School location is also a determinant factor in participating in private lessons. Beijing University students found that 55.7% of urban families, and 22.8% of students from rural families follow tutoring (Kim & Park, 2010; Liu & Bray, 2016), and students who take private lessons, in general, are the final level students at school (Kwo & Bray, 2014).

High and low socio-economic status of parents also affects the students who take private tutoring (Bray & Kwok, 2003). In Korea and Taiwan, a participation rate which is higher in private tutoring is a school of high quality, largely explained by student-level variables (educational aspirations, instrumental motivation, confidence, and education of the father) and school context variables (size of the community and SES school). While in the Philippines and Romania, the demand for private tutoring is equal in both high level and low level schools (Song et al., 2013). The development of private tutoring's rapid growth can have implications, or a great impact, on the community. This would increase inequality and social cohesion in various circles (Heyneman, 2011). Children who come from high-income families would be able to perform better in school, while children from lower income families could not follow with private tutoring, would be unable to compete with their friends, and would probably drop out of school at an early age (Bray, 2006). Only students who come from the high-income family can afford private tutoring so that they have a greater chance to pass to universities (Dawson, 2010). New technology, especially the Internet, has significantly changed the geographical space where the tutoring can be given. For example, there is a huge gap between the number of students in the USA who use internet and use a credit card and the number of students in India who do so. This will be harmful to the physical and mental well-being of children and the health of the nation (Bray, 2013), but have a positive impact for tutors in improving livelihood.

3 RESEARCH METHODS

The research looked at the participation of the students attending a tutoring program, with the socio-economic status of middle school students in Bandung. In this case, the focus of this study took a sample of public schools, with high and low socio-economic status, as well as private schools with high and low socio-economic status.

The dependent variable in this study is the participation of students in participating in tutoring among high school students in Bandung with high socio-economic status. Low-level indicators look at the students' motivation to follow the guidance of learning, combined with their learning achievement in school. The independent variable focuses on the socio-economic status of high school students in Bandung. The indicators include paren-

Table 1. The question tutoring research participation.

No	Question
1	What is your parent occupation?
2	What level of education are your father and mother graduated?
3	How is the income level of your parent?
4	How much money do you have every day?
5	How is the quality of learning in school?
6	How does your participation when learning in the classroom?
7	How are your achievements in school?
8	Does the school encourage you to follow the guidance of learning?
9	Does the school facilitate you to follow the guidance of learning?
10	Do you need guidance to learn in improve your performance in school?
11	What is your motivation in participating in tutoring?

tal education and the income of parents who would use one of the basic student tutoring programs.

This research uses a questionnaire which is also reinforced by field observations and literature studies. Table 1 presents some questions at the heart of research related to the participation of tutoring, seen from the socio-economic aspects of high school students in Bandung.

The research questions given in samples taken from public and private schools were classified as having the high and low socio-economic status, and therefore there are 130 respondents who are selected from each respective school. The sampling technique is stratified random to verify the hypothesis using linear regression analysis, to see the relationship between the socio-economic disparities that underlie participation in the tutoring program.

4 RESULTS AND DISCUSSION

4.1 Results

This research includes the participation of students in the tutoring program participation seen from the aspect of socio-economic status. Data is presented in Table 2.

Table 2 shows more than half of cluster tutoring high participation in the high category, with a number of 83 people or 63.8%, while for the medium category being 30.8% or 40 people, and low category with 5.4% or 7 people. In the lower cluster, it shows that more than half of the level of participation in tutoring cluster low in the medium category with a number of 94 people or 72.3%,

Table 2. Student participation in the tutoring program.

Participation rate	High cluster ($n = 130$)	Low cluster ($n = 130$)
Low	5.4%	2.3%
Middle	30.8%	72.3%
High	63.8%	25.40%

Table 3. Participation of students against intensity tutoring program.

Intensity of tutoring program	High cluster ($n = 130$)	Low cluster ($n = 130$)	Total
Yes	92	43	135
No	38	87	125

while 25.4% higher category or 33 people, and low of 2.3% or 3 people.

To show how intense the students in Bandung participate in the tutoring program, Table 3 shows that the students who do tutoring are about 135 people (51.9%), while those who do not do the tutoring are about 125 people (48.1%). More specifically, the intensity level of participation of tutoring tailored to schools with a high and low cluster the data is presented in Table 3 below.

Table 3 shows that the students who perform intense tutoring are about 135 people and those who are not intense in tutoring are about 125 people. A high cluster consists of 92 respondents participating in the intense tutoring program. 38 of them do not participate in the intense tutoring program. Unlike the lower cluster, as many as 43 respondents participate in tutoring programs. 87 of them do not participate in an intense tutoring program.

Then, in the calculation of the data analysis regression at school with a high cluster, t count obtained for cluster tutoring high participation was 5.873 of 130 total sample, then df = 130-2 = 128, and then the table was 1.338. It can be seen t > t count table (5.873 > 1.338), so that H1 is rejected and the H2 is accepted. This means that the partial participation in the cluster tutoring high-intensity significantly influences tutoring. So, the level of participation of tutoring in high school affects the intensity of the tutoring in general significantly. The influence of the participation of cluster tutoring high school against the intensity of the tutoring was 21.2%. Regression analysis of data on a cluster of low, obtained t count for the cluster tutoring low participation of 3.107 and 130 total sample, then df = 130-2 = 128, and then the table was 1.338. It can be seen t > t count table (3.107 > 1.338), so

that H1 is rejected and the H2 is accepted. This means that the partial participation in the cluster tutoring is significantly influential to low intensity of tutoring. So, the level of participation of tutoring in high schools affects significantly the intensity of tutoring in general. The influence of the participation of cluster school tutoring lows against the intensity of the tutoring is 7%.

Data shows the level of program participation tutoring can be categorized: the students who are in a cluster of socio-economic high have the participation that is high enough to follow the guidance of learning, compared to students in low socio-economic cluster who follow the guidance of learning at the low level.

4.2 Discussion

The trend in participation tutoring developed rapidly. Many people tried tutoring that offers the best deals to enhance student achievement in school. The student tutoring program is utilized as an alternative to academic guidance to progress their accomplishments (de Castro & de Guzman, 2012).

The participation of high school students in the city of Bandung against participation tutoring shows different results between schools with low and high socio-economic status. Many factors underlie the case. The most basic is to be viewed from the support of the parents (in this case, the earnings of parents). The parents with a high socio-economic status send their children to participate in tutoring to improve their child's achievement in school.

In line with Atalmis et al. (2016), tutoring has an important role in learning and students who have a high socio-economic status have a high chance to follow the guidance of learning to support learning. From the research, the selection of tutoring was adjusted to the ability of parents' income, which means that parents can be assured that more expensive tutoring will also help to increase their children's achievements. As expressed by Park et al. (2011), parents try to choose a tutoring for their children according to the needs and quality. Students who have a low socio-economic status are not too concerned for their child to follow the guidance of learning, as expressed by Park et al. (2011). In other words, the income of parents allows a child to engage in tutoring (Kozar, 2013).

However, indirect tutoring makes a good social inequality between high and low schools in terms of socio-economic status, in spite of the goal of increasing student achievement in school (Kozar, 2013). This inequality makes the existence of stratification in education so clear, strengthened by the results of studies conducted in both public and private schools. Schools with a higher socio-economic status tend to follow the guidance of learning facilities and guarantee them the best performance improvement, regardless of how much money they spend. In contrast, schools with low socio-economic status cannot not follow the guidance of learning. Their willingness to follow the guidance of learning still exists, but the capacity is limited due to economic aspects. This makes a significant difference in the participation of tutoring for students with high and low socio-economic status, especially in Bandung.

5 CONCLUSIONS

This research attempts to reveal the level of student participation in Bandung, which follows the tutoring. The results revealed that students who have high socio-economic status tend to have intense in participating in tutoring. On the other hand, it is proportional to the students who have low socio-economic status as they are not intense in following the guidance of learning. From what has been disclosed, indirectly, the role of parents is very important here as the higher socio-economic status of a family will increasingly support their child's participation in tutoring to increase their achievement. It is also a misnomer because the participation of tutoring indirectly creates a social stratification that is facilitated by education.

REFERENCES

Atalmis, E.H., Yilmaz, M. & Saatcioglu, A. (2016). How does private tutoring mediate the effects of socio-economic status on mathematics performance? Evidence from Turkey. *Policy Futures in Education, 14*(8), 1135–1152.

Aurini, J. & Davies, S. (2013). Supplementary education in a changing organizational field: The Canadian case. *International Perspectives on Education and Society,* 22. Emerald Group Publishing Limited.

Bray, M. (2006). Private supplementary tutoring: Comparative perspectives on patterns and implications. *Compare: A Journal of Comparative and International Education, 36*(4), 515–530.

Bray, M. (2013). Shadow education: Comparative perspectives on the expansion and implications of private supplementary tutoring. *Procedia – Social and Behavioral Sciences, 77,* 412–420.

Bray, M. & Kwok, P. (2003). Demand for private supplementary tutoring: Conceptual considerations, and socio-economic patterns in Hong Kong. *Economics of Education Review, 22*(6), 611–620.

Chung, Y.S. & Choe, M.K. (2001). Sources of family income and expenditure on children's private,
after-school education in Korea. *International Journal of Consumer Studies, 25*(3), 193–199.

Dang, H. & Rogers, F.H. (2008). How to interpret the growing phenomenon of private tutoring: Human capital deepening, inequality increasing, or waste of resources? *World Bank Research Observer, 23*(2), 161–200

Dawson, W. (2010). Private tutoring and mass schooling in East Asia: Reflections of inequality in Japan, South Korea, and Cambodia. *Asia Pacific Education Review, 11*(1), 14–24.

De Castro, B.V. & de Guzman, A.B. (2012). From scratch to notch: Understanding private tutoring metamorphosis in the Philippines from the perspectives of cram school and formal school administrators. *Education and Urban Society, 46*(3), 287–311.

Heyneman, S.P. (2011). Private tutoring and social cohesion. *Peabody Journal of Education, 86*(2), 183–188.

Ireson, J. (2004). Private Tutoring: How prevalent and effective is it? *London Review of Education, 2*(2), 109–122.

Jung, J.H. & Lee, K.H. (2010). The determinants of private tutoring participation and attendant expenditures in Korea. *Asia Pacific Education Review, 11*(2), 159–168.

Kim, J.H. & Park, D. (2010). The determinants of demand for private tutoring in South Korea. *Asia Pacific Education Review, 11*(3), 411–421.

Kozar, O. (2013). The Face of Private Tutoring in Russia: Evidence from online marketing by private tutors. *Research in Comparative and International Education, 8*(1), 74–86.

Kwo, O., Bray, M. (2014). Understanding the nexus between mainstream schooling and private supplementary tutoring: Patterns and voices of Hong Kong secondary students. *Asia Pacific Journal of Education, 34*(4), 403–416.

Liu, J. & Bray, M. (2016). Determinants of demand for private supplementary tutoring in China: Findings from a national survey. *Education Economics, 5292,* 1–14.

Park, H., Byun, S. & Kim, K. (2011). Parental involvement and students' cognitive outcomes in Korea: Focusing on private tutoring. *Sociology in Education, 84*(1), 3–22.

Song, K.O., Park, H.J. & Sang, K.A. (2013). A cross-national analysis of the student- and school-level factors affecting the demand for private tutoring. *Asia Pacific Education Review, 14*(2), 125–139.

Šťastný, V. (2016). Private supplementary tutoring in the Czech Republic. *European Education, 48*(1), 1–22.

Yi, J. (2013). Tiger moms and liberal elephants: Private, supplemental education among Korean-Americans. *Society, 50*(2), 190–195.

Zhang, Y. (2013). Does private tutoring improve students' National College Entrance Exam performance? – A case study from Jinan, China. *Economics of Education Review, 32*(1), 1–28.

Zhou, Y. & Wang, D. (2015). The family socioeconomic effect on extra lessons in greater China: A comparison between Shanghai, Taiwan, Hong Kong, and Macao. *Asia-Pacific Education Researcher, 24*(2), 363–377.

Ideas for 21st Century Education – Abdullah et al. (Eds)
© 2017 Taylor & Francis Group, London, ISBN 978-1-138-05343-4

Factors affecting the study completion time of Bogor Agricultural University's graduate students and its managerial implications

F. Siregar, D. Syah & N. Nahrowi
Bogor Agricultural University, Bogor, Indonesia

ABSTRACT: The aims of this study were to identify the factors that caused the length of the completion time of the studies of the graduate students at Bogor Agricultural University, and to formulate an adaptive managerial implication. Factors that may have influenced the study period of graduate students were examined and classified through the CART (Classification and Regression Trees) method, using the data of students who graduated between 2003–2013, consisting of 5,224 master's and 1,786 doctoral programs. A descriptive analysis method was also used to analyze the academic data related to the stage of study completion. Factors that affected the doctoral program were the faculty, age and master's GPA (Grade Point Average), while factors affecting the master's program were bachelor's GPA, faculty, profession, and the linearity of previous education. In-depth interviews also showed that student motivation was crucial to the completion time of their studies. In term of the management side, strengthening the administrative system so that it can support the study process of the students, starting from the registering process until graduation time, is essential. Therefore, the proposed managerial implications are the improvement of the admission system, improvement of the evaluation of the students' study progress, and optimization of the academic rules implementation.

1 INTRODUCTION

Timely completion of the studies of graduate students is an important outcome for the students, the host university and the economy (Pitchforth et al., 2012). A lot of research related to the process of study completion has been undertaken by the government, universities and graduate candidates themselves (Bourke, 2004), and these considered that the rate of failure of the completion of postgraduate study was quite high, and the time taken to complete studies was too long. In general, a number of research projects were conducted to determine the factors that cause the high dropout rate and the why the completion time is beyond the national standard, especially for the doctoral program. Several research projects were also conducted on the students who successfully complete their studies on time, in order to discover the factors that supported the success of their studies (Bain et al., 2009).

Although it has been discussed since the 1970s (Manathunga, 2005), this issue is still a complicated problem in a number of educational institutions now, including in Indonesia. Based on the data gathered by the Directorate General of Higher Education (2015), it is reported that the timely completion rates of graduate students in Indonesia are very low, especially for the doctoral programs. In the case of Bogor Agricultural University (IPB) Graduate School, the average percentage over the last 10 years of graduate students who graduated within 2 years is only 16.16%, while doctoral students who can complete their studies within 3 years is only 4.76%. On average, a doctoral student in IPB would complete his studies in more than five years, while a master's degree student takes between 2 to 2.4 years. This achievement is definitely still below the national standard, especially for doctoral programs. Therefore, this study aims to identify the factors that cause the length of study completion time of graduate students and formulate an adaptive managerial implication, based on student characteristics, in order to help them complete their studies in the optimal time.

2 RESEARCH METHODS

2.1 *Data*

The research was based on administrative data of 5,224 master's degree students and 1,786 doctoral students who graduated in the years 2003 to 2013. Administrative data used in the CART method consists of three types: demography (gender, age, marital status, profession), educational background (status of the previous university, previous study program and its accreditation, cluster of study program, place of previous university, GPA), and admission characteristics (funding resources, faculty, acceptance status).

2.2 Analysis method

The analysis was performed in two ways, based on the student's characteristics and the process of study completion. The first analysis used CART (Classification and Regression Trees) methods by Breiman et al. (1984), which describes the formation of a binary decision tree (Han & Kamber, 2006). This method was used to examine and classify the factors affecting the study completion time of graduate students.

Descriptive analysis was used to present, analyze and interpret data related to the process of study completion. Investigations were performed on the stages of study completion and academic evaluation. The data of students who graduated in 2011, 2012 and 2013 (50 students each year) with the longest completion times were used to examine the academic evaluation process. In-depth interviews were also performed with a number of graduates with certain characteristics that were obtained through the CART method.

3 RESULTS AND DISCUSSION

3.1 Graduates characteristics

The characteristics of graduates in the IPB Graduate School were dominated by male students, married, aged 33–49 years (S3), 26–35 years (S2), supported by scholarship, and the majority worked as lecturers. Associated with the study completion time, it was found that graduate students with certain characteristics, such as male, married, and those in the oldest age range, tended to take a longer time to graduate. Wright and Cochrane (2000) and also Wamala and Oonyu (2012) mention that gender does not always provide a real impact on the success of the study, but at least it can describe the IPB students' performance from the perspective of gender.

It was also found that the doctoral students who were not supported by scholarships took a longer time to complete their studies, but the opposite is true for the master's degree program. Bain et al. (2009) state that financial support also influences the success of the students studies. It was also found that lecturers need a longer time to complete their studies for doctoral programs, while for the master's degree program, freshman tend to graduate faster. On the faculty side, the longest times taken for doctoral students to complete their studies were those from the Faculty of Economics and Management (FEM), and from the Faculty of Human Ecology (FEMA) for the master's degree students. Based on the statistical test (α 0.05), factors that are not significant to the study period were only found in the doctoral program, which were gender, marital status and age.

3.2 Analysis of factors affecting the study completion time

Postgraduate students represent a significant range of diversity (Abiddin, 2011). The CART method was used to determine which factors influenced the study completion time of graduates in the Graduate School of IPB. For the doctoral program, the first variable that became a partition was faculty, which means that the selection of faculty is the most influential factor. The next influential factors were the master's GPA, followed by the student's age while pursuing a doctorate. The doctoral student's cohort who have the longest study completion time were characterized by a group of students who studied at the Faculty of Agriculture (FAPERTA), FEM, Faculty of Fisheries and Marine Sciences (FPIK), Faculty of Animal Science (FAPET), Faculty of Forestry (FAHUTAN), FEMA, and had master's GPA < 3.54 (class 3), as shown in Table 1. While the group of students who studied for a doctorate at the Faculty of Veterinary Medicine (FKH), Faculty of Agricultural Technology (FATETA), Formal Science and the Multidisciplinary Program had an average study period that was relatively faster than the students who chose other faculties. The other factors, such as funding resources, gender, profession, previous study program and its accreditation, status of previous university, and the acceptance status in the Graduate School of IPB, did not seem to affect the study period of the doctoral program.

Factors that affected the study completion time in master's degree programs were more varied than those in the doctoral programs, such as the bachelor's GPA, faculty, profession, and prior education cohort. The first variable that became partition was bachelors' GPA. In other words, bachelor's GPA is

Table 1. Identifier variables that affect the completion time of the doctoral program.

Class	N	Average completion time (year)	Identifier variables
1	302	4.85	Faculty (FKH, FATETA, IPA, Formal Science & Multidisciplinary Program)
2	217	5.30	Faculty (FEM, FAPERTA, FPIK, FAPET, FAHUTAN, FEMA), Master's GPA ≤ 3.54, Age ≤ 40.5 years
3	181	5.49	Faculty (FEM, FAPERTA, FPIK, FAPET, FAHUTAN, FEMA), Master's GPA ≤ 3.54
4	121	4.88	Faculty (FEM, FAPERTA, FPIK, FAPET, FAHUTAN, FEMA), Master's GPA > 3.54, Age > 40.5 years

the most influential factor for the study completion time of the master's degree program. The characteristics of the master's degree students who had the fastest study completion time were those who had bachelor's GPA $\geq$ 3.21 and took their master's program at FAPET, FAHUTAN, and Faculty of Mathematics and Science. The group with the longest study period were characterized as the students who had bachelor's GPA $\leq$ 3.21, took their master's degree program at FEM, FAPERTA, FATETA, FKH, FPIK, had a profession as lecturers and researchers, and whose prior education was not allied with the study program in the Graduate School. Other factors, such as age, funding resources, gender, previous study program and its accreditation, status of previous university, and the acceptance status in the Graduate School, were not found to have affected the study completion time of the master's program.

3.3 Analysis of completion time based on the stages of study completion and academic evaluation

There are a lot of stages of study that should be passed by graduate students while they pursue a master's or doctoral program. For doctoral students, the stages are more complicated compared to those of the master's degree students. The essential step that needs to be accomplished for both master's and doctoral students after finishing the courses stage is colloquium. Most of the students took a longer time to achieve this step. On average, over the last ten years, the doctoral students needed six semesters or more to perform the colloquium, as shown in Figure 1, while master's students needed four semesters or more. About 50% of the study period was used to achieve the colloquium stage, 30% to achieve the seminar, and the rest for the final examination.

Regarding the thesis and dissertation submission as the final work, some issues actually emerged. Based on the academic rules in IPB,

the students were given three months to revise their thesis or dissertation after the final examination before it was submitted. But, in fact, there were 14.6% of doctoral students who submitted their revised dissertation more than three months later, and this was 8.13% for the master's degree program. The total percentage may not be a large number, but in 2003 more than a half of the doctoral students submitted their final dissertation after more than three months. In 2012, there was even a doctoral student who finally submitted their dissertation after 106.07 months. The percentage was somewhat better for the master's degree program. In 2003, 19.94% of master's degree students submitted their thesis in an average of 5.6 months. However, the percentages were getting better during the following years.

Academic evaluations were held at the end of every semester, to evaluate the study progress of the students. On that evaluation process, the status of the students will be decided. Warning letters will be given gradually until the study period has ended. The final warning letter is a dropout warning letter (SPO). This should only be given once and, after that, if the student could not finish his studies, he will be dropped out from the university. But, from the records, in several cases it was found that SPO were given many times to the same students in large numbers. As an example, for the doctoral program in 2011 there were six students who received an SPO letter seven times. However, in 2013, this was reduced to only twice for 37 students (Figure 2). Similar cases also happened in the master's degree program.

3.4 Managerial implication

Faculty was found to be the first factor that affected the completion time in a doctoral program, while for the master's degree program it was bachelor's GPA. Thus it is very important for the candidates to select a suitable faculty for them, and also important for the university to select the best

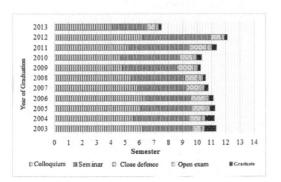

Figure 1. Stage of study completion for doctor's students.

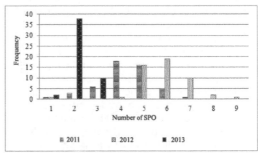

Figure 2. Frequency of "SPO" letter given to the same doctor's degree student.

intake. Based on the interviews with the graduates' cohort who had the fastest completion time, it was found that they undertook research about the faculty that they intended to join, and prepared advanced study and research plans, which were implemented accordingly. On the contrary, it was revealed that the graduates' cohort with the longest study completion time did not know very much about their study program, generally did not have a clear research plan, and faced a lot of disadvantages in the research process. Therefore, the university should start considering some improvements in the admissions system.

New requirements should be added to the selection process, in which the candidates are expected to find their prospective supervisors before they apply to the university, and prepare an advanced research synopsis. By implementing this step, the information system in the university should be strengthened at the same time. It is very important to ensure that the necessary information, such as the faculty and the lecturer's profile, is accurate and up to date for prospective applicants. Furthermore, the selection process should not only focus on academic requirements, but should also examine the research synopsis to find out its suitability with the expected supervisor. By implementing this step earlier, the supervising and research process can be started earlier; the supervisor can monitor their students earlier, which, in turn, can accelerate the study period. Manathunga (2005) claimed that the supervisors hold an important role in ensuring timely completions.

Generally, the students did not face a lot of constraints at the lecture stage. The constraints were faced in the phase of researching and writing the thesis/dissertation. Thus, there should be an improvement in managing the curriculum, for example managing the colloquium schedule more seriously, since it was found to be the crucial step for the research stage. It should be scheduled simultaneously on every study program, and consideration should be given to the consequences for the students who could not do the colloquium on time.

Another important factor found from the interviews was the motivation of the students themselves. A lot of effort to build student motivation can be done by creating a conducive atmosphere within the teaching and learning activities, for example by improving the quality of administrative services, creating seminars or providing training in writing scientific papers regularly, and improving student facilities.

It is also important to consider improvements on the academic evaluation side. The evaluation of students in the first and second semester usually only focuses on the student's GPA as a requirement to continue to the next semester. Therefore, the evaluation should also focus on the study progress of

the students from the beginning, by involving their supervisors. It was also found that the academic rules were not implemented firmly while holding the academic evaluation process. So in order to reduce the accumulation of problems later on, the academic rules should be implemented more optimally.

4 CONCLUSIONS

Factors affecting the doctoral program were the faculty, age and master's GPA, while those affecting the master's program were bachelor's GPA, faculty, profession, and the linearity of previous education. Student motivation was also crucial to the completion time of their studies. In terms of the management side, strengthening the administrative system so that it can support the study process of the students, starting from the registering process until graduation time, is essential. Therefore, the proposed managerial implications are an improvement of the admission system, improvement of evaluation of the students' study progress, and optimization of the academic rules implementation.

REFERENCES

Abiddin, N.Z. (2011). Attrition and completion issues in postgraduate studies for student development. *International Review of Social Sciences and Humanities, 1*(1), 15–29.

Bain, S., Fedynich L.V. & Knight M. (2009). The successful graduate student: A review of the factors for success. *Journal of Academic & Business Ethic, 3,* 1.

Bourke, S. (2004). *Attrition, completion and completion times of PhD candidates.* Paper presented at the AARE Annual Conference Melbourne, 28 Nov – 2 Dec 2004.

Breiman, L., Friedman, J.H., Olshen, R.A. & Stone, C.J. (1984). *Classification and regression trees.* New York (US): Chapman & Hall.

[Dikti] Directorate General of Higher Education. (2015). *Analysis of study completion time of BPPDN scholarship awarde.* Jakarta (ID): Dikti

Han, J. & Kamber, M. (2006). *Data mining: Concepts and techniques* (2nd ed.). San Francisco (US): Morgan Kaufmann Publisher.

Manathunga C. (2005). Early warning signs in Postgraduate Research Education: A different approach to ensuring timely completions. *Teaching in Higher Education, 10,* 219–233.

Pitchforth, J., Beames, S., Thomas, A., Falk, M., Farr, C., Gasson, S., Thamrin, S.A. & Mengersen, K. (2012). Factors affecting timely completion of a PhD: A complex systems approach. *Journal of the Scholarship of Teaching and Learning, 12*(4), 124–135.

Wamala, R. & Oonyu, J.C. (2012). Completion time dynamics for master's and doctoral studies at Makerere University. *Contemporary Issues in Education Research – Second Quarter 2012, 2*(2), 140–146.

Wright, T. & Cochrane, R. (2000). Factors influencing successful submission of PhD Theses. *Studies in Higher Education, 25,* 181–195.

Ideas for 21st Century Education – Abdullah et al. (Eds)
© *2017 Taylor & Francis Group, London, ISBN 978-1-138-05343-4*

The location analysis of junior high schools in West Java Coastal Zone

T.C. Kurniatun, E. Rosalin & L. Somantri
Universitas Pendidikan Indonesia, Bandung, Indonesia

A. Setiyoko
Remote Sensing Technology and Data Center, LAPAN Indonesia, Indonesia

ABSTRACT: This study aims to determine the location of Junior High Schools (JHSs) in southern coastal areas of West Java (Sukabumi and Cianjur Municipality). In Indonesia, coastal areas are generally geographically remote and in need of more school presence until 2020. Analysis of the location is based on the requirements of the minimum service standards of education. This location analysis method is based on GIS and in need of an analysis of the social demand of school-based planning approach. Sources of data are in the form of Pleiades, Worldview 2 and GeoEye1 satellite imageries. The results showed that the ideal locations to establish five school buildings are in the District of Agrabinta Cianjur Regency (Walahir village, Neglasari and Sukamulya village Sukamanah, Wanasari and Karangsari village). In Sukabumi, there are eight ideal locations for junior high school buildings to be established, based on the density of residential areas. In the plain area, the best locations are Cisolok, Karangpapak, and Pasirbaru villages. Further, for the hilly area, the best locations are Caringin village, Cikakak, and Gunung Tanjung village.

1 INTRODUCTION

In carrying out the educational planning approach based on social needs there are often obstacles, in particular related to the accuracy of demographic data as well as data flows students. Further problems also arise in the context of determining the location of the school, considering the geographical conditions, which vary greatly. Generally, coastal areas in Indonesia have topography characteristics that vary and are categorized as remote areas. It is necessary to identify needs based on this approach as well as demographic and GIS, especially in West Java. This is because the nine-year compulsory education program in West Java has not been successful yet.

2 RESEARCH METHODS

Explorative methods were used in this study by using the study of documentation, interviews and analysis of satellite imagery. The data source is a document from the Department of Education Cianjur and Sukabumi and Pleiades satellite imagery, Worldview 2 and GeoEye1 were acquired between the dates of 3-3-2013 until 19-4-2015.

3 RESULTS AND DISCUSSION

The results of this study were shown in the form of data and information on the needs of educational facilities and determining the location of junior high schools in the coastal areas of West Java, including:

3.1 *Sub district Cisolok Sukabumi*

It is estimated that the distribution locations of JHSs in Cisolok sub district, Sukabumi were based on image interpretation Pleiades 2014. Based on the distribution of Cisolok settlements in the district, there are many settlements in the villages, such as in Cimaja, Cisolok, Karangpapak, Cikahuripan, Pasirbaru, and Caringin. Further north, settlements were increasingly rare, such as in the villages of Cileungsing, Sirnarasa, Sirnaresmi, Cicadas, Gunungkramat, and Cikelat. Even in the villages of Sirnaresmi, Sirnarasa, and Cileungsing, the settlements were very rare for the third village that is side to side with Mist Mountain Salak National Park. The settlements in the District Cisolok were concentrated in the southern regions that form the plains, while getting to northern settlements was increasingly rare because of the topography of the high hills. The settlements were scattered in the south because of the easy accessibility due to impassable roads at Palabuhan ratu Lebak. Settlements in the south were scattered among rice fields. Further north, ridge settlements follow the contours of the hills. The use of land in the hills consists of shrubs and mixed farms. The center of population is commonly in the southern region bordering the sea.

There are eight locations that are ideal for the establishment of JHS buildings, based on the density of settlements, such as locations in the densely populated plain areas in Cisolok village, Karangpapak, and Pasirbaru. Each village is ideal for establishing JHS buildings. As for reaching the hilly areas, schools can be extablished in the densely populated settlements in the villages of Caringin, Cikakak, and Mount Tanjung. Further, to reach mountain areas, although still rare for settlements, the accessibility is too far from Cisolok village: around 10–15 Km. Thus, there needs to be a junior high school somewhere around the Cicadas and Simarasa border.

Thus, the possible locations for the District Cisolok SMP (Sekolah Menengah Pertama-Junior High School), Sukabumi District are as follows:

3.2 Sub district Agrabinta Cianjur

Determining the approximate distribution locations of JHSs in Agrabinta sub district, Cianjur is based on image interpretation and GeoEye's World View 2015. The distribution of settlements in the District of Agrabinta were widely spread in the villages of Puncak Wangi, Walahir, Sirnasari, Sukasirna, and Sukamulya. The settlements were numerous in the villages of Pusakasari, Neglasari, Sirnalaut, and Bojongkaso. However, many settlements are from the sea which makes it difficult for people to reach. Plantations consist of a rubber plantation and a chocolate plantation. Settlements in the village of Bojongkaso follow along the river channel Cibuni. The topography of the District of Agrabinta is hilly, with narrow plains in the south, so that the concentration of the population is widely spread throughout the countryside. The settlements are common in South Trans, which connects West Java Cianjur and Sukabumi in the southern coastal areas.

There are five ideal locations for Agrabinta school buildings. The first location is in the village of Walahir, because the dense settlement includes the village of Puncak Wangi Walahir, and will reach children of ages SMP in both villages. Then there are locations at Neglasari and Sukamulya. Although quite far, the schools are accessible. Then, a school can be built in the village of Sukamanah to reach the children in the villages of Sukamanah, Wanasari, and Karangsari. Furthermore, there are locations in the villages of Sirnasari and Bojongkaso, because both have passed the village of Trans Line South West Java. JHSs, which were established in the village of Bojongkaso, can also reach the children of the village Sinar Laut, bordering the sea and with a sparse population.

Thus, the possible locations for the District of Agrabinta SMP, Cianjur regency are as follows:

Figure 1. Settlement distribution map of sub district Cisolok Sukabumi regency.

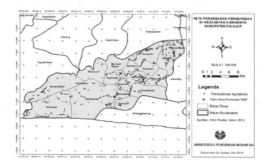

Figure 2. Settlement distribution map of sub district Agrabinta Cianjur regency.

4 CONCLUSIONS

Currently, the number of junior high schools in Sukabumi and Cianjur is still much smaller than the number of elementary schools. The average difference is > 80%. This illustrates that in order to achieve the target of compulsory education by 100%, facilities and adequate infrastructure still need to be pursued. The number of schools should be increased in anticipation of the number of primary school graduates who continue their education to junior high school. By 2020, both Sukabumi and Cianjur regencies still need to build new school buildings in accordance with the minimum requirements for education. Coastal areas in Sukabumi and Cianjur are still dominated by plantations and forests.

REFERENCES

Adekoya, S.O.A. & Gbenu, J.P. (2008). *Fundamentals of educational planning* (revised and enlarged). Lagos: Micodex Nig. Ltd.

Balai Konservasi Sumber Daya Alam, Kabupaten Sukabumi dalam angka 2011.

Balai Konservasi Sumber Daya Alam, Kabupaten Sukabumi dalam angka 2012.

Bappeda Kota Depok Tahun 2009 tentang Angka Partisipasi Murni Kota Depok.

Best, J.W. (1982). *Metodologi penelitian pendidikan*, Disunting oleh Sanafiah Faisal dan Mulyadi Guntur Waseso. Surabaya: Usaha Nasional.

Coomb, P.H. (1970). *What is educational planning*. Paris: United Nation, Education, Scientific, and Culture Organization.

Campbell, O.O. (2002). *Educational planning, management and school organization*. Lagos: Babs Olatunji Publishers.

Effendi, E. (2000). *Perencanaan berbasis sekolah*. Malang: FIP UM.

Fabunmi, M. (2007). *Perspectives in educational planning*. Ibadan: Odun Prints.

Gbadamosi, L. (2005). *Basic of educational planning*. Ogun State: Triumph Publishers.

Gbenu, J.P. (2012). The adoption of the principles of social demand approach (SDA) as a strategy towards ensuring the success of the UBE programme in Nigeria. *Knowledge review*, 24(1), 1–5.

Global Monitoring Report. (2008). *Education for all*. [Online]. Available at: UNESCO.ORG/Education/HomeEducationforAll.

Instruksi Presiden Nomor 5 Tahun 2006 tentang Gerakan Nasional Percepatan Penuntasan Wajib Belajar Pendidikan Dasar Sembilan Tahun dan Pemberantasan Buta Aksara.

Kementrian Pendidikan Nasional Tahun 2012 tentang APM Pendidikan

Longe, R. (2003). Introduction to educational planning. In J.B. Babalola (ed.), *Basic text in educational planning* (pp. 11–12). Ibadan: Awemark Industrial Printers.

Maswarita, M. (2010). *Teori perencanaan*. Teori Perencanaan/Pertama/11540(2).htm

Maqin, A. (2011). *Pengaruh kondisi infrastruktur terhadap pertumbuhan ekonomi di Jawa Barat*. Skripsi (diterbitkan). Bandung: Fakultas Ekonomi.

Rasheed, K.A. & Gamilym H.I.E. (2013). GIS as an efficient tool to manage educational services and infrastructure in Kuwait. *Journal of Geographical Information System, 5,* 75–86.

Timan, A. (2004). *Perencanaan pendidikan*. Buku Ajar. Malang: Jurusan AP FIP UM.

Udin, S.S. & Abin, M.S. (2005). *Perencanaan pendidikan*. Bandung: Remaja Rosdakarya.

UNESCO. (2008). *Application of the principles of Social Demand Approach (SDA) in the implementation of education*. [Online]. Tersedia: www.portal.unesco.org/education.

Vembrianto, S. (1985). *Pengantar pengajaran modul*. Yogyakarta: Yayasan Pendidikan Paramita.

Ideas for 21st Century Education – Abdullah et al. (Eds)
© *2017 Taylor & Francis Group, London, ISBN 978-1-138-05343-4*

The inclusion of gender issues in global education in contemporary Indonesia

E. Haryanti
UIN Sunan Gunung Djati, Bandung, Indonesia

ABSTRACT: This study aims to build a framework for the inclusion of gender issues in the Indonesian education system. Grounded from the ideas of global education, gender issues, framing by gender justice and equality, males and females as human beings and members of society can reach their fullest potential through education. As a qualitative case study research, it explores and analyzes the ideas and thoughts of selected Islamic university lecturers and Islamic high school teachers. Through delivering questionnaires, interviews, and providing focus group discussions, its results show the existing realities of the globalizing world and show that gender issues have been positively endorsed among the respondents. Equal access and opportunity in education, as well as equal treatment for both sexes, is not a problem at all. A little note emerges on the discussion of the gender division of labor, which should consider women's nature. However, the overall findings show the emergence of an Islamic moderate perspective on gender justice and equality in education.

1 INTRODUCTION

Today globalization is again becoming a world phenomenon, since the world is getting closer to one another and creating interdependence between each other. The causes are clearly visible due to the rapid development of sciences and various inventions of technology, such as the emergence of advances in telecommunications and transportation. Similarly, in the political and economic arenas, nations and countries have connected due to their similarities and uniformities, with individuals and groups building into a social system, which should actually have created world citizenship. As an example, the existing adoption of democracy in the Indonesian political system has been the result of the third global wave of democratization, which has developed since the 1990s (Indonesian Center for Civic Education ICCE, 2008). However, globalization can also create negative impacts. For one thing, it has reduced the traditional border of a country and national cultural-traditional values (Steiner-Khamsi, 2004).

In the field of education, globalization should be perceived to have a positive impact. Considering that the function of education has a strategic aspect in developing a human life, its roles point to it being a guardian of the future generation. Therefore, globalization in education should aim to produce professional graduates with international standards, promoting an educational system that enables graduates to compete with other

countries, provide qualified human resources with competitive capacities, provide changes to educational structures and systems, among other things. The aim of all these positive impacts should be to lead to the implementation of reformation in educational processes. The emphasis would be to create a more comprehensive and flexible educational system that can produce graduates who can contribute effectively to the building of world democratic global life (Idrus, 2009). This paper is aimed at building a framework for the inclusion of gender issues in the Indonesian education system, as the first attempt to further the implementation of global education. With the inclusion issues in global education, such as gender, multiculturalism, Human Rights, environment, and peace building issues, this paper chooses gender issues for this study. It considers that gender equality and justice have been widely introduced in Indonesian societies for about two decades, but its implementation seems to be stagnant, especially after the release of the Indonesian President Instruction No.9 Year 2000 on Gender Mainstreaming.

2 LITERATURE REVIEW

The inclusion of gender issues in global education are referred to in several sources. Firstly, they refer to a global movement and the commitments of various countries that declare support to the United Nations (UN) Convention resulting from the World Conference on Education for All held in 1990 and continued up to 2015. The campaign of making Education for All (EFA) has aimed to eliminate gender disparities and achieve gender equality in education; the focus clearly ensures that girls have full and equal access to and achievement in basic education of a good quality (the United Nations Educational, Scientific and Cultural Organization UNESCO, 2015).

Secondly, scholarly work has contributed to policies and programs and has widened gender discourse more comprehensively. This study, among other things, includes gender justice and equality in education in terms of the right to education of females; access and participation in education or equal access and opportunity; gender sensitive educational environments included in the gender division of labor; and processes and outcomes or equal treatment (Subrahmanian, 2005). At the same time, equal education is also widened by processes of gender justice and equality like perspectives of most Indonesian people.

Thirdly, gender justice and equality framed by the concept of global education is used to see the way that global issues can be combined with national and regional culture and tradition perspectives, thus there would be found a distinctive framework of gender justice and equality reflected by the respondents of Indonesian devout Muslims. For this frame, there are four components to developing issues included in global education: 1) conceptualization, which means that educational concepts should be articulated in what global education means to the concepts and their program; 2) national/international context, which means that educational concepts have to be participated in and understood in the national and international contexts of global education; 3) a local context, which means that educational concepts must be understood and taken advantage of by local contextual factors so that they can make an effective program of global education; and 4) controversy in the nature of global education (Merryfield & Harris, 1992).

Several definitions are chosen here in order to get certain indicators of relevance to this study. First of all, it is from an education perspective, which regards that people who live and interact in the world today are connected one to another in an increasingly globalized world. They are also trained to understand and discuss complex relationships of various issues so that they can come to new ways of thinking and acting (Corbin, 1989). In global education, learning can be about many problems and issues cutting across national boundaries and about the interconnectedness of various cultural, ecological, economic, political, and technological systems internationally. It is understood

and appreciated by neighboring countries who have different cultural backgrounds (Association for Supervision and Curriculum Development – ASDC 1991) as quoted by (Tye, 2003).

Gender as a concept is defined as the social differences constructed by women's and men's roles and responsibilities, which vary from culture to culture and change over time (Mansson & Farnsveden 2012). This understanding, then, develops in terms of the gender division of labor, which differentiates what roles women and men are expected to play in domestic and public spheres. Conventionally, women are supposed to be in the domestic sphere, taking care of family and household works, while men are expected to have the responsibility of income generation. Here gender justice and equality are analytically repositioned, using these conventional gender divisions of labor as interchangeably cultural roles, as women are considered to have equal capacities to play in the public sphere and vice versa.

Gender equality in education discusses the access and experience of girls and boys in education at the same levels of quality and outcomes of education time (Mansson & Farnsveden 2012). Furthermore, the concept also identifies terms such as equal access and opportunity, and equal treatment, which explores how boys and girls experience school and classroom activities. Further discussion is shifted to explore the possibilities of how gender equality can be implemented in the reformation of the curriculum; how it can be included in the curriculum and how to carry it out in the real field of education, i.e. schools and classrooms.

3 RESEARCH METHODS

This research was carried out among 8 Islamic graduate students whose backgrounds are, 4 university lecturers and 2 high school teachers, 6 males and 2 females. They were chosen to be respondents because of experiencing discussions on global education with gender equality issues for at least three meetings. The idea behind this study, as mentioned above, is to find out about the building of gender equality in education with Islamic perspectives. It is identified that global issues are most controversial in nature. Here, a global issue like gender equality is to see tolerance, acceptance, and even support reflected among respondents, considering that the issues often raised debates and conflicts and often refer to a threat to women's domestic roles, especially as wives and mothers.

This study uses a qualitative approach aimed to result in descriptive data that consists of the respondents' own written or spoken words, which related to their experiences, attitudes, beliefs, and thoughts (Denzin & Lincoln, 2011). The researcher used case studies because "...it tries to illuminate a decision or set of decisions: why they were taken, how they were implemented, and with what result" (Yin, 2006). Also, to get more people's realistic responses on why respondents reveal their views differently on the topic under discussion, having certain attitudes and opinions, and how they can build ideas of complementary gender issues, especially here in the field of education.

This study employed questionnaires and semi-structured interviews, in order to get more rich qualitative data that can bring clear research findings. It was also equipped by using Focus Group Discussion (FGD). By concentrating on aspects of gender equality, the FGD was to explore what respondents understood about gender concepts, how gender equality and justice can be implemented in schools and education, and what are their Islamic perspectives. All the gathered data was analyzed to follow the basis of the following steps: organize and prepare the data for analysis; reading through all the data; gaining a general sense of the information and reflecting on the overall meaning (Creswell, 2009). Furthermore, the main step was to conduct an analysis that focuses on the basis of the specific theoretical approach, which includes the connection between global education and gender equality and the use of the case study method. The researcher generated a description of the setting or people and identified themes, then searched for theme connections, such as how gender equality can be included in the school curriculum and more elaboration on equal opportunity and treatment in education. Finally, the researcher presented a research report leading to the development of the interpretation of the larger meaning of the data.

4 RESULTS AND DISCUSSION

4.1 Global education

Learning to understand other countries with different cultural backgrounds has reached 85% of agreement on the implementation of global education. It has provided positive impacts for shared perception on equality for all. It unites the concept, idea and its implementation with reference to equality that should be implemented at various levels so that existing discrimination and differences, especially those related to human rights, should be eliminated. This elimination should also include the protection of the human rights of women and men. This acknowledgment is also important for people who are predestined to live in a majority environment, where the people who are in the majority have differences in social, political,

economic, and technological ideas, as well as races, and statuses.

Respecting one another is not a novelty among Indonesians since they are respectful to each other, even to minorities. In recent times, when everything is borderless, people need to begin to widen their horizons and perspectives and have more knowledge; they need to be able to survive within the global community. An opinion can be shown of the importance of having a global education, "O mankind, indeed We have created you from male and female and made you peoples and tribes that you may know one another. Indeed, the noblest of you in the sight of Allah is the most righteous of you. Indeed, Allah is Knowing and Acquainted (QS Al Hujurat, 49:13)."

4.2 Gender division of labor

The inclusion of gender issues with the concepts of justice and gender equality, in which men and women are differentiated by their natural performance, leads to their social life having societal roles in which women's and men's roles generally can be interchangeable. Similar to the previous finding, the respondents also show their agreement to this with more than 83% support. Essentially there is no discrimination among men and women because they have equal status. The concept and the movement of gender equality has been changing and many people experience and observe the emergence of gender equality. The reality has shown that more and more women have public careers; this development also provides not only positive impacts on their families but also negative ones. The concept of justice has different forms, thus, several equality concepts that prevail among society need to be reconsidered. In other words, it needs to be advisedly filtered. All functions of gender roles should be selected first as the existence of women's nature; people in decision-making bodies should select the concept carefully.

4.3 Equal access and opportunity

Boys and girls have the same rights in attaining education in schools. Almost 100% of the respondents agreed about this equal access. Looking at life praxis recently shows that education cannot be differentiated between the sexes when dealing with attaining equal participation in the teaching and learning process. In this globalization era, when gender issues have spread around the world, it has answered the problems of equality and justice in education.

Respondents say that attaining an education is obligatory for every human being; boys and girls must have the same access to receiving an education. Whatever the conditions, the right to education between boys and girls should not only be allocated for primary and secondary schools but also until higher education. This is important for future life provision. If the right of girls or women is still questionable, it has led to the creation of violation against them. For a girl or a woman, the equal right to attain education means that there are no differences in intelligence, talents, and interests. The thing that makes them different is their different capacities due to their nature and choices. Boys and girls are given equal access; the same knowledge has to be given equally without coercion from parents, teachers, or anyone else. They actually have equal access and opportunity to decide what kind of future they wish to have.

4.4 Educational reform

Gender inclusion in the curriculum becomes significant to girls and women's empowerment to attain a better education. A female respondent said that gender equality should be included in the curriculum because there are so many types of discrimination, subordination, stereotypes and hegemony experienced by a certain sex, i.e. women. Human nature and religious doctrines are endorsed to justify that women's nature cannot be altered. This situation is actually a strategy for preserving hegemony, perpetuating domination and the limitation of human rights, and the impact often causes discrimination towards a certain sex by another sex. Through the curriculum, the inclusion of gender equality can be implemented in schools, considering that girls, in fact, have potential and interests that can be developed well.

4.5 Equal treatment

According to the respondents, girls should also be encouraged to participate actively in school activities. To produce better achievement, educators and parents should not provide discrimination towards girls and boys. If girls have the talent and quality then girls and boys should have equal roles. Girls can also have distinctive potential that can provide benefits for provision. Their potential is one of the human resources that cannot be eliminated because of the lack of opportunity given to girls. In competition, people cannot be differentiated against because of their gender. If a boy has the capacity to be a class leader, so does a girl. As proof, today there are more and more girls able to occupy leadership roles in various institutions and school activities. In many cases, women who have better capacities can surpass men because of their brilliant achievements.

4.6 Building religious framework

Religious teachings explain about gender equality in education. All the respondents agree that Islam gives equal status to women and men to pursue education. On the basis of the teaching, "seek knowledge and get education because of it is an obligation, an equal duty to men and women." There are many other Prophet Tradition and Qur'anic verses that strongly encourage Muslims to better achievement and continue a better life in the future.

Education in Islam can play a strategic role in educating women since women are believed to be a great asset of a nation in Islam. Women have the same right to learn and develop their potential and interests at schools, but it should be in accordance with the nature of women and *Sharia* – Islamic law. If women understand their role and obligations as mothers/housewives, men also have to understand their roles as fathers/heads of household and as husbands and being aware of their obligations as fathers and husbands. Women and men can play equal roles in educating and raising their children. To sum up, women and men must know and understand their rights and duties in accordance with what God has ordained in the Qur'an and *Sunnah* – Prophet Tradition, if this does not happen there would be imbalance and chaos in understanding the issues of gender.

5 CONCLUSIONS

Global education with the inclusion of gender issues has resulted in an Islamic fair moderate perspective, which is complementary with an amalgam of global and national/local perspectives. In many arguments, the respondents endorse much of the implementation of gender equality and justice implemented in education. On the other hand, smaller discussions still need to consider things contradictory to cultural religious values. Schooling is necessary for girls and women, their potential and interests should definitely be developed as well as those of boys and men; equal access and opportunity should be provided maximally to girls and women. However, women's nature and their distinctive roles as guardians of families cannot be forgotten. Equal treatment should also be prevailed to girls and boys. Girls should also be given various supports in classroom and school activities. Finally, implementation of gender equality in education should accommodate various angles: global and Islamic perspectives that would lead to the building of adaptable, flexible concepts to build an upcoming sustainable future generation.

REFERENCES

Corbin, D. (1989). What is global education? *Councilor*, 1–7.

Creswell, J.W. (2009). *Research design: Qualitative, quantitative, and mixed methods approaches* (3rd ed.). Los Angeles: Sage Publications, Inc.

Denzin, N.K. & Lincoln, Y.S. (2011). *The SAGE handbook of qualitative research*. California: SAGE Publications, Inc.

Haryanti-Kahfi, E. (2016). Women's representation in the parliament in the new Indonesia's democratization: SWOT analysis. In *Proceedings of Asian Studies 2016, 12–12 June, Toronto Canada* (pp. 72–78). Toronto: Unique Conference Canada.

ICCE. (2008). *Pendidikan kewarganegaraan (Civic Education): Demokrasi, hak asasi manusia, & masyarakat madani*. A.R. Ubaedillah, A., ed., Jakarta: Indonesian Center for Civic Education (ICCE) UIN Syarif Hidayatullah.

Idrus, A. (2009). *Manajemen pendidikan global: visi, aksi, dan adaptasi*. GP Press.

Mansson, A.B., & Farnsveden, U. (2012). Gender and skills development: A Review. Background paper for the EFA Global Monitoring Report 2012. *Prepared by Hifab International*, Sweden. New York: UNGEI.

Merryfield, M.M. & Harris, J. (1992). Getting started in global education: Essential literature, essential linkages for teacher educators. *School of Education Review, 4*, 56–66.

Steiner-Khamsi, G. (2004). Globalization in education: Real or imagined? *Global Politics of Educational Borrowing and Lending*, 1–6.

Subrahmanian, R. (2005). Gender equality in education: Definitions and measurements. *International Journal of Educational Development, 25*(4), 395–407.

Tye, K.A. (2003). Global education as a worldwide movement. *Phi Delta Kappa International, 85*(2), 165–168.

UNESCO. (2015). *Education for all 2000–2015: Achievements and challenges*. Paris: UNESCO.

Yin, R.K. (2006). Case study research-design and methods. *Clinical Research, 2*, 8–13.

Learning Teaching Methodologies and Assessment (TMA)

Ideas for 21st Century Education – Abdullah et al. (Eds)
© *2017 Taylor & Francis Group, London, ISBN 978-1-138-05343-4*

The relationship between metacognitive skills and students' achievement analyzed using problem based learning

B. Milama, N.A. Damayanti & D. Murniati
Universitas Islam Negeri Syarif Hidayatullah, Jakarta, Indonesia

ABSTRACT: The aims of this study were to reveal the effects of problem based learning on metacognitive skills and academic achievement, to analyze the metacognitive skills and academic achievement according to gender, and to discuss the correlation between metacognitive skills and academic achievement profiles. Students' metacognitive skills were measured with Metacognitive Activity Inventory (MCA-I) that was developed by Cooper and Urena (2008), while a chemistry achievement test was undertaken to measure the students' academic performance. A quantitative method was adapted to examine the results. Parametric t-test and Pearson Correlation Coefficient were used for data analysis. As result of analyzing the data, it can be concluded that problem based learning contributed positively both on students' metacognitive skills and students' academic achievement. Furthermore, significant correlation at the level of 0.01 exposed the relationship between metacognitive skills and students' achievement. According to gender, female students scored higher than males, both in MCA-I and the achievement test.

1 INTRODUCTION

High school students face an unpredictable future in which the skills and knowledge they require are constantly changing. This presents an interesting challenge to students and their teachers. Accordingly, a common theme throughout many statements about the purpose of education in the twenty-first century is the need to develop lifelong learners. Metacognitive skills play an important role in allowing students to become lifelong learners, because they affect how they can apply what they have learned to solve problems (Rompayom, 2010).

The basic definition of metacognition is a process of what somebody does to monitor their thoughts or their cognitive domain (Flavel, 1979). In fact, metacognitive includes how to reflect the known, how to analyze what is taught, how to solve what is analyzed, and how to apply what is learned (Tosun & Senocak, 2013). According to Livingston (2003), metacognitive is separated into three activities: planning, which has something to do with one's connection to their previous knowledge; monitoring, which involves one's self-checking at each stage of the task; and evaluating, which includes the learner's appraisal of the outcome and reflection on what new knowledge he or she has gained. These three aspects are commonly known as metacognitive skills.

Metacognitive skills are specifically defined as self-assessment, which is about the students' ability to access their own cognition, and self-management, which is related to the students'

ability to regulate their own cognitive development (River, 2001). Metacognitive skills play important roles in the learning process. They refer to students' cognition features while solving a problem. In the process of problem solving, students use many cognitive awareness behaviors, such as conceiving the problem, determining how to solve it, and analyzing and evaluating the process. Therefore, the process of problem solving improves students' awareness in the learning process (Gay & Howard, 2000). However, the more often that students use their metacognitive skills, the better academic performance they will achieve (Young & Fry, 2008).

Based on this fact, it is obviously required that educators recognize the importance of providing learning strategies that encourage students to develop their metacognitive skills. In the literature, it can be seen that student-centered learning has a relatively more significant effect on improving students' metacognitive skills compared to the traditional teaching methods (Ning & Downing, 2010). In a study carried out by Celiker (2015), Problem Based Learning (PBL) has a positive effect on the development of students' metacognitive skills.

PBL is the leading method among student-centered strategies and provides individuals with self-learning and lifelong learning skills, developing metacognitive skills and helping them to find alternative solutions to the problems they face or might face in daily life (Tosun & Senocak, 2013). The basis of PBL is rooted in Dewey's learning by doing and experiencing. The PBL is an active learning method, which enables the student to become aware of and determine his/her problem-solving ability and learning needs, to be able to make knowledge operative and to perform group works in the face of real life problems. It provides students with opportunities to consider how the facts that they acquire relate to the specific problem at hand. It also obligates them to ask what they need to know (Silver, 2004). Moreover, by having students learn through this experience of solving problems, they can learn both contents and thinking strategies. However, it will enable students to engage in complex tasks requiring higher-level thinking skills in order to support their academic achievements. Thus, metacognitive skills can be trained to the students to support their learning success.

According to the explanation above, this study clearly has three questions to answer: (1) Does problem based learning have any effect on metacognitive skills and academic performance? (2) Do metacognitive and academic performance posttest scores vary according to gender? (3) What is the correlation between metacognitive skills and academic performance?

2 RESEARCH METHODS

2.1 Research design

In this study, the research design employed is a post-test only with control group. In this design, the effect of problem based learning is tested on two different groups, which are separated into experimental and control groups. In the learning process, both of the groups had different treatment. PBL is used in the experimental group, while conventional learning is used in the control group (Creswell, 2012).

2.2 Study group

The research groups of the study comprised 76 eleventh-grade students of a state senior high school in Tangerang Selatan, in which both the experimental and control groups consisted of 38 students. Examining the gender distribution of the control group, 13 (34.21%) of them are males and 25 (65.79%) of them are females. While the gender distribution of the experimental group is not much different to the other one, it consists of 11 (28.95%) males and 27 (71.05%) females.

2.3 Assessment instrument

As part of the research to determine the metacognitive skills, the "Metacognitive Activity Inventory" test designed by Cooper et al. (2008) was used. The test has 27 questions, which includes

12 statements of planning skills, 13 statements of monitoring skills, and 2 statements of evaluation skills. On the other hand, the chemistry achievement test, which consists of 20 questions, was undertaken to examine students' cognitive and academic performance.

2.4 Data analysis

The data were analyzed via SPSS 21 packet program and Microsoft excel. To examine the normal distribution range in both the experimental and control classes, the lilliefors test was used and the L value of the test results were 0.11 and 0.09. Since values higher than 0.05 were interpreted as suitable, there was no deviance from normal distribution on this significance level; homogeneity test, promoting Fisher test, was used to examine students from both groups with no difference. The value found that the two sample groups were at the same level. According to the t-test result, the value was found to be 7.89, however, this result can be described that there are positive effects of PBL on dependent variables of the research. In other words, it can be concluded that null hypothesis is obviously rejected. The Pearson Correlation Coefficient was used to examine the relationship between metacognitive and academic achievement.

3 RESULTS AND DISCUSSION

The t-test results conducted on the significance of the distinction between the posttest scores of the experimental and control groups of metacognitive skills and the chemistry academic achievement test are given in Table 1.

This finding has shown that problem based learning has an important effect on increasing the metacognitive skills and academic performance.

Blakey and Spence (1990) determined metacognitive behavioral development strategies as defining what one knows and does not know, thinking by speaking, keeping a diary, planning, organizing and summarizing. Revising all the steps of this strategy it has shown that it is parallel in many ways to the problem based learning steps that are applied during

Table 1. T-test result of posttest scores for metacognitive skills and chemistry academic achievement test total score.

| Group | N | Mean | |
		Metacognitive skills	Academic achievement
Control	38	64.75	75.47
Experiment	38	81.86	83.00

this study. The implemented steps of this study were influenced by Arends (2012), which consisted of five steps. It began with making students familiar with the problem, organizing students to study, assisting independent and group investigation, developing artifacts (learning outcome), and analyzing and evaluating the problem-solving process (reflecting). However, through all these steps students are encouraged to develop their thinking skills process, implement previous information to solve the problems and to organize and remember new ways of creating knowledge by solving problems. This kind of opportunity becomes important because it will help them to enhance their metacognitive skills and improve their cognitive processes, which will further turn them into students that gained a better academic performance. Hence, the improvement in metacognitive skills will be followed by every skill in metacognitive itself, such as planning, monitoring, and evaluation. The average scores for planning, monitoring, and evaluation in the experimental group were higher than in the control group. However, this finding can be interpreted as being that problem based learning has a significant effect on the development of metacognitive skills in order to be aware of regulating their cognitive processes when solving a problem.

T-test results conducted on the significance of the difference between the experimental and control groups posttest average scores for the metacognitive skills test sub dimension, which consists of planning, monitoring, and evaluation skills, are shown in Table 2.

As a result of problem based learning there has been found to be a distinction between the average scores of the control and experimental groups. Planning, monitoring, and evaluation average scores in the experimental group are higher than in the control group. The implementation of PBL has really made a great impact on the metacognitive skills. It is based on the stage of PBL itself, where it can be a proper place for students to develop and be aware of their thinking skills. In stages 1 and 2 of PBL, students are encouraged to develop their planning skill by conducting plans to solve problems, recalling previous knowledge, and so on. In stages 3 and 4, students are intended to monitor what they have done to solve the problem, such as asking themselves if the way that they solved the problems are correct or not. And in the final stage, students will be able to evaluate and reflect on all

the processes they used in previous knowledge. However, evaluation skills will be developed.

Problem based learning has more effect on the development of metacognitive skills and academic performance of female students than of males. This result can be examined in Table 3. It shows that the difference between the posttest score mean of female students is higher than that of males.

The difference in the mean scores appearing in Table 3 between male and female students is influenced by the implementation of problem based learning. It can be said that PBL has more effect on improving the metacognitive skills and academic performance of female students than males. According to this fact, and supported by previous studies, students' achievement is believed as an output of their comfort level to engage in an interactive learning environment. PBL is one of the leading methods that is known for the interactive learning it involves. However, females more typically display more confidence in their achievement abilities than males. This confidence promoted their achievement and greater willingness to set further goals (Chisholm, 1999). However, because of the confidence inside female students, they can possibly reach positive achievements both in metacognitive skills and academic performance.

On the other hand, the Pearson Correlation Coefficient was used in this study. The value pointed at the 0.01 level, so it can be concluded that the correlation between metacognitive skills and students' achievement is significant. This finding can be seen in Table 4, which describes the Pearson Correlation Coefficient between posttest scores in the experimental class.

The results indicate that a positive relationship between metacognition and students' academic

Table 3. T-test results of posttest scores for metacognitive skills and academic achievement test according to gender.

Mean	Group			
	Experiment		Control	
	M	F	M	F
Metacognitive Skills	80.47	82.44	61.53	65.52
Academic Achievement	82.27	83.30	63.63	78.14

Table 4. Correlation between students' metacognitive skills and academic performance.

Variable	Pearson correlation	Sig.	N
Metacognitive Skills & Academic Performance	0.455	0.004	38

Table 2. T-test results of posttest scores for the metacognitive skills sub dimension.

Group	Planning	Monitoring	Evaluation
Control	67.38	64.12	76.32
Experimental	85.16	80.79	85.52

performance exists in the sample researched. It seems that students who do well in examinations are better on metacognition measures. Metacognition is no doubt important in one's learning process (Brown, 1978).

A student who has metacognition awareness understands himself as a learner, knows the best learning strategies that work for him, and knows when and why to use such strategies. More importantly, metacognitive students are very good at planning their learning, monitoring their progress and learning strategies and evaluating their learning strategies, learning output, self-strengths and self-weaknesses throughout the whole learning process. Metacognitive students have the ability to think about, understand and manage their own learning (Schraw & Dennison, 1994). Previous studies show that learners who score highly on measures of metacognition are more strategic, and generally outperform learners who score low on metacognitive measures (Garner & Alexander, 1989). Students who have metacognition also tend to be successful learners (Cooper et al. 2008). Hence, this finding has proved that students with a higher metacognitive skills score will have better academic performance during class and when facing problems.

4 CONCLUSIONS

Based on the explanations that have been discussed earlier, some conclusions were obtained that indicate that problem based learning has positive effects on metacognitive skills and students' achievement, females mean scores are higher than males, and metacognitive skills have a significant correlation with students' academic achievement.

However, because of its influence on students' academic performance, students' metacognitive skills are in high need of being developed by teachers as the responsible person in the learning processes. To encourage and be able to develop every dimension of metacognitive, teachers should apply learning strategies that could be able to direct students to empower their cognitive activity. It is necessary for teachers to implement student-centered learning in the future to develop students' thinking skills as well as metacognitive skills.

REFERENCES

Arends, R.I. (2012). *Learning to teach* (9th ed.). New York: The McGraw Hill Companies, Inc.

Blakey, E. & Spence, S. (1990). *Developing meta cognition*. New York: ERIC clearinghouse on information resources.

Brown, A.L. (1978). Knowing when, where, and how to remember: A problem of metacognition. In R. Glaser (Ed.), *Advance in instructional psychology* (pp. 77–165). Hillsdale, NJ: Erlbaum.

Celiker, H.D. (2015). Development of metacognitive skills: designing problem-based experiment with prospective science teachers in biology laboratory. *Academic Journal, 10*(11), 147–1495.

Chisholm, J.M. (1999). *The effects of metacognition, critical thinking, gender, and gender role identification on academic achievement in the middle years*. A thesis of Mount Saint Vincent University.

Cooper, M.M., Sandi-Urena, S. & Stevans, R. (2008). Reliable multi method assessment of metacognition use in chemistry problem solving. *Chemistry Education Research and Practice, 9*, 18–24.

Creswell, J.W. (2012). *Educational research: Planning, conducting and evaluating quantitative and qualitative research*. Boston: Pearson.

Davidson, J.E., Deuser, R. & Sternberg, R.J. (1995). The role of metacognition in problem solving. Retrieved from *http://springerlink.com*.

Flavell, J. (1979). Metacognition and cognition monitoring: A new area of cognitive-developmental inquiry. *American Psychologist Association, 34*(10), 906–911.

Garner, R. & Alexander, P. (1989). Metacognition: Answered and unanswered questions. *Educational Psychologist, 24*, 143–158.

Gay, G., & Howard, T.C. (2000). Multicultural teacher education for the 21st century. The Teacher Educator, 36(1), 1–16.

Livingston, J. (2003). *Metacognition: An overview*. Washington DC: ERIC.

Ning, H.K., & Downing, K. (2010). The impact of supplemental instruction on learning competence and academic performance. Studies in higher education, 35(8), 921–939.

River, W. (2001). Autonomy at all costs: An ethnography of metacognitive self-assessment and self-management among experienced language learners. *Modern Language, 85*(2), 279–290.

Rompayom, P., Tambunchong, C., Wongyounoi, S., & Dechsri, P. (2010). The Development of Metacognitive Inventory to Measure Students' Metacognitive Knowledge Related to Chemical Bonding Conceptions. *International Association for Educational Assessment, 1*, 1–7.

Schraw, G. & Dennison, R.S. (1994). Assessing metacognitive awareness. *Contemporary Educational Psychology, 19*, 460–475.

Silver, C.E. (2004). Problem-based learning: What and how do students learn? *Educational Psychology Review, 16*(3), 235–266.

Tosun, C. & Senocak, E. (2013). The effect of problem-based learning on metacognitive awareness and attitude toward chemistry of prospective teachers with different academic background. *Australian Journal of Teacher Education, 38*(3), 61–73.

Young, A. & Fry, J.D. (2008). Metacognitive awareness and academic achievement in college students. *Journal of the Scholarship of Teaching and Learning, 8*(2), 1–10.

Ideas for 21st Century Education – Abdullah et al. (Eds)
© *2017 Taylor & Francis Group, London, ISBN 978-1-138-05343-4*

Perception towards school physics learning model to improve students' critical thinking skills

N. Marpaung
Universitas Negeri Medan, Medan, Indonesia

L. Liliasari & A. Setiawan
Universitas Pendidikan Indonesia, Bandung, Indonesia

ABSTRACT: The objective of the study is to analyze responses from pre-service physic teachers and lecturers towards a School Physics learning model with arguments based on Multiple Representation (MR), its implementation, and how it may improve students' Critical Thinking Skills (CTS) after the learning. The study was conducted in one of the educational institutions of educational personnel in North Sumatra on pre-service physic teacher and lecturers. The study uses qualitative analysis with one group pretest-posttest. The instruments are questionnaire and MR-based CTS test. The results showed that the students' responses toward the learning model and implementation were generally positive. Lecturers' responses were also positive. The limited trial of learning model resulted in improvement of students' CTS at medium category. Thus the school physics learning model developed with the argument based on MR is regarded suitable to be implemented in larger scale.

1 INTRODUCTION

School physics is a compulsory expertise course to prepare teacher candidates in educational institution for educational personnel of Physics Education study program. The objective of School Physics (SP) is that after following the course, students are able to analyze and develop content of physics learning in high school or equivalent levels. The SP content has a very broad range since it covers the entire content material of physics in grade 10 until 12. To be able to analyze and develop the content in the school, students must be able to understand the concepts of the material and associate/integrate one concept to another. This ability to associate/integrate one concept to another in different situations is critical thinking skill (Tsapartis, & Zoller, 2003).

Critical thinking skills (CTS) in fact have long been a focus of concern in the world of education (Osborn, 2007, Tsapartis, & Zoller, 2003). One of the most important goals of all education sectors is CTS (Philip & Bond, 2004). Critical thinking enables a person to organize and articulate thoughts promptly and rationally, draw conclusions based on available evidence and be accurate (Nosich, 2012); to solve complex problems, find and select required information to solve these problems (Nickerson et al. 1985). Therefore, CTS enables a person to solve complex problems encountered in real life, be a responsible citizen and can live successfully in a democratic nature.

CTS cannot develop naturally, however it can be learned (Nickerson, 1985) through deliberate and planned learning. According to Brudvik, (2006) and Kuhn in Lubben et al. (2010) that argumentation can be used as a means of developing CTS. Through scientific arguments, students are involved in the process of proposing, supporting, evaluating, and criticizing using available and accurate evidence, and purify ideas about the learning and overcome misconceptions. Applying CTS in learning by teachers is very important to improve students' reasoning (Paul & Elder, 2007), so that students have deeper understanding of the concepts.

On the other hand, to explain the concepts of physics in a good learning, teachers must be able to represent the concepts in a variety of representations. This ability is a fundamental skill for physics teachers (McDermot, 1990). According to Etkina (2006), representing the concepts of learning in a variety of formats (multiple representations) is a scientific competence that must be mastered by good teachers.

Based on the characteristics of SP, these two skills are very strategic capabilities to be developed in SP, to prepare for teacher candidates that are qualified and competent. The reality on the field has not been as expected. In the preliminary study in one of the educational institutions of educational personnel in North Sumatera, it was found that CTS has not been an objective SP planning. Similarly, the MR concept has not been sought in learning. Mean while, both are very important skill to be given and developed for teacher candidates of physics.

Therefore, this study aims to solicit students' and lecturers' opinion on the learning models of SP that has been developed with MR-based arguments to improve the CTS of teacher candidates.

2 RESEARCH METHODS

This study was a limited trial on learning model with argumentation based on MR. The study used qualitative analysis with one group pretest-posttest design. The subjects of the trial were 12 students and 2 lecturers in one of LPTK in North Sumatera. The trial was conducted in 2 sessions (2×200 minutes). The instrument to acquire students' and lecturers' response was enclosed questionnaire (4 response options with a score of 1–4). Response score data of students and lecturers were analyzed based on the percentage distribution.

The instrument for CTS was a validated 30 item multiple-choice test with the context of dynamic and static electricity. The tests included CTS indicators: application of principles, determining the definition, identifying conclusions, interpreting, identifying relevant aspects, and giving a reason. The format of test representation included verbal, mathematical, pictorial, and graphic representations. CTS N-gain (<g>) was obtained interpreted according to Hake, (1998): high if (<g>) ≥ 0,70; medium if 0,70 > (<g>) ≥ 0,30; low if (<g>) < 0,30.

3 RESULTS AND DISCUSSION

3.1 Students' response towards learning model

Students' responses towards the SP learning model were collected using 16 statements and were grouped into five aspects. 1) Aspect of perception towards the implementation of the learning model; 2) motivational aspect regarding whether the lectures motivate students to prepare themselves better before the learning, to be active in learning, to do tasks optimally and acquire materials from other sources; 3) aspect of benefits regarding whether the learning help students to understand the concepts of learning, to have systematic thinking, and to solve problems; 4) aspect of opening the horizon and promote confidence regarding whether the learning offers a different view of the subject matter of learning, not hastily accept or reject

different opinions, give opinions based on available evidence, opens the horizon on the importance of independence in learning; 5) aspect of improving teacher competence regarding whether the learning equips/increases the competencies that a student/teacher candidate must have. Recapitulation of student responses cores is presented in Table 1. The mean score on the student response to five aspects of the learning model is 3,46 (very good).

Recapitulation level student consent to the implementation of the developed model is presented in Table 2. The average percentage of students towards learning model approval and its implementation is 86.89%, indicating approval of the students towards learning and implementation is very good or very positive.

Students' approval towards the learning model implementation increased from session I to session II (Table 2), indicated that the learning with arguments based on MR made students more comfortable in attending the learning. Generally stated that students' perceptions towards learning SP with MR-based argument was very positive, and gives a new perspective to the concept of learning.

Table 1. Recapitulation of students' response towards the learning model.

Aspect	Mean score			
	I	II	Mean	Criteria*
Perception towards lectures	3,50	3,67	3,54	Very good
Motivation for lectures	3,33	3,46	3,39	Very good
Benefits of lectures	3,31	3,52	3,42	Very good
Opening the horizon and confidence	3,33	3,54	3,43	Very good
Improving the competence of physics teacher	3,39	3,61	3, 50	Very good
Overall mean			3,46	Very good

*Criteria: 3,26–4,00 very good; 2,51–3,25 good; 1,76–2,50 poor; and 1,00–1,75 very poor (Ismet, 2013)

Table 2. Recapitulation of students' approval level towards the impelemation of learning model.

Aspect	Session		Mean (%)
	I (%)	II (%)	
Perception towards lectures	88,00	92,00	90,00
Motivation for lectures	84,33	86,46	85,62
Benefits of lectures	82,81	88,02	85,42
Opening the awareness and confidence	83,33	88,54	85,93
Improving the competence of physics teacher candidates	84,72	90,28	87,50
Overall mean			86,89

3.2 Lecturers' response

The score recapitulations of the 16 response items from the lecturers towards the learning model were grouped into three aspects presented in Table 3. Aspect of content is related to truth, depth and breadth of concepts; aspect of presentation is related to presentation structure, learning organization; and aspect of relevance to improve CTS based on MR.

In Table 3, the three observed aspects had a mean score of 3.64 > 3.26. It showed that the lecturers' response was very good or very positive towards learning model with MR-based arguments. In Table 4, the mean of lecturers' approval towards the implementation of learning was 91.15%. It can be stated that the lecturers strongly agree or give very positive response to the implementation of the learning model.

3.3 N-gain of critical thinking skills in physics

Students' pretest, posttest, and N-gain scores on CTS related to static-dynamic electricity content is presented in Figure 1. The mean of pre-test is 24.7% and post-test is 69.2%. Students' CTS improvement in 16.7% in the high category, 75.0% in the medium category, and 8.3% in the low category. The average of students' N-gain on CTS was 0.60. In general, the learning model trial increases students' CTS in the medium category. These results were consistent with Butchart, et al. (2009) that debate can enhance students' critical thinking skills. Student response also showed that

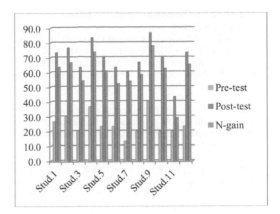

Figure 1. Chart of pre-test, post-test, and N-gain scores of students in percent (%).

MR-based learning helped them to solve physics problems. This was consistent with results from other studies (Adadan et al., 2009, Kohl & Finkelstein, 2006) that MR aids students in solving problems and improve the understanding of the concept.

4 CONCLUSIONS

The results of the limited trial on learning model of SP learning with MR-based arguments showed that students' perceptions of learning model was very positive. The perception of course lecturers on the learning model and its implementation was also very good, or very positive. The average N-gain of students' CTS was 0.60 indicated that the trial of SP learning model improved students' CTS in the medium category.

Therefore, it can be stated that SP learning model with MR-based arguments may be beneficial to be implemented.

Table 3. The score recapitulation of lecturers' reponse towards the learning model.

| Aspect | Mean score | | | |
	I	II	Mean	Criteria
Content	3.67	4.00	3,83	Very good
Presentation	3.38	3.50	3,44	Very good
Relevance	3.56	3.78	3,67	Very good
Overall mean			3,64	Very good

Tabel 4. Persentage rate of approval lecturer on the implementation of learning model.

| Aspect | Mean of questionnaire | | Mean |
	I (%)	II (%)	(%)
Content	91.67	100.00	95,83
Presentation	84.38	87.50	85,94
Relevance	88.89	94.44	91,67
Overall mean	88.31	93,98	91.15

REFERENCES

Adadan, E. Irving, K.E. & Trundle, K.C. 2009. Impacts of multi-representational instruction on high school students' conceptual understandings of the particulate nature of matter. *International Journal of Science Education*, 31(13): 1743–1775.

Brudvik, C. 2006. Assessing the impact of a structured argumentations board on the quality of students' argumentative writing skills. *Proceeding of the 14th international conference on computer in education*, 141–148, Amsterdam: IOA Press.

Butchart, S. Forster, D., Gold, I., Bigelow, J. Korb, K. & Serrenti, A. 2009. Improving critical thinking using web based argument mapping exercises with automated feedback. *Australasian Journal of Educational Technology*. 25(2): 268–291.

Costa, A.L. 1985. Goal for critical thinking curriculum. In Costa A. L (ed). Developing minds: *A Resourse Book for Teaching Thinking*. Alexandria: ASCD.

Etkina, E. 2006. Scientific abilities and their assesment. *Physic Review Special Topics-Physics education Research*. 2,020103.

Hake, R.R. 1998. Interactive-engagement versus traditional methods: A six-thousand-student survey of mecanics test data for introductory physics courses. *American Journal of Physics*. 66: 64–74.

Ismet, 2013. Pengembangan perkuliahan mekanika berbasis multiple representasi untuk meningkatkan kecerdasan spasial (spatial intelligence) mahasiswa calon guru. *Ph.D. Dissertation*. Indonesia University of Education: Bandung.

Kohl, P.B. & Finkelstein, N.D. 2006. Effects of representations on students solving physics problems: A fine – grained characterization. *Physical Review Special Topic-Physic Education Research* 2,010106.

Lubben, F. Sadeck, M. Scholtz, Z. & Braund, M. 2010. Gauging students' untutored ability in argumentation about experimental data: A South African case study.

International Journal of Science Education. 32(16): 2143–2166.

McDermot, L. C 1990. "Research and computer-based instruction: Opportunity for interaction". *American Journal of Physics*, 58: 452–462.

Nickerson, R.S. Perkin, D.N. & Smith, E.E. 1985. *The Teaching of Thinking*. New Jersey: Lawrence Erlbaum Associates, Inc.

Nosich, G.M. 2012. *Learning to think things through*. Fourth Ed. International. Boston: Pearson Education, Inc.

Osborne, J. Erduran, S. & Simon, S. 2004. Enhancing the quality of argumentation in science classrooms. *Journal of Research in Science Teaching*, 41(10): 994–1020.

Paul, R. & Elder, L. 2007. *A guide for educators to critical thinking competency standards*. [Online]. Available: http://www.criticalthinking.org [Marc 2015]

Philips, V. & Bond, C. 2004. Under graduates' experiences of critical tingking. *Higher Education Research & Development*. 23(3): 277–294.

Tsapartis, G. & Zoller, U. 2003. *Evaluation of higher vs. lower-order cognitive skills type examination in chemistry: implication for university in class assessment and examination. U. Chem. Ed.* 7, 50–57.

Ideas for 21st Century Education – Abdullah et al. (Eds)
© *2017 Taylor & Francis Group, London, ISBN 978-1-138-05343-4*

The implementation of 5E learning cycle model-based inquiry to improve students' learning achievements

A. Malik, Y. Dirgantara & A. Agung
Universitas Islam Negeri Sunan Gunung Djati Bandung, Indonesia

ABSTRACT: The research aimed at implementing the 5E learning cycle model-based inquiry to improve students' learning achievements on static fluid. This study used a pre-experimental research method promoting one group pre-test and post-test design. The samples were 30 students from a senior high school, using a purposive sampling technique. The data were collected by measuring the cognitive using essays, measuring for affective by using the Likert scale, and by using observation for psychometrics. The results showed that the means for teachers' and students' activities were 91% and 87%, respectively, which were classified as really good categories. The increase in the students' learning achievements was shown for cognitive with n-gain of 0.41 (medium category), for cognitive: 76% (good category), and psycho-motor: 71% (good category). The study used the statistical t-test: Paired Two Sample for Means with $t_{count} > t_{table}$ (15.529 > 2.042). Thus, the study concluded that the 5E learning cycle model-based inquiry can be used as an alternative choice to improve students' learning achievements on static fluid.

1 INTRODUCTION

Tiberghien (1998) discussed the role of teachers, particularly of physics, with a role of a physics scholar. This means that physics teachers need to be able to transfer their knowledge into an easy-to-understand learning materials so that students are not demotivated, but formulated to an easier material to learn.

The facts showed that physics is still considered to be a difficult subject for some students. According to Yesilyurt (2004), more explanations of the research have been provided to overcome the difficulties in science (physics, biology, and chemistry). Based on the research, Arief and Khoerul (2012) confirmed that the difficulties in understanding physics were due to factors of interests, abilities, motivation, facilities, supports, and activities.

The preliminary studies conducted in MAN (Madrasah Aliyah Negeri Islamic Senior High School) Subang concluded that almost all of the physics materials were difficult to understand for most students, in particular for those students who were less active and less enthusiastic in the classroom. However, the teacher had actually sought to apply various methods with the aim of the more active students understanding what was described by the teacher. In line with the declaration of the teachers, students also stated that they were less interested in the subject of physics. Further, students said that physics seemed abstract and they only memorized the concepts. This was consistent

with the student value, which was quite low. Direct observations made by the author while in the classroom also captured the same thing. Some students were not enthusiastic and not directly involved in the learning process. In addition, the teacher was still the center of learning and used the lecture method to deliver the subject, so that learning in the classroom became passive. What was said by physics teachers and students in MAN Subang was in compliance with what was happening in the classroom. The average value of students' achievements on the static fluid topic was the lowest compared to other topics. Thus, there was a need for a model of learning that could make students more involved in the classroom and obtain satisfactory academic results.

The 5E learning cycle model-based inquiry is a learning model centered on the student. The 5E learning cycle model is a realistic and constructivist method of leading students through a learning sequence where students were engaged in a topic, explored the topic, were given an explanation for their experiences, elaborated on what they have learned, and evaluated (Wilder & Shuttleworth, 2005).

Some research on the application of the 5E learning cycle model can improve the learning of concepts and the ability to apply concepts, students' achievement, students' understanding, learning outcomes, attitude of the students, students' academic achievement and the permanence of their knowledge (Piyayodilokchai et al., 2013;

Tuna & Kacar, 2013; Yadigaroglu & Demircioglu, 2012; Cepni, & Sahin, 2012; Uzunöz, 2011; Kaynar et. al., 2009).

2 LITERATURE REVIEW

2.1 *5E learning cycle model-based inquiry*

The 5E learning cycle model has five instructional stages, which are engage (intriguing), explore (exploration), explain (explain), elaborate (elaboration/develop), and evaluate (evaluation) (Bybee, 2004). Duran and Duran (2004) state that using a learning cycle approach in the classroom helps to facilitate inquiry practices because learning cycles focus on constructivist principles and emphasize the explanation and investigation of phenomena, the use of evidence to back up conclusions, and experimental design. The 5E instructional model (Bybee & Landes, 1990) can be used to design a science lesson, and is based upon cognitive psychology, constructivist learning theory, and best practices in science teaching.

2.2 *Learning achievement*

Learning achievement indicators for cognitive in this study were based on Krathwohl (2002), with domains of C_1 (remembering), C_2 (understanding), C_3 (applying), and C_4 (analyzing). Meanwhile, learning achievement indicators for affective and psychomotor were based on Bloom. Affective indicators were receiving, responding, valuing, organization, and characterization, whereas psychomotor indicators were imitation, manipulation, precision, articulation, and naturalization.

3 RESEARCH METHODS

The research method administered in this study was a pre-experimental method using one group pre-test post-test design. This research involved students in grade 11 who were in the science program (IPA) of MAN Subang.

The instrument used was an observation sheet to record the activities of teachers and students during the implementation of the 5E learning cycle model-based inquiry. For measuring the cognitive, affective, and psychomotor aspects, this study used an essay test, the Likert scale, and an observation sheet, respectively.

The data analysis of implementing the 5E learning cycle model-based inquiry was based on the percentage of teachers' and students' activities in every cycle rated by the observer. The increase rate of the students' learning achievements for cognitive was based on the n-gain (Cheng et al., 2004). The analysis of the affective was based on the interpretation of the percentage, and for psychomotor, it was based on the percentage of each psychomotor indicator shown by the students in the observation sheets.

4 RESULTS AND DISCUSSION

4.1 *Implementation of 5E learning cycle model-based inquiry*

The implementation of the 5E learning cycle model-based inquiry increased the activity of teachers and students in learning. After the first implementation, the average percentage of teachers' and students' activities reached 79% and 87%, respectively. These percentages increased even more after the second implementation of the 5E learning cycle model, after which the average percentage of teachers' and students' activities reached 96% and 92%, respectively, and classified as a really good category. After the third cycle, the average percentage of teachers' and students' activities reached 99% and 97%, respectively, and classified as a really good category.

It can be seen that there was an increase at every meeting, especially for students' activities. However, the increase was insufficient in the case of teachers' activity. This was due to the teachers being less able to manage the time. In general, the 5E learning cycle is considered as a "guided inquiry", where the teacher provides only the materials and problems to investigate and the students execute their own procedures to solve the problem under the guidance of the teachers (Martin-Hauser, 2002; Windschitl, 2003).

The interpretation of the teachers' and students' activities at every step of the model is summarized in Table 1.

Table 1. Mean implementation at every stage of the 5E learning cycle model-based inquiry.

5E learning cycle model-based inquiry	Mean activity		Interpretation	
	Teacher (%)	Student (%)	Teacher	Student
Engagement	92	88	Very Good	Very Good
Exploration	90	87	Very Good	Very Good
Explanation	92	83	Very Good	Very Good
Elaboration	91	90	Very Good	Very Good
Evaluation	93	89	Very Good	Good
Cover	96	88	Good	Good
Mean	91	87		

It can be concluded that the implementation of the 5E learning cycle model-based inquiry on the teacher's and students' activities was in the very good category. To optimize this model, time management became the most important point. Moreover, since the phase elaboration on this research category was still low, the teacher should be able to explain the theory associated with everyday life so that the concept is understood by the students.

4.2 *Improved learning achievements of students*

The students' learning achievements were improved by the implementation of the 5E learning cycle model-based inquiry, from a mean value of 12.33 (pretest) to 33.67 (posttest), with an overall n-gain value equal to 0.41 with moderate interpretation. Based on the analysis of the overall n-gain, the increase in the learning achievements of the students was better after learning with the 5E learning cycle model-based inquiry. This condition was caused by the 5E learning cycle model-based inquiry having associated significantly with the indicators of cognitive, affective, and psychomotor. From the total of 30 students who took part pretest and posttest, there were 3% of students with n-gain categorized as high, 77% of students with n-gain categorized as medium, and 20% of students with n-gain categorized as low.

With regards to the affective aspects, each indicator was improved after the implementation of the 5E learning cycle model-based inquiry at each meeting. The indicators of happy to follow learning process and writing important things obtained the highest mean value of 80%, with a good category, while the indicator of easily answered question became the lowest value compared to the others. It might be that the ability of students on mathematical operations was low However, when viewed as a whole, the mean of the affective learning outcomes of the students increased and reached 76% with both categories.

With regards to the psychomotor aspect, each indicator was improved after the implementation

of the 5E learning cycle model-based inquiry at each meeting. The indicator of capable in designing lab results into practical reports obtained the highest mean value, which was equal to 76% with both categories, while the indicator of can perform lab activities earned the lowest mean value, which was equal to 69% with both categories. This was due to, at the first meeting, the students' still having difficulty in doing practicals so that the psychomotor value had very little impact on the mean value. However, when viewed as a whole, the mean of the psychomotor aspect increased and reached 71% with both categories.

Based on the results of hypothesis test, t_{count} was 15.53 at the significance level of 0.05 and the t_{table} was 2.042. Thus, since the t_{count} is bigger than the t_{table}, Ho was rejected and Ha was accepted. This indicates that the implementation of the 5E learning cycle model-based inquiry can improve the learning achievement of students in a static fluid material at the subject case (class XI IPA 1 MAN Subang). Based on the data analysis, the cognitive domain of students in this study was placed in "medium" category, their affective domain was in "good" category and their psycho-motor domain was both in "medium" and "good".

The achievement of some indicators was still low, in particular in the aspects of analyzing (cognitive aspect), easily answer the question (affective indicator), and follow the practicum instructions (psychomotor indicator). To improve these three indicators, the teacher should be able to provide concrete experiences that correspond to daily life associated with an existing concept or theory that cognitive, affective, and psycho-motor should go hand in hand. The higher one aspect gets, the higher the others get, too. This was consistent with various studies that advocated the correct use of the 5E learning cycle instruction effectiveness for improving students' achievement in cell concept and scientific epistemological beliefs (Kaynar et al., 2009). In addition, Dorji et al., (2015) stated that a learning cycle approach can improve students' learning and awareness.

5 CONCLUSION

The implementation of the 5E learning cycle model-based inquiry was categorized as really good. The study confirmed that the implementation of the 5E learning cycle model-based inquiry improved the students' learning achievements in the cognitive domain, shown by normal gain index at a medium level; in the affective domain, shown with a good category; and in the psychomotor domain, shown by the means of observed rate for each session to the good category.

Table 2. Mean score and n-gain of cognitive aspects on pre-test and post-test.

| Aspect cognitive | Item number | Average score | | N-Gain | Interpretation |
		Pretest	Posttest		
C_1	1,5,7,11	3.63	9.63	0.49	Moderate
C_2	2,8,12,16	3.73	9.23	0.45	Moderate
C_3	3,10,13,15	2.43	7.50	0.37	Moderate
C_4	4,6,9,14	2.43	7.30	0.36	Moderate
Means		3.05	8.42	0.42	Moderate

REFERENCES

Arief, A. & Khoerul, M. (2012). Identifikasi kesulitan belajar fisika pada peserta didik RSBI studi kasus di RSMABI se-Kota Semarang. *Unnes Physics Education Journal*, *1*(2), 6–10.

Bybee, R. & Landes, N.M. (1990). Science for life and living: An elementary school science program from Biological Sciences Improvement Study (BSCS). *The American Biology Teacher*, *52*(2), 92–98.

Bybee, R.W. (2004). Scientific inquiry and science teaching. In L.B. Flick & N.G. Lederman. (Eds.), *Scientific inquiry and nature of science: Implications for teaching, learning, and teacher education* (pp. 1–14). Dordrecht, Netherlands: Springer.

Cepni, S. & Sahin, C. (2012). Effect of different teaching methods and techniques embedded in the 5E instructional model on students' learning about buoyancy force. *Eurasian Journal of Physics and Chemistry Education*, *4*(2), 97–127.

Cheng, K.K., Thacker, B.A., Cardenas, R.L. & Crouch, C. (2004). Using online homework system enhances students' learning of physics concepts in an introductory physics course. *American Journal of Physics*, *72*(11), 1447–1453.

Dorji, U., Panjaburee, P. & Srisawasdi, N. (2015). A learning cycle approach to developing educational computer game for improving students' learning and awareness in electric energy consumption and conservation. *Educational Technology & Society*, *18* (1), 91–105.

Duran, L.B. & Duran, E. (2004). The 5E instructional model: A learning cycle approach for inquiry-based science teaching. *The Science Education Review*, *3*(2), 49–58.

Kaynar, D., Tekkaya, C. & Cakiroğlu, J. (2009). Effectiveness of 5E learning cycle instruction on students' achievement in cell concept and scientific epistemological beliefs. *Hacettepe University Journal of Education*, *37*, 96–105.

Krathwohl, D.R. (2002). A revision of Bloom's taxonomy: An overview. *Theory into practice*, *41*(4), 212–218.

Martin-Hauser, L. (2002). Defining inquiry. *The Science Teacher*, *69*(2), 34–37.

Piyayodilokchai, H., Panjaburee, P., Laosinchai, P., Ketpichainarong, W. & Ruenwongsa, P. (2013). A 5E learning cycle approach–based, multimedia-supplemented instructional unit for structured query language. *Educational Technology & Society*, *16*(4), 146–159.

Tiberghien, A. (1998). *Connecting research in physics education with teacher education: making the results of researching physics education* Available to Teacher Educators. International Commission on Physics Education. Pan-American Copyright Conventions.

Tuna, A. & Kacar, A. (2013). The effect of 5E learning cycle model in teaching trigonometry on students' academic achievement and the permanence of their knowledge. *International Journal on New Trends in Education and Their Implications*, *4*(1), 73–87.

Uzunöz, A. (2011). The effects of the activities of current textbook and 5E model on the attitude of the students: Sample of the global effects of natural resources unit. *Educational Research and Reviews*, *6*(13), 778–785.

Wilder, M. & Shuttleworth, P. (2005). Cell inquiry: A 5E learning cycle lesson. *Science Activities: Classroom Projects and Curriculum Ideas*, *41*(4), 37–43.

Windschitl, M. (2003). Inquiry projects in science teacher education: What can investigative experiences reveal about teacher thinking and eventual classroom practice? *Science Education*, *87* (1), 112–143.

Yadigaroglu, M. & Demircioglu, G. (2012). The effect of activities based on 5E model on grade 10 students' understanding of the gas concept. *Procedia-Social and Behavioral Sciences*, *47*, 634–637.

Yesilyurt, M. (2004). Student teacher's attitudes about basic physics laboratory. *The Turkish Online Journal of Educational Technology*, *3*(4), 49–57.

Ideas for 21st Century Education – Abdullah et al. (Eds)
© *2017 Taylor & Francis Group, London, ISBN 978-1-138-05343-4*

Development and validation of creative thinking skills test in the project of laboratory apparatus modification

C. Diawati
Universitas Lampung, Lampung, Indonesia

L. Liliasari & A. Setiabudi
Universitas Pendidikan Indonesia, Bandung, Indonesia

B. Buchari
Institut Teknologi Bandung, Bandung, Indonesia

ABSTRACT: This study reports on the development and validation of test assessments designed to measure students' creative thinking skills in a Visible Spectrophotometer (VS) and Atomic Absorption Spectrophotometer (AAS) apparatus modification project. The test form was an essay that was arranged based on Torrance's Framework of creative thinking skill indicators. A preliminary draft of the test instrument was validated by three experts in the field of chemical education. The instrument was administered to third-year undergraduate students in the chemistry education study program in Province Lampung. The results showed that the internal consistency of the instrument was fairly good, which indicates that the instrument can be used to measure the acquisition of students' creative thinking skills. This result is highly recommended for future empirical research in developing a creative thinking assessment, especially on the topic of a particular subject matter.

1 INTRODUCTION

Creative thinking skills are a major goal of science education, because school leavers and graduates who think creatively will contribute positively to the personal, social, technological, and economic world that they will inhabit as adults in the 21st century (Wellestrand & Tjeldvoll, 2003; Diawati, 2016; DeHaan, 2009; Trnova, 2014). However, sufficient emphasis has not been given to the measurement of creative thinking skills, in particular in science domains such as chemistry. Creativity is very difficult to define and measure (Runco, 1993). To estimate creative thinking potential, divergent thinking tests are often used. The term and the measures of divergent thinking were invented by Guilford (Hong & Milgram, 2010; Hong et al., 2013). Currently, there are two types of creative thinking measurements that have been developed by researchers. The first is the measurement of creative thinking in the general domain; the second is a specific domain measurement.

In the beginning, the instrument that was most often used to measure creative thinking over the years was an instrument measuring creative thinking ability, that is, the divergent thinking test, which always includes the measurement of ideational fluency (for example, Torrance, 1974, 1999). In the measurment of general domain creative thinking, the problem posed to the respondents to be completed is very different from the type of problems that people encounter in everyday life. For example, when responding to the ideational fluency measure, respondents were asked to name all the ways in which to use a newspaper. Although they have long been considered reliable measures of creative thinking ability (Runco, 1990), divergent thinking tests have been criticized for their low correlation with real-world performance, because the generalization from general domain creative thinking scores based upon ideational fluency measures to creative thinking in practical life situations is not entirely justified (Hong & Milgram, 2010; Okudo 1991).

Measurements designed to assess specific domain creative thinking have been developed (for example, Okudo 1991). *Ariel Real Life Problem Solving*, for example, provides respondents with the opportunity to utilize their creative thinking abilities in a variety of specific domain real life situations (Hong & Milgram, 2010). However, the problems that are provided to the respondents on the specific domain creative thinking measurement are general real life problems to be solved, not those for specific science-oriented domains, such as chemistry. For example, a test item is as follows, "Your friend Teddy sits next to you in class. Teddy likes to talk to you a lot and often bothers you while you are doing your work. Sometimes the teacher scolds you for talking, and many times you don't finish your work because he is bothering you. What are you going to do? Remember to give as many answers as you can" (Okudo, 1991).

In relation to a visible spectrophotometer (VS) and atomic absorption spectrophotometer (AAS) modification practice that is being developed, we need a specific domain creative thinking skills test. Therefore, this article aimed to develop and validate the specific domain creative thinking skills test instruments of VS and AAS modification in project-based learning.

2 LITERATURE REVIEW

2.1 Creative thinking skills

Creative thinking is a skill to develop, to find, or to create new constructive combinations based on the data, information, or elements that already exist, with a different perspective that appears as a manifestation of their perceived problems, so as to produce a useful solution (Al-Suleiman, 2009; Lawson, 1979).

Although there are different concerns, creativity is considered as an essential life skill, which must be fostered through education (Shen & Lai, 2014). The previous research has shown that creative thinking is influenced by various circumstances, including whether collaboration works and the extent to which individuals are motivated to solve problems (Brophy, 2006; Zhou et al., 2010; Doppelt, 2009; Cheng, 2010). Most research suggests that there are differences in creativity and that students' gain achievement when the classroom environment is manipulated (Baker 2001; Sternberg, 2003).

2.2 Assessment of creative thinking skills

Many researchers have developed test assessments to measure creative thinking skills. The term and the measures of divergent thinking were originated by Guilford (Hong & Milgram, 2010; Hong et al., 2013) to measure the potential of creative thinking. The Torrance Creative Thinking Test (TCTT) has become a standard for assessing the ability of creative thinking. TCTT often requires considerable testing time, as it covers the figural and verbal forms. Torrance, Wu, and Ando created the Torrance Form Demonstration Test (D-TCTT), which requires less testing time, in 1980. The success of the short form when working with adults led to the current developments of the Abbreviated Torrance Test Adults (ATTA) (Shen & Lai, 2014). The development of measurements was designed to assess specific domain creative thinking (for example, Okuda 1991), which provides respondents with the opportunity to utilize their creative thinking abilities in a variety of specific domain real life situations (Hong & Milgram, 2010). Doppelt (2009) applies four layers as an assessment criteria of the Creative Thinking Scale (CTS) to assess the crea-

tive work of high school students. CTS includes, layer 1: awareness; layer 2: observation; layer 3: strategy; and layer 4: reflection.

3 RESEARCH METHODS

The method used in this research was descriptive. This article attempted to describe systematic, factual, and accurate information on the development and validation of the creative thinking skills test instruments in the chemistry domain during the VS and AAS modification project.

3.1 The development of test instruments

Tests were aimed to assess specific creative thinking skills in the VS and AAS modification project. Test items have been constructed so that such domain emerged. Tests were developed in an essay form, and based on creative thinking skills indicators using Torrance's Framework, that is, fluency, flexibility, originality, and elaboration (Al-Suleiman, 2009). Once the test was developed, then rubrics and scoring were created. The rubric was developed with four levels of gradation. The highest gradation level was scored 4, and the lowest level was scored 1.

3.2 Expert judgment

The test instrument was validated by three experts in the field of chemical education. Assessment aspects for the tests construction validation included: (1) the sentence is easy to understand, (2) does not waste words; (3) accordance with the concept scope; (4) the truth of concept, and (5) accordance with creative thinking skills indicators. Scores obtained from experts on the five aspects were analyzed using Intraclass Correlation (ICC) Two-Way Mix ANOVA that emphasizes the similarities of the assessment between raters. In addition, to examine the inter-rater reliability, the ICC correlation result was used to determine the validity of an assessment instrument based on the consistency of assessment among experts (ICC consistency).

3.3 Testing the instrument

A validated test instrument was administered to third-year students in the Program Studi Pendidikan Kimia in Province Lampung (N = 35). The test results were analyzed using Product Moment Pearson correlation: it is a type of correlation test to determine the empirical validity. Reliability was also analyzed using the Cronbach-Alpha formula.

4 RESULTS AND DISCUSSION

The developed test instrument consisted of 21 items. Examples of indicators and creative thinking skills test items are shown in Table 1.

Table 1. Examples of indicators and creative thinking skills test items.

The creative thinking skills indicators	Test items
Propose alternative ideas of VS/AAS component replacement and its modification (originality)	What are the VS components that can be modified? What is the reason? What is an alternative to its modification?
Describing work process flow of modified VS/AAS using images/charts in detail (flexibility and elaboration)	Describe work process flow of modified VS/AAS, using picture/charts in detail
Propose the idea of how to prove modified VS/AAS (fluency, originality)	How do you prove that the VS/AAS modified components work as expected?

Table 2. The results of the analysis of intraclass consistency between expert judgments.

Aspects		Intraclass correlation[b]	95% Confidence interval	
			Lower bound	Lower bound
The sentence is easy to understand	Single Measures	.110[a]	.001	.001
	Average Measure	.756[c]	.028	.028
Doesn't waste words	Single Measures	.071[a]	−.011	.821
	Average Measure	.658[c]	−.364	.991
Accordance to the concept scope	Single Measures	.075[a]	−.010	.825
	Average Measure	.669[c]	−.322	.992
The truth of concept	Single Measures	.234[a]	.045	.932
	Average Measure	.884[c]	.539	.997
Accordance to creative thinking skills indicators	Single Measures	.139[a]	.101	.888
	Average Measure	.801[c]	.208	.995

[a]The estimator is the same, whether the interaction effect is present or not.
[b]Type c intraclass correlation coefficients using a consistency definition-the between-measure variance is excluded from the denominator variance.
[c]This estimate is computed, assuming the interaction effect is absent, because it is not estimable otherwise.

The summary of the output analysis of the intraclass correlation coefficient (ICC) consistency between experts using SPSS 20.0 is shown in Table 2.

Based on the statistical analysis of the Two-Way Mixed ANOVA, the ICC consistency between experts on all the aspects is as follows: (1) the sentence is easy to understand is adequate (ICC = 0.756), (2) does not waste words is adequate (ICC = 0.658); (3) accordance with the scope concept is adequate (ICC = 0.669); (4) the truth of concept is good (ICC = 0.884), and (5) accordance with creative thinking skills indicators is good (ICC = 0.801). These analyses were conducted at the significance level of 95%. It indicates that the test instrument is valid and can be used to assess student creative thinking skills.

Of the results of the scores analysis obtained by students from the 24 test items of the preliminary version, 21 test items are valid and 3 test items are invalid. The three invalid test items were not used further. Therefore, the number of revised version test items is 21. Analysis using the Cronbach-Alpha formula at the significant level of 95% indicates that the test is a good internal consistency (α = 0899). These results indicated that the test instrument was valid and reliable, therefore, it can be used to assess student creative thinking skills.

The VS and AAS modification project requires students to apply knowledge and to train high-order thinking skills, such as creative thinking skills. Students formulated the problems, sought the replacement apparatus alternative, designed and constructed apparatus, tested and evaluated. These creative thinking activities are very specific, therefore they cannot be assessed using a creative thinking skills instrument of the general domain. Considering the importance of developing students' ability to think creatively in the specific domain of science, especially chemistry, researchers and practitioners should create test instruments

that are valid and reliable to evaluate the effectiveness of various learning efforts.

Both qualitative and quantitative data analysis was conducted and showed that the initial stage of this test instrument can be used to properly assess the students' creative thinking skills. Content experts were involved in reviewing the items during the item development stage, which provided evidence that the test items were clear and elicited the targeted specific domain creative thinking of VS and AAS modification.

Quantitative evidence showed that, in the stage of theoretical validation, the test instrument produced sufficient inter-rater consistency between the experts on aspects of ease of sentence to understand, do not waste words, and appropriateness with the concept scope; and good inter-rater consistency on aspects of the truth of concept and accordance to creative thinking skills indicators. This evidence indicates that the instrument is valid to assess student creative thinking skills.

In the empirical validation stage, quantitative analysis of the students' test scores showed that, of the 24 test items to have been developed, 21 test item were valid, so therefore the number of items on the revised version of the test instrument was 21. The analysis also showed that the test instrument for creative thinking skills produced a good reliability coefficient ($\alpha = 0.899$), which means that the test instrument has good reliability to assess students creative thinking skills.

The development of the creative thinking skills tests described in this study were largely in line with the recommended guidelines for the preparation of the test and other performance tests (for example, Adams & Wieman, 2011; Aydın & Ubuz, 2014; Benjamin et al., 2015; Tiruneh et al., 2016). Although the procedure is based on the guidelines established from previous studies, this study has proposed a framework for assessing creative thinking skills that can be used to measure creative thinking skills in the specific domain of chemistry. It is hoped that creative thinking skills tests can be used as a good basis for future empirical research as well as for teaching purposes assessment focusing on the integration of creative thinking skills in a particular subject matter instruction. The test can be used to answer the research questions involving the assessment of the effectiveness of learning on the acquisition of the specific creative thinking skills of chemistry.

5 CONCLUSION

The instrument for measuring creative thinking skills during the VS and AAS modification project has been developed. The test consists of 21 items. Intraclass correlation coefficient (ICC) consistency between experts was adequate and good. This indicates that the test instrument was valid. Analysis for the Cronbach-Alpha formula indicates that the test had a good internal consistency ($\alpha = 0899$). These results indicate that the developed test instrument is valid and reliable, therefore it can be used to assess students' creative thinking skills. This result is highly recommended for future empirical research in developing a creative thinking assessment, especially on the topic of a particular subject matter.

REFERENCES

Adams, W.K. & Wieman, C.E. (2011). Development and validation of instruments to measure learning of expert like thinking'. *International Journal of Science Education*, *33*(9), 1289–1312.

Al-Suleiman, N. (2009). Cross cultural studies and creative thinking abilities. *Journal of Educational and Psychology Science*, *1*(1), 42–92.

Aydın, U. & Ubuz, B. (2014). The thinking-about-derivative test for undergraduate students: Development and validation. *International Journal of Science and Mathematics Education*, *13*(6), 1279–1303.

Baker, M. (2001). Relationships between critical and creative thinking. *Journal of Southern Agricultural Education Research*, *51*(1), 173–188.

Benjamin, T.E., Marks, B., Demetrikopoulos, M.K., Rose, J., Pollard, E., Thomas, A. & Muldrow, L.L. (2015). Development and Validation of Scientific Literacy Scale for College Preparedness in STEM with Freshmen from Diverse Institutions.

Brophy, D.R. (2006). A comparison of individual and group efforts to creatively solve contrasting types of problems. *Creativity Research Journal*, *18*(3), 293–315.

Cheng, V.M.Y. (2010). Teaching creative thinking in regular science lessons Potentials and obstacles of three different approaches in an Asian context. *Asia-Pacific Forum on Science Learning and Teaching*, *11*(1).

DeHaan, R.L. (2009). Teaching creativity and inventive problem solving in science. *CBE Life Sci. Ed*, *8*, 172–181.

Diawati, C. (2016). Students' conceptions and problem-solving ability on topic chemical thermodynamics. In T. Hidayat et al. (Eds.), *Mathematics, science, and computer science education* (pp. 040002). *AIP Conference Proceedings* (Vol. 1708, No. 1, p. 040002). AIP Publishing.

Domain generality and specificity. *Creativity Research Journal*, *22*(3), 272–287.

Doppelt, Y. (2009). Assessing creative thinking in design-based learning. *International Journal of Technology and Design Education*, *19*, 55–65.

Hong, E. & Milgram, R.M. (2010). Creative thinking ability.

Hong, E. et al. (2013). Domain-general and domain-specific creative thinking test: Effects of gender and item content on test performance. *The Journal of Creative Behavior*, *47*(2), 89–105.

Lawson, A.E. (1979). *AETS. Yearbook the psychology of teaching for thinking and creativity. Clearing house for science, mathematics, and environmental education*. The Ohio State University College of Education.

Okudo, S.M. (1991). Creativity and finding and solving of real-world problems. *Journal of Psychoeducational Assessment*, *9*, 45–53.

Runco, M.A. (1990). The divergent thinking of young children: Implications of the research. *Gifted Child Today*, *13*, 37–39.

Runco, M.A. (1993). Divergent thinking, creativity, and giftedness. *Gifted Child Quarterly*, *37*, 16–22.

Shen, T. & Lai, J. (2014). Exploring the relationship between creative test of ATTA and the thinking of creative works. *Procedia Social and Behavioral Sciences*, *112*, 557–566.

Sternberg, R.J. (2003). Creative thinking in the classroom. *Scandinavian Journal of Educational Research*, *47*(3), 325–338.

Tiruneh, D.T. et al. (2016). Measuring critical thinking in physics: Development and validation of a critical thinking testing electricity and magnetism. *International Journal of Science and Mathematics Education, Advanced online publication* doi:10.1007/s10763-016-9723-0.

Torrance, E.P. (1974). *The torrance tests of creative thinking*. Bensenville, IL: Scholastic Test Services.

Torrance, E.P. (1999). *Torrance test of creative thinking: Norms and technical manual*. Beaconville, IL: Scholastic Testing Services.

Trnova, E. (2014). IBSE and creativity development. *Science Education International*, *25*(1), 8–18.

Welle-strand, A & Tjeldvoll, A. (2003). Creativity, curricula and paradigms. *Scandinavian Journal of Educational Research*, *47*(3), 359–372.

Zhou, C. et al. (2010). Creativity development for engineering students: Cases of problem and project based learning. *Joint International IGIP-SEFI Annual Conference 2010*. Trnava, Slovakia.

Ideas for 21st Century Education – Abdullah et al. (Eds)
© *2017 Taylor & Francis Group, London, ISBN 978-1-138-05343-4*

The implementation of guided inquiry learning to improve students' understanding on kinetic theory of gases

D. Nanto & R.D. Iradat
Syarif Hidayatullah State Islamic University, Jakarta, Indonesia

Y.A. Bolkiah
Senior High School, Madrasah Aliyah Pembangunan Syarif Hidayatullah State Islamic University, Jakarta, Indonesia

ABSTRACT: The kinetic theory of gases topic consists of many abstract concepts, which are not only hard for students to understand, but it also makes it hard for physics teachers to deliver the content knowledge. It makes teaching and learning harder for both teachers and students when the class is crowded. To solve the problems, we propose the guided inquiry learning method, which is observed by means of class action research in our school. We found that in the first and the second cycles of observation, students' activity in learning increased. There was an increase from 80% to 100% in activity learning. This study showed an improvement in the learning process of abstract lessons in physics, as students become more active and self-motivated. As a result, there is a significant improvement in student achievement in this particular topic.

1 INTRODUCTION

The purpose of the learning process in the classroom is a fundamental challenge for students in learning physics. The process becomes harder if the content has abstract characteristics. This is because students are unable to imagine the material in their daily context. Moreover, another factor that may cause difficulties in the learning process is controlling the classroom atmosphere, especially when it is not conducive for learning.

Shah and Inamullah (Khan & Mohammad, 2012) found that overcrowded classes could have a direct impact on students' learning. They not only affected students' performance, but also the teachers had to face different problems, such as discipline, behavioral problems, poor health and the poor performance of students, which put stress on teachers and increased the dropout rate of students.

The kinetic theory of gases is a matter of physics that describes the movement of gas particles in the microscopic and macroscopic studies. This material uses the ideal gas that meets the requirement of the laws of gas. Furthermore, these materials introduce the state formulated by the equation,

$$PV = nRT, \tag{1}$$

where P = Pressure; V = volume; and = the number of moles of gas; R = the universal gas constant; and T = the absolute temperature. A theoretical concept of kinetic energy in gas is also learned that satisfies the equation,

$$E = \frac{3}{2}nRT. \tag{2}$$

Equations (1) and (2) are samples, which students found difficult to understand. Pathare and Pradhan (2010) explained an overview of students' misconceptions about the basic idea of students' understanding of heat, temperature, pressure, heat transfer mechanism, particle properties of matter and the kinetic theory.

According to Wenning (2010), the pedagogy of an inquiry lesson is one in which learning activity is based upon the teacher slowly relinquishing charge of the activity by providing guiding, indeed leading, questions. It was expressed by Nwosu and Nzewi (Nworgu & Otum, 2013) that guided inquiry, which is employed in this study, has been defined as a set of activities characterized by a problem-solving approach in which students are placed in a problematic situation and surrounded by a lot of appropriate and suitable materials with which to explore their environment and solve problems. Guided inquiry creates an environment that motivates students to learn by providing opportunities for them to construct their own meaning and develop deep understanding, as Carol and Kuhlthau stated (Nivalainen, 2013). They offered that, through guided inquiry, students gain the ability to use tools and resources for learning as they are learning the contents of the curriculum. By guided inquiry, the activities concentrate on what students are thinking, feeling and doing, as they are learning throughout the inquiry process. Finally, the end product becomes a natural way of sharing their learning with the rest of the students in their learning community.

Wenning and Khan (2011) suggest that the inquiry learning stages include observation, manipulation, generalization, verification, and applications. We found an interesting class that was pretty complicated in the process of learning physics. Our classroom observations found some serious problems with the majority of the students, such as being unable to focus on learning, performing other activities that do not support learning, having difficulty in absorbing the learning material, giving negative responses when the teacher asks them to do exercises, showing low initiative, and having low average achievement.

Based on those problems, we introduced a guided inquiry learning method. We found that an interesting challenge of physical phenomena plays an important role in improving students understanding, motivation and achievement results. We describe our work in detail below.

2 RESEARCH METHODS

This study uses Classroom Action Research (CAR), a research design that refers to the Kurt Lewin model with a cycle that includes planning, acting, observing, and reflecting. The research was conducted in two cycles, with every cycle consisting of two meetings per week. The CAR was conducted at the Class XI Science 1 of Madrasah Aliyah Pembangunan. It is a private laboratory school under the Faculty of Education Sciences of Syarif Hidayatullah State Islamic University, Jakarta.

3 RESULTS AND DISCUSSION

3.1 Stages cycle I

3.1.1 Planning
At this stage we made data collection instruments, such as observation sheets, achievement tests, and student worksheets. We made a lesson plan in accordance with the method of the guided inquiry learning process on the material of the kinetic theory of gases.

3.1.2 Acting
The implementation stage of the guided inquiry learning method is presented in Table 1.

Table 1. Stages of learning by using guided inquiry method of cycle I.

No.	Phase	Learning activities
1	Finding the problem	The teacher gives a problem by means of doing a demonstration related to the concept of an ideal gas.
2	Introduce hypothesis	Each group proposed a hypothesis based on their observation.
3	Designing experiments	Students design experiments and consult the teacher to make sure whether the experiment is correct or not and take data based on students' worksheets.
4	Analyzing the data	Discussion with several groups of students guided by teachers to analyze the relationship between the concepts of an ideal gas with the applicable laws.
5	Make conclusion	Several student representatives to draw conclusions from the results of the experiment and correlate with the concept of ideal gas and ideal gas laws.

Based on the first cycle of learning activities, there is a change in the attitudes of students from passive to active with the application of the guided inquiry learning method. It can be seen from the activities of the students during the formulation of the problem and analyzing the data. At that stage, the other students have not been active in doing activities such as formulating hypotheses, experimenting with a discussion with the group, and drawing conclusions.

Students received some questions from the physics teacher before they did a demonstration, "Does a balloon could fit into a small container?" and "Is there a relationship between volume and pressure when you press the balloon?" Then, at the stage of analyzing the data, the students had a discussion led by the teacher. One student replied that the temperature is proportional to the pressure and he explained the lower the pressure of the balloon gets, the cooler the temperature is. The second stage had an important impact on the proceedings of the students, so that learning became more active.

3.1.3 Observing
We observed that the guided inquiry learning process occurs when students join the new concepts of temperature, pressure and volume. Some students argued that the temperature is directly proportional to the volume. This idea came from their experiments. They transfer heat on to balloons that consist of water, which made the balloon

expand. This phenomenon was observed when the temperature increased; the volume would increase while the pressure was inversely proportional to the volume.

The data observation activity of students in the study, performed by the observer, can be seen in the following table.

In the first cycle, the liveliness of Class XI Science 1 was included in the category of active in the learning process, using the guided inquiry method, with the achievement of 80% of the indicators set.

The learning atmosphere observation can be seen in the following table.

In the first cycle, the learning atmosphere of the students of the Class XI Science 1 category

Table 2. Indicators of student activity observation first cycle.

| Indicator | Implementation | |
	Yes	No
Have a problem-solving discussion.	√	
Perform various experiments to solve the problem with the tools and materials available.	√	
Initiative to understand the subject matter deemed not understood by asking the teacher or friend group.	√	
Linking the physics of matter that is relevant to the experiments performed.	√	
Enthusiasm during the experiment and group presentations.		√

Table 3. Indicators of observational learning environment first cycle.

| Indicator | Implementation | |
	Yes	No
The majority of students are busy with group activities, such as looking for ideas, designing experiments, and concludes.	√	
Class conditions to support the learning process: the seating position is patterned, purposeful activity, and interaction in two directions.	√	
Meaningful dialog (student—student and student—teacher).	√	
Availability completeness learning support.		√
Seen in the laboratory and bustle of hectic discussions in the classroom.		√

was conducive to the learning process, with the students' achievement of 60% of the indicators set.

3.1.4 Reflecting
Based on the results of the first cycle of reflection, it is necessary to have a remedial action on the first cycle as shown in Table 4 below.

3.2 Stages cycle II

3.2.1 Planning
Planning the second cycle is a reflection of the first cycle. At this stage the physics teacher motivated students to engage in activities that make them more active.

3.2.2 Acting
The implementation stage of the guided inquiry learning method is presented in Table 5.

Based on the second cycle of learning activities, there was a change in the attitude of the students from the first cycle to being more active and there was conducive implementation of the guided inquiry learning methods. It can be seen from the activities of the students when they formulate the problem, propose hypotheses, conduct experiments, and analyze data.

Table 4. Reflection action on the first cycle.

No.	Results reflection cycle I	Corrective action
1	There are some students who are not cooperative with activities other than ordered.	Every students' activity will be monitored and evaluated so that they can perform better in the next meetings.
2	Student worksheets have some instructions that were not understood by the students. Therefore many students could not perform laboratory work well.	Students will get regular briefings prior to every lesson in order that they understand each activity. Any worksheet will be clearly explained and clarified.
3	There still exist some students who did not pay attention to instructions from the teacher.	Physics teacher gave clarifying information in each group in order to avoid misconceptions to do instructions given.
4	The student position in class of the group also determines the state of student learning in an orderly manner.	Physics teacher changed student learning class position to enable better learning to take place.

191

Table 5. Stages of learning by using guided inquiry method in cycle II.

No.	Phase	Learning activities
1	Finding the problem	The physics teacher provides the video about the phenomenon of tubes/cans exploding. Then asked to associate and make tentative conclusions of these two things.
2	Introduce hypothesis	Students are asked to discuss with members of their group to propose a hypothesis based on the results of the video that aired.
3	Designing experiments	Teacher explains that all allegations must be proven by experiment. Then groups of students conducted experiments using the tools provided in the form of a model kinetic theory of gases, which was accompanied by the student worksheet to guide them to collect the experimental data.
4	Analyzing the data	Each group was given the opportunity to present the conclusions of the group, guided by the teacher. At this session will be reunited several groups who have different concepts so that the heated discussions will occur between groups.
5	Make conclusions	Physics teachers with students concluded the material provided. At this session, the teacher showed a video related to the material, to provide the correct concept.

At the stage of formulating the problem, each group shared their thoughts about the video shown by the physics teacher. We found the idea that when a bottle of perfume is burned, there is an increase in temperature resulting in particles becoming unstable and explosions. Students ultimately connect the relationship between the temperature of the particles and motion. This happened when they formulated hypotheses.

At this stage of the experiment, there were some students who tried to advise other students. It became a positive condition for them to encourage their colleagues to express their ideas. Students observed the motion of particles in a prop and they commented that when the temperature was too high then the motion of particles was high, and when pressed the motion of the particles also became faster. The connection with the energy of motion is kinetic energy.

At this stage of the discussion, there were 16 students who gave opinions on the results of the experiments conducted. Meanwhile, other students provided arguments only to the group's friends. The physics teacher then asked the students about the meaning of Eq. (2). Students have the opinion that the number "3" indicates a freedom to move in 3 axes dimensional of the particles. Students assumed that if gas was occupying a space, then there were three coordinate axes, namely $(x, y, and z)$.

In the second cycle, the proceedings occurred when students conducted experiments using props, as shown in Figure 1. There were some students who could construct a connection among the variable amounts of the kinetic theory of gas. Besides the balloon experiment, the physics teacher showed another experiment that is the motion of marbles in a shake jar. Then, the teacher asked the students what the relationship would be between the marbles speed and their temperature? Students saw that the balls moved fast and then their temperature rose. This might happen because the many collisions among the marbles made the temperature increase. This indicates that the students succeeded in connecting the relation between the higher speeds of marbles in a jar with greater kinetic energy.

Stages of learning from the first cycle and the second cycle increased the activity of the learning process. This was also expressed by Nivalainen et al. (2013), who stated that there is a change in some of the participants' attitudes to the use of practical work, as noted. An open-guided inquiry laboratory course promoted the acquisition of positive attitudes to the use of inquiry in teaching for a majority (21 out of 32) of the research participants. A similar study has been reported by Wee et al. (2007). They studied the professional development of participants in a program dealing with the use of inquiry. Once again, it is important to remember that we are examining a group of pre-service students who do not have an adequate level of theoretical understanding. The result may be the opposite in the case of a different subject or group in a different context.

Figure 1. Kinetic theory experiment of gases using the tool.

3.2.3 *Observing*

The data observation activity of the students in the study, performed by the observer, can be seen in Table 6 below.

In the second cycle, the liveliness of Class XI Science 1 was included in the category of active in the learning process, using the guided inquiry method, with the achievement of 100% of the indicators set. The guided inquiry method has led students to discover concepts independently due to the learning process and hence they become more active. This is supported by Sarwi et al. (2016), who states that guided inquiry can help students become more attentive during learning, more cooperative within the group, answering questions, and asking or responding. Moreover, Ural (2016) stated that in an inquiry guided learning environment, students are required to think, map concepts, and actively participate in learning. To think and look for a difficult process, for students who are used to obtaining information that has been packaged, is important.

The learning atmosphere observation can be seen in the following table.

In the second cycle, the learning atmosphere of the students of Class XI Science 1 was conducive to the learning process, with the achievement of 80% of the indicators set. In the fifth indicator, it was not easy to achieve the expected goals. It requires a long time to make students more disciplined; this is supported by the statement of Ural (2016) that guided inquiry learning cannot take a short time to pass a good scientific action. It needs to be a longer process so that students are accustomed to it.

3.2.4 *Reflecting*

Based on the observations of the second cycle, there should be a reflection to see both the weaknesses

Table 6. Indicators of observation of student activity cycle II.

Indicator	Implementation	
	Yes	No
Have a problem-solving discussion.	√	
Perform various experiments to solve the problem with the tools and materials available.	√	
Initiative to understand the subject matter deemed not understood by asking the teacher or friend group.	√	
Linking the physics of matter that is relevant to the experiments performed.	√	
Enthusiasm during the experiment and group presentations.	√	

Table 7. Indicators of observational learning environment second cycle.

Indicator	Implementation	
	Yes	No
The majority of students are busy with group activities, such as looking for ideas, designing experiments, and concluding.	√	
Class conditions to support the learning process: the seating position is patterned, purposeful activity, and interaction on two directions.	√	
Meaningful dialog (student—student and student—teacher).	√	
Availability completeness learning support.	√	
Seen in the laboratory and bustle of hectic discussions in the classroom.		√

and the successful implementations of the second cycle. The results of the second cycle of reflection are explained below:

1. In the implementation of the second cycle students' self-management in the laboratory was good. These things affect the conducive learning atmosphere. Some groups started to experiment independently. Interestingly, the students discussed with the group's friends while they were conducting experiments and testing hypotheses.
2. Student worksheets can already be understood by the students. Video appearance in early learning gave students a preliminary description of the concepts that they will get through the trial.
3. Almost all of the students began to pay attention to the teacher's instructions and explanation. Students also posed many questions, showing that they did not understand the concept and provided scientific statements to the teacher when conducting experiments.
4. The seating position of the groups also began to be changed, so room for discussion with friends in the group can be done in an orderly and conducive way. Observers saw that the atmosphere of learning in the laboratory and the implementation of discussion in the classroom has been providing comfort to the learning process.

Another result of the study was a reflection from the distributed questionnaire. Here are some of their comments:

Student 1: *"I feel more obedient on the hypothesis because it has been proved by experiment."*

Student 2: *"It is fun and not too seriously also more fun to learn in the lab."*
Student 3: *"Learning through practice more interesting and exciting than just work on the problems."*
Student 4: *"I am happy to learn in the lab. New atmosphere, indeed I would understand if demonstrated rather than explanation only."*

3.3 Student understanding of the material kinetic theory of gases

During the learning process, by implementing a guided inquiry learning model in Class XI IPA 1, an increase in the students' mastery of the concepts on the subject of the kinetic theory of gases was seen from the first cycle to the second cycle, as shown in Figure 2.

It is apparent that in the first cycle, the average value of the students' mark point in mastery of concepts is still lower than 67, in which the thoroughness of the class was only 50%. A number of students scored below the passing grades of 70 and only 50% of students who completed the study scored concept mastery of at least 70.

In the second cycle there was a significant increase, where the average value of the mastery of concepts in Class XI Science 1 is 73 and the class learning completeness reached 78%. This shows that, by implementing the model of guided inquiry *(guided inquiry)*, it can improve the students' mastery of the concepts on the subject of the kinetic theory of gases.

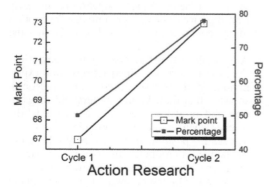

Figure 2. The average value of students in the first cycle and the second cycle, and percentage of success in the first cycle and the second cycle.

4 CONCLUSION

We successfully implemented guided inquiry learning of the kinetic theory of gases in two cycles of classroom action research in a crowded class. The first cycle to the second cycle of observation shows a significant increase of student activity learning. Moreover, the guided inquiry learning has improved not only the students' performances but also improved their active involvement and self- motivation. Some challenging questions may play an important role in improving students understanding, motivation and performances, because they succeed in constructing their own understanding.

ACKNOWLEDGMENTS

We would like to thank Drs. Rusli Ishaq, M.Pd, and the principal Islamic Senior High School Pembangunan Jakarta, for his support to this study. This research was funded by Syarif Hidayatullah State Islamic University.

REFERENCES

Khan, P. & Mohammad, I. (2012). Overcrowded classroom: A serious problem for teachers. *Elixir International Journal, 9*, 10162–10165.
Nivalainen, V., Asikainen, M.A., & Hirvonen, P.E. (2013). Open guided inquiry laboratory in physics teacher education. Journal of Science Teacher Education, 24(3), 449–474.
Nworgu, L.N., & Otum, V.V. (2013). Effect of Guided Inquiry with Analogy Instructional Strategy on Students Acquisition of Science Process Skills. Journal Education and Practice 4,(27) 35, 40.
Pathare, S.R., & Pradhan, H.C. (2010). Students' misconceptions about heat transfer mechanisms and elementary kinetictheory. *Physics Education, 45*(6), 629.
Sarwi, S., Sutardi, S., & Prayitno, W.W. (2016). Implementation of guided inquiry physics instruction to increase an understanding concept and to develop the students'character conservation. *Jurnal Pendidikan Fisika Indonesia, 12*(1), 1–7.
Ural, E. (2016). The effect of guided-inquiry laboratory experiments on science education students' chemistry laboratory attitudes, anxiety and achievement. *Journal of Education and Training Studies, 4*(4), 217–227.
Wenning, C.J. & K. Khan. (2011). Experimental inquiry in introductory physics courses. *Journal Physics Teacher Education, 6*(2), 2–8.
Wenning, C.J. (2010). Levels of inquiry: Using inquiry spectrum learning sequences to teach science. *Journal Physics Teacher Education, 5*(3), 11–20.

Ideas for 21st Century Education – Abdullah et al. (Eds)
© *2017 Taylor & Francis Group, London, ISBN 978-1-138-05343-4*

Creativity assessment in project based learning using fuzzy grading system

A. Ana, A.G. Abdullah, D.L. Hakim, M. Nurulloh, A.B.D. Nandiyanto & A.A. Danuwijaya
Universitas Pendidikan Indonesia, Bandung, Indonesia

S. Saripudin
Politeknik TEDC, Bandung, Indonesia

ABSTRACT: Project Based Learning (PBL) is a learning approach adopted from the concept of constructivist-based learning. The characteristics of PBL are stressed as authenticity, student-centered, and work on product-based projects. Therefore, alternative assessment is used in PBL. The innovation of the assessment tool designed in this study is carried out to meet the demands of technological developments and the demands of using alternative assessments that help lecturers to apply the assessment. One assessment that is needed is an assessment of creativity. The variables of creativity used as an assessment input in the PBL model include fluency, flexibility, and novelty. The method used was a quasi-experimental design with the posttest Only Design with Nonequivalent Groups. The subjects were 82 students. The Fuzzy Grading System-based (FGS) creativity assessment tool is potentially used as an alternative assessment method in schools, polytechnics or universities.

1 INTRODUCTION

Project Based Learning (PBL) is an approach to learning that is adopted from a constructivist-based learning concept. PBL is an appropriate model for vocational technology education in responding to issues of improving the quality of vocational and technological education and the major changes occurring in work fields.

Assessment in PBL is an integral part of learning. Teachers, as project planners, must describe how to measure the achievement of student learning and the final outcome of the project that has been completed.

Alternative assessment in PBL is recommended to assess performance because the performance is assessed by using the rubric and the task, which are considered to be more objective and reliable.

The performance assessment is an assessment of the acquisition, application of knowledge and skills, which demonstrates the ability of students in both the process and the product (Marzano & Pickering, 1994; Stiggins, 1994; Zainul, 2001). The assessment was based on certain standards, which are required to clearly identify what students should know and what students should do. The standards are known as the performance criteria or rubric (Marzano, 1992; Zainul, 2001).

The application of the assessment process that is authentic in schools is still a fundamental problem experienced by educators/teachers. It is difficult to resolve and teachers often encounter the problems of subjectivity, injustice, missing assessment components and evaluation parameters. The assessment process undertaken by teachers is also rigid, and this leads to a final decision that is less flexible, reliable, or authentic. This, of course, is very detrimental to the students because they get final results that do not correspond with their actual capabilities gained from the learning process. These issues will result in unachieved learning objectives that have been formulated (Suranegara et al., 2014).

The use of the fuzzy expert system in the process of the assessment of learners has been undertaken in some countries and has produced a more authentic assessment tool compared to classical or conventional assessment (Yadav et al., 2014). Fuzzy logic is widely used because the concept is easy to understand by using the basic set theory, which is very flexible to adapt to the changes. The fuzzy logic also has tolerance for inappropriate data (Kusumadewi & Purnomo, 2010). In this paper the authors will examine an assessment tool to measure creativity by using fuzzy logic.

2 RESEARCH METHODS

To test the effectiveness of the creativity assessment tool, this study used a quasi-experimental design: the posttest Only Design with Nonequivalent Groups.

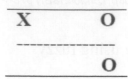

Figure 1. Post-test only design with non-equivalent groups.

The experiment was tested on a math class in education electrical engineering UPI and involved 82 students.

3 RESULTS AND DISCUSSION

The use of fuzzy sets in the evaluation process begins by determining the assessment variables in accordance with the purpose of learning. Determining the variables of fuzzy-based assessment was done by converting the variables determined to be the inputs of the fuzzification process. Furthermore, the assessment rubric based on the fuzzy set was developed to determine the scale of scoring. After the fuzzification was completed, the next step was to start the inference, that is, providing fuzzy algorithm rules followed by a process of defuzzification, which was necessary to get the appropriate output to the input and the rules that have been made (Suranegara et al., 2014; Peji et al., 2013; Ingoley & Bakal, 2012; Yadav et al., 2014; Abdullah et al., 2014).

The creativity assessment of the students was assessed by three variables: fluency, flexibility, and novelty. The variables were inputted one by one in to the mat lab software and the value of creativity appeared instantly.

After learning the scenarios of fuzzy and the rules of FGS, which have a smooth surface viewer and high levels of consistency on the final decision issued, the next step was to make a judgment of creativity using conventional methods by taking the average value of the three input variables of creativity (Sakthivel et al., 2013). The experimental results show the creativity assessment using conventional methods and using FGS.

Table 1 shows ten creativity assessment data ranging from student 1 to student 10. It shows the results of the assessment of the students' creativity using conventional methods by calculating the average value of the three assessment variables or constant mathematical rules. To see the difference between the final decisions of creativity assessment, conventional methods were used with creativity assessment using the FGS, as presented in Table 1. It contains a comparison between the conventional method and the FGS method, which is the scenario of fuzzy-3 rules.

Table 1 shows the difference between the final decisions of creativity assessment using FGS against those of conventional methods. From this table it is seen that the difference between the FGS final decision and the conventional method has a large range. For example, the final decision by student number four has a value range of 74.8 to 85 with 10.2 points difference. To further clarify the analysis of the differences between the two, Figure 2 shows a comparison between the final decisions by taking the example of two types of scenarios of FGS compared with conventional methods.

Figure 2 shows the difference between the final decision of the creativity assessment tool that uses FGS and that of the conventional evaluation method. When the results of the final decision were juxtaposed, there was a significant difference between the two in making the final decision.

Fuzzy set, which allows for a value that is more than one domain, suggests that this assessment is flexible and not rigid, in contrast to conventional methods that do not have a degree of membership.

Table 1. The comparison of the final results using FGS creativity assessment and conventional methods of assessment.

S. No.	Fluency value	Flexibility value	Novelty value	Fuzzy-3 The rule above	The rule below	Conventional method
1	80	75	80	75.5	74.8	78.33
2	25	70	25	50	25.2	40
3	20	0	20	45.5	11	13.3
4	90	75	90	80.7	74.8	85
5	50	40	50	50	41.2	46.66
6	50	0	50	50	26.5	33.33
7	40	0	40	41.2	23.4	26.66
8	100	25	100	78.4	73.5	75
9	100	75	100	89.2	74.8	91.66
10	80	0	80	55.7	51.7	53.33

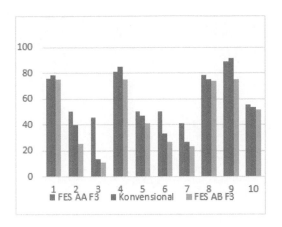

Figure 2.

The difference between the two ratings can be seen on the final decision obtained by student 1 and student 8 (Suranegara et al., 2014).

REFERENCES

Abdullah, A.G., A. Ana & Hakim, D.L. (2014). Perakitan modul latih otomasi industri melalui project-based laboratory dengan penilaian kinerja berbasis fuzzy grading system. *Prosiding konvensi nasional asosiasi pendidikan teknologi* dan kejuruan (APTEKINDO) ke-7 ISBN: 978-602-72004-0-1.

Ingoley, S.N. & Bakal, J.W. (2012). *Students' performance evaluation using fuzzy logic.* Nirma University International Conference on Engineering, Nuicone. IEE 987-1-4673-1719.

Kusumadewi, S. & Purnomo, H. (2010). Aplikasi *logika fuzzy untuk pendukung keputusan.* Yogyakarta: Graha Ilmu.

Marzano, R.J. (1992). *A different kind of classroom: Teaching with dimensions of learning.* Verginia: Association for Supervision and Curriculum Development.

Marzano, R.J. & Pickering, D.J. (1994). *Assessing student outcomes: Performance assessment using dimensions of learning model.* Alexandria: Association for Supervision and Curriculum Development.

Peji, A., Stani, P.M., Pletl, S. & Kiss, B. (2013). Fuzzy multicriteria analysis for student project evaluation. *IEEE 11th International Symposium on Intelligent Systems and Informatics* (pp. 409–413).

Sakthivel, E., Kannan, K.S. & Arumugam, S. (2013). Optimized evaluation of students performances using fuzzy logic. *International Journal of Scientific and Engineering Research, 4*(9), 1128–1133.

Stiggins, R.J. (1994). *Student-centered classroom assessment.* New York: Macmillam College Publishing Company.

Suranegara, G.M., Abdullah, A.G. & Saputra, W.S. (2014). Alat penilaian kinerja pembelajaran kooperatif teknik digital berbasis logika fuzzy. *Prosiding konvensi nasional asosiasi pendidikan teknologi dan kejuruan* (APTEKINDO) ke-7 ISBN: 978-602-72004-0-1.

Yadav, R.S., Soni, A.K. & Pal, S. (2014). A study of academic performance evaluation using Fuzzy Logic techniques. *Computing for Sustainable Global Development (INDIACom), 2014 International Conference on, (April)* (pp. 48–53). http://doi.org/10.1109/IndiaCom.2014.6828010

Zainul. A. (2001). *Alternative assessment.* Jakarta: Ditjen Dikti.

Ideas for 21st Century Education – Abdullah et al. (Eds)
© *2017 Taylor & Francis Group, London, ISBN 978-1-138-05343-4*

Students' attitude towards mobile-assisted language assessment: A case of speaking class

A.A. Danuwijaya, A.G. Abdullah & A.B.D. Nandiyanto
Universitas Pendidikan Indonesia, Bandung, Indonesia

ABSTRACT: The use of mobile devices has been widely known to provide potential for language learners to improve their language skills. This present study investigates the attitude of English learners to using mobile devices, particularly in assessing speaking skills. This study used a case study approach, and 34 university students participated in the study. Within a semester, each student volunteered to record a two-minute speech about one argumentative topic and shared the recording to a LINE group application consisting of class members as well as the teacher. At the end, students were welcome to give feedback and comments. The results showed that this practice improves both their skills and motivation to speak English better. Some language aspects, such as pronunciation, can be improved, and students feel less anxious when practicing their speaking. Some problems were also discussed.

1 INTRODUCTION

The growing use of mobile devices has made it possible for people to change their landscape of e-learning in many ways. Instead of being substitutes for existing learning devices, mobile devices serve as an extension for learning in a new environment (Kukulska-hulme, 2009; Palmer & Yann, 2015), which is spontaneous, informal, and personalized (Miangah & Nezarat, 2012). Panagiotidis (2012) suggests that this environment has affected students' learning habits, particularly at the tertiary education level. In personalized learning environments, students have the opportunity to form their own personal way of working, using appropriate applications and tools to achieve their purpose.

There are a large number of studies investigating the use of mobile devices in language learning, such as the iPad or MP3 player for listening and speaking (Demouy et al., 2010), podcasting for presentation (Abdous et al., 2009; Chan et al., 2011), and PDA for reading comprehension (Chang & Hsu, 2011). Apart from language learning, a recent study shows that mobile devices can also be useful when implemented in language assessment. Tarighat and Khodabakhsh (2016) introduced the term Mobile-Assisted Language Assessment (MALA) as the integration of Mobile-Assisted Language Learning (MALL) and language assessment. They further argue that MALA can provide space for students to practice their speaking skills while being assessed. In the study, students showed mixed attitudes towards MALA in assessing speaking. As the study of MALA is still limited, this present study aims to further investigate students' perceptions towards MALA using social network applications in speaking classes.

2 LITERATURE REVIEW

Learning mediated by handheld devices has opened up a new way of learning that is available anywhere and at any time (Kukulska-hulme & Shield, 2008). The learning is characterized by the two main characteristics of mobile devices: connectivity and portability. For connectivity, a mobile system must have the capability of being connected to and communicating with the learning website, using the wireless network of the device to access learning material, including Short Message Service (SMS) and mobile e-mail, and portability enables learners to move mobile devices and bring learning materials (Miangah & Nezarat, 2012). In short, mobile learning provides students with the experience of feeling a sense of freedom in terms of time management and places to learn languages.

MALL is basically defined as the use of research, which has provided many benefits for improving students' learning. In speaking and listening, mobile devices could successfully support learning activities that promote collaboration and interaction among students (Kukulska-hulme & Shield, 2008).

The study of using mobile devices in a speaking context was carried out by Salamat and Pourgharib (2013). This experimental study shows the positive effects of mobile learning, particularly in changing students' conception of learning. Mobile learning offers opportunities for language learners to be more familiar with the application students can carry around at all times. It also helps them to reduce their stress and improve their self-confidence in speaking. In the area of assessment, mobile devices can also be used as a means of helping students to practice speaking, while at the same time being assessed by their peers and the teacher.

Assessment can be defined as an action to collect and use information about one's knowledge and skills with the aim to improve learning (Berry & Adamson, 2011). According to McConnell & Doolittle (2011), within the classroom context, one of the assessment principles is that assessment must be directed to see the improvement of student learning and focus on three main issues: process, evaluation, and purpose. They include how the information is gathered from students: test or non-test; how the information is evaluated and can feedback be used to modify teaching and learning; and how assessment improves learning by modification in understanding.

One way to get information about students' improvement in learning is to find out how much students achieve of their learning goal in some tasks. A task provides students with opportunities to use their linguistic and non-linguistic resources to complete the activity, such as to fulfill their needs to convey information and to express opinions (Ellis, 2009). According to Brown (2004), speaking assessment tasks are categorized into five categories: imitative, intensive, responsive, interactive, and extensive. The task in this study was extensive, in which students made a monologue on a certain topic. This task was characterized by complex and lengthy stretches of discourse, deliberative use of language style, and use of formal language (Brown, 2004). The assessment tasks were mostly in the form of presentations and monologues in front of the class. However, the expansion of technology means that this can now be involved in the task, even application-based activities in which students complete the task and are assessed at the same time using an application.

Some studies have examined the use of mobile devices for language assessment purposes. An exploratory study conducted by Wong et al. (2006) investigates a prototype to assess students' English grammar using a mobile phone. Based on the trial, the participants gave positive comments on the design and the instant feedback to responses. In speaking skills, a study conducted by Tarighat and Khodabakhsh (2016) investigated learners' feasibility and attitudes in speaking through MALA, through the use of a social networking application by Iranian advanced learners. The results showed that practice with MALA provides opportunities for self-correction, peer-correction and teacher feedback. For students, MALA can be a fairer way, which gives them the choice to assess their peers, and the teacher can use it as an alternative to assess students' speaking skills. The present study investigates how students perceive the practice of MALA in speaking classes at university level.

3 RESEARCH METHODS

This present study was triggered to provide new experiences for students to improve their speaking skills. Most of the sessions in speaking class were dominated by speaking practice in pairs, in groups or by giving a short speaking practice in front of the class. Students practiced their speaking and got feedback from their peers and a teacher. For some students, this practice was challenging, and sometimes intimidating. They felt stressful when speaking in front of others. In this case, students needed a new way of practicing their speaking skills by using a mobile device with a social networking application

This present study used a case study design, as it aimed to investigate the phenomenon of using a mobile device and a social network application to assess speaking skills. The study involved twenty second-year students majoring in English in a public university in Bandung. The samples were selected based on the convenience of them being available and willing to participate in the study. The data were collected through interview, with open-ended questions asking their attitude towards using a mobile application for speaking class. The interview questions cover not only the students' preferences and concerns about using a mobile device for speaking practice, but also students' attitudes towards assessment, its benefits, and its drawbacks. The data were analyzed using content analysis.

4 RESULTS AND DISCUSSIONS

Based on the interviews, the results show that the use of mobile devices for speaking practice and assessment has some potential.

The practice of assessing speaking using a mobile device was a new experience for students. Students think that the practice of assessing their peers' using a mobile device for recording provides a more comfortable avenue in terms of flexibility in preparing their speaking practice and giving comments. They had more freedom to listen to their friends' work wherever they were and to give comments whenever they had the time. This sense of freedom was mentioned by some students.

"I can do my work anytime and anywhere."

"It is quite flexible for me to listen to my friends' recording and give comments at any time."

"I can prepare my speaking first, then record it. If I don't like it, I record it again. I can do it at home or at the library. It's not easy, but I have more time to do it anywhere."

"I'm quite confident to do the task because I have flexible time to prepare it."

The excerpts above indicated that students felt a sense of flexibility and freedom when preparing and doing their speaking task. As the task was a monologue, which is quite complex and requires them to do preparation (Brown, 2004), students seemed enthusiastic to perform the task. Some students felt that this practice improved their confidence as well. It corroborates the study by Salamat and Pourgharib (2013), suggesting that the use of mobile applications reduces stress and improves confidence.

In addition to new experiences, assessment practice using mobile devices offers opportunities for students to do self-assessment and get feedback from peers, which allows them to be more critical when practicing speaking. In this learning context, students are expected to submit their speaking tasks, provide comments to other students, and learn from the feedback of others. To complete the task, students need to prepare their speaking topic and record and upload their speaking to the application. This process allows students to monitor their progress in providing a good speaking task.

"I have time to think about the topic and words that relevant to my speaking. It makes me think and rethink about the organization and the flow of ideas before I record it. It's good because I have time to make sure that my points are covered in my task."

The excerpt above shows a student's efforts to complete the task. There are steps where he needs to recheck his work before submitting it to the application. This shows how the student tried to achieve the learning goal. Black et al. (2004) argue that, when students can understand what learning goal they need to achieve and they can assess what they need to do to achieve the learning goal, students are developing their capacity for self-assessment and monitoring their own work.

Apart from self-assessment, the use of MALA can be helpful for students to improve their speaking practice from peer feedback. Peer feedback can function to complement self-assessment and students may accept criticisms from their friends (Black et al., 2004). Figure 1 shows an example of feedback provided by another student via the application. The student comments on another student's

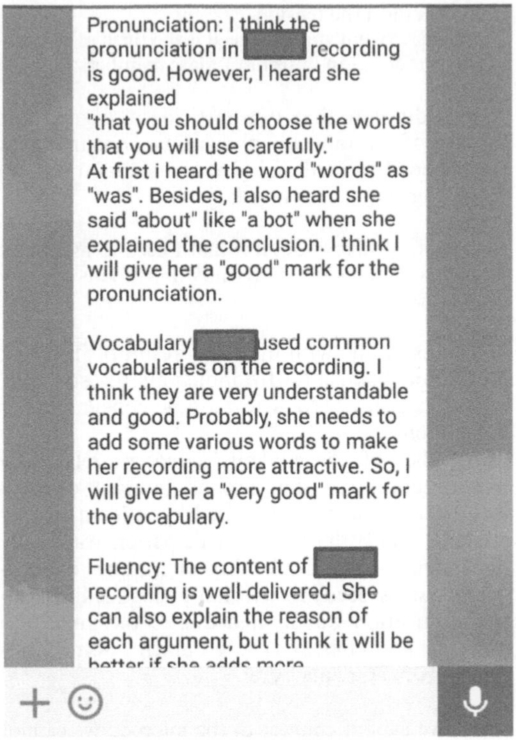

Pronunciation: I think the pronunciation in ███ recording is good. However, I heard she explained "that you should choose the words that you will use carefully." At first i heard the word "words" as "was". Besides, I also heard she said "about" like "a bot" when she explained the conclusion. I think I will give her a "good" mark for the pronunciation.

Vocabulary ███ used common vocabularies on the recording. I think they are very understandable and good. Probably, she needs to add some various words to make her recording more attractive. So, I will give her a "very good" mark for the vocabulary.

Fluency: The content of ███ recording is well-delivered. She can also explain the reason of each argument, but I think it will be better if she adds more

Figure 1. Peer-feedback on speaking task.

recording on some aspects of language, such as pronunciation, vocabulary, fluency, and organization. To improve vocabulary, for example, the student comments that his friend should use a wider range of vocabulary to improve his/her speaking.

This type of feedback allows students to be more critical and cautious when speaking, particularly in pronunciation and vocabulary. This practice is useful to improve their monitoring of the task, as shown in the excerpt below.

"Sometimes we don't know about what is wrong with our speaking. But the assessment given by peers makes us aware about our mistakes. This helps us in improving our speaking."

For many students, the use of a mobile device in speaking assessments provides many benefits. However, some students feel that this practice reduces the real tension of the nature of speaking, such as speaking in front of the class. The use of the application replaces the real tension and challenge of what real speaking is. One student argued that he felt unchallenged when speaking via an application. As shown in the excerpt below, the student thought that the use of an application lowered the tension and thrill. Two participants mentioned:

"I don't feel the tension."
"I believe what students need is the thrill of delivering presentation in front of class members."

In addition to that, two participants argued that the use of an application was time consuming as they needed to read and give comments to others work. One participant said:

"What I dislike about it is that I need to listen and give comments on my friends' work. It is a waste of time."

This is in line with the study result of Tarighat and Khodabakhsh (2016) indicating the students' views on using MALA to be time consuming for the students.

Another of the students' concerns related to this practice was that limited internet connection becomes a problem faced by students. Most students think that their participation was really dependent on the availability of internet connection. Besides, students felt that giving feedback on the application was limited due to the limitation of space and characters to comment. Some comments from students were:

"If we cannot connect to the internet, we cannot participate it."
"I want to give more feedback for my friends, such as giving corrections. But I can't write more words."

5 CONCLUSION

The main focus of the study was that the use of a mobile device for assessment purposes could potentially be used to improve students' learning, particularly language skills. It provided more opportunities for students to experience flexible learning with better preparation to improve speaking skills. For students, it has supported learning with collaboration and more interaction with others, and it has reduced the level of tension in practicing their speaking. Despite some limitations, the use of a mobile phone helped students to experience new ways of practicing their skills and, for teachers, this practice can be an alternative that can be used to collect information about students' speaking performance and learning.

REFERENCES

Abdous, M., Camarena, M.M. & Facer, B.R. (2009). MALL technology: Use of academic podcasting in the foreign language classroom. *ReCALL Journal, 21*(1), 76–95.

Berry, R. & Adamson, B. (2011). Assessment reform past, present and future. In R. Berry & B. Adamson (Ed.), *Assessment reform in education: Policy and practice.* New York: Springer.

Black, P., Harrison, C., Lee, C., Marshall, B. & Wiliam, D. (2004). Working inside the Black Box: Assessment for learning in the classroom. *Phi Delta Kappan, 86*(1), 8–21.

Brown, H.D. (2004). *Language assessment: Principles and classroom practices.* New York: Pearson Education.

Chan, W.M., Chi, S.W., Chin, K.N. & Lin, C.Y. (2011). Students' perceptions of and attitudes towards podcast-based learning: A comparison of two language podcast projects. *Electronic Journal of Foreign Language Teaching, 8,* 312–335.

Chang, C. & Hsu, C. (2011). A mobile-assisted synchronously collaborative translation—annotation system for English as a foreign language (EFL) reading comprehension. *Computer Assisted Language Learning, 24*(2), 155–180.

Demouy, V., Eardley, A., Kukulska-Hulme, A. & Thomas, R. (2010). Using mobile devices for listening and speaking practice in languages : The L120 Mobile Project. In *Eurocall 2010, Language, Cultures and Vurtual COmmunities, 8-11 September 2010.* Bordeaux, France.

Ellis, R. (2009). Task-based language teaching : Sorting out the misunderstandings. *International Journal of Applied Linguistics, 19*(3).

Kukulska-hulme, A. (2009). Will mobile learning change language learning ? *ReCALL, 21*(2), 157–165.

Kukulska-hulme, A. & Shield, L. (2008). An overview of mobile assisted language learning: From content delivery to supported collaboration and interaction. *ReCALL Journal, 20*(3), 271–289.

Miangah, T.M. & Nezarat, A. (2012). Mobile-assisted language learning. *International Journal of Distributed and Parallel Systems, 3*(1), 309–319.

Palmer, R. & Yann, R.T.Y. (2015). How can a mobile device be effectively utilised for studying in an e-learning context? *Hirao School of Management Review, 5,* 81–97.

Panagiotidis, P. (2012). Personal learning environments for language learning. *Social Technologies, 2*(2), 420–440.

Salamat, A. & Pourgharib, B. (2013). The effect of using mobile on EFL students speaking. *International Research Journal of Applied and Basic Sciences, 4*(11), 3526–3530.

Tarighat, S. & Khodabakhsh, S. (2016). Mobile-assisted language assessment : Assessing speaking. *Computers in Human Behavior, 64,* 409–413.

Wong, C.C., Sellan, R. & Lee, L.Y. (2006). Assessment using mobile phone: An exploratory study. In *Proceedings of IAEA 2006 Conference, Grand Copthorne Waterfront Hotel, Singapore City, Singapore.*

Student's understanding consistency of thermal conductivity concept

I.S. Budiarti, I. Suparmi & A. Cari
Universitas Sebelas Maret, Surakarta, Indonesia

V. Viyanti
Universitas Lampung, Lampung, Indonesia

C. Winarti
Universitas Islam Negeri Sunan Kalijaga, Yogyakarta, Indonesia

J. Handhika
IKIP PGRI Madiun, Indonesia

ABSTRACT: Preliminary observation in three universities revealed that its focus remains on the level of knowledge, understanding, and application. The aims of the research were to explore and to describe the consistency of students' understanding of thermal conductivity concept, whereas the testing instruments were adapted from HTCE (Heat and Temperature Concept Evaluation). The study was carried out in three higher learning institutions: IKIP PGRI Madiun, UIN Sunan Kalijaga Yogyakarta, and Lampung University with a sample of 145 students. Descriptive research design was employed, while the purposive random sampling technique was used to obtain the sample. Data were obtained using multi-representation tests and in-depth interviews. Based on the data analysis, it was found that 87.59% students were inconsistent to answer test of multi-representation concept.

1 INTRODUCTION

Learning Physics is an endeavor to comprehend Physics concepts and to discover how to gather fact information and induce the principles underlying such facts as well as the attitude of the Physicists toward their discoveries. It is through studying Physics that students can discover the meaningfulness of the concepts under study and as such can apply those concepts in their daily lives. Rosser (1984) stated that a concept was an abstraction representing a class of objects, events, activities or relationships sharing the same attributes. It may be the case that the concepts the students have evolved from daily encountered phenomena. However, such intuitive comprehension is completely different from scientific concepts. It is, therefore, important to analyze how students' concepts differ from scientific explanation (Alwan, 2011).

The concepts of Temperature and Heat is offered in as early as Elementary School years through High School to Higher Learning Institution. Sozbilir (2003) the concepts of Temperature and Heat which were still abstract or vague upon learning resulted in the various interpretation or perception on the part of students. Thomas et al. (1995) found that students had difficulties accepting or internalizing such

concept as "the final temperature of two different objects will be the same upon being exposed to the same temperature in a particular environment." This result consistent with the research findings observed by Baser (2006), in that the students had difficulties solving the problem of heat transfer using conduction taking place in a seat belt, partly made of metal and non-metal. The comprehension of concepts by students can observe from their correct answer consistency. Ainsworth (2006) stated that "the consistency of the students' responses in comprehending Physics concepts demands a deeper understanding on the part of the students themselves to perceive equality of Physics problems presented in various or multiple ways." A more in-depth understanding will make a student consistently adhere to what he or she perceives, understands and believes. An implication of such consistency is attested inconsistency of representation in that what they perceive may not be scientifically true. It takes an understanding of the equivalence of the multiple representations and scientific consistency which demand scientific comprehension to perceive it consistently (Nieminen, Savinainen, & Viiri, 2010). Students' consistency will take them to a higher level of understanding in perceiving different Physics concepts represented in various problems.

A student's ability to succeed in learning is determined, among others, by his or her thinking skills he or she possesses. The most important is the ability to solve problems encountered during learning processes. By putting his or her thinking skills into practice, a student can train and develop his or her cognitive intelligence he or she possesses, as well as relate various facts or information to the knowledge previously gained to formulate a prediction of the outcome. Thinking ability is not restricted to defining. Thinking is considered an intellectual process or higher-order cognitive process (Wilson, 2000). Thinking ability is thus a skill and strength to be internalized in learning subject matters to enhance performance and lessen learners' weaknesses (Heong, 2011). Teaching-learning activities should involve specific thinking ability, thereby facilitating categorizing thinking ability based on the open framework (Kong et al., 2012).

Senk et al. (1997) stated that higher-order of thinking showed the characteristics of solving unusual problems and the solutions were a result of logical thinking. The characteristics of mental activities of higher-order of thinking processes often involve complex thoughts, non-algorithmic in nature, as well as independence in thinking processes and being able to provide various applicable solutions (Resnick, 1987) upon encountering unusual situations filled with problems, faced with options requiring him or her to make a decision (King, Godson & Rohani, 1998).These high order thinking skills in this study refer to the three top domains of Bloom's Taxonomy. Zoller (1993) stated that the three highest cognitive domains of Bloom's Taxonomy, i.e. analyzing, evaluating and creating, which demands higher-order thinking skills.

Based on the theoretical and case studies the researchers conducted the study to explore and describe the consistency of students' understanding of thermal conductivity concept. The research question proposed in this study was how consistent the Students' conceptual understanding consistency of conductivity is?

2 RESEARCH METHODS

The method of study was descriptive in nature to scrutinize and describe the consistency relationship between the students' comprehension of the concept of conduction and their higher-order thinking skills. The sample was taken using purposive random sampling from three higher learning institutions, namely IKIP PGRI Madiun, with twenty-five students participating; UIN Sunan Kalijaga Yogyakarta, with fifty-two students; and Universitas Lampung, with sixty-eight students. The test of

consistency of representation concept can also use as a tool for identifying the cause of misconceptions (Cari, et.al, 2016). The data were collected using multiple-choice test items with reasoning as well as interviews. Four multiple-choice items were those regarding heat transfer using conduction. The interviews were conducted to investigate the students' underlying comprehension in yielding the answers, thereby revealing their understanding of the concept line with their higher-order thinking skills.

3 RESULTS AND DISCUSSION

The percentage of the students' understanding of thermal conductivity concept based on their answers can see in Table 1 below. Based on Table 1, the average of consistently correct answers yielded by the students was 1.38%, this indicates that their comprehension of the concept of heat transfer using conduction was low, meaning that they had difficulties to see equivalence in problems presented in multiple or various ways. The students should have had a better understanding if they had been consistent toward what they perceive to be true. The multiple representations of problems demand a scientific comprehension on the part of the students. However, as a result of inconsistent comprehension resulting from different perspectives employed in perceiving the problems, consistently incorrect answers occurred. Based on consistency theory, students will be brought to a higher level of comprehension in perceiving various Physics concepts presented in multiple representations.

About the data on the percentage of consistency level obtained from the students in the three higher-learning institutions taken as a sample in this study, in the following is described the students' consistency of concept comprehension regarding the subject matter of heat transfer using conduction.

Table 1. The percentage of the consistency of the students' comprehension of concept.

Consistency level	Percentage (%)			
	UIN sunan kalijaga	Universitas lampung	IKIP PGRI madiun	Average
Correctly Consistent	1.92	1.48	0	1.38
Consistenly with the wrong answer	7.70	11.76	16	11.03
Inconsistent	90.38	86.76	84	87.59

3.1 *The consistency of students' understanding of thermal conductivity concept*

Based on the data, the students who yielded low-quality responses clear gave an explanation on the microscopic mechanism about conduction taking place in metals. The majority of the students' responses about conduction mainly in solid objects was not supported by a satisfying argument since they could not state the correct reason. The student's answer is shown in Figure 1 below.

Based on Figure 2, it can see that the student was hesitant in deciding between the choice A and D before finally deciding on the latter. Considering the reason stated, it could identify that the student already had the correct concept regarding Temperature and Heat. However, this particular student's answer not in line with the problem presented. The problem presented in this particular item is the transfer of heat. As such, the correct answer yielded should have been A.

It can see from the Figure 4 above that the student did not comprehend the concept, the student's answer which does not reveal the relationships between Temperature, Heat and other units related to heat transfer using conduction.

Based on Figure 5 above, it can see that the student was able to solve the problem well, as observed in the correct answer. The in-depth interview conducted revealed that the student could explain the relationship between the rate of heat transfer and

Wellem mengambil dua penggaris, berbahan logam dan kayu yang berada di tempat yang sama. Dia menyatakan bahwa penggaris logam terasa lebih dingin daripada penggaris kayu. Apakah penjelasan yang tepat mengenai situasi tersebut?

a. Energi kalor merambat dari tangan ke logam lebih cepat daripada kayu
b. Kayu secara alami merupakan bahan yang lebih panas daripada logam
c. Penggaris kayu mengandung lebih banyak kalor daripada penggaris logam
d. Logam merupakan radiator panas yang baik daripada kayu
e. Suhu logam lebih rendah dari pada suhu kayu
Alasan: ...

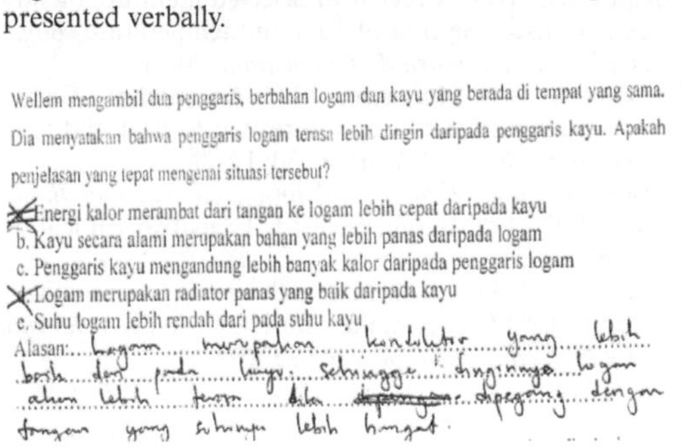

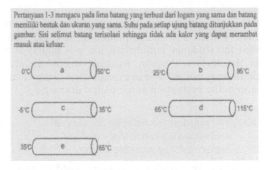

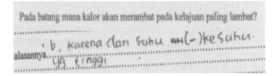

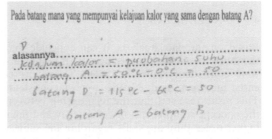

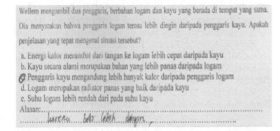

Figure 1. A student's incorrect answer to a problem presented verbally.

Wellem mengambil dua penggaris, berbahan logam dan kayu yang berada di tempat yang sama. Dia menyatakan bahwa penggaris logam terasa lebih dingin daripada penggaris kayu. Apakah penjelasan yang tepat mengenai situasi tersebut?

a. Energi kalor merambat dari tangan ke logam lebih cepat daripada kayu
b. Kayu secara alami merupakan bahan yang lebih panas daripada logam
c. Penggaris kayu mengandung lebih banyak kalor daripada penggaris logam
d. Logam merupakan radiator panas yang baik daripada kayu
e. Suhu logam lebih rendah dari pada suhu kayu
Alasan: logam merupakan konduktor yang lebih baik dari pada kayu. sehingga dinginnya logam akan lebih terasa bila dipegang dengan tangan yang suhunya lebih hangat.

Figure 2. A hesitant student's answer to a problem presented verbally.

Pertanyaan 1-3 mengacu pada lima batang yang terbuat dari logam yang sama dan batang memiliki bentuk dan ukuran yang sama. Suhu pada setiap ujung batang ditunjukkan pada gambar. Sisi selimut batang terisolasi sehingga tidak ada kalor yang dapat merambat masuk atau keluar.

Figure 3. A problem on conduction presented visually.

Pada batang mana kalor akan merambat pada kelajuan paling lambat?
b. karena dari suhu (-) ke suhu yg tinggi
alasannya...

Figure 4. A student's incorrect answer to a problem presented visually.

Pada batang mana yang mempunyai kelajuan kalor yang sama dengan batang A?
D
alasannya...
kelajuan kalor = perubahan suhu
batang A = 50°C − 0°C = 50
batang D = 115°C − 65°C = 50
batang A = batang D

Figure 5. A student's correct answer to a problem presented visually.

change in temperature. As the rate of heat transfer using conduction is positively correlated to change in temperature, rod A has the same rate of heat transfer as rod D, since the temperature change in rods A and D are the same. The student's answer that "rate of heat transfer = temperature change" carries the meaning that the rate of heat transfer is positive correlate with temperature change. As such, it can say that the majority of the students was inconsistent in comprehending the concept of conduction as can be seen in their answers to both verbally and visually presented problems.

3.2 *Higher-order thinking skills*

The higher-order thinking skills in this study referred to the three top or highest domains of Bloom's Taxonomy. Zoller (1993) asserted that the three top cognitive domains in Bloom's

Taxonomy—analyzing, evaluating and creating—demand higher-order thinking skills. Based on their written and oral explanation, it could state that the students' comprehension was mainly still in the analysis domain. In other words, the students were not yet able to express their comprehension in the evaluation and creation domains.

3.3 The relationship between the concept understanding consistency and the higher-order of thinking skills

Based on the students' answers yielded during the tests and in-depth interviews, it could say that the relationship between the concept comprehension consistency and higher-order of thinking was attested. Several students' answers regarding conduction which were consistently correct. Based on these consistently correct answers, the researchers were able to detect the presence of higher-order of thinking skills. However, the correct answers yielded by those particular students revealed that their understanding was still in the analysis domain.

4 CONCLUSIONS

Based on the data analysis, it can conclude that the answers yielded by 1.38% of the students were consistently correct, 11.03% were consistently incorrect, and 87.59% were inconsistent. This result revealed that the relationship between the concept comprehension consistency and their higher-order thinking skills regarding the subject matter of conduction attested in their answers yielded in the tests and in-depth interviews. Apart from that, another factor at play was the tendency among the students to learn Physics merely using memorizing without fully comprehending the concepts. Furthermore, still another factor contributing to this low consistency was the daily practice of Physics learning which mainly dealt with how to solve problems without a deeper understanding of the details. As a result, the low mastery of Physics concepts rendered the students' comprehension weak, as attested in the higher-order of thinking skills regarding the concept of conduction.

Further factors triggering this lack of consistency learning everyday physics emphasize on how to resolve the matter, without understanding the issues in detail, consequently weak mastery of physics that make less profound understanding experienced by students, especially high-level thinking skills regarding conduction concept.

ACKNOWLEDGEMENT

This research supported by Mandatory grand UNS with contract number 632/UN27.21/LT/2016

REFERENCES

Alwan. A.A. 2011. The misconception of Heat and Temperature among Physics Student. *Procedia Social and Behavioral Sciences*. 12: 600–614.

Ainsworth, S.E. 2006. DeFT: A Conceptual framework for considering learning with multiple representations. *Learning and Instruction Journal.* 16: 183.

Baser, M. 2006. Effect of conceptual change oriented instruction on students' understanding of heat and temperature concept. *Journal of Maltese Education Research*. 4(1): 64–79.

Cari, et. al., 2016. Student's preconception and anxiety when they solve multi-representation concepts in newton laws, and it is application. *Proceeding ICOPIA*.

Heong, Yee Mei, WidadBinti Othman & Jailani Bin MdYunon, 2011, The level of Marzano higher order thinking skills among technical education students. International Journal of Social Science and Humanity. 1(2).

King, F.J., Goodson, L. & Rohani, F. 1998. Higher-order thinking skills: definitions, strategies, and assessment. URL:http://www.cala.fsu.edu/files/higher_order_thinking_skills.pdf.

Nieminen, P., Savinainen, A. & Viiri, J. 2010. *Force Concept Inventory based Multiple-choice Test for Investigating Students' Representational Consistency.Physical Review Special Topics*. Physics Education Research 6,020109. Retrieved from http://prstper.aps.org/abstract/PRSTPER/v6/i2/e020109.

Resnick, L. 1987. Educational and learning to think. Washington D.C.: National Academy Press.

Senk, S.L., Beckmann, C.E. & Thompson, D.R. 1997. *Assessment and grading high school mathematics. journal for research in mathematics education*. 28(2): 187–215.

Sozbilir, M. 2003. A review of selected literature on student's misconception of heat and temperature. *Bogazici University Journal of Education*. 20(1).

Thomas, M.F. & Marquiset. 1995. An attempt to overcome alternative conception related to heat and temperature. *Physics Education*. 30: 19–26.

Wilson, V. 2000. *Educational forum on teaching thinking skills*. Edinburgh: Scottish Executive Education Department.

Zoller, U. 1993. Are lecture and learning compatible? Maybe for LOCS: Unlikely for HOCS (SYM). *Journal of Chemical Education*. 70: 195–197.

Ideas for 21st Century Education – Abdullah et al. (Eds)
© *2017 Taylor & Francis Group, London, ISBN 978-1-138-05343-4*

Students' science literacy skills in ecosystem learning

M. Arohman

SPS Universitas Pendidikan Indonesia, Bandung, Indonesia

ABSTRACT: Science literacy skill is the capacity to use scientific knowledge, to identify the question and to draw conclusions based on facts and data to better understand the nature and changes that happen due to human activities. This research is conducted to understand the science literacy skills of MTs students in ecosystem learning. Science literacy skill is measured by the Test of Scientific Literacy Skills (TOSLS), which is adopted from Gormally et al. (2012). The test was given to 17 students in MTs Nurul Ikhsan Belawa Cirebon. The results obtained by the use of the TOSLS test were in the intermediate category.

1 INTRODUCTION

Nowadays, the global community has entered a new era, an era where the acceleration of change happens in different areas, including education. The demand in the 21st century is to make the educational system relevant and to correspond to changes in this era. As quoted by Correia et al. (2010):

"The relevance of such educational issues is confirmed by the United Nations, which declared the years between 2005 and 2014 to be the 'Decade of Education for Sustainable Development.'"

The relevance of the educational issue was confirmed by the United Nations (UN) when it declared the years from 2005 to 2014 to be the decade of education for sustainable development. Attitudes towards the new challenges of the post-industrial society are a direct consequence of scientific and technological development, an explosion of knowledge, and globalization. Support from respected stakeholders is needed to achieve a condition that corresponds to development in the 21st century.

It has become very important for science literacy to be developed by students as preparation for facing the challenge of 21st century development. This is consistent with Treacy et al. (2011):

"Scientific literacy is directly correlated with building a new generation of stronger scientific minds that can effectively communicate research science to the general public."

Referring to the aforementioned quote, science literacy is directly correlated with building a new generation who have strong scientific thinking and attitudes and effectively communicate the science and results of the research to the general community. Someone with science literacy is a person who utilizes science concepts, having science-process skills to evaluate making daily life decisions when dealing with other people, society and the environment, including social and economic development.

The findings of PISA's research (Programme for International Student Assessment), which has been conducted since 2000, do not show satisfactory results, because students' average scores are still far below the international average, which reached a score of 500. The average value obtained by Indonesian students was 371 in 2000, 382 in 2003, and 393 in 2006, with the reading average score of 405. The result, indeed, indicates a significant difference with the average score of international students. Based on the aforementioned result, most of the Indonesian students' science skills are still at the level of knowing some basic facts, but they are still unable to communicate and relate with skill the various topics in science, let alone to implement the scientific concepts (Toharudin, et al., 2011).

The environment is an important issue, which has been much highlighted lately. The reclamation of Jakarta bay has become one of the issues concerning damage to the environment. The environmental impacts of this activity are the loss of some important ecosystems, rising sea levels that possibly leads to more severe flooding, as well as many other environmental impacts (Anggraeni, 2014). Based on the explanation above, a measurement of students' science literacy in ecosystem learning is needed.

2 RESEARCH METHODS

This research is a descriptive research, which is aimed at identifying students' science literacy in ecosystem learning. The subjects of analysis are 8th grader students in Madrasah Tsanawiyah

Table 1. Category of indicator and sub-indicator in TOSLS.

Indicator	Sub-indicator
I. Understanding the method of inquiry that leads to scientific knowledge.	1. Identifying correct scientific argument (I1)
	2. Using an effective literature search (I2)
	3. Evaluation in using scientific information (I3)
	4. Understand the design element of the research and how it impacts the scientific discoveries (I4)
II. Organize, analyze and interpret quantitative data and scientific information.	5. Create a graph to represent the data (I5)
	6. Read and interpret data (I6)
	7. Problem solving using quantitative capabilities including statistics of probability (I7)
	8. Understands and is able to interpret the basic statistics (I8)
	9. Presenting conclusions, predictions based on quantitative data (I9)

Nurul Ikhsan Desa Belawa, Kecamatan Lemahabang, Kabupaten Cirebon.

The instrument utilized is the science literacy skill test developed by the writers, which corresponds with the indicators provided in the developing test devices TOSLS (Test of Scientific Literacy Skills) by Gormally et al. (2012).

The following is an indicator and sub-indicator table of TOSLS.

3 RESULTS AND DISCUSSION

Based on this research, data regarding the students' literacy skill is presented in Figure 1.

The average value of the science literacy skills obtained in this study is 42.35%. This value is in the intermediate category. This indicates that strengthening and learning activities that can improve students' science literacy skills are needed; this can be done by the provision of a supportive learning approach. The lowest value in the sub-indicator of science literacy is contained in the sub-indicator of understating the design element of research and how it impacts towards the scientific discoveries (I4), with an acquired value of 27.45%. Students are still not familiar with activities that contain scientific measures, so they need to be exposed to learning with a scientific approach. As for the

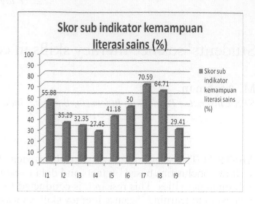

Figure 1. Graph of students' science literacy. *score of sub-indicator of students' science literacy.

Table 2. Scores per indicator of the ability of science literacy.

Indicator	Score indicator of the ability of scientific literacy (%)
Understanding inquiry method that leads to scientific knowledge	37.7425
Organize, analyze, and interpret qualitative data and scientific information	51.178

highest score of the sub-indicators, this lies on the sub-indicator of science literacy skills at solving problems using quantitative capabilities including statistics of probability (I7), with a score value of 70.59%.

From the assessment results obtained using the test tool of TOSLS, the average value of each indicator can be seen in Table 2.

Based on the table above, the students tend to be more capable of processing quantitative data compared to understanding the scientific method. This is due to the fact that, in school, the skills of processing, analyzing, and interpreting quantitative data are more often used as an integrated part with other subjects (for example, mathematics). As for the understanding of scientific measures, it is less acquired by the students. According to Aisyah (personal interview, July 11 2016) the practicum is rarely conducted.

4 CONCLUSIONS

The conclusion of this study is that the students' science literacy skill in MTs Nurul Ikhsan is in the intermediate category (42.35%). There is a need to implement teaching methods that can support the improvement in students' scientific literacy.

ACKNOWLEDGMENTS

To all respected parties that supported this study, the researchers would like to say thanks, especially to all the staff of MTs Nurul Ikhsan Belawa Village district. Lemahabang Kab. Cirebon.

REFERENCES

Anggraeni, L. (2014). Mengkaji: *Dampak Reklamasi Pantai di Teluk Jakarta*. Retrieved at (http://kompasiana.com/post/read/650460/3/mengkaji-dampak-reklamasi-pantai-diteluk-jakarta.html).

Correia, P. R. M., Valle, B. X., Dazzani, M., Malachiasm, M. E. & Infanta, M. E. (2010). The importance of scientific literacy in fostering education for sustainability: Theoretical considerations and preliminary findings from a Brazilian experience. *Journal of Cleaner Productions, 18*, 678–685.

Gormally, C., Brickman, P. & Lutz, M. (2012). Developing a Tests of Scientific Literacy Skills (TOSLS): Measuring undergraduates evaluation of scientific information and argument. *CBE-Life Science Education, 11*, 364–377.

Toharudin, U., Hendrawati, S. & Rustaman, A. (2011). *Membangun literasi sains peserta didik*. Bandung: Humaniora.

Treacy, D. J., Collins, M. & Kosinski, M. S. (2011). Using the writing and revising of journal articles to increase science literacy and understanding in a large introductory Biology laboratory course. *Atlas Journal of Science Education, 1*(2), 29–37.

Ideas for 21st Century Education – Abdullah et al. (Eds)
© *2017 Taylor & Francis Group, London, ISBN 978-1-138-05343-4*

Developing historical thinking skills in learning history through teaching and learning methods

E.M. Karima
Pasundan University, Bandung, Jawa Barat, Indonesia

D. Supardan & A. Zainul
Universitas Pendidikan Indonesia, Bandung, Jawa Barat, Indonesia

ABSTRACT: This article attempts to discuss historical thinking skills of high school students. To investigate the matter, the present study focuses on the impact of conventional learning methods commonly used by history teachers; more specifically, it is concerned with the effects of lecturing method, discussion method, assignment method, and question and answer method on historical thinking skills. Senior high school students across Bandung were chosen randomly as the participants of this study. The method used was an analytical survey with a quantitative approach through linear regression techniques for analyzing the influence of the independent variables on the dependent variable. The findings show that lecturing, discussion, assignment, and question and answer methods had influence on historical thinking skills with a total score of 38.5%, while the rest was influenced by other factors not discussed in this study. Teachers might use appropriate learning methods to improve students' historical thinking skills and explore the knowledge in interpreting history.

1 INTRODUCTION

Teaching and learning history does include remembering some dates and names, but what is most important is to get a basic understanding of the time period in which the events occurred. A good understanding of teaching materials will be a meaningful learning for life in the society. This is in line with Hasan's argument (2012) that "education histories need not limit themselves to the science alone in which it also basically has the ethical aspects and other valuable aspects" (p. 25). Historical thinking skills for learners can be improved by the teacher through various learning processes, and one of them is the learning method. Moore noted (1989, p. 304), "As teachers we feel that we cannot be replaced by a textbook or a videotape." Conducted in state senior high schools in Bandung in the history subject, this study therefore aims to find how lecturing, discussion, and question and answer, and assignment methods in teaching historical thinking skills are implemented using the survey method. The researchers chose the sample randomly to see the effects of teaching methods on learners' historical thinking skills; furthermore, the researchers analyzed which method had significant influence on the students' historical thinking skills. "Survey method is a method of collecting primary data by providing questions to respondents"

(Jogiyanto, 2014, p. 3). The approach used in this study was a quantitative survey method. Generalization of this method can therefore represent a large population. This method takes a lot of data from respondents using a questionnaire. Unlike interview, a survey is conducted without having direct communication with the respondents. "The questionnaire survey is used to communicate with the respondents" (Jogiyanto, 2014, p. 4). Through questionnaire, the researchers obtained answers to the proposed questions. The hypothesis of this study is there are significant influences of the independent variables, namely the lecturing method (X1), the discussion method (X2), assignment (X3), and question and answer method (X4) on the dependent variable, which is historical thinking (Y). It is expected that history teaching will help pupils develop important and broadly applicable skills. Moreover, Hamalik (2007, p 31) noted that the learners' skills of understanding historical items might be linked with the result of historical study, patterns and values, mastering definitions, attitudes and appreciation, and skills with curious and enquiring minds.

Relevant research has been conducted by Roaet Waedakae. His research was conducted using a survey to see each influence on learning outcomes and also the influences of all factors on the learning outcomes of students. Through a survey study, the result showed that there was a positive influence, so that teachers need to improve their competence so as to improve the students learning outcomes. Secondly, there was research conducted by Richardson Dilworth. The purpose of the study was to compare three modes of historical thinking as a warehouse of analogues, a set of historical institutionalist models of stability and change, and as a "stream" in terms of the likelihood that they provide useful skills. The results of the study showed that historical institutionalist models possess the greatest potential for skill building and historical analogizing.

2 RESEARCH METHODS

2.1 Learning methods

Learning method is the way in which the teacher carries out historical items to successfully meet the subject's goals. According to Kropp (1973), "A statement of objectives would help focus teaching. It contains expectations which are held for him, and it would provide him with references by which to assess his progress" (p. 758). With a statement of objectives, the learning that takes place will become more meaningful since the teachers respectively share and deliver the lesson objectives and the expected progress of the undertaken learning.

2.2 Historical thinking

Historical thinking refers to the skills of learners to look at the past and use their mind to go beyond the past, so that the interpretation of the learners nowadays probably will have an influence from the past. Keirn & Martin (2012) demonstrated that "historical thinking is a phrase that is becoming a standard in conversation about teaching history. Not necessarily a new idea—it calls for teaching historical habits of mind to go back at least a century, but there has been an explosion of resources in the past two decades that support making history classroom sites of analysis, interpretation, and questioning, rather than of memorization" (p. 489). In addition, Drake and Brown (2003, p. 466) described that a good teaching history is done through stimulating learners to look for other sources related to the definable subject matters in the classroom. The learning process developed by Gagne is based on the distribution of stimulation received by the five senses to the central nervous system as a form of information. Placing the functions of memory in the sequence of encoding, storage, and retrieval, provides a simple way of dividing it up (Green, ed., 2003, p. 53).

2.3 Historical thinking of Sam Wineburg

Historical thinking is not the skills to memorize names, dates, or flow of historical events, but learners are required to be able to understand the meaning of each historical important event: "Actually it goes against the grain of how we ordinarily think, one of the reasons why it is much easier to learn names, dates, and stories than it is to change the basic mental structures we use to grasp the meaning of the past" (Wineburg 2001, p. 7). Lee also noted (2005), "It is vital for the student to have opportunities to evaluate and discriminatingly choose sources". Teachers need to interact with learners and the learning process will be more active; hence, it will encourage many questions on either the teachers or fellow learners.

2.4 Research method

The objective of this study is to see how the condition of teaching history in Bandung through lecturing method, discussion method, and question and answer method, and students' assignment on historical thinking skills is. The approach used in this study was a quantitative survey method. Generalization of this method can therefore represent a large population. This method takes a lot of data from respondents using a questionnaire. The population used in this study was the students of state senior high schools (SMAN) across Bandung. From 37,296 students, 412 were chosen randomly. Collecting the necessary data in this study was done using a questionnaire designed by the researchers. It contains a list of questions regarding the learning methods used to improve the students' historical thinking. The questionnaire used the ordinal and Likert scale techniques. After obtaining the data, the researchers conducted statistical tests using multiple regression analysis.

2.5 Analysis method

The analysis used in this present study was multiple regression analysis or linear regression using SPSS 21.0 for Windows. This analysis was used to determine whether or not there were influences of the independent variables on the dependent variable. This study draws attention to four independent variables, namely students who received lectures, discussions, assignments, and frequently asked questions, and one dependent variable called the skills of historical thinking. Conducted to 412 respondents of high school students in Bandung, the survey distributed questionnaires randomly and used a cross sectional approach. Finally, this investigation was done only once without any further treatment to the respondents.

3 RESULTS AND DISCUSSION

3.1 Classical assumption test

Before conducting the linear regression analysis, the classical assumption tests were done. The tests consisted of normality, multicollinearity, autocorrelation and heteroscedasticity tests. The classical assumption tests were done as the requirement to undertake the regression model. Furthermore, as stated by Gani & Amalia (2015), "The result of regression model can meet the standards so that the parameters obtained become logical and reasonable" (p. 123).

Data were analyzed using parametric statistics. The distribution patterns of the histogram are as follows: (a) a bell shape means that the distribution of data is particularly normal. Tests were also done using a normal probability plot; (b) to detect whether or not the data used might be distributed normally. The residual score is normally distributed, for the data distribution is located around the diagonal line and follows the diagonal direction which is from left side to the upper right.

The score of the standard error of each regression coefficient might also be infinite. Calculations were done by looking at variance inflation factor (Variance Inflation Factor/VIF). If it is more than 10 (VIF > 10), then multicollinearity occurs; meanwhile, when VIF is less than 10 (<10) then it indicates that there is no multicollinearity or tolerance score is getting close to 1. VIF score in X1 is 2.987, which means that the score is less than 10 (<10) and tolerance score in X1 is 0.335. A similar range of scores is also observed in X2, X3 and X4 variables, in which each had a VIF score of 2.511, 1.701, and 1.825 respectively. Because the score is smaller than 10 m, so it can be inferred that there is no multicollinearity.

Figure 2 shows that there is no heteroscedasticity due to the scatter plot. The data distribution tends to be random and does not draw particular meaningful patterns.

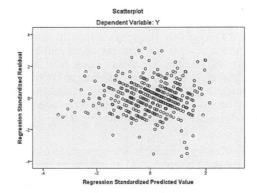

Figure 2. The result of heteroscedasticity assumption test.

Table 1. The Result of multicollinearity assumption test.

Model	Collinearity statistics	
	Tolerance	VIF
X1	0.335	2987
X2	0.398	2511
X3	0.588	1701
X4	0.548	1825

Coefficients[a]
a. Dependent Variable: Y.

Table 2. The result of autocorrelation assumption test.

Model summary	
Model	Durbin-Watson
1	1.226

a. Predictors: (Constant), X4, X2, X3, X1.
b. Dependent Variable: Y.

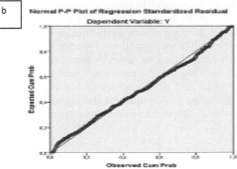

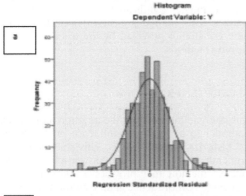

Figure 1. (a) Dependent variable histogram of historical thinking skills, (b) Normal probability plots.

If Durbin-Watson's score is less than 1 and greater than 3 (<1 and >3), then the criteria for autocorrelation are violated. The Durbin-Watson's score from the independent variables of lecturing teaching, discussion, assignment, and question and answer methods to the dependent variable of historical thinking is 1.226. This means that there is no autocorrelation beyond the regression model. This is obvious because 1.226 is greater than 1 and less than 3.

3.2 The influence of learning methods on historical thinking skills

Table 3 shows the results obtained from the calculation of multiple linear regression. Question and answer method has the highest impact compared to other conventional methods. Active learners in the learning activities might become one of the influences in developing the skills of historical thinking. "Through encouraging students to formulate educated reviews for their responses and express opinions, teachers are able to assess how familiar or interested they are in the material" (Critelli & Tritapoe. 2010, p. 2). Lecturing, discussion, and assignment methods have smaller influences on the development of historical thinking skills.

3.3 History learning in developing historical thinking skills

The learning method with the greatest influence on students' historical thinking skills was question and answer method. Other useful methods were lecturing method, discussion and assignment method, respectively. The zero-order correlation (r) for the question and answer variable (X4) is 0.543; the score of correlation (R) is 0.543, and the contribution of R2 is 0.295. Unique distribution percentage is 29.5%, which means the influence of question and answer method (X4) on historical thinking is not affected by other variables. The result of the zero-order correlation (r) for assignment variable

Table 3. Learning method regression of historical thinking.

| Variable | Correlation | | | |
	r	R	R^2	Percentage unique distribution
Question and Answer (X4)	0.543	0.543	0.295	29.5%
Assignment (X3)	0.466	0.613	0.375	1.1%
Discussion (X2)	0.48	0.62	0.385	1%
Lecturing (X1)	0.525	0.603	0.364	6.9%
Total				38.5%

(X3) is 0.466; the score of correlation (R) is 0.613; the contribution of R2 is 0.375; and unique distribution percentage is 1.1%, which means the amount of influence of the assignment variable (X3) on historical thinking is not influenced by other variables. The score of Zero-order correlation (r) for discussion variable (X2) is 0.480; the score of correlation (R) is 0.620; the contribution of R2 is 0.385; and the unique distribution percentage is 1%, which means the influence of discussion variable (X2) on historical thinking is not determined by other variables. Finally, the score of Zero-order correlation (r) for lecturing variable (X1) is 0. 525; the score of correlation (R) is 0.603; the contribution of R2 is 0.364; and the unique distribution percentage is 6.9%, which means the influence of lecture variable (X1) on historical thinking is not determined by other variables.

Therefore, all of the methods, namely question and answer, assignment, discussion, and lecturing methods, had influence on historical thinking. The total score is 38.5%, while the rest 61.5% was due to other factors not discussed in this study. The understanding is processed within the learners' mind in the form of further information on the entries in both short-term memory and long-term memory. Historical thinking then is a skill possessed by students based on the results of the processing, and this might be indicated through the behavior changes. This change is clearly demonstrated, for there are stimuli provided by the teacher through learning methods.

4 CONCLUSIONS

This study found out that the question and answer method had a significant influence on the learners' historical thinking skills. This study thus recommended the question and answer method as an effective method that provides a stimulus to the learners to be more active in the classroom activities. On the other hand, lecturing, discussion, and assignment methods had smaller influences on the historical thinking skills than the question and answer method. The participation of students in asking or answering questions develops their skills of historical thinking. The learners' constructive thinking will develop, and they start making interpretation of knowledge. Finally, learners also argue their opinion based on the sources that they comprehend.

REFERENCES

Critelli & Tritapoe. 2010. Effective questioning techniques to increase class participation. *Department of Teacher Education Shippensburg University.* 2(1): 1–7.

Drake, F. & Brown, S. 2003. A systematic approach to improve students historical thinking. *The History Teacher.* 36(4): 465–489.

Gani, I & Amalia, S. 2015. *Alat analisis data: aplikasi statistik untuk penelitian bidang ekonomi dan sosial.* Yogyakarta: Andi.

Green, V. 2003. *Emotional Development in Psychoanalysis, Attachment Theory and Neuroscience: Creating Connections.* New York: Taylor and Francis Group.

Hamalik, O. 2007. *Proses belajar mengajar.* Jakarta: Bumi Aksara.

Hasan, S. 2012. *Pendidikan sejarah indonesia: isu dalam ide dan pembelajaran.* Bandung: Rizqi Press.

Jogiyanto, H. 2014. *Pedoman: survei kuesioner (mengembangkan kuesioner, mengatasi bias, dan meningkatkan respon).* Yogyakarta: BPFE-Yogyakarta.

Keirn, T. & Martin, D. 2012. Historical thinking and preservice teacher preparation. *The History Teacher.* 45(4): 489–492.

Kropp, R. 1973. Teaching method. *American Journal of Agricultural Economics.* 55(4): 757–761.

Lee, P. 2005. Putting Principles into practice: Understanding History. In S. Donovan & J. Branford (Eds.) *How Students Learn: History Mathematics and Science in the Classroom.* Washington: National Academic Press. 29–78.

Moore, D. 1989. Teaching history. *Cultural Literacy.* 62(7): 303–305.

Wineburg, S. 2001. *Historical thinking and other unnatural acts: charting the future of teaching the past.* Philadelphia: Temple University Press.

Ideas for 21st Century Education – Abdullah et al. (Eds)
© *2017 Taylor & Francis Group, London, ISBN 978-1-138-05343-4*

The effect of the outdoor learning model on biology learning motivation in SMAN 2 Bekasi on biodiversity matter

E. Suryani
Universitas Indonesia, Indonesia

ABSTRACT: The purpose of the study is to investigate the effect of the outdoor learning model on biology learning motivation on biodiversity matter. The study applied a quasi-experimental design, divided into an experimental group and a control group. In order to achieve the purpose of the study, the experimental group applied the outdoor learning model and the control group applied speech and discussion learning. Data was collected through a pretest and posttest learning motivation score. The data analysis of an independent t-test shows that there is a difference between the posttest means for both the control and experimental groups. Findings confirm that the outdoor learning model can motivate students in learning biology. Teachers are expected to apply this model to biology learning at their respective schools to achieve better learning motivation.

1   INTRODUCTION

Learning biology is an integrated process that deals with how to systematically find out and understand nature (Depdiknas, 2003). Therefore, the biology learning process in high school is not just a collection of knowledge in the form of facts, concepts, or principles, but also a process of discovery. The learning process emphasizes providing direct experience to develop competencies in order to understand the surrounding nature scientifically (Mulyasa, 2007). The learning model that is needed in biology learning is a learning model that puts students in direct contact with the object to be studied by direct observation. Selection of the learning model will affect the students' motivation (Sutrisno, 2008). Motivation is one of the factors that determines the level of success or failure of the process of learning (Slavin, 2008). Therefore, teachers are required to implement an innovative learning model, which will be able to raise the motivation of students to enrich their learning experience, and facilitate students to interact with their environment as a learning resource.

Learning models that involve learning through experience are at the core of the model of learning outside the classroom (Dumouchel, 2003). This learning experience is expected to provide a better understanding for the students by being confronted directly with the object being studied.

2   LITERATURE REVIEW

2.1   *Outdoor learning mode*

The learning model is a plan or pattern that can be used to shape the curriculum (long-term learning plan), design learning materials, and guide learning in the classroom or other location (Joyce & Weill, 2000). The learning model can be used as a reference for learning, meaning that teachers can choose an appropriate learning model that is efficient to achieve education goals.

Outdoors, in the context of education, is an education that emphasizes the process of experiential learning, as shown in the following passage: "outdoor education is an experiential process of learning by doing, which takes place primarily through exposure to the out-of-doors" (Priest & Gass, 1997 in Dumouchel, 2003). This was saying that outdoor education is a process that is gained through the experience of learning by doing things and that the experience is gained in the outdoors.

Things can be learned outside of the classroom, and the outdoors is an effective setting in which to teach students about environmental issues (Martin, 2003). The outdoors is an effective place to teach learners about things related to the environment. Outdoor learning can be concluded as a learning process through the experience gained from learning by doing in an outdoors environment, which can be adapted to learning topics such as those relating to the environment.

The outdoor learning model has four stages. Each stage has a different function and purpose. These stages are:

1. Ice breaking

Ice breaking plays an important role in order to stimulate curiosity and build the concentration of participants, as this event serves as a means of introduction between participants and facilitators, to evoke the spirit of the participants, as well as to minimize the passivity of the participants.

2. Material

Generally, this stage raises the curiosity of the participants to the core of learning.

3. Evaluation

This stage can be done with an interesting activity. For example, make a poem, story, painting, and so on. It can leave memorable moments for them even after leaving school.

4. Sharing

This stage is the exchange of experiences after the site visit and is very important. Not all participants know or see something that is considered to be of interest to them. Some like insects, but some like the plants, and so on. The exchange of experiences after returning from the field is expected to broaden the participants' knowledge (Widyandani, 2008).

2.2 *Learning motivation*

Motivation and learning are two things that are mutually inclusive. Motivation is the basic impulse that drives a person to behave (Mulyasa, 2007). Learning is a step change in the behavior of individuals all over the relative settlements as a result of experience and interaction with the environment that involves cognitive processes (Syah, 2006).

Motivation to learn is the overall driving force within the students who lead learning activities, which ensures continuity of learning activities and provides direction on the learning activities; then the purpose desired by the students is reached (Winkel, 1991 in Uno, 2006). It can be concluded that the motivation to learn is a basic urge that drives students to learn.

Many of the strategies, techniques, approaches or learning models are used by teachers to improve students' motivation. Students will be motivated to learn and will be eager to learn and avoid boredom if educators are clever and raise the motivation to learn. However, students will be lazy, bored, and listless learners if teachers are less able to raise the students' learning motivation (Mulyasa, 2007).

Teachers are required to raise the motivation to learn. Some of the ways that teachers can improve the motivation of learners are, among others:

(1) clarify the objectives to be achieved; (2) generate interest in students; (3) create a pleasant atmosphere in the study area; (4) give reasonable praise to every student who has success in learning; (5) give the assessment; (6) give comments on the results of student jobs; (7) create competition and cooperation (Sanjaya, 2008).

Indicators of motivation to learn can be classified as follows: (1) their desires and wishes for success; (2) lack of motivation and learning needs; (3) their hopes and ideals for the future; (4) lack of respect in the classroom; (5) the existence of interesting activities in learning; (6) the existence of a conducive learning environment, allowing somewhere that students can learn well (Uno, 2006).

3 RESEARCH METHODS

The research design used in this study is a pretest and posttest two group design. Sampling was done by purposive sampling where two classes out of eight were chosen. There was one experimental class and one control class. The number of samples in the experimental class and control class were respectively 20 students. The techniques used to collect the data in this study include: instruments motivation and observation sheet. The data gathered were analyzed by using SPSS.

4 RESULTS AND DISCUSSION

At the beginning of the study, in order to check the student's motivation, a pretest was administrated to the student at the beginning of the lesson. Table 1 presents the mean, standard deviation, and standard error of mean for the participants in the pretest.

According to the statistics displayed in Table 1, the mean difference of the pretest for the two groups is 0.1, which is not statistically significant. In order to make sure that the difference between the pretests of the experimental group and control group is not significant, an independent sample

Table 1. Descriptive statistic for score of learning motivation of the experimental and control groups in the pretest.

Code	N	Mean	Std. Deviation	Std. Error Mean
Pretest				
Control	20	15,40	3,43971	−76914
Experimental	20	15,50	3,15394	−70524

Table 2. The result of the t-test for the experimental and control groups in the pretest.

| Group | Mean | T-Test | | | | | | |
		T	df	P	Mean Difference	Std. Error Difference	Lower	Upper
Experimental	15.5	0,096	38	0,924	0,1	1,04	−2,01	2,21
Control	15.4							

Table 3. Descriptive statistic for score of learning motivation of the experimental and control groups in the posttest.

Code	N	Mean	Std. Deviation	Std. Error Mean
Pretest				
Control	20	14,90	2,78908	−62366
Experimental	20	16,85	2,87045	−64185

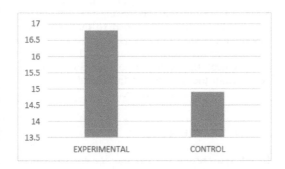

Figure 1. Graphical representation of the means for hypothesis.

t-test was run between the pretest scores of the two groups. Table 2 illustrates the result of the t-test.

Table 2 illustrates the summary of the result of the t-test. It indicates that the observed t for the experimental and control groups in the pretest is 0,096 (obs.t = 0,096) while the critical t (df = 38, α = 0, 05) is 1, 68 (crit.t = 1, 68), which is higher than the observed t (obs. T < crit t); therefore in the pretest the experimental and control groups were not significantly different.

As is clearly shown in Table 3 and Figure 1, the posttest mean for the experimental group is 16,8500, which is higher than that of the control group, being 14,9000, and having a mean difference of 1,95. Therefore, it can be said that the two groups have different performances compared to the pretest. In order to ascertain that the difference between the posttests of the experimental group and the control group is significant, an independent sample t-test was run between the posttest

scores of the two groups. Table 4 illustrates the result of this t-test.

Table 4 illustrates the summary of the result of the t-test. It indicates that after implementing the treatment and measuring the statistics, the observed t for the experimental and control groups in the posttest was 2,179 (obs.t = 2,179), while the critical t (df = 38 and α = 0, 05) is 1, 68, which is not higher than the observed t (obs.t > crit.t); as a result, there was a statistically significant difference between the experimental and control groups performances on the posttest.

After implementing the outdoor learning model, the progress of both the experimental and control groups in learning motivation was compared using a t-test. The result indicated that the posttest scores were significantly increased.

The increased motivation of the experimental class showed that learning outside the classroom (outdoor learning) has a positive influence on the motivation of high school students to learn biology. Learning outside the classroom (outdoor learning) is learning that involves the philosophy of experiential learning or learning through experience (Priest and Gass, 1997 in Dumouchel, 2003).

In the selected experiment in teaching methods, models of outdoor learning were used where students could find and associate learning experiences with the material learned. The selection of the learning model will affect the students' motivation (Sutrisno, 2008). Results showed that lessons fulfillment was 97%. This study fulfilled the included criteria very well according to the fact that the average increase in student motivation in the experimental class is higher than that in the control class. It has been proven that outdoor learning can make the subject matters being taught easier to understand and increase students' motivation.

At the icebreaking stage, the teacher raised student motivation by clarifying the objectives and aroused the interest of the students through games. Students' understanding of the learning objectives can cultivate their interest in learning, which in turn can increase their motivation to learn (Sanjaya, 2008). The clearer the goals to be

Table 4. The result of the t-test for the experimental and control groups in the posttest.

| Group | Mean | T-Test | | | Mean Difference | Std. Error Difference | Lower | Upper |
		T	df	P				
Experimental	16,85	2,179	38	0,036	1,95	0,89	0,13	3,76
Control	14,9							

achieved, the stronger the increased student motivation will be. In the outdoor stages of learning, the teacher raised the motivation of the students to create a good atmosphere for learning and created competition and cooperation. The pleasant learning atmosphere was gained by learning with a lively atmosphere that was fresh and free from tension (Depdiknas, 2008). Students gained a new atmosphere from the usual classroom atmosphere that they feel, as saturation and tension is not visible when the students' attention is focused on outdoor learning activities. In addition, students were grouped into small groups, which is expected to foster cooperation and healthy competition between groups. Healthy competition can provide a good influence on the success of the students' learning process. Student competition was made possible through trying earnestly to get the best results (Depdiknas, 2008).

At the evaluation stage, the teacher gave students the opportunity to present their observations, provided an opportunity to file a response and give a response to the feedback, and gave students' their opinions with compliments and comments. At the sharing stage, the teacher evaluated learning activities, gave students the opportunity to share experiences, to accommodate suggestions and feedback, and provided a verbal assessment on the outdoor learning activities. Reasonable compliments were given for every student success, and assessments and comments were provided on the results of the students' work, which is an aspect that can increase student motivation (Sanjaya, 2008).

Stages for models of learning outside the classroom (outdoor learning) make sure that students are in a condition to have a free sense of curiosity in learning and get experience with the lesson materials directly. Such circumstances increase students' motivation for learning and are used to gain an understanding of the material that has been given. This proves that the amount of knowledge gained during the students' learning will be reflected in attitudes, behavior, mindset, and motivation (Widayanti, 2003).

In the control class, the success of the lesson was 80%; however, there is a limitation of the study particularly regarding the implementation of outdoor learning. It is reflected from the lesson plan that the implementation of outdoor learning was not optimum.

The learning experience in the experimental class does not reflect the typical experiences of the model of learning outside the classroom (outdoor learning), this was due to the lack of references to the learning model outside the classroom (outdoor learning) so that the nature of the learning model was not reached. In the lesson plan (RPP) for the model of learning outside the classroom (outdoor learning) the experimental class still described the RPP as learning in a classroom, where the learning should be more developed so that the activities are not so stiff and formal. Student worksheets for outdoor activities learning also do not reflect a worksheet that can support outdoor learning activities in the material of biological diversity, in both content and appearance.

5 CONCLUSIONS

Based on the results of this study, it can be concluded that there was a significant model of learning outside the classroom (outdoor learning) on the motivation to learn biology for students at SMAN 2 Bekasi on material biodiversity. The model of learning for outside the classroom (outdoor learning) can be used in the process of learning biology to improve students' motivation. The use of the model of learning for outside the classroom (outdoor learning) adjusted to the subject matter to be studied by considering the effectiveness in achieving the learning objectives.

ACKNOWLEDGMENT

My acknowledgment is given to LPDP (Indonesia Endowment Fund for Education) for taking part in providing me with the financial support on my scholarship.

REFERENCES

Depdiknas. (2003). *Kurikulum 2004 SMA: Pedoman khusus pengembangan silabus dan penilaian.* Direktorat Pendidikan Menengah Umum.

Depdiknas. (2008). *Proses pembelajaran di kelas, laboratorium, dan lapangan.* Jakarta: Direktorat Tenaga Kependidikan.

Dumouchel, D. (2003). *Experiential practice: Outdoor, environmental and adventure education.* Accessed on November 8, 2009 at 9:57 pm from http://www.newhorizons.org/strategies/environmental/dumouchel.

Holsinger, K. (2007). Functional diversity. Accessed on September 18, 2009 at 10:40 GMT from http://www.DiversityandStability.com.

Joyce, B. & Weill, M. (2000). *Models of teaching.* Boston: Allyn and Bacon.

Martin, S.C. (2003). *The influence of outdoor schoolyard experiences on students' environmental knowledge, attitudes, behaviors, and comfort levels,* Accessed on 10 January 2010 at 11:45 pm. from http://eric.ed.gov/PDFS/EJ852422.pdf.

Mulyasa, E. (2007). *Menjadi guru profesional: menciptakan pembelajaran kreatif dan menyenangkan.* Bandung: Remaja Rosdakarya.

Sanjaya, W. (2008). *Strategi pembelajaran berorientasi standar proses pendidikan.* Jakarta: Kencana.

Slavin, R.E. (2008). *Educational psychology: Theory and practice.* Jakarta: PT index.

Sutrisno, J. (2008). *Pengaruh metode pembelajaran inquiry dalam belajar sains terhadap motivasi belajar siswa.* Jakarta: Erlangga.

Syah, M. (2006). *Psikologi belajar.* Jakarta: Raja Grafindo Persada.

Uno, H.B. (2006). *Teori motivasi dan pengukurannya: analisis di bidang pendidikan.* Jakarta: Bumi Aksara.

Widayanti, N. (2003). *Efektivitas pembelajaran geografi melalui metode outdoor study dalam upaya meningkatkan minat belajar siswa.* Accessed on December 29, 2009 at 11:20 from http://pakguruonine.pendidikan.net.

Widyandani, S.B. (2008). *Belajar bersama alam.* Accessed on October 24, 2009 at 9:32 pm from http://bintangbangsaku.com/artikel/2008/03/.

Ideas for 21st Century Education – Abdullah et al. (Eds)
© *2017 Taylor & Francis Group, London, ISBN 978-1-138-05343-4*

Spatial thinking in frame-based learning of plant anatomy and its relation to logical thinking

E. Ermayanti
Universitas Pendidikan Indonesia, Bandung, Indonesia
Universitas Sriwijaya, Palembang, Indonesia

N.Y. Rustaman & A. Rahmat
Universitas Pendidikan Indonesia, Bandung, Indonesia

ABSTRACT: A study on the framing of spatial thinking in a plant anatomy course was conducted to investigate spatial thinking in frame-based learning of plant anatomy and its relation to logical thinking. This research used a pre-experimental research design. A number of biology education students (n = 42) were involved as participants. Data were collected using instruments of observation, a spatial thinking test, and a Test of Logical Thinking (TOLT). The data were analyzed quantitatively. Research results show that the spatial thinking of students in the frame-based learning of plant anatomy involved: (i) generating a representation in 2D and 3D; (ii) maintaining the characteristics of tissue in working memory to construct 3D structures; (iii) scanning the 2D and 3D representations; and (iv) transforming the representations. These were factors that improved students' logical thinking on each indicator. The students' logical thinking before and after the frame-based learning instruction resulted in the following pretest and posttest scores: (1) proportional reasoning (42.9 and 64.3); (2) controlling variables (3.6 and 11.1); (3) probabilistic reasoning (9.5 and 15.5); (4) correlational reasoning (11.9 and 27.4); and (5) combinatorial reasoning (17.9 and 45.2). Analysis of the relationship between spatial thinking and logical thinking showed significant correlation. It is concluded that frame-based plant anatomy learning improves students' spatial thinking and logical thinking.

1 INTRODUCTION

Spatial ability is an important skill in various activities in daily life and some carriers depend heavily on spatial ability. Therefore, it is a fundamental ability in the 21st century (Diezmann & Lowrie, 2012). The importance of spatial thinking was also shown in the work of the National Research Council (2006), which stated that "learning to think spatially" is a key skill in various educational curriculums. The importance of spatial ability in science learning is related to one's ability to solve spatial problems, particularly mental rotation, which comprises the ability to manipulate and transform 3D objects in the brain (Brownlow et al., 2003). Visual representation, in the form of 3D objects, is a very important part of understanding the phenomena in biology and mechanics and in solving spatial problems (Bolotin & Nashon, 2012).

Logical thinking is a skill, which is determined in the period of abstract process in Piaget's cognitive development phase. Logical thinking is a mental operation used by individuals when they solve specific problems (Piaget, 1969). There are five different modes of formal logical thinking: proportional reasoning, controlling variables, probabilistic reasoning, correlational reasoning and combinatorial reasoning (Tobin & Capie, 1981). The students solve the problems by undertaking various mental practices or rules or by doing some abstraction and generalization. This activity is related to spatial thinking and logical thinking.

Spatial thinking is also very important in the plant anatomy course. The demands of the plant anatomy syllabus requires students to understand the structures and functions of the cells, tissues, or organs of plants, which are three-dimensional (3D) structures. Students need to recognize the characteristics of plant tissue (for example, the shape, size, positions, cell wall thickness, air cavity and another characteristics) and to relate it to its function. In order to understand the structure of plant anatomy, spatial thinking is much needed.

The results of the preliminary test in plant anatomy give the information that the plant anatomy course strongly requires spatial thinking ability in each student, because cell structure and plant tissue are 3D structures and are abstract, whereas the pictures contained in the student's handbook and the results of microscopic observation are two-dimensional (2D) structures. So to know the characteristics of plant tissues, students should be able to observe carefully and make representations in 2D and 3D. But in reality, the students' spatial thinking in plant anatomy was less than satisfying, especially for the parts related to thinking about three-dimensional structures, positions and knowing every part of the cellular structure of a tissue or an organ. In general, students find difficulties in constructing representations from 2D into 3D and transforming representations (for example, creating a new perspective) to understand the structure of plant anatomy as a whole.

Students' difficulties in understanding structure and function at a cellular level was also found in earlier studies (Lazarowitz & Naim, 2013). Spatial visualizations of 2D or 3D models might help to resolve the spatial difficulties encountered when learning anatomy (Hoyek et al. 2014). It is much needed to be able to understand spatial concepts in a better way (Hoffler 2010). The previous research on spatial concepts did not focus on the types of spatial thinking in framing based plant anatomy and its relation to logical thinking. The previous research instead focused on content and students' spatial ability (Hoffler 2010, Lazarowitz & Naim 2013), and the role of visualizations of 2D or 3D models to resolve spatial difficulties (Hoyek et al., 2014; Jones et al., 2011). Research on how to frame the cognitive processes in spatial thinking for the plant anatomy course and its relation to logical thinking is not yet available.

Based on the analysis of the previous studies, it is clear that studies on how to understand the cognitive processes of spatial thinking on a framing based plant anatomy course and its relation to logical thinking have never been done. This paper focuses on the discussion of the four cognitive processes in spatial thinking that seem to describe a student's spatial thinking in plant anatomy and the relation of the student's spatial thinking to their logical thinking ability.

2 LITERATURE REVIEW

In this section, we described two points from the research literature that were related to this research. The first point describes the concept of spatial thinking and the second point describes the framing concept.

Spatial thinking can be developed based on: (1) the concept of space; (2) the representation; and (3) the process of reasoning (National Research Council, 2006). Spatial thinking involves several cognitive processes, such as visualizing relations; imagining between one scale transformation and another scale; remembering locations of objects, their shapes, and moves; rotating the objects to see other sides; creating a new perspective; transforming object orientations, and others (National Research Council, 2006). Visualizations of 2D or 3D models might help to resolve the spatial difficulties encountered when learning anatomy (Hoyek et al., 2014). Moreover, involving students in constructing the 3D structure models of the cell will improve the students' understanding of the cellular structure and function (Lazarowitz & Naim, 2013).

Framing is a dynamic and ongoing process, where people continue to constantly frame and reframe how to understand "what is happening" in a small adjustment of the scheme (Berland & Hammer, 2012). Frame is an individual feeling about "what is it that's going on here?" (Goffman, 1974). Framing in the biology lesson and class and the social reality, particularly in the biology learning process, influences the ability of scientific argumentation of the students. (Berland & Hammer, 2012; Boerwinkel et al., 2014), resolves the cognition pressure (Autin & Croizet, 2012), recalls and transfers information stored in the memory (Engle et al., 2011) and builds the ability to explain (Boerwinkel et al., 2014).

3 RESEARCH METHODS

This research was pre-experimental research with one group of pretest-posttest designs. The instruction of the plant anatomy course was generally consistent with framing.

3.1 Participants

This research was conducted at the Biology Education Study Program at a state university in South Sumatra, Indonesia. This research involved 42 students (41 females and 1 male), who were enrolled in the third semester and taking the plant anatomy course.

3.2 Instruments and procedures

The instruments used in this research were a spatial thinking test and a Test of Logical Thinking (TOLT) instrument. The spatial thinking test was self-developed based on the spatial thinking cognitive process (Kosslyn, 1978). The spatial thinking test was specifically designed for the research and it was validated by experts through field testing. The test instrument employed four indicators, namely: (1) generate a representation; (2) manage and maintain the representation in working memory;

(3) scanning the maintained representation in working memory; (4) transform a representation with rotation or view the object from a different perspective (Kosslyn, 1978). Whereas, the logical thinking test instrument consisted of 10 test items, in the form of multiple choice questions with four options with reasons (Tobin & Capie, 1981).

The dynamic of framing was adopted and modified from framing terms in earlier studies (Engle et al., 2011; Autin & Croizet, 2012, Goffman, 1974). Framing was especially designed into: concepts questions, spatial-related concepts questions, directing sentences, and also examples of 2D and 3D plant anatomy pictures that had been constructed well (worked examples). Framing directed the students to think spatially about plant anatomy concepts.

The instruction at each stage of learning is consistent with framing that directs students to think spatially. Students work in a group to solve spatial problems and concepts of plant anatomy. After solving the problems related to the concepts of plant anatomy, the participants worked on the posttest.

3.3 Data analysis

A test was administered at the beginning and at the end of the frame-based plant anatomy learning. Quantitative data were obtained by calculating the average or percentage in each spatial thinking and logical thinking indicator. The criteria of the test scores were classified by the referencing and modifying of Bao et al. (2009), namely: ≤ 34 (very low); 35–50 (low); 51–65 (moderate); 66–80 (high); ≥ 81 (very high). The improvement in students' spatial thinking and logical thinking was measured with an average N-Gain (Meltzer, 2002). Statistical analysis was performed with SPSS version 22 for windows. The descriptive analyses were used to explain the students' cognitive processes in spatial thinking during the frame-based plant anatomy learning process.

4 RESULTS AND DISCUSSION

The results showed that there was an improvement in the students' spatial thinking activities. The spatial thinking activities were observed by focusing on the spatial activities that were expected to emerge at every step of the frame-based plant anatomy course. The analysis of the students' spatial thinking activities in learning with frame-based instruction involved several cognitive processes that supported the students' spatial thinking, such as recognizing the shape, size, positions, cell wall, cellular air space and other characteristics of plant tissue. Students

identified and scanned the characteristics of each tissue from a microscopic slide and created a 2D representation to help them to keep the concepts in their working memory. Students scanned a 2D representation and maintained in their working memory the relative shape, size, positions, cell wall, cellular air space and other characteristics of plant tissues. They focused their attention on some parts to construct these into 3D. By scanning the picture from the examples, students can construct the 2D pictures into 3D or vice versa.

The visualization of the tissue structures in 2D and 3D representations gave students complex information about the shapes and locations of one of the tissues or one of the various tissues. In addition, the students created the representation with multiple anatomical views (for example, an analysis of a microscopic slide from a cross section, longitudinal section or radial section) from different perspectives. These cognitive processes support students' spatial thinking.

This result was supported by research findings that show an improvement in the spatial thinking and logical thinking of students after having experienced framing based learning with an average N-Gain of 53.7 and 17.3 respectively (Table 1). The posttest score was significantly different when compared to the pretest score, with $\alpha < 0.05$. The percentage of spatial thinking in each indicator can be seen in Table 2.

Based on Table 2, it can be seen that there was an improvement of spatial thinking in each indicator.

Table 1. The mean score pretest-posttest spatial thinking and logical thinking.

	Score			
	Spatial thinking		Logical thinking	
	Pretest	Posttest	Pretest	Posttest
Average	28.11	67.27	17.14	31.67
Normality	0.20	0.16	0.07	0.08
N-Gain	53.7		17.3	

Table 2. Percentage of spatial thinking in each indicator.

Spatial thinking	Pretest	Posttest	Criteria
Generating a representation	30.00	73.95	High
Maintaining a representation	22.21	69.79	High
Scanning representation	20.00	71.94	High
Transforming representation	15.00	64.46	Moderate

It gives the information that frame-based plant anatomy learning can facilitate students in spatial thinking. Framing resolves the students' difficulties in thinking spatially. This result was also supported by the previous result, which showed that framing can resolve cognitive pressure in solving task difficulties, so it enhances working memory capacity (Autin & Croizet, 2012). Besides, framing in a learning context can enhance the ability to recall what is stored in the memory and sharing the knowledge with the students (Engle et al., 2011).

The framing of spatial thinking in plant anatomy also improved logical thinking ability (Tables 1 and 3). The posttest score was significantly different when compared to the pretest score, with $\alpha < 0.05$. The percentage of logical reasoning can be seen in Table 3.

Based on the posttest scores shown in Table 3, it can be seen that the highest score of the logical reasoning was proportional reasoning, with 64,29 (moderate), and the lowest was controlling variable, with 11,09 (very low). Framing based plant anatomy learning had more trained proportional, combinatorial and correlation reasoning compared to other reasoning. This is due to framing using trained cognitive spatial processes to create spatial visualization (for example, 2D representation and constructing 3D representation). It is directly related to students' ability to combine and analyze the proportion of 2D representations. Students had analyzed a 2D representation to its component parts and then combined these parts to construct a new 3D representation. These are strongly related to logical reasoning (for example, proportional reasoning, correlation reasoning and combinatorial reasoning).

The analysis of the relationship between spatial thinking and logical reasoning showed r (42) = 0.69 ($p < 0.01$) (significant correlation). Students with high spatial thinking have high logical reasoning. Improvements in spatial thinking will improve logical reasoning. This result is also supported by previous studies, which showed that visualizing the structure of 2D into 3D requires spatial perception, which is related to logical reasoning of cognitive aspects in formal situations (Lazarowitz & Naim, 2013). Students with a concrete operational cogni-

tive stage were not masters in formal operational skills (Shemesh & Lazarowitz, 1988), because they could not conceive concepts at an abstract level (for example, constructing a 2D microscopic structure into 3D) (Yenilmez et al., 2005).

5 CONCLUSIONS

Based on the pretest-posttest scores and N-Gain, there was improvement in spatial thinking and logical thinking after having experienced framing based learning. This research also indicated that students have spatial thinking activities during the frame-based learning of plant anatomy: (i) creating a representation in 2D and 3D; (ii) maintaining a representation in working memory; (iii) scanning the representation; and (iv) transforming the representation. These were factors that improved students' logical thinking on each indicator. These results showed that framing based plant anatomy learning led to more trained proportional, combinatorial, and correlation reasoning compared with other reasoning. Analysis of the relationship between spatial thinking and logical thinking showed significant correlation.

ACKNOWLEDGMENT

The authors would like to express gratitude to the head of the Biology Education Program, Faculty of Teacher Training and Education, Sriwijaya University, who has greatly helped in this study.

REFERENCES

Autin, F. & Croizet, J.K. (2012). Improving working memory efficiency by reframing metacognitive interpretation of task difficulty. *Experimental Psychology*, *141*, 610–618.

Bao, L., Cai, T., Koening, K., Fang, K., Han, J., Wang, J. & Nianle, W. (2009). Learning and scientific reasoning education. *Education Forum*, *232*, 586–587.

Berland, K.L. & Hammer, D. (2012). Framing for scientific argumentation. *Research in Science Teaching*, *49*, 68–94.

Boerwinkel, D.J., Swierstra, T. & Waarlo, A.J. (2014). Reframing and articulating socio-scientific classroom discourses on genetic testing from an STS perspective. *Science & Educ*, *23*, 485–507.

Bolotin, M.M. & Nashon, S.M. (2012). The essence of student visual-spatial literacy and higher order thinking skills in undergraduate biology. *Protoplasma*, *249*(1), 25–30.

Brownlow, A., McPheron, T.K. & Acks, C.N. (2003). Science background and spatial abilities in men and women. *Science Education and Technology*, *12*(4), 371–380.

Table 3. Percentage of logical reasoning.

Logical reasoning	Pretest	Posttest	Criteria
Proportional reasoning	42.86	64.29	Moderate
Controlling variable	3.57	11.09	Very low
Probability reasoning	9.52	15.48	Very low
Correlation reasoning	11.90	27.38	Very low
Combinatorial reasoning	17.86	45.24	Low

Diezmann, M.C. & Lowrie, T. (2012). Learning to think spatially. What do students "SEE" in numeracy test items? *International Journal of Science and Mathematics Education, 10*, 1469–1490.

Engle, R.A., Nguyen, P.D. & Mendelson, A. (2011). The influence of framing on transfer: Initial evidence from a tutoring experiment. *Instructional Science, 39*, 603–628.

Goffman, E. (1974). *Frame analysis an essay on the organization of experience.* Boston: Northeastern University Press.

Hoffler, T.N. (2010). Spatial ability: Its influence on learning with visualization a meta-analytic review. *Education Psychology, 22*, 245–269.

Hoyek, N., Collet, C., Di Rienzo, F., De Almeida, M. & Guillot, A. (2014). Effectiveness of three-dimensional digital animation in teaching human anatomy in an authentic classroom context. *Anatomical sciences education, 7*, 430–437.

Jones, M.G., Gardner, G., Taylor, R.A., Wiebie, E. & Forrester, J. (2011). Conceptualizing magnification and scale: The roles of spatial visualization and logical thinking. *Research in Science Education, 41*, 357–368.

Kosslyn, S.M. (1978). Measuring the visual angle of the mind's eye. *Cognitive Psychology, 10*, 356–389.

Lazarowitz, R. & Naim, R. (2013). Learning the cell structures with three-dimensional models: Students achievement by methods, type of school and questions' cognitive level. *Journal of Science Education and Technology, 22*, 500–508.

Meltzer, D.E. (2002). Normalized learning gain: A key measure of student. *Learning American Journal of Physic, 70*(6), 639–654.

National Research Council (NRC). (2006). *Learning to THINK SPATIALY.* Washington: The National Academies Press.

Piaget, J. (1969). *The origins of intelligence in children.* New York: International University Press.

Shemesh, M. & Lazarowitz, R. (1988). Formal reasoning skills of secondary school students and their subject matter preferences. *School Science and Mathematics, 88*(5), 376–389.

Tobin, K.G. & Capie, W. (1981). The development and validation of a group test of logical thinking. *Educational and Psychological Measurement, 41*, 413–423.

Yenilmez, A., Sungur, S. & Tekkaya, C. (2005). Investigating student's logical thinking abilities, the effects of gender and grade level. *Hacettepe Universitesi Egitim Facultesi Dergisi, 28*, 219–225.

Hypnoteaching and learning motivation enhancement

F. Fauzan & L. Indriastuti
UIN Syarif Hidayatullah, Jakarta, Indonesia

ABSTRACT: This study aimed to investigate the effect of hypnoteaching to increase students' learning motivation in civic subject. The method used in this research was the Classroom Action Research (CAR). Methods of data collection were observation, questionnaire, and documentation. The results showed that the learning process using hypnoteaching method in the first cycle increased most of student motivation. The percentage of student who had a high motivation increased from 13.79% to 68.87%. The students' motivation increased more in the second cycle, and reaches 83.45%.

1 INTRODUCTION

In the learning process, teachers have a very strategic role. Teacher is a central figure in term of the effort to make changes and form a maturity of learners. Besides teaching, the other main task of teacher is to educate, guide, direct, facilitate, mediate learners to interact, be educative and has implications for the change of attitude (social, spiritual), knowledge, and skills. Therefore, the choice of approaches, strategies and methods of teaching-learning becomes a necessity. However, there are still many teachers who are less precise in using a variety of strategies and methods, resulting in the lack of students' motivation and their unsatisfied learning. Preliminary research on civics class showed that the average value of the motivation of 29 students, consisting of 18 boys and 11 girls, was 60.01, out of 100. The students who already have the motivation to learn were only 13.79%. Most of students have not been able to concentrate, to be responsible, to be enthusiastic, to ask the teacher, and to follow the directions given by teachers.

Teachers need to have miraculous ways, which can lead students through language and attitude, and using appropriate methods. Teacher also should be able to make learning more attractive. Hopefully, with the right method, students have a high learning motivation to participate in the learning process.

Hypnoteaching method is a combination of teaching that involves the conscious and the subconscious mind of students. In contrast, the existing methods involve only the student conscious mind. Hypnoteaching method is a unique, creative, and imaginative way of teaching, in which students can learn in a fresh state. In this method, teachers also need to consider the emotional and psychological aspects of students, so that students are motivated in learning activities.

2 LITERATURE REVIEW

2.1 *Learning motivation concept*

Motivation can be referred to the process of encouragement, direction, and persistence of behavior, so the motivated behavior is full of energy, focus, and last a long time (Syaifurahman & Ujiati 2013). Motivation can also be a change in a person's energy, in the form of a real and physical activity. Because someone has a specific purpose in the activity, then someone has a strong motivation to achieve with every effort (Djamarah, 2011). Motivation can be defined as urge that arise in a person, consciously or unconsciously, to perform an act with a specific purpose. Motivation also can be defined as businesses that can cause a person or a particular group of people motivated to do something because they want to achieve their goals (Asrori 2009).

Student motivation can arise within the students themselves, called intrinsic motivation. It can also arise from outside the student i.e., coming from a pleasant environment or activity created by the teacher; the gifts prepared by teachers as a reward for students who successfully complete the task well. Those things can be a motivation for students to be able to follow the lesson well. Students may be forced to perform an act, but he might not be forced to live action, as appropriate. Teachers can impose learning materials to students, but it is not possible to impose to learn in the truest sense. Therefore, teachers strive to make students want to learn and have the desire to learn continuously. Motivation to learn is the power (power motivation), the driving force, the willingness tool builder, and the strong desire for self-learners to be active learning, creative, effective, innovative, and fun in order to change behavior, both in cognitive, affective, and psychomotor (Hanafi & Suhana 2012).

There are two types of motivation often affects human activities, namely: intrinsic motivation and extrinsic motivation. Intrinsic motivation is the motives that become active or function without stimulated from the outside, because inside every individual have the urge to do something. Extrinsic motivation is the active motives that become function due to the push from outside.

To determine the presence and absence of motivation in students, there are indicators of motivation that exist in every person. Those are: 1) diligently complete the task (can work continuously for a long time, never stopped before completion), 2) tenacious in the face of adversity (not easy to despair), 3) demonstrate an interest in a variety of problems, 4) more happy to work independently, 5) quick bored on tasks routine (things that are mechanical, repetitive course, making it less creative), 6) maintain the opinion (if it is convinced of something), 7) not easy to release things that they believed, 8) glad to locate and troubleshoot problems (Sardiman 2012). In other reference, it also mentioned the indicators of students who are already motivated. Those are: (1) have desire to do activities, (2) urge for action, (3) have hopes and ideas, (4) appreciate and respect themselves, (5) in a good environment, and (6) exist in interesting activities (Uno 2008).

2.2 Hypnoteaching method

Hypnos is a state of consciousness that is very easy to receive various advice or suggestion. In this condition, the role of critical area (temporary data container to be processed based on analysis, logic, aesthetics, and others; for each person is different) get minimal.

To be able to achieve its intended purpose, and make learning more interesting and not boring, a teacher must have a correct method in performing their duties during the process of learning and teaching. Nowadays, many methods have been developed and implemented in schools. One method that was developed is hypnoteaching learning methods; learning methods in the delivery of the material where teachers will use more of the languages of the unconscious, which can grow its own interest to students. Hypnoteaching improvisation can be regarded as a method of learning. This method provides a new conceptual approach in the field of education and training (Yustisia 2012). In classroom, teachers do not have to euthanize his

portage, when giving suggestions. Teachers simply provide persuasive words as a communication tool in accordance with the expectations of the students. So, hypnoteaching is a learning method that uses the power of words and attitudes from a teacher. Students can easily follow the learning process that will affect the atmosphere and learning outcomes through growing motivation of the student.

Teachers who apply this method, will get some benefits, such as: 1) learning become more fun and more exciting for students and for teachers; 2) learning can attract the attention of students through a variety of creative games that are applied by the teacher; 3) teachers become better to manage his emotions; 4) education can foster a harmonious relationship between teachers and students; 5) teachers can cope with children who have learning difficulties through a personalized approach; 6) teachers can cultivate the spirit of the students in learning through games of hypnoteaching. Through the method of hypnoteaching, teachers use some other learning methods with games and certain approaches to students with learning difficulties.

3 RESEARCH METHODS

The method used in this research was the Classroom Action Research (CAR), developed by Kurt Lewin, which is consisting of four components, namely: 1) (planning), 2) (acting), 3) (observing), 4) (reflecting). With this model the researchers could observe the process of learning firsthand. The learning process included student activities, student motivation, and level of understanding of students. The subject of this study was 29 students, consisting of 18 male students and 11 female, at MIS Nuroniyah Gunung putri, Bogor. Methods of data collection were observation, questionnaire, and documentation.

4 RESULTS AND DISCUSSION

The learning process using hypnoteaching method was performed in the civics subject, about living in harmony. The learning process using hypnoteaching in the first cycle increased most of student motivation. The percentage of student who had a high motivation increased from 13.79% to 68.87%. The students' motivation increased more in the second cycle, and reaches 83.45%. The rise in students' motivation to learn civics on the second cycle was very good. Thus the achievement of the target of research that has been presented before has been reached.

The implementation of hypnoteaching method in the learning process changed the student behavior.

There were no rowdy student during the learning process and students seemed to focus on the lesson and can complete the tasks given well. Students seemed enthusiastic and the atmosphere of learning was fun for students and teachers. Even the interaction among students and interaction with the teacher looked pleasant as expected in this study.

The learning motivation of students was observed very high both at the time the observation was made and after the application of the hypnoteaching method. Students seemed excited and happy during learning activities. The high motivation to learn civics could be seen in the observation of students and teachers where the researchers also witness in the learning process.

5 CONCLUSION

Implementation of hypnoteaching method in learning process increased students' learning motivation. Moreover, the method also changed the students' attitude. The change can be achieved through the example given on the method of hypnoteaching by providing words of suggestion and words of motivation to be able to achieve the expected competencies in learning civics.

REFERENCES

Asrori, Mohammad. 2009. *Psikologi pembelajaran.* Bandung. Wacana Prima.

Djamarah, Bahri, S. 2011. *Psikologi belajar.* Jakarta. Rineka Cipta.

Hanafiah & Suhana, C. *Konsep strategi pembelajaran.* Bandung. Refika Aditama.

Sardiman. 2012. *Interaksi & motivasi belajar mengajar.* Jakarta. Raja Grafindo Persada.

Syaifurrahman, Ujiati, T. 2013. *Manajemen dalam pembelajaran.* Jakarta. Indeks.

Yustisia. 2012. *Hypnoteaching seni ajar mengeksplorasi otak peserta didik.* Jokyakarta. Ar-Ruzz Media.

Ideas for 21st Century Education – Abdullah et al. (Eds)
© *2017 Taylor & Francis Group, London, ISBN 978-1-138-05343-4*

The development of an Augmented Reality (AR) technology-based learning media in metal structure concept

F.S. Irwansyah, I. Ramdani & I. Farida
UIN Sunan Gunung Jati Bandung, Indonesia

ABSTRACT: This research was motivated by the importance of developing students' abilities concerning the material representation of submicroscopic metallic structures, so that the necessary media were able to visualize metal structures by the simple, efficient and useful devices that are already available. This study aimed to describe the stages of manufacture of the technology-based learning media AR on the material metal structures, to develop students' capabilities in submicroscopic representation, and to analyze the feasibility of AR technology-based learning media on the material of metal structures. The research and development had produced products in the form of AR technology-based learning media on the concept of metal structures. The stages of the research were carried out by the analysis of metal structural concept development, design, validation, due diligence, and limited testing. Validation and limited testing were done in order to obtain feedback in the form of recommendations for the improvement and the assessment of the learning aspect, the substance of the concept, the visual communications, and the software engineering, as well as the feasibility of its use for the purpose of making the products. In general, the test results obtained $r_{calculation} = 0.8$–1.0 feasibility and an average value of 72.5 to 88.33% was obtained at the stage of testing of a limited percentage of eligibility. It showed that the presentation of AR technology-based learning media on the concept of metal structures was feasable for use as a learning resource for students to gain the ability to develop the submicroscopic representation.

1 INTRODUCTION

Chemistry includes three levels: macroscopic, submicroscopic, and symbolic levels. The problem was that learning chemistry usually emphasized the symbolic level and problem solving only, while the visualization of both the macroscopic and submicroscopic levels were also required (Daviddowitz & Chittleborough, 2009) to make students understand the whole concept of chemistry. The metal structure was formed by the order of the same atoms tightly packed in a crystal. Atoms, molecules, and ions were theoretical models that underlaid the dynamic explanation of particle levels closely related to the representation of the submicroscopic. Learning abstract concepts and abstract concepts with concrete examples were hard to do in the laboratory, although these phenomena could be observed visually. However, animations were required for further explanation to illustrate the phenomena on a molecular basis. Research showed that many high school students, college students and even some teachers found it difficult to transfer from one level to another level of representation (Chittleborough & Treagust, 2007). In understanding chemical phenomena, both teachers and textbooks did not emphasize the differences

and linkages between all three levels of representation. This was because students were considered to be able to distinguish and relate the three levels of representation (Chittleborough & Treagust, 2007). Students experienced difficulties, especially at submicroscopic and symbolic levels, because the representation was invisible and abstract while their thinking relied on sensory information. Based on these explanations, one of the alternatives that could be used to develop the three levels of representation in chemistry learning, and to help students in understanding chemistry, was using the tools to build a coherent mental representation from the material presented in the form of instructional media, such as words, pictures, and animations (Mayer, 2003). Due to the development of science and technology today and the demands of the future, computer technology was widely used in various fields, such as in the areas of information, education, business and communication, associated with educational computer technology or developed in learning. Moreover, there was a computer technology that was currently being developed, called technology-based Augmented Reality (AR) (Furht et al., 2010). The advantages of AR technology itself could be implemented widely in various media, as an application in a smartphone, in a product package, and even in the print media, such as books, magazines, or newspapers, so that it was user friendly for the low cost tools and facilities producing awesome learning media. Therefore, AR had many opportunities to develop, and could support educational facilities. In this decade, progress in the development of pedagogical concepts, applications, and technologies, the decrease of hardware costs, and the use of small-scale AR technology in educational institutions made it possible that it could be used as interesting learning media. The purpose of this research was to analyze the feasibility of AR technology-based learning media on the material of metal structures to develop the students' capabilities of submicroscopic representation (Kaufmann, 2002).

2 METHOD

The method used in this research was Research and Development (R & D). The research and development method was a research method used to produce particular products, and to test the effectiveness of these products. The subjects of this research were expert lecturers instructed to examine the feasibility of AR technology-based learning media, including experts in media education and learning, and some students of chemistry education at UIN Sunan Gunung Djati Bandung. In conducting this research, the researchers focused on the making of AR technology-based learning media. The stages of making these AR technology-based learning media refered to the CAI (Computer Assisted Instruction) tutorial design model, which had been modified. Therefore, in general, the stages of making these media consisted of the analysis phase and the design development phase, described as follows:

a. The analysis phase
b. Design development phase
c. The making visualization

3 RESULTS AND DISCUSSION

3.1 *The analysis phase*

At this stage, an analysis of the concept was conducted and the concept map of the metal structures was created based on the curriculum. It aimed to produce a material suitable with metal structure and instructional media created. The analysis results of the material concept of metal structures can be seen in Table 1 below.

After analyzing the concept, the next stage was arranging the learning indicators that would be used in the making of the AR technology-based

Table 1. The analysis resume of the concept of metal structures in general.

No.	Concept was analyzed	Type of concept
1	The crystal structure of solids	Abstract
2	The crystal lattice	Abstract
3	Cell unit	Abstract
4	Metallic crystals	Abstract
5	Ionic crystal	Abstract
6	Molecular crystal	Abstract
7	Covalent crystals	Abstract
8	Close-packed structure	Abstract
9	Cubical close-packed structure	Abstract
10	Hexagonal close-packed	Abstract
11	Body centered cube	Abstract
12	Face centered cube	Abstract
13	Cube	Abstract
14	Tetragonal	Abstract
15	Trigonal	Abstract
16	Hexagonal	Abstract
17	Orthorhombic	Abstract
18	Monoclinic	Abstract
19	Triclinic	Abstract

learning media. The indicators of the metal structure learning media are presented in table. The following table visualized the indicators of metal structure learning (Table 2).

3.2 *Design development phase*

For the development of AR technology-based learning media on the metal structure studies, in any design, the development should consider the workflow design or the informational processing flow based on the flow chart and the storyboard. Moreover, after completing the analysis and the development phase, the visualization of AR technology-based learning media for metal structure studies was made. One of the instructional media visualizations can be seen in the figures below.

The learning media uses a smartphone with AR technology-based learning media for students' worksheet. While learning using these AR technology-based learning media, the students were asked to answer questions on a worksheet, in which indicators had been designed and developed, which led to the development of the students' ability with regards to submicroscopic representation in the material of metal structures.

In general, displays in this learning media consisted of: 1) home display, which was a display containing links and the identity of the material, 2) the display of learning objectives, containing learning objectives on the concept of metal structure, 3) materials display, containing the concept of

Table 2. Worksheets indicators of students learning about metal structures.

No.	Label concept	Learning indicators
1.	Solid structure	Through the crystalline structure of solids and amorphous solids, students could determine the structure of solids and amorphous solids accurately.
		Based on the marker containing the structure of solid crystals and amorphous solids, students could explain the differences in crystalline solids and amorphous solids.
		Based on the marker containing the structure of solid crystals and amorphous solids, students could accurately describe the structure of amorphous and crystalline solids.
2.	The seven basic crystal systems	Based on the marker presented, students could accurately describe the shape of the crystal lattice structure.
		Based on presented markers, students could accurately determine the basic parameters of the three-dimensional crystal.
		Based on the markers presented, students could determine the cell unit with the correct marker.
3.	Close-packed structure	Students could accurately determine the structural pattern of Hexagonal Closest Packing (HCP).
4.	The metal structure	Based on the data on the object marker, students could accurately calculate the atomic radius in nm (nanometers).
		Based on presented markers, students could calculate the volume of the metal crystal cell unit accurately.
		Based on presented markers, students could calculate the density of the metal appropriately.
		Based on the structure of the object marker, students could calculate the volume of the unit cell in cm^3 (cubic centimeters) accurately.

Figure 1. Example marker.

Figure 2. Example of an object on a marker.

metal structures based on learning objectives, which refered to the submicroscopic indicators, describing the material consisting of 3D-shaped crystal solids and amorphous, seven basic crystal systems, and metal structures, 4) the constituent profile page, containing information about the constituents.

The due diligence of the AR technology-based learning media was done in two stages: 1) validation test consisting of the validation test of the learning aspect, the aspects of the material substance, the aspects of visual communication, and software engineering of learning media on metal structure, 2) trial limited test of the student group, consisting of ten randomly selected students of chemistry education. The results can be seen in the table below:

The results of $r_{calculation}$ on each criterion in the learning aspects had the highest result of a feasibility value of 1.0 or was valid on the indicator of the accuracy of the use of learning strategies, so that the use of the correct learning strategies could help students to improve their comprehension and attractively and reliably present data (Arsyad, 2007). Meanwhile, the lowest result of $r_{calculation}$ was in the completeness and quality indicator of learning materials, with $r_{calculation}$ 0.8, which showed that an improvement in the quality of the materials was needed, since the quality of the teaching materials could improve the quality of the learning outcomes so that it could be well integrated (Arsyad, 2007). The result of the validity of the media stated that teaching materials were valid if the value of $r_{calculation}$ was above the value of r_{critic}, which was 0.30. Therefore, it can be concluded that the learning aspect in this research was valid and feasible to be used as a teaching material (Sugiyono, 2011).

Table 3. Results of validator expert of aspects of learning, the material substance, visual communication and software engineering.

No.	Aspect	$r_{calculation}$	Result
1.	Aspect of learning	0.90	Valid
2.	Aspects of material	0.80	Valid
3.	Aspects of visual communication	0.84	Valid
4.	Aspects of software engineering	0.95	Valid
	Average	0.87	Valid

Table 4. Average ratings of students on AR technology-based learning media.

No.	Aspect of the material content	Percentage (%)
1.	The relevance of the learning objectives	88.33
2.	The time efficiency of the use of the products	72.50
3.	Effectiveness to solve the limitations of media learning	74.16
4.	The flexibility of the media usage	80.00
5.	Media display	85.00
6.	Increase student motivation in learning	77.50
7.	The ability to encourage students to learn more	75.75
8.	Prospecting other similar media development	82.55
	Average	79.50

The result of $r_{calculation}$ on each of the criterion in the software engineering aspects, on the indicators of effectivity, efficiency, compatibility, and usability, had a value of 1.0 or was valid. It proved that the use of technology in learning not only became a tool but also delivered a learning message (Sadiman et al., 2009). Therefore, the validation of the AR technology-based learning media on the material of metal structures in the aspect of software engineering was valid.

According to the results of the stages of making the learning media, including the stages of concept analysis, indicator analysis, and design development, all three stages produced instruments in the form of flow charts and storyboards, which were used as a reference in the production of AR technology-based learning media on the material of metal structures. It showed that the storyboard was the explanation of the flow that had been designed in the form of a flow chart and used as a reference in the making process of learning media (Darmawan, 2012).

The next stage of making the AR technology-based learning media was gathering three-dimensional objects fitted with the storyboard on a metal structure material in the Google SketchUp application, after it created a marker of the media, using corel draw X.5. Furthermore, the marker was registered to vuforia developers site, so it could then be used to create an AR technology-based media that combined the unity of 3D applications so that virtual objects could be projected in real time (Roedavan, 2014).

After the AR technology-based learning media was completely made, the validation phase of the media products was done by three lecturers of chemistry education, as validation experts. This validation was conducted on the learning aspect, the aspect of material substance, visual communication aspects and aspects of software engineering. The result generally had a feasibility value (r) between 0.8–1.0, or was valid. These validation results indicated that every aspect was valid and feasible. Therefore, it can be concluded that the AR technology-based learning media on the material of metal structures on all aspects of the supporting elements of learning devices was valid and feasible for learning media.

4 CONCLUSIONS

According to the aspects of learning, conceptual substances, visual communications, and software engineering of AR technology-based learning media developed, the feasibility assessment conducted in this research resulted in a value with $r_{calculation}$ 0.8–1.0 or which had an interpretation of high feasibility, from an expert assessment or validator. This result showed that the AR technology-based learning media on the concept of metal structures was feasible to be used. In addition, the results of the feasibility test based on the responses of ten students showed good responses of 72.5 to 88.33%. Therefore, it could be concluded that AR technology-based learning media on the concept of metal structures could be used as learning tools or media.

REFERENCES

Arsyad, A. (2007). *Media Pembelajaran*. Jakarta: PT Raja Grafindo Persada.
Chittleborough, G. & Treagust, D. F. (2007). The modelling ability of non-major chemistry students and their understanding of the sub microscopic level. *Journal Chemistry Education Research and Practice*, 8, 274–292.
Darmawan, D. (2012). *Teknologi Pembelajaran*. Bandung: PT Remaja Rosda Karya.

Daviddowitz. B. & Chittleborough, G. D. (2009). *Linking macroscopic and submicroscopic level*. Dordrecht: Springer.

Furht, B. et.al. (2010). Augmented reality technologies, systems and applications. *Journal Department of Computer and Electrical Engineering and Computer Sciences*, *51*, 341–377.

Kaufmann, H. (2002). Collaborative augmented reality in education. *Journal, Institute of Software Technology and Interactive Systems: Vienna University of Technology*, 188, 9–11.

Mayer, R. E. (2003). The promise of multimedia learning: Using the same instructional design methods across different media. *Journal, Department of Psychology*, *13*, 125–139.

Roedavan, R. (2014). *Unity tutorial game engine*. Bandung: Informatika Bandung.

Sadiman, A. S. et. al. (2009). *Media Pendidikan: Pengertian, Pengembangan dan Pemanfaatannya*. Jakarta: Rajawali Pers.

Sugiyono, S. (2011). *Statistika Untuk Penelitian*. Bandung: Alfabeta.

Treagust, D. F., Chittleborough, G. & Mamiala, T. L. (2003). The role of submicroscopic and symbolic representations in chemical explanations. *International Journal of Science Education*, *25*, 1353–1368.

237

Ideas for 21st Century Education – Abdullah et al. (Eds)
© *2017 Taylor & Francis Group, London, ISBN 978-1-138-05343-4*

The effectiveness of the local culture-based physics model in developing students' creative thinking skills and understanding of the Nature of Science (NOS)

I.W. Suastra
Universitas Pendidikan Ganesha, Bali, Indonesia

ABSTRACT: The purpose of this study was to test the effectiveness of the local culture-based physics model of teaching in developing students' creative thinking skills and understanding of the NOS. The study was conducted through a quasi-experimental posttest design. The subjects are 72 students in classes X MIPA.1 and X MIPA.2 SMA Negeri 3 Singaraja, Bali. The data regarding the students' creative thinking skills was measured by Torrance Tests of Creative Thinking (TTCT). In addition, the students' understanding of the NOS was evaluated by questionnaire. The data regarding the creative thinking and understanding of the NOS were analyzed using statistical and descriptive methods, and Multivariate Analysis of Variance (MANOVA) to verify the hypothesis. Statistical analysis shows the significant difference in the students' creative thinking and understanding of the NOS between the group of students who learned through the Local Culture-based Model of Teaching (LCBMT) and those who learned through the Conventional Method of Teaching (CMT). It can be concluded that the culture-based physics model of teaching is more effective than the conventional (regular) model of teaching in developing creative thinking skills and the understanding of the NOS in students.

1 INTRODUCTION

The low ability of students' to solve problems in their daily life, which is also driven by the low development of their nation, leads to the low of their creative thinking skills. The Trend International Mathematics Science (TIMSS) in 2007 reported that the average score in science in the cognitive domain of Indonesian students was ranked 36th out of 49 countries throughout the world (Gonzales et al., 2008). Indonesia gets 425 for its knowing score, 426 for its applying score, and 438 for its reasoning score. All of these scores, respectively, are below the TIMSS average, which is 500.

In this era of industrialization and globalization, with increasingly tougher competition, mastering knowledge and technology is important. This challenge calls for Indonesia's human resources to be reliable and qualified, people who are not only able to master knowledge and technology, but who are also able to develop a strong character. Gardner (2007) states that to meet the challenges of a future (towards the 2045 generation) that is increasingly more complex, we need five types of mind for the future, encompassing: disciplined mind, synthesizing mind, creating mind, respecting mind, and ethical mind. Furthermore, Tilaar (2012) states that globalization has to "be fought" by developing creativity and entrepreneurship through critical transformative pedagogy in national education.

One of the models of innovative teaching that can challenge the students in developing creative thinking is the local culture-based teaching model (Suastra, 2012). The steps are (1) exploration of the students' local culture (knowledge and beliefs) that is relevant to the physics teaching at the time of learning, (2) focusing (focus of inquiry), (3) inquiry from various perspectives (scientific, sociocultural, and historical), (4) elaboration, and (5) evaluation (Suastra, 2010). In this model, the teacher plays an important role as "a cultural broker" who bridges the gap for students in crossing the two cultures; that is the students' local culture and scientific culture (Western culture). Thus, the understanding of the students of the NOS can be hypothesized as something that can be developed. The purpose of this study was to test the effectiveness of the local culture-based physics model of teaching in developing creative thinking skills and understanding of the NOS in senior high school.

2 LITERATURE REVIEW

The students' creative thinking is a High Order Thinking (HOT) that should become the main objective in thinking and be given an emphasis in the school curriculum (Fisher, 1999). Furthermore, Hadzigeorgiou et al. (2012) state that science (physics) education has a very high probability for developing creative thinking, such as in the slogans "creative science", "creative problem solving", and "creative inquiry". Creative thinking is a novel way of seeing and doing things that are characterized by four components: (a) fluency (generating ideas), (b) flexibility (shifting perspectives easily), (c) originality (consisting of something new), and (d) elaboration (building on existing ideas) (Anwar et al., 2012).

The Nature of Science (NOS) is an important factor that determines the direction of science teaching in the future. Wenning (2006) describes the nature of science as empirical, creative, imaginative, theoretical, sociocultural-context bound, and with tentative characteristics. Bell and Gilbert (1996) state that the NOS encompasses some concepts that are defined in a simple way, in terms of the ontology, epistemology, and axiology of science. The three aspects are explained as, 1) ontology: knowledge in the fields of articulation, sociology, and history; 2) epistemology: knowledge as the way to reach understanding, insight, and wisdom; and 3) axiology: knowledge that puts more emphasis on the benefits of knowledge for society and the environment. To understand the NOS is an important part of science literacy (Bavir, 2012). The American Association for the Advancement of Science and National Research Council stresses the important role in increasing the students' NOS. NOS becomes important because it is needed in making, managing, and processing the objects of science and technology, informing the decision makers concerning socio-scientific issues, respecting science values as contemporary culture, developing understanding of the norms of the scientific community to materialize moral commitment that has a general value for society, and facilitates the main problems in science teaching (Hardianty, 2015).

3 RESEARCH METHODS

This quasi-experimental study was done in the ninth grade of SMA Negeri 3 Singaraja involving 72 students dispersed in 2 parallel classes, that is, Class X MIPA.1 and Class X MIPA.2. Class X MIPA.1 was selected randomly as the experiment class and Class X MIPA.2 as the control one, thus this study used Posttest Only Control Group Design. Creative thinking was evaluated using Torrance Tests of Creative Thinking (TTCT). The TTCT was developed within an educational context to test creativity (Anastasi, 1976). The data concerning the understanding of the students of the NOS were collected using questionnaires. The data regarding the creative thinking and the NOS were analyzed using descriptive statistics, such as

percentage, mean, and standard deviation, and one-way MANOVA was used to verify the hypothesis at a 5% level of significance.

4 RESULTS AND DISCUSSION

The results of the descriptive analysis of this study can be seen in Table 1. There are four aspects of creative thinking being evaluated in this study, namely: fluency, flexibility, originality, and elaboration. Based on the result of measurement the means for the students' creativity, which have been converted into 100, can be seen in Table 1.

Based on Table 1 it can be seen that the originality aspect of the LCBMT group has the qualification very good, while that of the CMT group has the qualification good. In terms of fluency and flexibility, the LCBMT group has the qualification of sufficient for both, while in terms of flexibility, both the LCBMT and CMT groups have the qualification of very insufficient. Viewed from the mean in creative thinking, the LCBMT group has a mean of 58.62, better than the CMT group, which is 56.30, and falls into the qualification of insufficient. It can be inferred that the students who learned through the LCBMT have a higher level of creativity than those who learned through the CMT. However, these data need to be given serious attention in the teaching and learning process in the future so as to ensure a higher level of student creativity, particularly in terms of flexibility, since it falls into the qualification of very insufficient for both groups.

The t-test analysis found that there is a difference in creative thinking and in the NOS between the group of students who learned through LCBMT and those who learned through CMT, as shown in Table 2 below.

Table 1. Means for each aspect of creative thinking. N = 72 students.

No	Aspect of creative thinking	Mean			
		LCBMT	Qualification	CMT	Qualification
1	Fluency	67.22	Sufficient	64.00	Sufficient
2	Flexibility	13.89	Very Insufficient	12.87	Very Insufficient
3	Originality	87.72	Very good	83.99	Good
4	Elaboration	65.64	Sufficient	64.33	Sufficient
	Mean	**58.62**	**Sufficient**	**56.30**	**Sufficient**

Note: LCBMT: Local Culture-Based Model of Teaching. CMT: Conventional Model of Teaching.
Source: Suastra, 2015.

Table 2. Recapitulation of the result of one-way MANOVA of the result of hypothesis testing.

	Value	F	Hypothesis df	Error df	Sig.
Pillai's trace	.45	21.23[a]	2.00	51.00	.01
Wilks' lambda	.55	21.23[a]	2.00	51.00	.01
Hotelling's trace	.83	21.23[a]	2.00	51.00	.01
Roy's largest root	.83	21.23[a]	2.00	51.00	.01

Based on the recapitulation of the result of the one-way MANOVA analysis, shown in Table 2, it can be interpreted that the respective level of significance for Pillai's Trace, Wilk's Lambda, Hotelling's Trace, and Roy's Largest Root is lower than 0.5, respectively, so that Ho is rejected. Thus it can be said that there is a significant difference simultaneously in students' creative thinking and the NOS between the group of students who learned through LCBMT and those who learned through CMT (F = 21.23; p < 0.05). The further result of the LSD test shows that there is a difference in the mean score for students' creative thinking and the understanding of the NOS for students who learned through LCBMT and those who learned through CMT. $\Delta\mu = [\mu$ (LCBMT) $- \mu$ (CMT) = 8.20 at a lower than 0.05 level of significance. Therefore, the mean scores for the students' creative thinking and the understanding of the NOS of students who learned through LCBMT and those who learned through CMT differ significantly at the 0.05 level of significance. It means that LCBMT gives a better effect than CMT in developing the students' creative thinking and the understanding of the NOS.

The difference in the students' creative thinking was caused by many factors, as follows. First, the LCBMT starts the teaching activities by unearthing the students' prior knowledge and beliefs concerning the material to be learned. Ausubel's view (Suastra, 2013) is that "the most important single factor influencing learning is what the learner already knows. Ascertain this and teach him accordingly". Furthermore, the teacher focuses the students' attention on the readiness to start an inquiry into their prior knowledge. However, the CMT prior knowledge and ideas/beliefs of students received less attention.

The next step in LCBMT is the students making the inquiry from various perspectives. In this step, the students make the inquiry, both from the scientific perspective and the historical/socio-cultural perspectives. If the concept is related to the scientific concept, then the inquiry takes the

form of a scientific inquiry (Trawbridge & Bybee, 1990). However, if it is related to socio cultural concepts, then it can be investigated from the sociocultural perspective, which also includes the historical perspective. CMT uses scientific inquiry in perspectives.

Another effect of LCBMT is the improvement of the students' understanding of the NOS, which is better than the effects of CMT. This is due to the fact that the steps in the inquiry come from various perspectives, both from scientific as well as from socio cultural perspectives, which causes the students to understand the characteristics of science (physics) not only from the empirical but also from the socio cultural perspective. To the same effect, Wenning (2006) states that the NOS is the understanding of the characteristics of science that are related to the specific properties of science, such as empirical, creative, imaginative, theoretical, socio culturally contextual, and tentative. Therefore, science lessons will be understood by the students holistically. This is also supported by Minner's view (in Suastra, 2015), which states that inquiry-based science teaching, in addition to its capacity to improve the students' understanding of concepts, can also develop a high order thinking (thinking creativity) and the students' responsibility for learning. Kind and Kind (in Hadzigeorgiou et al., 2012) state that scientific theories are creative products (ideas) made by scientists.

The integration of local wisdom (cultural), especially Balinese local wisdom into physics teaching, gives a different nuance in teaching. By integrating the values of local wisdom into physics teaching, it makes physics teaching not as "scary" as the students first thought, as physics is often seen as a difficult subject, but it will make them become closer to the natural environment and their socio cultural environment.

5 CONCLUSIONS

Based on the results of the data analysis, it can be concluded that the culture-based physics model of teaching is more effective than the conventional (regular) model of teaching in developing creative thinking skills and the understanding of students of the NOS. These results indicate that the developed model can be implemented to improve students' creative thinking skills and students' understanding about science.

REFERENCES

Anastasi, A. (1976). *Psychologycal testing* (4th ed.). New York: Macmillan Publishing.

Anwar, M. N. et al. (2012). Relationship of creative thinking with the academics achievements of secondary school students. *International Interdisciplinary Journal of Education, 1*(3).

Bell, V. & Gilbert, J. (1996). *Teacher development: A Model from science education.* Falmer Press.

Cakici, Y., & Bayir, E. (2012). Developing children's views of the Nature of Science through role play. *International Journal of Science Education,* 34(7), 1075–1091.

Fisher, R. (1999). Thinking skill to thinking school: Ways to develop children's thinking and learning. *Early Child Development and Care,* 153(1), 51–63.

Gardner, H. (2007). *Five minds for the future* (Alih Bahasa Tome Beka). Gramedia Pustaka Utama.

Gonzales, P., Williams, T., Jocelyn, L., Roey, S., Kastberg, D. & Brenwald, S. (2008). *Highlights from TIMSS 2007: Mathematics and science achievement of U.S. fourth and eighth grade students in an international context.* Washington DC: Institute of Education Sciences.

Hadzigeorgiou, Y., Fokialis, P. & Kabouropoulou, M. (2012). Thinking creativity in science education. *Scientific Research Journal, 3*(5), 603–611.

Hardianty, N. (2015). *Nature of science: Bagian penting dari literasi science.* Prosiding Seminar Nasional Inovasi dan Pembelajaran Sains.

Suastra, I. W. (2010). Model pembelajaran sains berbasis budaya lokal untuk mengembangkan kompetensi dasar sains dan nilai kearifan lokal di SMP. *Jurnal Pendidikan dan Pengajaran, 43*(1).

Suastra, I. W. (2012). *Model pembelajaran fisika berbasis budaya lokal untuk mengembangkan kreativitas berpikir dan karakter bangsa berbasis kearifan lokal.* Proceeding Konvensi Nasional Pendidikan Indonesia. 7.

Suastra, I. W. (2013). *Pembelajaran sains terkini.* Singaraja: Penerbit Undiksha.

Suastra, I. W. (2015). The effectiveness of local culture-based physics model of teaching in developing physics competence and national character. *Proceeding in International Conference Implementation and Education of Mathematics and Science (ICRIEMS).*

Tilaar, H. A. R. (2012). *Pengembangan kreativitas dan enterpreneurship dalam pendidikan nasional.* Jakarta: Penerbit Kompas.

Torrance, E. P. (1963). *Education and the creative potential.* Minn.Mn: University of Minn Press.

Trawbridge, L. & Bybee, R. W. (1990). *Becoming a secondary school science teacher.* London: Merril Publishing Company.

Wenning, C. J. (2006). A framework for teaching the nature of science. *Journal of Physics Teacher Education Online. 3*(3), 3–10.

Developing creative thinking ability and science concept understanding through SCSS problem solving oriented performance assessment teaching at primary schools

I.N. Jampel & I.W. Widiana
Universitas Pendidikan Ganesha, Bali, Indonesia

ABSTRACT: This was a classroom action research which was aimed at improving creative thinking ability and science concept understanding of primary school students. This study was conducted in two cycles in which each cycle consisted of four stages: planning, action, observation/evaluation and reflection. The study was conducted at SD 2 Kampung Baru with 18 subjects/students (11 males and 7 females). The data were collected with an essay test that had been validated before. The data were analyzed using quantitative descriptive statistical analysis technique. The results showed that there was an increase in creative thinking ability by 12.4% with an average of 69% (creative enough) in cycle 1 becoming 84.5% (creative) in cycle II. Based on this result it can be concluded that the implementation of SCSS problem solving oriented performance assessment teaching can improve creative thinking ability and science concept understanding. Science concept understandings that can be easily developed are remembering and repeating as well as identifying and selecting concepts, principles, and procedure. Creative thinking abilities that can be easily developed are related to the indicators of fluent thinking skill, flexible and original thinking ability.

1 INTRODUCTION

In this globalization era, we cannot deny that success and welfare of the society and country depend on creative contributions in the form of new ideas, new inventions and new technologies from the society. The creative contributions of the society, especially in science and technology are now undergoing rapid development, that they demand an increase in high quality and high reasoning human resources. The increase in high quality human resources is one of the main tasks in educational sector. This is because education can produce and develop humans who have logical thinking ability, critical thinking, creative thinking, and have and initiative and are adaptive to changes and developments in science and technology (Suastra, 2006). Science education is one of activities in education that has the potential to play a strategic role in progress and development in science and technology. The potential of science education includes and ability to develop a creative and innovative society.

Gardner (1999b) states that the general objective of education should be oriented to the achievement of understanding to master various disciplines. Understanding is mental process that occurs in which adaption and transformation of knowledge.

In some teaching taxonomies, understanding takes different cognitive level positions. Based on Gagne's taxonomy understanding is at verbal information level, according to Bloom's taxonomy it is at comprehension level, according to Anderson taxonomy it is at declarative knowledge level, in Merrill's taxonomy it is at remember paraphrased level, and according to Reigeluth's taxonomy it is at understand relationship level (Reigeluth & Moore, 1999). This explanation indicates that understanding needs knowledge prerequisite at high levels such as application, analysis, synthesis, evaluation, insight, and wisdom.

The students' science concept understanding and science teaching are very important since their low level of understanding will cause their misunderstanding or misconception. Understanding in science teaching is an ability: (1) to remember and repeat concepts, principles and procedures, (2) to identify and select concept principles and procedures, (3) to apply concept principles and procedures. The three dimensions in this study are basic thinking skills in thinking ability ladder (Krulik & Rudnick, 1995).

Understanding is a basic thinking skill that forms the basis for achieving creative thinking ability. Creative thinking ability is an organized process that involves mental process which includes problem solving, decision making, analysis, and science inquiry activities (Ennis, 1985). Creative thinking is the basis for analyzing arguments and producing insights into every meaning and interpretation. This thinking framework develops a cohesive, logical, reliable, brief, and convincing reasoning (Ennis, 1985). A person who has creative thinking ability can act normatively, is ready to reason about something that he or she sees, hears, or thinks and is able to solve the problem that he or she faces (Redhana, 2003a). According to Santyasa (2006), the characteristics of a person who has creative thinking competence is careful, likes to classify, is open, emotionally stable, ready to take actions when the situation requires, likes to demand things, appreciates feelings and opinions of other people. According to Munandar (2004) creative thinking ability is a person's ability to combine his or her last experiences with a new experience to think and find an appropriate solution that is reflected in his or her fluency, flexibility, originality and elaboration in thinking. Thus, the person who has creative thinking ability will always be ready to develop or find ideas/results that are original, esthetical, constructive, especially in using information and materials to produce or explain them with an original thinker's perspective.

Creative thinking ability is not given enough serious attention and has not been developed well. The teachers tend to focus their attention on giving information that is readily made, such as memorizing science concepts and principles that are found in the students text books. Similarly, problems given to students in the formative test, summative test, and daily quiz require students to memorize or repeat information in their text books. As the result their achievement of the essential objectives in science education becomes a failure. This is shown by the low quality in the teaching process and achievement in science at elementary school. The indicator is that the average score of the final examination for science at SD 2 Kampung Baru in the school year 2004/2005 was 5.63; in the school year 2005/2006, 5.91 and in the school year 2006/2007, 5.95.

The low science achievement at SD 2 Kampung Baru cannot be separated from various factors that have caused it, one of which is the teaching process. After doing an observation, an interview and a reflection of the learning process and the assessment procedure of the student learning achievement it was found out that the process of learning up to the time when the study started was characterized as follows.

1. It had not given an emphasis on the development of the student's ability in solving problems (problem posing and problem solving)
2. It tended to be oriented to cognitive strategy to achieve learning objectives, and
3. It had not been oriented to development of creativity and thinking productivity (creative and productive thinking) to achieve a high level of understanding (in-depth understanding).

Also the lecturing method so far has given more emphasis to students active learning, for example, through cooperative learning to present sub topics individually in turn, it turns out that the effects on problem solving ability and mathematical reasoning, and mathematical communications of the students is not yet apparent. This, of course, has a bad effect on the students' achievement level in understanding science materials, that finally has an effect on the students' low learning achievement and the high number of the students who do not pass science subject.

Based on this problem, we need to reorient science teaching toward developing students' creative thinking ability development in the future. Article 19 No. 1 of the Government Regulation No. 19 of 2006 about education process standard that becomes the basis for developing the School—Based Curriculum, the government instructs that the teaching process at school should be run in interactive, inspiring, joyful, challenging, encouraging manner for the students to participate actively and that gives them enough chances to initiate and to be creative, in accordance with their aptitude, interest and psychological development. This instruction is supported by Calg Sagan's opinions (Yasa, 2007) that defines science teaching process as a way of thinking to find out how nature works and its rules, not only as group of knowledge or theories about nature. Thus to be able to produce meaningful learning in the teaching process, we need to transform education from learning as memorization to learning as a thinking process. This can be realized if the teaching approach used is more varied and is student—centered.

Based on the basic concept of school-based curriculum as the curriculum at the school level, any teaching model as an innovative teaching process has to refer to the instructed process standard which gives learning experience that involves thinking skill that will develop the students' creativity (Yasa, 2006). As consequence, the learning strategy has to be student centered. In other words, the students are fully involved in the learning process.

The teacher as an educator is responsible to give the best alternative to the students to learn science. The alternative intended here has to be challenging, stimulating a thinking habit, and doing activities that are related to problem solving and their dairy life. One of the innovative teaching strategies that can stimulate students' thinking skill is problem solving strategy.

Problem solving is an integral part of the learning and teaching process in science at school as well at university (Tao in Suma,2004). In general, the aim is to increase a high order thinking skill. According to Tao (In Suma, 2004) problem solving is a means to deepen concepts and principles and understanding and to help students to apply them to various problems. If the students are able to solve problems that represent problems in new events, then they will be involved in thinking behavior. One of the teaching models that is related to problem solving is a search, solve, create and share (SSCS) teaching model.. The SSCS teaching model is one of the innovative teaching models that is very appropriate to be applied in teaching science.

This is also supported by a review of literature done by Rahmi (2011) who states that the advantage of the SSCS model is that the students can learn more meaningfully, since they are directly involved in finding concepts. Apart from this, they will become more active since they are trained to formulate a problem, to design a solution, to formulate the result and to communicate the result that they obtained. The SSCS teaching model puts more emphasis on the students' role as the center of teaching. In teaching by using this model the students do not only rely on the existing knowledge, but pay more attention to the process of acquiring the knowledge. By focusing on the process, the students are expected not only to memorize knowledge, but to understand more deeply so that the knowledge acquired will continually be retained. Thus, their concept understanding can be increased.

Seeing the completeness of the SSCS teaching model and performance assessment, then in this study an integration process was carried out. The integration of performance assessment and the SSCS teaching model will have an effect on the characteristics of the teaching. The teaching process will be adjusted to the time and the process of assessment done. The teaching will be more like the real constructivism. The teacher asks the students to be autonomous during the teaching process. In science teaching, the students' autonomy can be seen from how their performance is when they are solving problems in science. The feedback from the performance assessment given by the teacher will be taken as a motivation for the students to process in learning. The SSCS teaching model with performance assessment make the students able to strengthen, widen and apply knowledge and skill they possess in solving problems in science. Thus, the problem solved in this study is: Can the SSCS type problem solving—oriented performance assessment teaching develop creative thinking ability and science concept understanding at primary school?

2 METHODS AND ANALYSIS

This study was a classroom action research. The action in the teaching that was implemented was divided into two cycles. Every cycle consisted of four stages, that is, planning, action, observation/evaluation and reflection. The process of implementation in each cycle was done in four meetings. That is, 3 meetings for the implementation of action and once as an evaluation that was done that the end of

Table 1. The guideline to convert creative thinking and science concept understanding to a 5 point scale.

Criterion (%)	Level of science concept understanding	Level of creative thinking ability
85–100	Understands Very Much	Very Creative
70–84	Understands	Creative
60–69	Understands Sufficiently	Sufficiently Creative
40–59	Understands Insufficiently	Insufficiently Creative
0–39	Understands Very Insufficiently	Very Insufficiently Creative

the cycle. This study was conducted at SD 2 Kampung Baru with 18 subjects/students (11 males and 7 females). The data were collected by an essay test that has been validated beforehand. The data that were collected were analyzed using descriptive analysis technique. The result of analysis of the students' creative thinking ability obtained was then converted into the criterion referenced evaluation scale as shown in Tables 2 and 3 as follows.

3 RESULTS AND DISCUSSION

3.1 Result of creative thinking ability development

The analysis of the data on the students' thinking creativity in the third meeting in cycle 1 showed that the average percentage of the students' creative thinking ability was 69%. If this is converted into a 5 point criterion referenced evaluation then the students' creative thinking ability in cycle I was still within sufficiently active category (65%–79%). The result of data analysis obtained in cycles I and II showed that there was a development and improvement in teaching quality, both in creative thinking ability and science concept understanding. The result of data analysis in the students' creative thinking ability during cycles I and II can be seen in the following graph as follows.

Figure 1 shows that the students' creative thinking ability in cycle I has not optimal, that is it was within sufficiently creative category (64.5–69%). The result of analysis in cycle II shows that it has developed better and falls into very creative category (80%–89%). The indicators that dominantly developed according to Munandar' creative thinking ability (1999) were: in fluent thinking skill in which the students were able to express many ideas in solving the problem; to give many ways or suggestions to do various things; to work faster and to do more than the others. In flexible thinking ability the students were able to produce more ideas

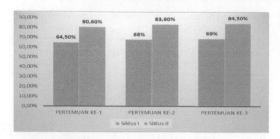

Figure 1. Graph showing creative thinking ability development in Cycle I.

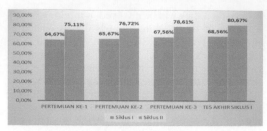

Figure 2. Graph showing concept understanding Development in Cycle II.

for solving a problem or more varied answers to a question, can see a problem from different perspectives, and present a concept using different ways. In original thinking ability the students were sufficiently able to give new ideas in solving a problem, making uncommon combinations of the parts or elements. In the ability or skill of elaboration the students were not yet able to develop or enrich ideas, to add or give details of an idea to improve the quality of the idea. In the ability or skill to evaluate, the students still had difficulties in determining a truth of a question or a truth of a plan in solving a problem, were able to solve a problem and could implement it correctly, and had a reason that could be accounted for to reach a decision.

First, in cycle I there were only some of the characteristics that occurred, then every meeting after the first meeting showed a significantly creative ability development so that in there was a significantly consistent and sustainable ability improvement.

3.2 Results in science concept understanding development

The data on the students' science concept understanding can be seen in Figure 2 as follows.

Figure 2 shows that the students' concept understanding in cycle I was 61% (sufficient category). The result of data analysis of the students' science concept understanding in cycle II was 94.4% (understands very much). The indicators for students' science concept understanding in science that developed were: having the ability to explain concepts, principles, and procedures, having the ability to identify and select concepts, principles, and procedures, not yet maximally having the ability to apply concepts, principles and procedures.

3.3 Unique findings during the lessons

The findings that were obtained during the implementation of cycle II were as follows: (1) During the teaching process optimal interactions occurred, both between a student an another student and between the students and the teacher. The students were more motivated to ask questions and to give their opinions. This can be seen in the fact that they had an initiative to ask or give opinions without waiting for the teacher to appoint them. In the interaction process, the students could appreciate each other. The students kept giving appreciation to other students who gave opinions or answered questions although the answers were not correct. (2) In the discussion process, every group showed good enthusiasm by doing tasks in the student's worksheet seriously. Thus, all groups could do the tasks in the student's worksheet correctly and on time. (3) the students' creative thinking ability had been evenly distributed among themselves. The students with a higher ability tried to help other students who had less ability as an effort to distribute and make equal understanding of the problem being discussed so that all the members in the group could participate actively at the time they solved problems in the student's worksheet and at the discussion time, after presentation and other stages in teaching. Apart from that, the students were seen to be focused at the time when they did activities in the classroom.

Compared with conventional model of teaching, teaching with the SSCS type problem solving oriented performance assessment is more in improving the students' learning achievement. This is also supported by the results of Kadir's study () that showed that the metacognition of the students who were given performance assessment was higher than those who were given written assessment. The interesting findings at the time of teaching using the SSCS type problem solving oriented performance assessment teaching were:

1. after the value of information obtained by the students was communicated by the teacher through qualitative feedback, the teacher felt that they were motivated to improve the quality of their learning based on their needs,
2. after the teacher related the new information and knowledge that the students had acquired, their interest to learn science was very high.

The students were more enthusiastic to listen to the teacher's explanation in the classroom.

3. after the teacher asked the students to interpret all of science materials that had been learned through an inspirative story, the students became more creative and had a high level of imagination (understanding meanings).

4. after the teacher involved the students actively in the teaching process, the students' activity was very high, no students wasted their time any longer during the teaching period, such as what had so far occurred and even tended to be boring. On this occasion, the lesson became interesting and joyful.

5. after the teacher asked the students to use the learning time more for discovering, probing, discussing, thinking critically, or doing a project and solving problems, the students' creativity in trying and observing and even investigating was very high. Many products from their projects were very interesting and imaginative.

6. after the teacher asked the students to build learning concepts out of their self-awareness and intrinsic motivation, they became more autonomous in learning. This autonomy was shown by the students' activities in doing practicum by themselves before the teacher came into the teaching process.

4 CONCLUSIONS

Based on this result it can be concluded that the implementation of SCSS problem solving oriented performance assessment teaching can improve creative thinking ability and science concept understanding. Science concept understandings that can be easily developed are remembering and repeating as well as identifying and selecting concepts, principles, and procedure. Creative thinking abilities that can be easily developed are related to the indicators of fluent thinking skill, flexible and original thinking ability.

REFERENCES

Anderson, Lorin W. & David R. Krathwohl. 2011. *A taxonomy for learning teaching and assesing, a revision of Blomm's taxonomy of educational objectives*. New York: Adison Weasley Longman.

Ennis, R. H. 1985. Goal Critical Thinking Curriculum. According to Costa, A. L. (Ed): *Developing minds: A resourse book for teaching thinking*. Alexandria, Virginia: *Association for Supervision and Curriculum Developing (ASCD)*. 54–57.

Herman, J. L., Aschbacher, P. R. & Winters, L. 1992. *A practical guide to alternatif assessment*. New York.

Association for Supervision and Curriculum Development.

Munandar, Utami. 1999. *Creativities and talent: incarnating creative potential and talent strategy*. Jakarta: PT Rineka Cipta.

Munandar, Utami. 2004. *Creativities development of talented children*. Jakarta: PT Rineka Cipta.

National Department of Education. 2002. *Contextual approach: contextual teaching and learning*. Jakarta: Dirjen Dikdasmen.

National Department of Education. 2002. *Curriculum and learning outcome*. Jakarta: Curriculum Center.

National Department of Education. 2004. *Physics competence standard*. Jakarta: PT. Rineka Cipta.

Pantiwati, Y. 2010. *The effect of biology assessment in TPS (Think Pair Share) cooperative learning toward high school student's cognitive, critical thinking, creative thinking and metacognitive awareness abilities in Malang city*. Dissertation. Malang State University. 2010. 155.

Redhana, I. W. 2003. Increasing student's critical thinking by cooperative learning based on problem solving strategy. *Education and Instruction Journal* IKIP Negeri Singaraja. 3(33): 11–23.

Reigeluth, C. M. & Moore, J. 1999. *Cognitive education and the cognitive Domain*. Indiana University.

Santyasa, I. W. 2004. Reasoning and problem solving learning model. *IKA IKIP Negeri Singaraja Journal*. 2(2): 26–43.

Santyasa, I Wayan. 2007. Conceptual based of learning media" a working paper that presented in learning media workshop for high school teachers.

Santyasa, I. W. 2004. *The effect of learning model and setting toward high school student's remidiation, misconception, concept comprehension and physics learning outcome*. Dissertation (unpublished). Malang State University Post Graduate Program Learning Technology Study Program.

Santyasa, I. W. 2005. *The Implementation of Innovative Learning in Teaching Practice*. Paper. Presented in early purchasing of miter compete grand LPPL IKIP Negeri Singaraja with Sekolah Laboratorium IKIP Negeri Singaraja, 18–20 of July 2005 in Singaraja.

Santyasa, I. W. 2006. *Accommodating student's paradigm alteration in learning: professorship oration in physics education at mathemathics and science faculty*. Presented in Open Session Ganesha University of Education Senate, Monday 28 of August 2006.

Suastra, I. W & Kariasa, I. N. 2001. *Developing Student's Creative Thinking through Erudition in Elementary School*. Lecturer Research Report. Unpublished. IKIP Negeri Singaraja.

Suma, K. 2004. *Developing Thinking Process and Problem Solving Skill in Physics Learning*. Paper. Physics Education Department Mathematics and Science Faculty IKIP Negeri Singaraja.

Widoyoko, S. Eko Putro. 2012. *The evaluation of learning program*. Yogyakarta: Pustaka Pelajar.

Yasa, P. 2007. Problem based learning strategy to increasing student's physics basic competence at grade VIII Junior High School 2 Singaraja. *Education and Instruction Journal UNDIKSHA*. 3: 622–637.

Ideas for 21st Century Education – Abdullah et al. (Eds)
© *2017 Taylor & Francis Group, London, ISBN 978-1-138-05343-4*

Identification of consistency and conceptual understanding of the Black principle

C. Winarti, A. Cari, I. Suparmi, J. Budiarti, H. Handhika & V. Viyanti
Universitas Sebelas Maret, Surakarta, Indonesia

ABSTRACT: The concept of physics can be understood not only in mathematical language but through a variety of representations. By understanding the concept, someone will be consistent in addressing the problems, even though they are presented in different representations. Consistent answers to the questions in physics will show the level of understanding of the concepts. The purpose of this research was to identify the consistent answers about the Black principle. This research used the descriptive qualitative approach. The instrument used in this research was nine multiple choice questions with reasons. The essential concept is the Black principle. Some questions were adopted from the Heat and Temperature Concept Evaluation (HTCE). The sample for this research was 145 students of the Physics Education Program of Sunan Kalijaga State Islamic University, Institute of Teachers' Training and Education of PGRI Madiun, and State University of Lampung. The results showed that 81.74% of students were inconsistent, 1.13% of students were consistent with the correct answers, and 17.03% of students were consistent with the wrong answers when responding to the test of multi-representation concepts.

1 INTRODUCTION

Physics is a science that studies the natural phenomena. The natural phenomena that occur should be studied more deeply in order to get the causal occurrence. It requires a mature understanding to have reasoning when solving the problems that arise from the occurrence of the natural phenomena. Understanding the concept gives students the ability to not only remember but also to re-explain the definition, special features, nature, essence, and content using their own words, without changing the meaning of the content of the information they receive. According to Wospakrik & Hendrajaya (1993), physics is a branch of science that aims to study and provide a quantitative understanding of the various symptoms or processes of nature and the nature of the substance and its application. Physics is not only realized in mathematical language. Someone's understanding of the concept can be judged from the language of communication, the language of physics, the language of math, the language of intuition, and the language of definitions/provisions. When someone does not understand one of these languages, there will be no understanding of the concept and there could even be misconceptions (Search, 2016). Natural phenomena and processes in the universe have certain consistent patterns that can be studied and assessed systematically. According to Meltzer (2004), physics can also be translated

into four representations, as follows: verbal representation, diagram or images representation, mathematical representation or mathematical symbols, and graphic representation. Thus, solving physics problems can be conducted by representing the questions of physics in various forms: verbal, graphic, images or pictorial, and math in the form of formulas. A concept can be built through the problem-solving task, which also involves use of the system of representation. The students' knowledge of the concept would be more significant if they were able to reveal the differences in the representation of the concepts being learned.

Various representations have the power that can be used to solve the problems appropriately. For instance, the problem is inadequate if it is being solved only by the verbal representation. But if it is solved by the good graphic representation, it would provide the most appropriate solution. It requires the proper consideration in using the representations for solving the problems. By viewing the level of scientific consistency of the students' answers, the students' understanding of the concepts could be measured. The scientific consistency is based on the number of correct answers for each concept, measured by different modes of representations. According to Nieminen et al. (2010), the consistency of the students will lead them to a better level of understanding in viewing various concepts of physics as outlined in various problems. In line with Ainsworth

(2006), the consistency of response of the students in understanding the concept of physics requires a deeper understanding by the students to view the equivalence of the problems of physics as outlined in various ways. The results of preliminary research show that there are problems with the temperature and heat concept at the middle school level, not only in the low national test scores, but also in the daily tests. Based on the interviews with the physics teacher, it is found that the lack of understanding of the concept is because the concepts of temperature and heat are close to being an everyday phenomenon. In another study on the concept of the Black principle, the students have difficulties in distinguishing between $Q_{deliver}$ and $Q_{receive}$ (Winarti et al., 2015). Many students were confused about the concepts of heat and temperature due to the same thing, the perception of temperature is only about hot and cold, and temperature can be transferred. The students memorized this concept and were not able to make a connection between their knowledge and the physics phenomena in everyday life (Winarti et al., 2016). It needs to examine the failing process of knowledge transfer in the class. Based on the interviews with the students, the data shows that, when teaching the subject of temperature and heat, the teacher only asks the students to memorize the subject and does not embed the concept of thinking in solving the problems. It can possibly happen to students majoring in physics education. According to Prince (2011), misconceptions and incomprehension in college greatly affect the performance of teaching. As the institution producing the physics teachers, it is important for the college to determine the readiness of the physics teacher candidates to jump into the field in order not to bring the misconceptions and incomprehension to the students. Difficulty in the understanding of the concepts of heat and temperature have been investigated by many educational researchers. Many students have not understood even misconceptions about the concept of temperature and heat (Wiser & Carey, 1988; Lee, 2007). Students come to class with an understanding that is not empty; the students come with a variety of knowledge from everyday life. According to Arnold & Billion (1994), students' understanding of the concept of temperature and heat comes from the experience that they get from everyday life. The concept of temperature and heat is directly related to the environment and daily life (Ericson, 1979).

Therefore, it is important to 1) identify the consistency of understanding from the answers of the students, and 2) analyze the understanding and the errors that occur in the students associated with the concept of the Black principle.

2 RESEARCH METHODS

This research used the descriptive qualitative approach. The research was based on the depiction of the phenomena or events captured by the researchers with the existing facts. Qualitative research is the method used to examine the condition of natural objects (Bao et al., 2002). The research was a preliminary study to see the profile of the students' reasoning. The instrument used in this research was nine multiple choice questions with reasons. The essential concept tested in this test was that of temperature and heat. Some questions were adopted from the questions of the Heat and Temperature Concept Evaluation (HTCE).

The research was conducted at three universities, which are: Sunan Kalijaga State Islamic University, Institute of Teachers' Training and Education of PGRI Madiun, and State University of Lampung. For the sample, there were 52 students of the Physics Education Study Program of Sunan Kalijaga State Islamic University, 25 students of the Institute of Teachers' Training and Education of PGRI Madiun, and 68 students of the State University of Lampung. The data analysis was conducted by:

- Finding the average of the correct and incorrect answers of each concept being tested.
- Determining the fraction of students who answered correctly or those who answered incorrectly out of the total students.
- Determining the distribution of the consistency of answers from the group of respondents to each concept to determine the tendency of consistency of the respondents.
- Determining the conception of the group of respondents to each concept to determine the tendency of not knowing the concepts and the misconceptions.

3 RESULTS AND DISCUSSION

The results of the research showed that the level of consistency of the students' answers is as shown in Table 1.

It shows the values of the consistency of the students in answering the questions on the temperature and heat concept. It can be seen that the greatest value from the three universities is the inconsistency of the students' answers. The average of the inconsistency of the students' answers from the three universities is 81.74%. The average of the consistency with the correct answers is only 1.13%. When facing the same question, but in a different form of representation, the students become confused.

On the question of the Black principle, eighteen students correctly answered the multiple choice

Table 1. Percentage of the level of consistency.

Level of consistency	Percentage (%)			
	Sunan Kalijaga State Islamic University (N = 52)	State University of Lampung (N = 68)	Institute of Teachers' Training and Education of PGRI Madiun (N = 25)	Average
Consistent with the correct answers	1.92	1.47	0	1.13
Consistent with the wrong answers	26.92	16.17	8	17.03
Inconsistent	71.15	82.24	92	81.74

Elemen pemanas diletakkan dalam gelas A dan B yang berisi sejumlah air yang sama, sehingga terjadi perpindahan kalor untuk menjaga gelas pada temperatur yang ditunjukkan. Jawaban mana yang dapat menggambarkan kelajuan kalor yang harus dipindahkan untuk menjaga temperatur yang ditunjukkan?

a. Gelas A akan membutuhkan kalor dengan kelajuan sekitar lima kali kelajuan B
b. Gelas A akan membutuhkan kalor dengan kelajuan sekitar tiga kali kelajuan B
c. Gelas B akan membutuhkan kalor sedikit lebih cepat dari A
d. Kedua gelas akan membutuhkan kalor dengan kelajuan yang sama
X Gelas B akan membutuhkan kalor sekitar tiga kali kelajuan A

Alasan: ...

Figure 1. Example of a student's answer on the subject of the Black principle.

question but gave the wrong reason. Figure 1 shows that the student only analogized it and took into account mathematically the calorific needs of each glass without using or relating to the concept of the Black principle and the rate of energy. It should be possible to use the equation to solve the problem. Glass B needs three times more heat than glass A because the heat is proportional to the rate of heat ($Q \sim H$), so glass B needs three times as much heat. The most common error seen from the answers of the students was that they only explained the difference in temperature and the thermal temperature alone, but they did not explain the rate of heat that occurs in glass A or glass B.

Based on the interviews, the students stated that the rate of heat concept was abstract and they could not find it in everyday life, so they answered

incorrectly the questions related to the rate of heat. If the students really understood the concept, they should consistently answer the questions about representations of images and graphics correctly. A student's answer when working on the question about the representation of the graphic is shown in Figure 2.

In answering the question on the representation of the graphic in Figure 2(a), the student answered correctly because he was able to link the question with the concepts of physics. In solving the question, the student should determine the relationship between the changes in temperature (ΔT) and time (t). The students should also link the case with the concept of thermal equilibrium. In the case of glass A, it has a temperature of 35°C, and when filled with 100 grams of water it has a space temperature of 25°C. If it stays still at a certain time it will reach the thermal equilibrium. When reaching the thermal equilibrium, the graphic corresponding to the case is graph E. But for this case, actually glass A and glass B will experience the same thing; each will already reach the thermal equilibrium so the right choice is graph C, where the temperature will be inversely proportional to the time. The students who answered as in Figure 2(b) have not understood the concept because, based on the interviews, the students stated that the final temperature of glass A will be the same because they thought there

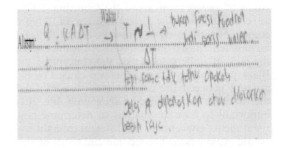

Figure 2(a). Result of writing the description of understanding the Black principle.

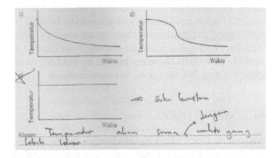

Figure 2(b). Wrong representation of the relationship of temperature and time.

251

was no change in the room temperature that affects the water temperature in glass A. The result of the answers of the interview on Figure B is as follows "the student stated that the temperature reached by an object depends on the time".

The results of the interviews show that the students sometimes used their intuition to answer the questions. This happens because the concepts presented in physics and in the questions are very close to daily happenings. Unfortunately, the analysis conducted by the students was less detailed in viewing the other factors that influence the case. The students only view one side, based on the visible variables, without thinking of the connection between the variables and the actual concepts. This is in accordance with the opinion delivered by Alwan (2010) in his research, who states that the students' concept develops based on their daily experiences, but sometimes their intuitive understanding is different from the scientific concepts. It is in line with Leura et al. (2005), who stated that the concepts embedded in the students' minds comes from the interpretation of ideas derived from their daily experiences (Leura et al., 2005).

4 CONCLUSIONS

Based on the research results and the data analysis, it shows that the level of consistency of the students' answers consists of the following: 1.13% are consistent with the correct answers, 17.03% are consistent with the wrong answers, and 81.80% are inconsistent. Based on the results of the interviews, the lack of understanding of the concept of the Black principle by the students was caused by the mixture between the correct understanding and the intuition encountered in their daily life. It is important to have an analysis to determine how the students' conceptions are different from the scientific explanation. The recommendations of this research are that, when getting to know the understanding or even the misconceptions of the university students or other students, we should keep doing the interviews to explore the reasons for the students' answers. It is very profitable to be able to investigate the problems that are caused by their incomprehension and to know what treatment is suitable to handle these problems, both in terms of learning and for the questions provided.

ACKNOWLEDGMENT

We thank UNS. This research was supported by Mandatory Grand UNS with contract number 632/UN27.21/LT/2016.

REFERENCES

Ainsworth, S. E. (2006). A conceptual framework for considering learning with multiple representations. *A Framework for Learning with Multiple Representations*. Retrieved from http://www.nottingham.ac.uk

Alwan. A. A. (2011). Misconception of heat and temperature among physics student. *Procedia Social and Behavioral Sciences, 12*, 600–614.

Arnold, M. & Millar, R. (1994). Children's and lay adult views about thermal equilibrium 'work' and 'heat': on a road toward thermodynamics. *International Journal of Science Education, 16*, 131–144.

Bao, L., Hogg, K. & Zollman, D. (2002). Model analysis of fine structures of student models: An example with Newton's third law. *American Journal Physics, 70*(7), 77–82.

Lee, O. (2007). Urban elementary school teachers' knowledge and practices in teaching science to English language learners. *Journal Science Teacher Education, 25*, 733–756.

Leura, G. R., Otto, C. A. & Zitzewitz, P. W. (2005). A conceptual change approach to teaching energy & thermodynamic to pre-service elementary teacher. *Journal Physics Teacher Education Online, 2*(4), 3–8.

Meltzer, D. E. (2004). Investigation of students' reasoning regarding heat, work, and the first law of thermodynamics in an introductory calculus-based general physics course. *American Journal of Physics, 72*(11), 1432–1446.

Nieminen, P., Savinainen, A., & Viiri, J. (2010). Force Concept Inventory-based multiple-choice test for investigating students' representational consistency. *Physical Review Special Topics-Physics Education Research, 6*(2), 020109.

Vigeant, M. A., Prince, M. J., & Nottis, K. (2011). The use of inquiry-based activities to repair studen misconceptions related to heat, energy, and temperature. *Proceedings from American Association for Engineering Education, Vancouver, BC.*

Winarti, W., Cari, Widha S. & Edi I. (2015). *Analyzing Skill dan Reasoning Skill Siswa Madrasah Aliyah di Kota Yogyakarta*. Prosiding Seminar Nasional Pendidikan Sains UNS. 19 November 2015. ISSN: 2407–4659.

Winarti, W., Cari, Suparmi, Widha, S. & Edi, I. (2016). *Development two tier test to assess conceptual understanding heat and temperature*. Proceedings of the International Conference on Science and Applied Science. UNS.

Wiser, M. & Carey, S. (1988). The differentiation of heat and temperature: History of science and novice expert shift. In S. Strauss (Ed.), *Ontogeny, phylogeny and historical development*. Norwood, NJ: Ablex Publishing Corpora.

Wospakrik, H. J., & Hendrajaya, L. (1993). Dasar-dasar Matematika untuk Fisika. Jakarta: Ditjen Dikti Depdikbud RI Proyek Pembinaan Tenaga Kependidikan Pendidikan Tinggi.

Ideas for 21st Century Education – Abdullah et al. (Eds)
© *2017 Taylor & Francis Group, London, ISBN 978-1-138-05343-4*

Relationship between vocational/senior high school educational background and the generic medical ability of midwifery students on a microbiology course

Y. Saparudin, N. Rustaman & A. Fitriani
Universitas Pendidikan Indonesia, Bandung, Indonesia

ABSTRACT: The aim of this research was to investigate the relationship between the senior high school/ vocational high school background of midwifery students and their Medical Generic Skills (KGM) on a microbiology course. The method used was a descriptive correlational method, with cross-sectional analytic study. This study involved students in the second semester of the 2015/2016 academic year. The independent variable was the background of their program in senior high school/vocational high school and the dependent variable was the average KGM score. The results showed that students with a natural science background at high school/vocational high school had higher scores compared with the ones that had social science, pharmaceutical and chemical analyst backgrounds. The relationship between the students' backgrounds and their KGM scores was significant at the 5% test level.

1 INTRODUCTION

Implementation of health education at college, including the delivery of midwifery education, should be based on national standards, namely professional competence and the competence of graduates. The implementation of midwifery education should start to adjust to global competence, formulated by the American College of Nurse-Midwives. Competence has no relationship with the subject of microbiology and largely overlaps with pathogenic microbes, which cause clinical problems. Therefore, the curricula of health workers, especially in microbiology courses, should include material on the microbial world, microbes that affect health, and concepts of microbes associated with the prevention, infection control, and sanitation.

The learning process should be student-centered and link the content of materials studied with the problems that exist in the real world. The process must be able to respond to scientific and technological progress, to participate in resolving clinical issues (Ricon et al., 2010; Schmidt et al., 2011; Wood, 2003) using thinking skills (Ault et al., 2011; Azer et al., 2013; Gasper & Gardner, 2013). One of the skills that must be developed in students is their ability in medical generic skills (Wood, 2003; Ault et al., 2011; Azer et al., 2013; Gasper & Gardner, 2013). The assessment of ability in medical generic skills (KGM), developed for this study, includes consideration of: (1) planning and organization, (2) critical evaluation of literature, (3) teamwork, (4) communication, (5) time management, and (6) self-directed learning and the use of resources (Wood, 2003; Razzaq & Ahsin, 2011; Murdoch & Whittle, 2012). Efforts to improve the ability of medical generic skills would be difficult to be realize, if learning outcomes of generic skills in medical microbiology by college students were unsatisfactory (Saparudin, 2013; Saparudin, 2014; Saparudin, 2016).

Previous research performed within the Nursing, Midwifery, and Public Health study program showed that the average generic capabilities of the students were below 50 (Table 1) (Saparudin, 2013; Saparudin, 2014). In addition, average understanding of the theory of microbiology by the students was below 70. These results were inadequate.

Table 1. Average value theory of microbiology and generic capabilities.

No	Student program	Average value theory of microbiology	Average value of generic capabilities
1	Nursing	69.20 (±4.57)	45.27 (±8.81)
2	Midwifery	67.34 (±4.12)	39.23 (±5.81)
3	Public health	69.72 (±5.13)	44.12 (±9.32)

Based on the above, research has been conducted to analyze the relationship between the senior high school/vocational high school background of students and the average value of theory and practice of nursing, midwifery, and public health students studying the microbiology course. The results showed that, in general, the students who took a natural science program during their vocational/high school education had a higher average score for both theory and practice in the microbiology course, than students who took a social science program.

2 RESEARCH METHODS

The target population of this study was students in the second semester (academic year 2015/2016) at a midwifery school (*STIKES*) in West Java. The sampling technique of the entire student population was used to select 30 students.

This study was a cross-sectional analytic study. The researchers conducted a correlation analysis between senior high school/vocational senior high school background and KGM. The independent variable of the study was the educational background of the students, while the dependent variable was the KGM of the students. The average KGM was measured when the students carried out practical problem solving, based on the material effects of temperature on microbial growth.

In this research, the data were analyzed using univariate and bivariate analysis (Chi Squared test). Univariate analysis was used to look at the percentage scores achieved during vocational/senior high school and the KGM values of students.

3 RESULTS AND DISCUSSION

Two programs offered in senior high school: a natural science and social science program. Two programs were offered in vocational high school: chemical analyst and pharmaceutical chemistry. The frequency of students who took the programs during vocational/high school and then entered midwifery school education is shown in Table 2.

Table 2. Frequency distribution of senior high school/vocational senior high school background programs.

Background program	Amount	
	N	%
1. Natural sciences	16	53.30
2. Social science	8	26.70
3. Chemical analyst	3	10.00
4. Pharmaceutical chemistry	3	10.00

More than 50% of midwifery students had a natural science program background.

Analysis of the KGM showed that the percentage of midwifery students who were in the good category (53.3%) was similar to the amount of students who were in the low category (46.7%) (Table 3).

Bivariate analysis on the relationship between the senior high school/vocational high school background of the students and their KGM showed that all students with a natural science background were categorized as good (100%) (Table 4). In contrast, the KGM of students with social sciences, pharmaceutical and chemical analyst backgrounds all (100%) had low KGM values. The relationship between the program background and KGM was significant at the 5% test level (Table 4). Interestingly, the students who had good KGM values all had natural science backgrounds.

Students with a natural sciences background had studied more on natural science subjects, such as biology, chemistry, physics and mathematics, than students with a social sciences background at senior high school. Therefore, the understanding of scientific concepts and high-level thinking skills would be more in depth for the natural science background students. In addition, the hours and the depth of scientific-subject learning (mathematics, biology, physics and chemistry) of students at vocational senior high school (chemical analyst and pharmaceutical chemistry programs) were lower than those for the students with natural science backgrounds at senior high school. For example, in the pharmaceutical chemistry program at the vocational high school, biology was only studied in grade X and XI, for just 2 hours per week. Therefore, the understanding of science and the students' abilities to solve problems would be better for senior high school students on the natural science program compared with the high school students on the social sciences program, or vocational high school students on the pharmaceutical chemistry and chemical analyst programs.

Students with a natural science background at senior high school received more hours of education on natural science subjects. They were therefore better equipped, with more in depth understanding, for learning activities that can

Table 3. KGM students.

KGM students	Amount	
	N	%
Good	16	53.30
1. Less	14	46.70

Table 4. Relationships between senior high school/vocational senior high school background and KGM value of midwifery students.

| No | Variable | KGM | | | | | | P_{value} | 95% CI |
| | | Good | | L | | Amount | | | |
		n	%	n	%	N	%		
1.	Background program:								
	a. Natural Sciences	16	100.0	0	0	16	100.0	0.000	30.000
	b. Social science	0	0	8	100.0	8	100.0		
	c. Chemical Analyst	0	0	3	100.0	3	100.0		
	d. Pharmaceutical Chemistry	0	0	3	100.0	3	100.0		

encourage the development of the implementation of the model syntax problem solving. The learning process carried out at senior high school in the natural sciences program generally centered on students, encouraging them to pursue opportunities and facilitate the development of their own knowledge, so that they could gain deep understanding, and ultimately improve their scientific knowledge. Learning to use the, relatively innovative, student-centered learning, and adhering to the constructivist-learning paradigm, students play an active role in constructing the concepts they have learned.

4 CONCLUSION

Based on the research results and discussion, it can be concluded that the KGM of students with natural science backgrounds from high school/vocational school was higher (categorized as good) than that of students with social sciences, pharmaceutical chemistry and chemical analyst backgrounds (categorized as low). The relationship between the vocational/senior high school educational background and KGM was significant at the 5% test level.

REFERENCES

Ault, J. F., Renfro, B. M., & White, A. K. (2011). Using a molecular-genetic approach to investigate bacterial physiology in a continuous, research-based, semester-long laboratory for undergraduates. *J. Microbiol. Biol. Educ*, 12, 185–193.

Azer, S. A., Hasanato, R., Al-Nassar, S., Somily, A., & AlSaadi, M. M. (2013). Introducing integrated laboratory classes in a PBL curriculum: impact on student's learning and satisfaction. *BMC medical education*, 13(1), 71.

Gasper, B. J., & Gardner, S. M. (2013). Engaging students in authentic microbiology research in an introductory biology laboratory course is correlated with gains in student understanding of the nature of authentic research and critical thinking. *J. Microbiol & Biol. Educ.*, 14(1), 25–34.

Murdoch-Eaton, D., & Whittle, S. (2012). Generic skills in medical education: developing the tools for successful lifelong learning. *Medical education*, 46(1), 120–128.

Razzaq, Z., & Ahsin, S. (2011). PBL Wrap up sessions: an approach to enhance generic skills in medical students. *J Ayub Med Coll Abbottabad*, 23(2), 162–165.

Ricon, T., Rosenblum, S., & Schreuer, N. (2010). Using Problem Based Learning in Training Health Professionals: Should it Suit the Individual's Learning Style?. *Creative Education*, 1(1), 25–25.

Saparudin, Y. (2013). *Profile execution problem solving models syntax stikes student: A case study results* (Unpublished magister dissertation). Universitas Pendidikan Indonesia, Bandung.

Saparudin, Y. (2014). *Profile lectures microbiology student stikes: Results of field study* (Unpublished magister dissertation). Universitas Pendidikan Indonesia, Bandung.

Saparudin, Y. (2016). Relationship between the educational backgrounds with syntax model of learning problem solving midwifery stikes students. *Prosiding SN-DPK 2016* (pp. 406–412). PPPM STIKes Bakti Tunas Husada, Tasikmalaya.

Schmidt, H. G., Rotgans, J. I., & Yew, E. H. (2011). The process of problem-based learning: what works and why. *Medical education*, 45(8), 792–806.

Wood, D. F. (2003). Problem based learning. *BMJ: British Medical Journal*, 326(7384), 328.

Other Areas of Education (OAE)

Ideas for 21st Century Education – Abdullah et al. (Eds)
© *2017 Taylor & Francis Group, London, ISBN 978-1-138-05343-4*

Identification of scientific literacy aspects of a science textbook for class V of elementary school

S.S. Nurfaidah
Universitas Pasundan, Bandung, Indonesia

ABSTRACT: Directed scientific literacy is important for learners because it is a basic competence in understanding environmental and economic problems in modern society. One effort to improve scientific literacy can be initiated by the identification of the scientific literacy aspects of textbooks. This study presents an analysis of the aspects of scientific literacy present in a classroom textbook for the fifth grade of elementary school. The aspects of scientific literacy considered in this study included knowledge of science, science as a means of investigation, science as a way of thinking, and the interaction of science, technology and society. The form of this research was descriptive research. The data was collected in the form of documentation analysis. The whole of one book was analyzed, and presented the various aspects of scientific literacy to different extents, namely: 73.5% in terms of aspects of scientific knowledge; 19.5% for aspects of science as a means of investigation; 5.8% for aspects of science as a way of thinking; 1.2% for aspects of the interaction of science, technology and society. Scientific knowledge consistently represented the highest proportion of any scientific literacy aspect in each chapter of the book, while the aspect of science as a means of investigation showed a smaller proportion, varying from 13% to 32% in any given chapter. These findings indicate that this leading textbook put more emphasis on factual content (aspects of scientific knowledge) and only provided low emphasis on mastering scientific process and context (i.e. science as a means of investigation).

1 INTRODUCTION

Life in the modern industrial state cannot be separated from the 'culture' of science and technology, where daily lives are significantly influenced by science (Miller, 1996, in Yuenyong & Narjaikaew, 2009). One indication is the rapid increase in the number of products of the science and technology community in everyday life. Thus, in the era of fierce international economic competition, the producers (graduate users) require skilled scientific staff, that is, those who have a good understanding and ability in science, to be able to meet the demands of the times. Therefore, learning about science and technology is very important for everyone in the world (Office of the National Education Commission, 2003, in Yuenyong & Narjaikaew, 2009).

Skills associated with the use of science and technology are described by the term 'scientific literacy'. Mastery of scientific literacy is expected to enable students to adapt to the advances of science and technology in the future. Therefore, factors affecting the level of scientific literacy become interesting to understand. One of these is textbooks that are available and to which students can relate. Yusuf (2008) has also stated that textbooks are one tool for

improving scientific literacy. Based on this, a study of the textbooks required is one way to improve the quality of science education in Indonesia.

The purpose of this research is to establish information regarding the presentation of scientific literacy in each chapter of a science textbook, and the textbook as a whole, which covers aspects of scientific knowledge and science as a means of investigation. This research was conducted at the level of elementary school, as previous studies have been conducted in middle and high school education. Class V was selected because at this level the student must be prepared to take the final exam, which will be done in the next class, meaning that students should be able to bring understanding and maturity to the material.

2 LITERATURE REVIEW

2.1 *Scientific literacy*

Scientific literacy is defined as the capacity of researchers to use knowledge and scientific capabilities to identify questions and draw conclusions based on existing evidence and data, in order to understand and help to make decisions about the

Table 1.	Scientific literacy domains.
Context	The personal, local/national and global issues, both today and in the past, that require some understanding of science and technology.
Science knowledge	Understanding of facts, concepts and theories that form the basis of scientific knowledge. This includes knowledge about natural and technological artifacts (content knowledge), knowledge of how ideas are produced (procedural knowledge) and an understanding of the underlying rationale and justification they use (epistemic knowledge).
Science competence/ process	Ability to explain scientific phenomena, evaluate and design scientific investigations, and interpret the data and scientific evidence.
Attitude	Attitude toward science is characterized by an interest in science and technology; appreciation of scientific inquiry approach, as well as perception and awareness of environmental issues.

natural world and human interaction with it (Rustaman et al., 2000, in Sains Edutainment, 2012).

The Programme for International Student Assessment (PISA) is a worldwide study by the Organisation for Economic Cooperation and Development (OECD) of 15-year-old school pupils' performance in mathematics, science, and reading. In its 2000 and 2003 reports, PISA identified three major domains of scientific literacy and, in 2006, added a fourth domain concerning students' attitudes toward science (OECD, 2013), as described in Table 1.

2.2 Textbooks

According to various experts, textbooks are compulsory reference books that have been standardized for teachers and learners in a particular field of study, compiled by experts in the field for the purpose of fulfilling instructional objectives, and which are used as a learning tool to support the learning process in schools and colleges (Tarigan & Tarigan, 2009; Departemen Pendidikan Nasional, 2008; Razak, 2012).

Textbooks have an important role in learning. Despite the increasingly rapid development of digital books, however, textbooks still have a very strong grip on teachers and, unfortunately, the result is that learning implementation is often fully oriented to the textbook without recognition of the curriculum. This is in line with Amalia (2009) who argues that many science teachers

believe so strongly in textbooks that it may lead to misconceptions in science, in addition to which, many science textbooks overemphasize terms and vocabulary.

Textbooks are a manifestation of the curriculum and competencies that must be achieved by students. Textbooks also contribute to the development of student competence in applying scientific knowledge. Therefore, textbooks that contain aspects of scientific literacy will affect the development of competence and increase scientific literacy of students (Riadiyani, 2009). Textbooks are one of the tools by which scientific literacy can be increased (Yusuf, 2004, in Amalia, 2009).

3 RESEARCH METHODS

The subject used in this research was a science textbook most widely used in the fifth grade of elementary schools. The form of research was descriptive research using documentation or document analysis methods. In using this method, the researcher utilized a checklist to find predefined variables. The data obtained was then processed by counting the number and percentage of presentations of these indicators of two aspects of scientific literacy for each chapter in the textbook used.

The book consisted of seven chapters:

1. *Fungsi Alat-alat Tubuh* (Function Body Tools);
2. *Tumbuhan* (Plants);
3. *Cara Makhluk Hidup Menyesuaikan Diri dengan Lingkungannya* (How to Adjust the Living Environment);
4. *Sifat Bahan Penyusun Benda* (Properties of Material Components);
5. *Gaya dan Pesawat Sederhana* (Style and Simple Aircraft);
6. *Cahaya dan Alat Optik* (Light and Optics);
7. *Tanah, Air & Alam Semesta* (Land, Water and the Universe).

Each chapter was of a different length, and the total number of pages in the book was 162. Each page in the book was analyzed, except for those that contained only review questions or vocabulary, and including the objectives and aims of learning. The total analyzed was 135 pages.

4 RESULTS AND DISCUSSION

4.1 Science knowledge

Based on the findings, the domain of scientific literacy that was most prominent was the aspect of science knowledge, which accounted for 73.5% of the indicators (i.e. facts, concepts, principles, laws, hypotheses, theories, and models, and a statement asking students to remember the knowledge or information). This indicated that the material in The Book presented more aspects of science knowledge than the other scientific literacy domains, so we can say that this book places more emphasis on the concepts of factual science, which sees science as a product.

These findings are consistent with the results reported by previous studies (Chabalengula et al., 2008). Moreover, the same result was shown by research conducted in Indonesia (Amalia, 2009; Riadiyani, 2009; Utami, 2008). This shows that the competences students need to acquire do not fully support them in facing the rapid changes in the development of science and technology.

The aspects of scientific knowledge that were identified consisted of three indicators of scientific literacy. The numbers of statements associated with these three indicators are shown in Table 2.

Indicators of the presentation of facts, concepts, principles and laws appeared in the largest numbers (and percentages) in every individual chapter.

4.2 Science as a means of investigation

The research findings indicated that the average incidence in the textbook of the science competence/ process domain (i.e. science as a means of investigation) of scientific literacy indicators was 19.5% (see Table 3). This domain was much less prominent than that of science knowledge. Material in the textbook that was included in the domain of science as a means of investigation included: requiring students to answer questions through the use of materials; requiring students to answer questions through the use of charts, tables, and others; requiring students to make calculations; requiring students to explain answers; and requiring students to engage in experimenting or thinking activities.

Table 2. Scientific literacy indicators of scientific knowledge aspects.

Indicator of scientific knowledge	Chapter						
	1	2	3	4	5	6	7
Presenting the facts, concepts, principles, and laws	216	49	65	61	98	89	108
Presenting hypotheses, theories, and models	11	4	13	4	6	13	2
Asking students to recall knowledge or information	34	32	35	40	33	44	54
TOTAL	261	85	113	105	137	146	164

Table 3. Scientific literacy indicators of means of investigation aspects.

Indicator of science as a means of investigation	Chapter						
	1	2	3	4	5	6	7
Requires students to answer questions through the use of materials	1	6	0	0	10	11	17
Requires students to answer questions through the use of charts, tables, and others	18	18	14	8	18	2	4
Requires students to make calculations	1	0	0	0	0	0	0
Requires students to explain answers	34	9	9	4	2	7	10
Statements involve students in experiments or thinking activities	2	7	0	7	15	13	14
TOTAL	55	40	23	19	45	33	45

Not all of these indicators of the means of investigation were always present in each chapter of the book. In fact, it can be said that the indicators by which students are required to make calculations was not present in the book at all, because there was only one statement present (in Chapter 1 only). Within this domain, the most prominent indicator was that requiring students to answer questions through the use of charts, tables, and others, which accounted for 32% of the scientific literacy indicators of this aspect.

The results of the analysis in Book X did not show a balanced proportion for every aspect of scientific literacy. This phenomenon suggests that students would not be fully able to obtain from the textbook the competence required to support them in the face of changing times due to the rapid development of science and technology. Based on the results found in this study, Book X placed more emphasis on the presentation of material that was the result of science, so the ability demanded of students was of lower thinking skills, such as memorization. However, students require scientific literacy skills with more emphasis on process capability and context.

Textbooks have greatly contributed to the development of student competence through the acquisition, investigation, and application of knowledge, and the process of thinking, which would eventually form students who have the competencies expected. Therefore, determining the quality of textbooks is becoming a necessity in supporting learning. Science is not just a collection of facts, concepts, principles, laws, and matters related to the content, but is rather the development of curiosity and skills in the scientific process, the fostering of environmental awareness, life skills development, learning that involves both hands on and minds on, and the development of creativity, among other things (Firman & Widodo, 2008).

5 CONCLUSIONS

The presentation of scientific literacy in Book X in the domains of 'knowledge of science' and 'science as a means of investigation' were 73.5% and 19.5%, respectively. These findings indicated that the book placed far more emphasis on fact-laden content (aspects of scientific knowledge) and gave only little support to mastering process and context (aspects of science as a means of investigation).

Suggestions for further research should include the expansion of the range of aspects of scientific literacy covered in textbooks, by determining the direct relationship between textbooks and the level of scientific literacy of students, and determining the relationship between learning materials and the presentational aspects of scientific literacy.

REFERENCES

Amalia, S. (2009). *Analisis buku ajar biologi SMP kelas VIII di kota Bandung berdasarkan literasi sains* (Unpublished thesis).

Chabalengula, V.M., Lorsbach, T., Mumba, F. & Moore, C. (2008). Curriculum and instructional validity of scientific literacy themes covered in Zambian high school biology curriculum. *International Journal of Environmental & Science Education, 3*(4), 207–220.

Departemen Pendidikan Nasional. (2008). *Permendiknas Nomor 2 Tahun 2008 Tentang Buku.* Jakarta, Indonesia: Departemen Pendidikan Nasional.

Firman, H. & Widodo, A. (2008). *Panduan pembelajaran ilmu pengetahuan alam SD/MI.* Jakarta, Indonesia: Pusat Perbukuan.

Lahiriah, R.S. (2008). *Analisis buku ajar biologi SMA kelas X di kota Bandung berdasarkan literasi sains* (Unpublished thesis).

OECD. (2006). Literacy skills for the world of tomorrow. [Online]. Paris, France: Organisation for Economic Cooperation and Development. Retrieved from http://www.oecd.org/edu/school/2960581.pdf

Razak, R. (2012). Manfaat dan fungsi buku teks dalam pembelajaran. [Web log post]. Retrieved from http://rahmanabdulrazak80.blogspot.com/2012/11/man-faat-dan-fungsi-buku-teks-dalam.html

Riadiyani, E. (2009). *Analisis Buku Ajar Biologi SMA Kelas XI di Kota Bandung Berdasarkan Literasi Sains* (Unpublished thesis).

Sains Edutainment. (2012). Definisi Literasi Sains [Web log post]. Retrieved from http://sainsedutainment.blogspot.com/2012/12/definisi-literasi-sains_23.html

Tarigan, H. G. & Tarigan, D. (2009). *Telaah Buku Teks Bahasa Indonesia.* Bandung, Indonesia: Angkasa.

Utami, A.A. (2008). *Analisis buku ajar biologi SMA kelas XII di kota Bandung berdasarkan literasi sains* (Unpublished thesis).

Yuenyong, C. & Narjaikaew, P. (2009). Scientific literacy and Thailand science education. *International Journal of Environmental & Science Education, 4*(3), 335–349.

Yusuf, S. (2008). Perbandingan gender dalam prestasi literasi siswa Indonesia. Retrieved from http://uni-nus.ac.id/tampil/data/data_ilmiah/Suhendra%20Yusuf%20-%20Makalah%20untuk%20Jurnal%20Uninus.pdf

Ideas for 21st Century Education – Abdullah et al. (Eds)
© *2017 Taylor & Francis Group, London, ISBN 978-1-138-05343-4*

Arung Masala Uli-e: The idea of the leader in Buginese myth

A.B.T. Bandung
Universitas Hasanudin, Makassar, Indonesia

ABSTRACT: The Buginese have different kinds of myths. Some people assume that myth is only a fairy tale. In fact, myth has social and moral messages that can be used as a guide in life. Lévi-Strauss (1978) argued that myth is a collective human product such that by understanding the structure of meaning in myths, the structure of human thinking can be revealed. One Buginese myths is that of *Arung Masala Uli-e*. This myth tells of various forms of Buginese structure, in relation to ordinary people, an *adat* leader, and a king. The method used in analyzing the deeper meaning of the myth followed Lévi-Strauss, namely, by dividing the myth into several parts or episodes. In each episode, the relationships can be established and then interpreted according to ethnographic data. The results of the study indicate that the glory and success of Buginese leadership is based on deliberation, especially when an important activity is to be done or decided. A Bugis leader or king has to be very careful and always be fair and firm in handling his children or close relatives.

1 INTRODUCTION

Myths study by Lévi-Strauss is a study conducted by experts of the Humanities Sciences in Indonesia. The myths can be used as a way of understanding a society's culture. Claude Lévi-Strauss was a French anthropologist who reviewed around 800 myths. He argued that by understanding the structure of meaning in myths, the structure of human thinking can be revealed, because the myth is not individual in nature, but is a collective human product (Lévi-Strauss, 1978).

In this article, the first step is taken in learning to examine the structure and meaning of the myth in a Bugis story, *Arung Masala Uli-e*. This article seeks to represent the Buginese cultural dynamics of this story. In addition to interpreting the *Arung Masala Uli-e* myth, some ethnography books about the Buginese culture in South Sulawesi are recognized as raw data for completing the interpretation of a myth that has taken place in a dialectical movement between ethnographic data and the myth itself. The *Arung Masala Uli-e* myth was taken from a book by A. Rahman Rahim (Rahim, 1992). The leadership system in Bugis society was comprehensively explained by Mattulada (1985); it was a very strong and unique system. For example, the king was very accountable, fair and firm in enforcing the law, even when it concerned his own son.

The method used to explore the underlying meaning of the myth follows Lévi-Strauss (1978) and Ahimsa-Putra (1994), namely, by dividing the

myth into several parts or episodes. In each episode, the relationships can be established and each relationship is then interpreted according to ethnographic data.

The myth of *Arung Masala Uli-e* concerns the diseased skin of a king's daughter; it is a myth that exists among the Buginese, and many people show their appreciation of the story by eating the forbidden albino buffalo.

The method can be divided into several steps: understanding each fragment of the myth story; analyzing and interpreting the relationships between these fragments; deciding upon the relationships among the fragments, leading to conclusions about the meanings of those relationships.

2 LITERATURE REVIEW

2.1 *Structural analysis and interpretation*

The first step in the analysis is to understand the entire story in the chosen myth, and the next step is to analyze and interpret the meaning of the myth. As mentioned in the introduction, this myth will be analyzed by following the method used by linguists in analyzing a language. Lévi-Strauss (2009) considers that myth can be included in the category of language because myths have a message, either explicitly or implicitly. Myth is a story that is linguistically a piece of prose or set of words from which the content appears to emerge through coincidence. Lévi-Strauss wanted to investigate whether this was just coincidence or whether the myth had its own pattern, just as language has its grammar. However, language users are not aware of the rules of the language used.

To analyze the myth, the myth will be cut into pieces and then from these fragments it will be determined whether the relationships are the same or different. The arrangement of the relationships will be interpreted, so that the hidden meaning is visible behind the myth. This myth will be divided into two episodes, with each episode representing an important event.

Episode I

> Character of Mappajungge (The King of Luwu)

This section describes how a king of Luwu, in the land of Bugis, faces and resolves problems in Buginese society.

A complex problem was faced by the king when his daughter was affected by a skin disease which was getting increasingly worse. The king was despondent in the face of this reality. People feared contamination by such diseases. The custom of Luwu was consensus, and the consensus was that there were two options that would be presented to the king in the form of a preference for one egg or many eggs. If the king preferred one egg, then the people would emigrate to other areas. However, the king chose many eggs (people). Thus, he was finally willing to exile his own daughter for the sake of upholding *ade'*, or traditional custom.

> Character of Arumpone (The King of Bone)

This section describes the anxiety of Arumpone, a king of Bone, and his wife on witnessing their sick son, Arummaloloe. The king summoned his slaves and asked them why his son was sick, but only Anreguru knew. Anreguru told of the meeting of Arummaloloe with a princess of Luwu, which caused him pain (having fallen in love). Arumpone decided to ask Qadi and Aruppitu to be an ambassadorial team for the purpose of procuring the marriage of his son to the princess. Arumpone instructed Qadi and Aruppitu that if the purpose was accepted then they should just determine the wedding day, but if it was refused then they should take up arms and declare war.

> Mappajunge > his daughter has skin disease > consensus > exiled
>
> Arumpone > his son is sick (fallen in love) > consensus > helped

Episode II

> The Princess of Luwu

This section shows the fortitude of the King of Luwu's daughter in accepting her misfortune, namely a skin disease. The decision of the *ade'* consensus that she be discharged into the river on a raft was never complained about by her, just as she had not complained of her disease. She simply complained to Allah. In her exile, the disease was cured by the licking of an albino buffalo (*tedong puleng*).

> The Prince of Bone

This episode tells of the Prince of Bone's travel to hunt deer. He and his entourage ran out of supplies, so that all of them felt hunger. The prince created some groups of food seekers, and one of the groups eventually found a village inhabited by the Princess of Luwu, along with her slaves. The princess helped by providing a very orderly arrangement of many foods, such as rice, chicken and eggs. The Prince of Bone felt indebted, so he wanted to meet her to express his gratitude. In a meeting between the prince and princess, the prince felt upset seeing her beautiful face. He was also fascinated to see how the setting and service of the food were arranged so appropriately. He kept his feelings to himself and he just complained to Allah. He simply cried while covering himself when he was asked by his parent why.

> Princess of Luwu > skin disease (outside) > smell > complaint to Allah > healed by albino buffalo
>
> Prince of Bone > love pain (inside) > no smell > complaint to Allah > healed by human

In order to grasp the structural meaning and depth of this myth requires considerable ethnographic data about South Sulawesi; in particular, the writer has to understand the leadership system of the Buginese.

2.2 *The Buginese leadership system*

The Buginese consider the king to be a descendant of To-Manurung (those who come from the over world). They are placed or treated as the best and noblest creatures and above other human beings. However, in the systems of power, they put a man at the same level because the authority of the king's leadership always holds to the certainty of state law provisions that have been agreed together.

According to Mattulada (1985), it has been stated that the country will be threatened, and even destroyed, if the actions of the kings are ruled by their own desires. Conversely, a small country will be victorious and stronger when the state upholds the correct rule of law, harmony in law, and the principle of consensus. Furthermore, Mattulada asserted that an honest king will be assisted by nature. The meaning of honesty in this context is not letting the citizens be affected by disaster. Many instances are told of how a king acted in good faith for the sake of his citizens' prosperity. An example: once upon a time, in the reign of La Pagala Nenek (Nene') Mallomo (1546–1654) in Sidenreng, the harvest was unsuccessful for three consecutive years. The citizens immediately consulted to look for the reason, especially the royal family and the traditional council. People were becoming desperate because they had not found the reason. In a confused and anxious atmosphere, the son of Nene' Mallomo suddenly came and knelt in front of his father and told him what he had been hiding all this time. He said that, three years ago during the plowing season, some tips of a *salaga* (plowing tool) were broken: *'I took a piece of wood belonging to neighbors without asking permission, and until now I have not asked for his consent'*. Nene' Mallomo said, *'Apparently, you broke pemali, my son, so God sent down a warning that befell the people and the land of Sidenreng; for the sake of honesty, you should face the traditional council'*. The decision taken with regard to him was the death penalty. *Ade'e temmakkeanak, temmakkeeppo*, which means that the law does not know child or grandchild (Rahim, 1992). Following this decision, the harvest was successful and abundant.

The hope of people is placed in the king. He exists for the sake of the people, and from the people he takes life. For the people, *siri'* is everything in life, so anything that is asked by the king within the limits of cultural tradition (*pangadereng*) will be provided with all sincerity (Mattulada, 1985). The king is seen as *siri'*s State; in other words, he is personified as the dignity of the country and, therefore, from this point of view, the people's loyalty to their country is reflected in loyalty in serving the king. His esteemed position is expected to enable him to protect the society. If he cannot, the people might depose him or they will leave him by emigrating to other areas.

Another thing that is emphasized in the leadership system of the kings of South Sulawesi is the purity of blood. Expertise in leadership is not the main criterion for being a king; the principal one is blood purity rather than political cleverness in the interest of government. Thus, Ahimsa-Putra (1988) described how, in order to maintain the purity of this blood, the king should marry a girl who had pure noble blood.

2.3 The interpretation of myth's meaning (episode I)

In this story, we can observe how the Buginese play out their lives, whether it is a king, the son of a king, the servants of a king or as common people. In their daily activities, they have *ade'* (rules of life). But in fulfilling *ade'* they sometimes meet many obstacles, so it seems very difficult to achieve. For example, the experience of the King of Luwu (Mappajunge) when his daughter contracted skin disease. He had a deep conflict of the soul, torn between loving his child and obeying the people's will. He exiles his daughter for the sake of upholding *ade'*, making for a heavy decision to take.

Another such instance befell the King of Bone (Arumpone) after he discovered that the sickness of his son was the result of falling in love with the Princess of Luwu, who lived near the Walanae river of Wajo and was also a daughter of Pattola. Arumpone decided to form an ambassadorial team (for the purpose of proposing marriage). Both of these events can be interpreted as showing that if the king always holds on to *ade'* that has been agreed in consensus, even though the sanction was on his son, then the king and his people always get assistance from nature and the Almighty.

2.4 The interpretation of myth's meaning (episode II)

Episode II is symbolic of the idea that if a young Buginese man or woman experiences hardship or suffering, they feel ashamed to reveal it publicly, especially for a young man like the son of Arumpone, even though he is a strong man. He is riding a horse and bravely hunting deer in the forest. But when he falls in love with a girl and there is a desire to marry her, he is not brave enough to tell her frankly but just acts 'strangely', so that his parents pay attention to him.

In this episode, we can also understand that the way in which food is laid out and a guest is accommodated indicates that the host belongs to a noble family that has high standards.

2.5 Messages derived from consensus

Islam first entered South Sulawesi in 1605 when embraced by the king of Tallo, named Manyonri I Mallingkaan Daeng. It was then followed by the king of Gowa-14, named I Manga'rangi Daeng Manrabia on 20th September 1605. Both of these kings spread Islam, so that over the course of six years all the kings of South Sulawesi, either great or small, had been successfully converted to Islam. This was a very significant influence on the culture of Bugis-Makassar or, in other words, the contact between Bugis-Makassar culture and Islam formed a new system pattern. One aspect of this was the model of consensus in Buginese society, which is described in this myth and is a tradition having Islamic heritage (Mattulada, 1998).

In this myth, we can see two major kingdoms in South Sulawesi giving priority to consensus in facing a problem. They believed that with consensus the assistance of nature and the pleasure of Allah would surely come. Perhaps they were practicing a hadith of the Prophet Muhammad, which more or less has the meaning: 'If there is consensus in a village then, when Allah would bring down His mercy, the mercy will be magnified, but when Allah would bring down reinforcements or disaster, so they will be suspended or transferred to other areas that have not yet attained consensus'.

When the King of Luwu was suffering because his daughter was experiencing skin disease, he could not do much except call on the shaman and healer, but the disease had not been healed, rather it was getting worse. He was in a frenzy. The elders and the Luwunese people made consensus as to what must be presented to the king; the people agreed to give options to the king of whether he preferred one egg or many eggs, and he eventually chose many eggs. The consensus was executed and the exiled princess got help from nature, inasmuch as she got a place where the land was fertile, near to the river, and her disease was cured by the licking of an albino buffalo. In her healing, the princess was successful in leading her slaves and their generation created a prosperous region. In their community, they always made consensus to deal with problems.

When the son of the King of Bone wanted to see the princess, because he felt indebted and wanted to express his gratitude, the slaves conducted a consensus to determine whether he and his entourage were or were not allowed to see the princess. They finally decided to agree that he could meet the princess. As a result, the son of the King of Bone fell in love and eventually married the princess and, according to the story, the result of this marriage was the generation who next became the kings of the Bugis region.

In this myth we also see that the King of Bone practiced the consensus. When his son would not eat or drink, and just shut himself in his room, Arumpone and his wife consulted with the slaves. In fact, his son had fallen in love and wanted to marry the daughter of the King of Luwu. An ambassadorial team was created and the marriage proposal was accepted without there being a war.

3 CONCLUSIONS

The relationship that is built up from episode to episode above shows two kings of different regions, and both their son and daughter, taking the same approach in facing the problems of society; namely, they consulted when making decisions. When faced with their children's problems, the approach is especially hard to determine for those two kings but there is a consensus, the result of which is a powerful law that must be executed or there will be a traditional sanction. It can be interpreted that the biological children or close relatives of a leader in South Sulawesi can often become obstacles to seeing honesty and firmness in that leader, and may affect whether he can be fair in following the provisions of custom or acts arbitrarily. Finally, it can be said that, indirectly, the children of kings or leaders in South Sulawesi sometimes become the reason why he does not uphold justice. In handling their child, a leader or a king needs to be cautious, so that he can apply wisdom to them fairly.

Therefore, although the myth or story happened in the past, the message remains relevant to current conditions and it can even help predict the future: if any leader faces biological children or relatives, they need to be cautious so that their decisions cannot rebound on themselves like a boomerang.

REFERENCES

Ahimsa-Putra, H.S. (1988). *Minawang, hubungan patron-klien di Sulawesi selatan.* Yogyakarta, Indonesia: Gadjah Mada University Press.

Ahimsa-Putra, H.S. (1994). *Analisis STRUKTURAL dan makna mithos orang Bajo* (Research report). Yogyakarta, Indonesia: Gadjah Mada University Press.

Levi-Strauss, C. (1978). *Myth and meaning.* London, UK: Routledge & Kegan Paul.

Levi-Strauss, C. (2009). *Antroplogi struktural.* Yogyakarta, Indonesia: Kreasi Wacana.

Mattulada, A. (1985). *Latoa: Satu lukisan terhadap antropologi politik orang Bugis.* Yogyakarta, Indonesia: Gadjah Mada University Press.

Mattulada, A. (1998). *Sejarah, masyarakat dan kebudayaan Sulawesi Selatan.* Makassar, Indonesia: Hasanuddin University.

Rahim, A.R. (1992). *Nilai-nilai utama kebudayaan Bugis.* Ujung Pandang, Indonesia: University Press.

Ideas for 21st Century Education – Abdullah et al. (Eds)
© *2017 Taylor & Francis Group, London, ISBN 978-1-138-05343-4*

The effectiveness of educational qualifications in organizational career development for education staff

A.Y. Rahyasih & D.A. Kurniady
Universitas Pendidikan Indonesia, Bandung, Indonesia

ABSTRACT: The effectiveness of educational qualifications should be reflected in the ability of educational staff to utilize their qualifications to develop their career. However, most employees with high levels of qualifications have not obtained corresponding career development in the organizational structures of universities in Indonesia. This study investigates the effectiveness of educational qualifications in developing the careers of education staff. A qualitative method was used and respondents from one public university in Bandung were involved in the study. The results showed that the organizational structure in the University had largely fulfilled the needs analysis (78.9%), and employee career development and placement was generally based on the educational qualifications of the employees (84.2%). To accommodate the needs of the University, placements were suited to the qualifications and background disciplines of the employees. Career development was determined by qualifications, experience, and the selection and simulation process used in relation to promotion. In addition, job performance and competence were also determining factors in career development. However, a true merit system could not be implemented in the University as the recruitment process was controlled by the limited availability of suitable candidates, and not based on needs analysis as required by formal regulation.

1 INTRODUCTION

The role of the individual within an institution is very influential in the development of organization within the institution due to the existence of career engagement and attachment between individuals and organizations. The development of the organization can be achieved through the effectiveness of the work of employees in that organization. Indirectly, the concept of career plays an important role in the reward and control systems of the organization (Kaswan, 2012).

An organization progresses and develops just as a career does (Halmard, 2008). Wigand, Picot, & Reichwald (1997) argue that career development within an organization is synonymous with increased education, training, job transfer and promotion. With new regulations concerning organizational structure in universities, it is necessary that employees improve their job performance, being expected to be more professional, more disciplined, and able to fulfill their jobs well and correctly (Fathonah & Utami, 2011).

Article 1 of Law No. 20 of 2003 on the Indonesian Education System stipulates that educational staff are members of the community that are recruited and committed in support of the educational administration. According to Article 39 of the same law, educational staff are responsible for managing the administration of, and developing, monitoring, and providing technical services to support, the educational process in education institutions.

Effectiveness is the ability to carry out the tasks and functions (the execution of program activities or objectives) of an organization without pressure or tension in implementation. The effectiveness of educational qualifications is reflected in their use by education personnel to obtain career development. However, in practice, there are many employees with higher education qualifications who have not received promotion in the organizational structure, which affects the quality of managerial implementation.

Career development is one of the functions of career management. The process identifies potential employee careers and characteristics and involves the implementation of appropriate ways to develop this potential. In general, the career development process begins with evaluating the performance of the employees. On the basis of the evaluation, a plan of career development based on training and non-training tracks will be established. Through the application of career development, it is expected that the career path of each person will be more focused and structured, growing the commitment of every employee to be successful, providing the career planning capability needed by employees and the development of expertise to support their career progress and helping them to achieve successful careers.

2 RESEARCH METHODS

This study used a qualitative descriptive case study design. It was conducted in one public university in West Java, Indonesia. The respondents of the study were (1) heads of divisions, (2) heads of subdivisions at the University of Indonesia, and (3) heads of sections at the University, who were selected according to whether their functions could provide the data necessary for the study. The data were collected through participative observation of several activities in the field to obtain relevant data and to analyze the effectiveness of educational qualifications in the development of the University, as well as regular and controlled (semi-structured) interviews, and general observation. Data were also collected by studying documentation in the form of written documents officially or unofficially issued by the university and the government regarding the career development of educational staff.

3 RESULTS AND DISCUSSION

In the era of global competition, employment has been an attractive market especially for those who believe to have certain competences, knowledge, expertise, skills, and proficiency. The challenges of life make it increasingly hard for people to make decisions that are important for their future. The speed of development of technology, digitalization, and even information in cyberspace, is increasingly rampant, and how to respond to it positively, informing and inspiring our education, is important so that developments in technology are able to change minds. Such developments are not limited by space and time, especially in terms of receiving and conveying important information as tools in the achievement of organizational goals.

As a college that has the task of educating the nation, and has the nation's competitiveness in its hands, the university must be able to accelerate in all aspects. Increasingly, high-quality education is affected by the quality of the human resources (HR). It is assumed that the higher one's education level, the more able one is to provide a high quality of service to one's customers. On the basis of these ideas, those educators who have pursued an increase in educational qualifications through further study are expected to improve the quality of services in the organizational units that deal directly with students and professors. Low-quality human resources will affect the quality of their institution's performance, and thus it is expected that educators contribute to the objectives of their institution through the performance indicated by the results of their formal education. By increasing educational qualifications, an individual gains knowledge, skills and the ability to apply their knowledge and to be responsible for their work.

The services of education personnel may involve academic administration services, which includes managing new student registration, the preparation of class schedules, drawing up course participant lists, or the preparation of a lecture hall for the benefit of every program in the environmental studies faculty or other campus areas. In addition to these services, such personnel must also provide lecture room facilities equipped with devices or media learning tools, including setting up LCD screens, which can certainly be used to provide good facilities for students and faculty during lectures. Another important area is managing finances, from planning and the operational budget, to financial reporting and accountability. This requires mastery and expertise, particularly in the rules of proper financial accountability.

Effective communication is important in achieving success and improving the quality of the workforce, especially for improving the quality of college education personnel in a relatively short time with refreshment through education and training appropriate to the areas the college is servicing. However, to catch up with and accommodate the latest developments in science and technology, as well as high-level managerial understanding, requires knowledge of the approaches of individuals and groups to aspects of thinking, the power of creative analysis of situations, and the state of higher education demands. Colleges are made up of courses that develop the various fields of science, which are continuously evolving, changing and demanding acceleration in providing services to students.

Developments in the marketplace and the world of work are closely associated with preparing college graduates. Based on these circumstances, the human resources in a university, especially education personnel, must follow a pattern of change in mindset that has not been effectively implemented, either in the form of policies or regulations, at any university. One example is the provision of a database of students, ranging from new student enrollment, selection, and graduation results, to the determination of selection results, which should be traced and well documented from the preparation of new admissions all the way up to graduation. All resources, assets and facilities for academic services, and other services and activities should be managed. The management of a university has the image of competitiveness that should be encouraged. Finances that are managed in a transparent and accountable way are strongly needed to support the university to have better teaching, research, and community services quality.

To meet the competencies required for the various purposes described above, and to understand how the educators in each of these areas are positioned, further academic and professional studies are needed. HR management in a university may have different types of personnel with different functions that should, in essence, contribute to the institution's achievement of its vision and mission. A university seeks to promote excellent services to its stakeholders. Each academic and non-academic service requires knowledge, soft skills and hard skills. In the process of education, interaction occurs between students and faculty, and also between the university and the community, fellow colleges, and relevant institutions to build a network between local and international educational institutions and agencies. The need for competency in managing the university should be translated and described through tasks and functions.

The educational level and qualifications of education personnel are expected to help them carry out tasks that meet the requirement of national standards so that the university is ready for betterment. Educational personnel, as human resources, support the implementation of the management college, and their role and existence determine the success and achievements of the university in the present and the future. In addition, the educational background of education personnel should support the implementation of the university's tasks and the development of its organization. The principles for the development of this organization should also be the basis and foundation for education personnel, such as honesty, discipline, commitment, loyalty, consistency and having the ability to communicate and maintain good relationships with others.

Career development for education personnel should be implemented on an ongoing basis. Negara, Tripalupi, & Suwena (2015) states that through the development of career, employees are encouraged or motivated to perform duties effectively and efficiently. Thus, it can be concluded that career development can create a productive atmosphere because it requires educators to work effectively and efficiently.

The relevance of educational qualifications to the position occupied in the university structure is significant because the position prepared for the personnel demands an appropriate level of managerial ability, so that they can fulfill the necessary tasks. However, personnel who are promoted to a particular position should be made fully capable by virtue of managed induction to the new position. In addition to educational qualifications, experience is an important factor, as suggested by Kuijpers and Scheerens (2006) who state that the more support the employee experiences, the more the employee manages his own work and learn-

ing process. Moreover, mobility opportunities and intentions to vary work within the current employment situation contribute to career control.

Based on our observations, what has been done in the University of Indonesia in West Java in relation to the promotion of one position still has a weakness because the promotion path for education personnel to a particular position is not clearly defined. The requirements and standards expected by the institution are not available. The University needs to prepare human resources to step up and address the challenge of how a university educates the nation to compete. To be a leading university, plans should be prepared in the form of programs relating to improvement of the quality of human resources, and this should be a priority associated with structural competence of officials at the operational level. In addition, educators must have high self-efficacy to achieve career goals, as argued by Yakushko et al. (2008, p. 366) who state that, 'Conversely, a person who has more positive self-efficacy regarding related educational or career goals, In addition to the role of self-efficacy, Also emphasizes the role of a person's outcome expectancies on educational and career goal formation'.

In addition to the managerial competence that must be possessed by a leader at any level, professionalism needs to be demonstrated through labor productivity, which has been the goal of the organization. The relevance of educational qualifications and the placement of a person can be seen from the indicators of planning, implementation, and performance obtained as part of the institutional performance.

Another important thing is the spirit of leadership, to take decisions according to their capacity and to accept the associated risks. There are other aspects to be considered, starting with academic knowledge and professional abilities, and leaders must have endurance and toughness in facing and solving problems in related organizations, and the ability to analyze situations that arise both inside and outside the workplace where they are assigned.

The organizational structure and work procedures are a form of structure that describes the functions of each level in carrying out duties and responsibilities in line with their authority. The policies and regulations for positions in the organizational structure of the University require appropriate human resources to occupy them and to carry out leadership tasks. Therefore, the managerial model that must be owned by a person should be based on competence, excellent performance, and the ability to manage areas of their jobs such that the vision and mission of the University can be translated into their duties and functions.

In 2015, the University obtained its new status as a legal university entity. This has impacts on the existing organizational structure, especially in the service units that directly support the university management system. Some significant changes have occurred, especially in the directorates that have development functions, but in fact the services were more in the nature of administrative services. The directorates with development functions tend not to accommodate career paths for education personnel. Therefore, the highest career mapping for education personnel is the position of head of a bureau, which are relatively small in number.

It is important for the University members that in mapping a good career for the post of director directorate and bureau chief in regulation MWA No. 03. In 2015 there is no discrimination towards either lecturers or educators. All are welcome to compete and have the same rights as long as they meet the requirements. Therefore, educators must use their competence and the opportunities to contribute to the University. One of the career development plans that was developed from the University policy is that academic staff, including human resources, are prepared to occupy positions in accordance with their capacity, capability, commitment and managerial competence.

The analysis of the needs for positions in the organizational structure of the University has been conducted in accordance with BKN Regulation No. 19 Year 2011 on General Guidelines for Preparation of Civil Servants needs. The University has established equality in each of these positions so that it becomes a basis to determine class positions.

The needs analysis of positions within the university emphasizes the urgent need for universities to improve the managerial qualities as well as the academic qualities of the service provided to students. The quality improvement ranges from career planning and policy implementation, to the determination of positions to be considered carefully to avoid waste and not to burden the university. The purpose of the analysis of the needs of the organization is to minimize the problems that increase the burden in terms of both budget and the effectiveness and efficiency of the university. In certain positions, there are still human resources who do not meet the standards for educational qualifications due to limited human resources in that field.

Once the University has autonomous status in the fields of human resources management, finance, assets and facilities, it can become more transparent and provide greater opportunities to qualified employees. The positions occupied by employees in the organizational structures of the University need no longer consider the echelon described by Government Regulation No. 100 of 2000, which stated that officials within the echelon structure should first follow and complete the necessary training before being promoted. The policy implemented by the University provides ample opportunity for the staff of the University to develop their career and accelerate the organization's performance.

4 CONCLUSIONS

In general, universities have had career development policies for education personnel. One indicator is that the career development meets the needs of the organization in accordance with educational qualifications and the organizational structure. Placement for positions has been in accordance with educational qualifications possessed. This can be seen from the disciplines and qualification requirements that accommodate the needs of the organization. Career development at the University of Indonesia in West Java is determined by the educational qualifications and experience possessed by each employee. There is the task of selection and simulation in terms of promotion. In addition to education and experience, work performance and competency can also be decisive factors in career development. Nevertheless, a merit system cannot be achieved because the hiring of academic staff is based on the availability of existing positions, not based on needs analysis required under the formal legal guidelines issued to universities. The career development of education personnel has not been effectively implemented by the University because there is no clear career development planning and no true merit system.

REFERENCES

Fathonah, S. & Utami, I. (2011). Pengaruh kompensasi, pengembangan karir, lingkungan kerja, dan komitmen organisasi terhadap kepuasan kerja pegawai sekretariat daerah kabupaten karanganyar dengan keyakinan diri (self efficacy) sebagai variabel pemoderasi. *Excellent: Jurnal Manajemen, Bisnis, dan Pendidikan, 1*(1), 1–20.

Halmard, G. (2008). *Development of HRM in work*. New York, NY: John Wiley & Sons.

Kaswan, M.M. (2012). *Manajemen sumber daya manusia untuk keunggulan bersaing organisasi*. Yogyakarta, Indonesia: Graha Ilmu.

Kuijpers, M. & Scheerens, J. (2006). Career competencies for the modern career. *Journal of Career Development, 32*(4), 303–319.

Negara, N.M.C.M.A., Tripalupi, L.E., & Suwena, K.R. (2015). Pengaruh Pengembangan Karir terhadap Kinerja Pegawai pada PT. Pos Indonesia (Persero) Kabupaten Jembrana Tahun 2014. *Jurnal Jurusan Pendidikan Ekonomi, 4*(1), 1–11.

Wigand, R.T., Picot, A., & Reichwald, R. (1997). *Information, organization and management: Expanding markets and corporate boundaries*. Chichester: Wiley.

Yakushko, O., Backhaus, A., Watson, M., Ngaruiya, K. & Gonzalez, J. (2008). Career development concerns of recent immigrants and refugees. *Journal of Career Development, 34*(4), 362–396.

Adventure-based counseling model to improve students' adversity intelligence

N. Rusmana & K. Kusherdyana
Universitas Pendidikan Indonesia, Bandung, Indonesia

ABSTRACT: This study is based on the poor adversity intelligence of students and a lack of adventure-based counseling implemented in universities in Indonesia. The study aimed to create an adventure-based counseling model to improve students' adversity intelligence. This study used a research and development approach consisting of four main stages, namely, an introductory study, hypothetical model design, model validation, and field testing. The targets of the study were 410 students of Tourism Colleges in Indonesia (STP Bandung, STP Bali, Akpar Medan, and Akpar Makasar), selected by incidental sampling technique. The study findings show that: (1) the existing adversity intelligence level of most Tourism College students is in the *camper* category; (2) there is no difference in students' adversity intelligence on the basis of gender, study program, Grade Point Average, or parents' salary; (3) an adventure-based counseling model can be assumed to be appropriate as an intervention model to improve students' adversity intelligence; and (4) the adventure-based counseling model is proved to be effective in improving the adversity intelligence of Tourism College students in Indonesia. Therefore, it is recommended that the adventure-based counseling model be applied as part of the development of the counseling program in Tourism Colleges in Indonesia in order to improve students' adversity intelligence.

1 INTRODUCTION

Individuals typically experience up to 30 difficulties (*adversities*) per day (Stoltz & Weihenmayer, 2006). Having a higher Intelligence Quotient (IQ) or Emotional (intelligence) Quotient (EQ) does not guarantee someone the capability to handle such obstacles. Recent studies have shown that there is another contributing factor, named *adversity intelligence* (Enriquez & Estacio, 2009; Zainal et al., 2011).

According to Huijuan (2009), adversity intelligence is defined as a numerical figure (AQ) that represents how well an individual deals with difficulties and challenges. It provides an indicator of how well an individual is likely to cope in facing unpleasant situations (Phoolka & Kaur, 2012).

In this context, Stoltz (1997) categorized people into *quitters*, *campers*, and *climbers*. *Quitters* are less ambitious, often choose to escape from problems, and avoid obligation; *campers* go as far as possible to avoid unpleasant situations; and *climbers* welcome challenges no matter what.

An individual's adversity intelligence score is composed from four dimensions, abbreviated as CORE: *Control, Ownership, Reach*, and *Endurance* (Stoltz, 1997; Stoltz & Weihenmayer, 2006). *Control* relates to empowerment and influence; *ownership* refers to the responsibility to solve problems; *reach* deals with the way people handle difficulties; *endurance* relates to the perception of the duration of the difficulties.

As adults, college students have to be able to handle problems effectively. Otherwise, they will impede their study, career development and future lives. Hence, an effort to improve the general level of students' adversity intelligence is beneficial.

Adventure-based counseling is one such approach. It is a combination of experiential learning and outdoor education that uses group counseling techniques (Fletcher & Hinkle, in Cale, 2010). Ringer (in Hans, 2000) and Alvarez and Stauffer (2001) define it as 'a class of change-oriented, group-based experiential learning processes that occur in the context of a contractual, empowering, and empathetic professional relationship'. Newes and Bandoroff (2004) state that adventure-based counseling provides positive reinforcement and attractive activities.

2 RESEARCH METHODS

The method used in this study was that of research and development. The stages adopted were: (1) preparation; (2) hypothetical model design; (3) hypothetical model validation; (4) hypothetical model modification; (5) model testing; and (6) final model design. The data collection instruments used pertain to the CORE dimensions developed by

Stoltz, and took the form of a semantic differential scale.

The target population was 410 tourism students of the class of 2013/14 in four institutions: *Sekolah Tinggi Pariwisata* (STP) Bandung, STP Bali, *Akademi Pariwisata* (Akpar) Medan, and Akpar Makasar. A two-stage sampling technique was employed. The students of Class A (control group - 34 students) and Class B (experimental group - 32 students) of the Patisserie Study Program (third semester) at STP Bandung were selected as the research sample.

The model advisability analysis was carried out by involving counseling experts and adventure-based counseling practitioners, while the analysis of students' adversity intelligence differences based on biographical background was tested by one-way ANOVA (analysis of variance), and model effectiveness was tested using ANCOVA (analysis of covariance).

3 RESULTS AND DISCUSSION

3.1 *Adversity intelligence profile of tourism college students in Indonesia*

As indicated in Figure 1, 106 (25.85%) students were classed as *climbers*, 205 (50.00%) as *campers*, and 99 (24.15%) as *quitters*; most students from the four tourism colleges belonged to the *camper* category.

Figure 2 suggests that most respondents belonged to the average category in all of the adversity intelligence dimensions: 53.66% in the *control* dimension, 50% in the *ownership* dimension, 48.76% in the *reach* dimension, and 50.97% in the *endurance* dimension.

These findings from the introductory study showed that, based on total score, most respondents' adversity intelligence was in the *camper* category. This is in line with the study findings of Villaver (2005), Amliati (2012), Bakare (2014) and Hasanah (2010), which stated that most of their respondents showed average adversity intelligence.

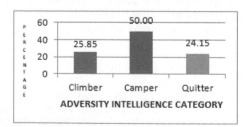

Figure 1. Adversity intelligence profile of Indonesian Tourism College student class of 2013/2014, based on total scores.

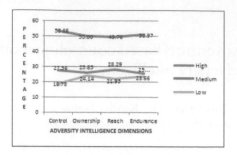

Figure 2. Adversity intelligence profile of Indonesian Tourism College students in academic year 2013/2014, based on CORE dimensions.

3.2 *The difference in students' adversity intelligence based on biographical background*

Hypothesis A1: *Male students have higher adversity intelligence than female students.*

According to the one-way ANOVA testing result, a score of F = 0.434 was obtained, with probability 0.510. This supports the null hypothesis, indicating that there is no difference in students' adversity intelligence based on gender.

This finding is in line with studies conducted by Huijuan (2009) and Kanjanakaroon (2012) which concluded that gender does not influence adversity intelligence. In the campus setting, the treatment given to male and female students is equal. However, the finding is not in line with the studies by Dweck (in Stoltz, 1997) or Bintari (2000) that concluded that, on average, the male adversity intelligence rate is higher than the female.

Hypothesis A2: *The students of the Hospitality Study Program have higher adversity intelligence than students of other study programs.*

According to the one-way ANOVA testing result, a score of F = 0.150 was obtained, with probability 0.698. This supports the null hypothesis, indicating that there is no difference in students' adversity intelligence based on the study program into which they have been accepted.

This study finding is in line with those of Cura and Gozum (2012), who found no difference in students' adversity intelligence based on their program of study. It is assumed that the two relevant study programs across the four Tourism Colleges in Indonesia provide similar challenges and difficulties for the students.

Hypothesis A3: *The students who have high Grade Point Averages (GPAs) have higher adversity intelligence than low-GPA students.*

According to the one-way ANOVA testing result, a score of F = 0.053 was obtained, with probability

0.984. This supports the null hypothesis, indicating that there is no difference in students' adversity intelligence based on GPA achieved.

This finding is in line with study findings by Bintari (2000), Tjundjing (2001) and Hasanah (2010), which found no significant difference in adversity intelligence based on learning achievement. Nevertheless, the hypothesis was based on the studies conducted by Huijuan (2009) and Setyaningtyas (2012), which concluded that there is a positive and significant correlation between students' adversity intelligence and their learning achievement. In relation to this, Stoltz (in Crawford & Tee, 2000) also emphasizes that there is a correlation between adversity intelligence and academic achievement.

Hypothesis A4: There is a difference in students' adversity intelligence based on parent's salary.

According to the one-way ANOVA testing result, a score of $F = 0.685$ was obtained, with probability 0.562. This supports the null hypothesis, indicating that there is no difference in students' adversity intelligence based on parent's salary.

This condition is presumably a result of similar challenges and difficulties faced by the student sample at home or school, and also similar parents' salary averages. It is in line with studies conducted by Tarapurwala (in Jain, 2013) and Villaver (2005), but is not in line with the findings of the study conducted by Schoon et al. (2004, p. 79), which concluded that subjects who had lower socioeconomic backgrounds actually have higher adversity intelligence.

3.3 Formulation of adventure-based counseling operational model to improve students' adversity intelligence

This model contains formulations of the associated rationale, purpose, intervention target, assumptions, counselor role, organization mechanism, counseling stages, program structure, service unit and assessment mechanism. It is important to have this model in order to identify problem-solving alternatives. Specifically, this model is worthwhile as it has satisfied a range of examinations, validations, draft improvements and testing.

3.4 Adventure-based counseling model effectiveness in improving students' adversity intelligence

Statistical analysis resulted in the following hypothesis testing results.

Hypothesis B1: The adventure-based counseling model is effective in improving students' adversity intelligence.

According to the ANCOVA testing result, a score of $F = 44.141$ was obtained with $p = 0.0001$. Given that $p < \alpha$, this means that the null hypothesis (H_0) is rejected. Thus, the average total score of the experimental group's adversity intelligence following the intervention is significantly higher than that of the control group and we conclude that adventure-based counseling is effective in improving the students' adversity intelligence.

This study finding is in line with those of the following similar studies: (1) respondents' resilience and resilience protective factors increased significantly after adventure-based counseling (Gillespie & Allen, 2009); (2) low-salary minority male students' resilience scores increased significantly (Green et al., 2000); (3) all participants showed positive changes in resilience (Neil & Dias, 2001); and (4) an adventure-based counseling program had positive and significant effects on teenagers' self-efficacy and resilience (Walsh, 2009).

Hypothesis B2: The adventure-based counseling model is effective in improving students' control dimension.

According to the ANCOVA testing result, a score of $F = 28.833$ was obtained with $p = 0.0001$. With score $p < \alpha$, this means that H_0 is rejected. Thus, the average score of the experimental group's *control* dimension is significantly higher than that of the control group and we conclude that adventure-based counseling is effective in improving the students' *control* dimension.

Hypothesis B3: The adventure-based counseling model is effective in improving students' ownership dimension.

According to the ANCOVA testing result, a score of $F = 13.525$ was obtained with $p = 0.0001$. With score $p < \alpha$, this means that H_0 is rejected. Thus, the average score of the experimental group's *ownership* dimension is significantly higher than that of the control group and we conclude that adventure-based counseling is effective in improving the students' *ownership* dimension.

Hypothesis B4: The adventure-based counseling model is effective in improving students' reach dimension.

According to the ANCOVA testing result, a score of $F = 22.828$ was obtained with $p = 0.0001$. With score $p < \alpha$, this means that H_0 is rejected. Thus, the average score of the experimental group's *reach* dimension is significantly higher than that of the control group and we conclude that adventure-based counseling is effective in improving the students' *reach* dimension.

Hypothesis B5: The adventure-based counseling model is effective in improving students' endurance dimension.

According to the ANCOVA testing result, a score of F = 22.420 was obtained with p = 0.0001. With score $p < \alpha$, this means that H_0 is rejected. Thus, the average score of the experimental group's *endurance* dimension is significantly higher than that of the control group and we conclude that adventure-based counseling is effective in improving the students' *endurance* dimension.

The intervention conducted in this study was effective because the program has been evaluated by experts, making it well-planned and systematic; it had clear objectives and was comprehensive in implementation; it provided active and fun learning activities. This is in line with the success criteria described by Durlak and Weissberg (in Christian, 2013, p. 39), who argued that effective adventure-based counseling programs have the following characteristics:

First, the programs were sequential; they intentionally ordered activities to address specific topics. Second, successful programs utilized active forms of learning. Third, the programs were focused and included specific components that addressed what they assessed. Finally, the successful programs were explicit in how they targeted behaviors for interventions.

In line with this, Neill and Dias (2001) put forward some components influencing the effectiveness of adventure-based counseling. First, appropriate program design for the participants' needs. Second, participants' motivation to get involved; Ewert (in Sibthorp & Skye, 2004, p. 33) stated that motivation significantly influences the effectiveness of adventure-based counseling. Third, a longer program means more of the target behaviors change.

Furthermore, friendship and familiarity can facilitate participants' emotional, social and learning adjustment and will influence high-quality development in interpersonal relationships (Bishop & Inderbitzen, 1995, p. 480).

Another finding shows that the counselors/lecturers in colleges start to understand that adventure-based counseling is effective. The advantages include: (1) longer-term improvement compared with a more traditional counseling intervention (Davis-Berman & Berman, 1994, p. 49); (2) students are better able to translate their learning into daily life (Glass & Benshoff, 2002, p. 270); (3) it gives the opportunity to explore and construct values and attitudes (Attarian, 1996, p. 44); (4) it has healing potential and (5) it has both immediate and long-term effects and would improve later (Hattie et al., 1997, p. 74).

To conclude, an adventure-based counseling model intervention has proved effective in improving students' adversity intelligence in all dimensions. Thus, this counseling model is afforded both theoretical and empirical credibility for implementation in colleges, especially State Tourism Colleges in Indonesia.

4 CONCLUSIONS

- The adversity intelligence level of most Tourism College students of the 2013/2014 academic year in Indonesia fell into the *camper* category. *Campers* are people who tend to enjoy hard work during unfinished climbing activities but do not use all of their potential to handle challenges. In the adversity intelligence dimensions, namely *control, ownership, reach,* and *endurance*, most of the students belong to the average category.
- There is no difference in students' adversity intelligence based on gender, study program, GPA or parent's salary.
- The adventure-based counseling model contains formulations of rationale, purpose, intervention target, assumptions, counselor role, organization mechanism, counseling stages, program structure, service unit and assessment mechanism. Validation has shown that the model is worthwhile as an adversity intervention model for students.
- An adventure-based counseling model has proven effective in improving the adversity intelligence, and all dimensions thereof, of students in the Tourism College class of 2013/2014.

5 RECOMMENDATIONS

5.1 *For future researchers*

- The effectiveness of this counseling model has only been proved in tourism college students. To strengthen the generalization of the findings, it is suggested that this model be applied to other students.
- It is recommended that the number of experimental groups be increased to two or more to give more empirical evidence of behavioral change as a specific effect of model implementation.
- It is recommended that *random assignment* be used in selecting research subjects so that the study findings have better internal and external validations.
- It is recommended that other related parties be involved, such as relatives or significant others, to provide more information and perspective in relation to the participants being counseled.

5.2 *For institutions*

Adventure-based counseling implementation needs new environments and particular devices to support the success of counseling and achieve the designated targets. Thus, it requires the associated institution(s) to provide facilities/devices and financial support.

REFERENCES

Alvarez, A.G. & Stauffer, G.A. (2001). Musings on adventure therapy. *Journal of Experiential Education, 24*(2), 85–91.

Amliati, W.O. (2012). *Studi deskriptif adversity quotient (AQ) remaja sekolah menengah atas (SMA) di Semarang* (Thesis). Post-Graduate Program, Diponegoro University, Indonesia.

Attarian, A. (1996). Integrating values clarification into outdoor adventure programs and activities. *Journal of Physical Education, Recreation & Dance, 67*(8), 41–44.

Bakare, B.M., Jr. (2014). *Students' adversity quotient® and related factors as predictors of academic performance in the West African senior school certificate examination in South-Western Nigeria* (PhD thesis). The International Centre For Educational Evaluation (ICEE), Institute Of Education, University of Ibadan, Nigeria. Retrieved from http://peaklearning.com/documents/PEAK_GRI_bakare.pdf

Bintari, R.D. (2000). *Hubungan antara adversity quotient dengan prestasi akademik pada mahasiswa fakultas teknik dan fakultas psikologi UI* (Thesis). Psychology Faculty, Indonesia University.

Bishop, J.A. & Inderbitzen, H.M. (1995). Peer acceptance and friendship: An investigation into their relation to self esteem. *Journal of Early Adolescence, 15,* 476–489.

Cale, C. (2010). *A case study examining the impact of adventure based counseling on high school adolescent self-esteem, empathy, and racism* (Dissertation). University of South Florida, USA. Retrieved from http://scholarcommons.usf.edu/etd/1585/

Christian, D.D. (2013). *Adventure based counseling: Exploring the impact of ABC on adaptive functioning in high school* (PhD dissertation). University of North Texas, USA. Retrieved from https://digital.library.unt.edu/ark:/67531/metadc283835/m2/1/high_res_d/dissertation.pdf

Crawford, L.E.D & Tee, T.C. (2000). Promoting adversity quotient among Singaporean school children. *National Institute of Education, Singapore, 1,* 10–14.

Cura, J. & Gozum, J. (2011). *Correlational study on adversity quotient® and the mathematics achievement of sophomore students of College of Engineering and Technology in Pamantasan ng Lungsod ng Maynilaw* [Online]. Retrieved from http://www.peaklearning.com/documents/PEAK_GRI_gozum.pdf

Davis-Berman, J. & Berman, D.S. (1994). Research update: Two-year follow up report for the wilderness therapy program. *Journal of Experiential Education, 17,* 48–50.

Enriquez, J.M. & Estacio, S.D.L. (2009). *The effects of mentoring program on adversity quotient of selected freshmen college students of FAITH* (BA dissertation). First Asia Institute of Technology and Humanities, Batangas, Philippines. Retrieved from http://peaklearning.com/documents/PEAK_GRI_enriquez.pdf

Gillespie, E. & Allen, S.C. (2009). The enhancement of resilience via a wilderness therapy program: A preliminary investigation. *Australian Journal of Outdoor Education, 13*(1), 39–49.

Glass, J.S. & Benshoff, J.M. (2002). Development of group cohesion through challenge course experiences. *Journal of Experiential Education, 25,* 268–278.

Green, G.T., Kleiber, D.A. & Tarrant, M.A. (2000). The effect of an adventure based recreation program on development of resiliency in low-income minority youth. *Journal of Park and Recreation Administration, 18*(3), 76–97.

Hans, T.A. (2000). A meta-analysis of the effects of adventure programming on locus of control. *Journal of Contemporary Psychotherapy, 30*(1), 33–60.

Hartung, J.G. (2010). Leading in tough times: Developing resilience. *Across the Board: A bulletin to assist, educate and communicate with volunteer board members, 14*(3), 1–56.

Hasanah, H. (2010). *Hubungan antara adversity quotient dengan prestasi belajar siswa SMUN 102 Jakarta Timur* (Essay). Psychology Faculty, Universitas Islam Negeri Syarif Hidayatullah, Jakarta, Indonesia.

Hattie, J., Marsh, H.W., Neill, J.T., & Richards, G.E. (1997). Adventure education and Outward Bound: Out-of-class experiences that make a lasting difference. *Review of educational research, 67*(1), 43–87.

Huijuan, Z. (2009). *The adversity quotient and academic performance among college students at St. Joseph's College, Quezon City* (BSc thesis). Faculty of The Departments of Arts and Sciences, St. Joseph's College, Quezon City, Philippines. Retrieved from https://www.termpaperwarehouse.com/essay-on/The-Adversity-Quotient-And-Academic-Performance/41900

Jain, P. (2013). *Development of a programmme for enhancing adversity quotient® of Std VIIIth students* (Masters dissertation). Department of Education, S.N.D.T. Women's University, Mumbai, India. Retrieved from http://peaklearning.com/documents/PEAK_GRI_priyankaJain.pdf

Kanjanakaroon. J. (2012). Relationship between adversity quotient and self empowerment of students in schools under the jurisdiction of the office of the basic education commission. *The International Journal of Learning, 18*(5), 7–10.

Neill, J.T. & Dias, K.L. (2001). Adventure education and resilience: The double edged sword. *Journal of Adventure Education and Outdoor Learning, 1*(2), 35–42.

Newes, S. & Bandoroff, S. (Eds.) (2004). *Coming of age: The evolving field of adventure therapy.* Boulder, CO: Association of Experiential Education.

Phoolka, S. & Kaur, N. (2012). Adversity quotient: A new management paradigm to explore. *Research Journal of Social Science and Management, 2*(7), 67–78.

Schoon, I., Parsons, S. & Sacker, A. (2004). Socioeconomic adversity, education resilience, and subsequent levels of adult adaption. *Journal of Adolescent Research, 19,* 383–403.

Setyaningtyas, E. (2012). *Hubungan adversity quotient (AQ) dengan prestasi belajar mahasiswa program studi kebidanan universitas sebelas maret* (Paper). Universitas Sebelas Maret, Surakarta, Indonesia.

275

Sibthorp, J. & Skye, A. (2004). Developing life effectiveness through adventure education: The roles of participant expectations, perceptions of empowerment, and learning relevance. *The Journal of Experiential Education, 27*(1), 32–50.

Stoltz, P.G. (1997). *Adversity quotient: Turning obstacles into opportunities*. Hoboken, NJ: John Wiley & Sons.

Stoltz, P.G. & Weihenmayer, E. (2006). *The adversity advantage: Turning everyday struggles into everyday greatness*. New York, NY: Simon & Schuster.

Tjundjing, S. (2001). Hubungan antara IQ, EQ, dan AQ dengan prestasi studi pada siswa SMU. *Anima, Indonesian Psychology Journal, 17*, 69–92.

Villaver, E. (2005). *The adversity quotient levels of female grade schoolteachers of a public and a private school in Rizal Province* (PhD thesis). Retrieved from http://peaklearning.com/documents/PEAK_GRI_villaver.pdf.

Zainal, S.R.M., Nasrudin, A.M. & Hoo, Q.C. (2011). The role of emotional intelligence towards the career success of hotel managers in the Northern States of Malaysia. *International Conference on Economics, Business and Management, 22*, 123–128.

Ideas for 21st Century Education – Abdullah et al. (Eds)
© *2017 Taylor & Francis Group, London, ISBN 978-1-138-05343-4*

The effectiveness of implementing an experience-based counseling model in reducing the tendency of students towards bullying behavior

N. Rusmana, A. Hafina & I. Saripah
Universitas Pendidikan Indonesia, Bandung, Indonesia

ABSTRACT: The research was conducted into the widespread phenomenon of violence among teenagers, especially the violence that occurs in the form of bullying. The objective of the study was to implement an experience-based counseling model for reducing the tendency of students toward bullying behavior. The study used the R&D (Research & Development) method. The subjects of the study were senior high school students in Bandung who were identified as committing bullying actions in low, moderate or high intensity. Data was collected using the instruments of bullying behavior scale and bullying behavior check lists. The data was analyzed using statistical tests including Central Tendency and T-Test to compare the values of the means from two samples. The results show that the implementation of the experience-based counseling model proved to be effective for reducing the symptoms or the characteristics of students' bullying behavior.

1 INTRODUCTION

Lately, the phenomenon of bullying has become more prominent after there were victims who died and the media exposed this issue widely. Some studies have shown that bullying is an international problem that occurs in almost schools (Novianti, 2008). The team of Psychology Faculty of Indonesia University (2008) discovered that bullying was quite common among senior high school students in Indonesia, especially those who study in big cities, such as Jakarta, Bogor, and Bandung.

Each educational institution needs to know about the existence and the impact of bullying and also to try preventing that from happening. If the incidence of bullying is taken for granted or the bullying actions still occur, the students will experience abuse or acts of violence. As a result, the students who are bullied will experience psychological stress and they will suffer for a lifetime. Bullying victims at secondary and higher education are believed to be the victims who suffered most since they have had bad experiences at the moment where they should have acquired the best ones. More worryingly, children who bully others are more likely to grow up to become criminals than those who do not.

To totally eliminate bullying in schools is indeed not possible, but it can be minimized. In an educational institution, school in this case, everyone can be a victim or even become the bully. A peer or peer group is the most potential side to become the bully in the school environment. The occurrence of bullying in schools is a process of group dynamics, where there are role divisions among the teenagers themselves (Salmivalli et al., 2005). Therefore, an

effort is required to inculcate noble values in the students in order to steer them away from bullying-related actions and situations, so they do not become neither perpetrators of bullying when they are teenagers nor the bullies who have the potential to turn into criminals when they grow up.

Some empirical facts about the phenomenon of bullying at school, with all its psychological implications, suggest the need for an effective treatment and intervention against the perpetrators of bullying. In general, the bully has a lot of energy, which is usually channeled aggressively against the target or victim (Olweus, 1993). Tasks in experience-based counseling model were designed ranging from physically and mentally light and easy exercises up to complex and challenging exercises in order to improve the ability of emotion regulation, channel the energy in a positive form, develop self-concept, self-efficacy and social skills (Newman-Carlson, & Horne, 2004). Thus, there is a need to design an intervention program through experience-based counseling to help the perpetrators of bullying in order to channel their energy into more positive things, such as developing good interpersonal skills, responsibility, empathy, self-control, anger management skills, and reducing their aggressive impulsive behavior.

There are several reasons underlying the importance of this study as the implementation of an experience-based counseling model in reducing the students' tendency towards bullying behaviors. The main reasons are as follows:

First, bullying is actually no longer a mere child's game, but a frightening thing that is experienced daily by almost all students in all levels of education. Over the past decade, bullying has become more deadly and more routinely, compared to the previous two decades (Adair, 1999). In addition, it is more distressing since the teenagers who like to bully others are likely to grow as criminals than those who do not. Therefore, we need an alternative solution to handle the problem of bullying behavior in teenagers, especially junior high school students, senior high school students and vocational school students.

Second, the experience-based counseling model for reducing the tendency toward bullying behavior has the potential to be implemented by several parties without demanding high cost. This is because the outdoor and indoor game activities in this counseling model can be implemented by many people, including non-experts.

Third, the experience-based counseling model for reducing students tendency toward bullying behavior is expected to become operational guidelines for schools, non-formal educational institutions and related NGOs that have concerns on how to reduce bullying.

Fourth, experience-based counseling can help the bullies acquire social skills, attain the ability to work together and take responsibility, and foster their empathy and tolerance towards others.

Fifth, this study provides a conceptual contribution to the fields of counseling and psychotherapy science, especially in dealing with the students who have a tendency to become bullies or who have been found guilty of bullying.

Therefore, the main focus of the problem in this research is to implement an effective experience-based counseling model for reducing bullying behavior tendency of high school students.

2 RESEARCH METHODS

2.1 *Approaches and methods of the research*

The approach used in this study is a mixed methodology consistive of qualitative and quantitative research methods. Overall, the study uses the method of R&D (Research & Development).

2.2 *Subject of the research*

The subjects of the research were senior high school students with the following characteristics: a) those who are in the age range 15–18 years (midteen); b) identified as having a tendency toward bullying behavior in the category of low, medium or high; and c) showing the need for experience-based counseling services to reduce bullying behavior of senior high school students.

2.3 *Location of the research*

The research was carried out in different places at every stage of the research. Data collection sites included several senior high schools in Bandung, namely SMAN 6 Bandung, SMAN 1 Bandung, and SMA Laboratorium Percontohan UPI. The location of the intervention provision was in Karang Tumaritis Campground, Cisarua, West Bandung Regency.

2.4 *Data collection instruments*

For the purpose of need assessment, the study used the problem identification instrument in the form of a check list of the characteristics of bullying. The form of the instrument is a Likert scale, which uses the following categories: Inappropriate (TS), Slightly Inappropriate (KS), Uncertain (RR), Appropriate (S), Absolutely Appropriate (SS).

3 RESULTS AND DISCUSSION

3.1 *Results*

3.1.1 *The implementation of an experience-based counseling model for reducing the tendency of students towards bullying behavior*

3.1.1.1 Initial stage

The strategy used in this stage was the use of ice breaking games. Outdoor activities undertaken at this stage were group games, including "This is Titi's Ball" (Ini Bola Titi), "Animal Family" (Keluarga Binatang), "Arranging Lines" (Susun Baris), and "Yelling the Group Slogans" (Yel-yel Kelompok). The implementation of group games at any stage in the experience-based counseling model was using the Socratic Method. This method uses four-step activities, namely: a) experiencing, b) identifying, c) analyzing, and d) generalizing.

Based on the observations during the initial stages, it was found that at the beginning of the implementation, there were several students who did not know each other. The situation was apparently awkward among the students in the beginning. However, after they were asked to introduce themselves through the *"Ini Bola Titi"* game, the students began to know each other's names. Before performing the next game, the facilitator split the students into six groups. Each group consisted of four people. After playing the game *"Keluarga Binatang"* and *"Susun Baris"*, the awkwardness began to fade and the students started to blend in with the other members in the group. This was proved by the students' cooperation when they made the group slogans.

3.1.1.2 Transition stage

The strategy used in this stage was playing games that had a warming-up and energizing effect. Outdoor activities undertaken at this stage were group games including: "Hula Hoop" game, "Dancing Logs" (Kayu Goyang) game, "Cap Guricap"

Figure 1. "Susun Baris" game.

Figure 2. "Hula Hoop" game.

game, "Papanjang-Panjang" game, and "Blind Trust Walk".

The second phase of the experience-based counseling is the transition phase, which contains outdoor activities such as group games that can strengthen group cohesiveness and foster the spirit of the group. Values of cooperation appeared in the transition phase since there were certain given values, signed roles, and several enforced rules. Based on the observations, the activities in this stage went smoothly.

3.1.1.3 Work stage

There were indoor and outdoor activities undertaken in this stage. The indoor activities were creating and presenting anti-bullying posters, watching the movie "Happiness for Sale", followed by reflecting and discussing the movie "Happiness for Sale". While the outdoor activities were "Spider Web", "Chocolate River", "Circle Fall", "Toxic Waste" (Limbah Beracun), and "Leaked Dish" (Gentong Bocor).

The work stage in experience-based counseling was packed with a series of outbound activities and indoor activities. Outdoor activities contained activities that provided the opportunity for participants to experience by themselves (letting the experience speak for itself), describe the experience (speaking about the experience), reflect on the experience (debriefing the experience), and explain the experience (framing the experience).

The indoor activities were in the form of a conventional encounter, which contained activities to explain and discuss specific subjects; complete a number of tasks; and reflect on the task. These activities aimed at providing an introduction for the participants about the experience-based counseling model in order to develop the communication and cooperation of a group in order to solve a particular problem.

All students were also actively involved in indoor activities. The students drew and colored the anti-bullying poster designs and attentively watched the "Happiness for Sale" movie. The film tells of the struggle of a victim of bullying in making friends with those who had bullied her. The film also shows that a perpetrator of bullying can change and be friends with the victim. Thus, based on the general observations, the work phase was progressing well.

3.1.1.4 Final stage

The final stage in experience-based counseling became the sentimental stage as the students were affected by the issue of farewell. Nevertheless, this stage ran smoothly. All activities at this stage were done indoors. The students started by playing a "Permainan Sarung" game, in which a student was asked to guess the names of the other students hiding in the holster (sarung). After positive energy began to be created, the counselors distributed a worksheet in the form of number puzzles and a letter maze to develop the students' skill in problem-solving. Then, the students were asked to give positive feedback to each member of their group. The facilitator also gave positive feedback to the students and provided rewards for the three best groups and the best participant. At the end of the activity, the facilitator showed slideshow photos of the students when they were engaging in the activities of experience-based counseling. This playback of the slideshow triggered an emotional atmosphere among the participants and facilitators. The session was closed by collecting the activity journals and taking a photo all together.

3.1.2 *The effectiveness of experience-based counseling in reducing the tendency of students towards bullying behavior*

In order to measure the effectiveness of the counseling process, an assessment mechanism of the counseling process and outcomes was carried out. The assessment mechanism used to assess the suc-

Figure 3.　"Spider Web" game.

Figure 4.　"Gentong Bocor" game.

cess of the counseling process, was to carefully observe the process of implementation, from the initial stage to the final stage. Journals of activities were given shortly after students attended each session of activities, in this case after the students played games and entered into a joint reflection session activity with the counselor or facilitator.

Based on the votes on the counseling camp, the following data shows the decreased number of bullying cases.

Chart 1 shows that all values of the indicators have decreased significantly. In other words, each of the symptoms decreased and the bullying behavior of the students was reduced. This shows that the intervention of experience-based counseling model is effective towards all aspects of behavioral indicators of students' bullying, such as lack of empathy, poor interpersonal skills, lack of anger management skills, weak self-control, lack of responsibility and impulsive-aggressive behavior.

3.2　Discussion

A series of activities in experience-based counseling can help students improve their empathy through a variety of games that require teamwork. When there was a member of the group who was injured, other members also felt her/his suffering so that they worked together to take care of the wounded member. In addition to empathy, students' interpersonal skills were also trained through games that help developing team-building and collaboration capabilities. Strict game rules also allowed students to play in a fair and

Figure 5.　Indoor activity: Giving positive feedback.

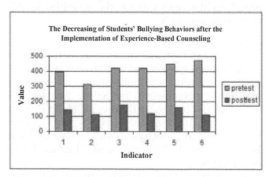

Chart 1.　The decrease in the students' bullying behavior after the implementation of experience-based counseling.

sportsmanlike manner, thus it promoted no harm among groups.

Games in experience-based counseling help students channel their excess energy in a positive way. Channeling this energy is very good for students with anger management problems. Some games demanding focus and calmness of the students, such as the game of *"Limbah Beracun"*, make the students more focused and relaxed and affect the ability to control anger in a positive way.

As for the ability of self-control, games in experience-based counseling were designed to help students with poor self-control to keep their internal locus control. Through the reflection activity, the students were encouraged to express their thoughts, experiences, feelings, and expectations held during and after the games with their group. The willingness of students to follow the instructions of the facilitator, discuss and cooperate with their friends, as well as trying to complete the task in each game as best as they could reflected an increase in their internal locus of control and a sense of responsibility of the students.

The pattern of impulsive aggressive behavior was successfully controlled in the implementation of experience-based counseling activities. In the initial stage, there were some students who made rude jokes about their friends. Also, there were students who seemed unable to accept defeat and began screaming and destroying objects or hitting something. But with the counselors' skills in providing guidance and reflection for each participant, the impulsive behavior of some participants could be controlled. The distribution of energy in a positive way and the increase of good characteristics in students (empathy, responsibility, self-control) were expected to help minimize impulsive aggressive behavior patterns in students. It is obviously true that the pattern of impulsive aggressive behavior often requires clinical efforts.

According to Drosopoulos (2008), bullying actions do not happen by accident or just happen for no reason. Instead, bullying behavior emerges through the learning process and being learned. Therefore, bullying behavior can be reduced through certain values inserted in a new process of learning. This is consistent with the main purpose of experience-based counseling, namely the transfer of learning into new patterns of behavior, so that there will be changes in the counselee's behavior (Gillis & Gass, 2004).

Adventure activities can develop values in shaping a positive self-concept, including self-image, self-respect, self-satisfaction, and self-realization in achieving success when facing new challenges. Several studies have demonstrated the efficacy of experience-based counseling in dealing with

juvenile inmate's aggressive behavior, developing the skills of employee teamwork, and overcoming post-traumatic stress (Mitchell, Finkelhor, & Wolak, 2004).

According to Mitchell, Finkelhor and Wolak (2004), there are some advantages that can be obtained from the experience-based counseling in an outdoor setting, namely counselee can:

1. have knowledge about nature and love of outdoor life;
2. maintain the balance of nature and appreciate the interdependence of all living things;
3. awaken spiritual awareness of God by way of appreciating the universe. Such situations arise when we look at the high mountains, silent valleys, the amazing colors when the sun sets, and the flashes of sunlight on the surface of a lake;
4. understand the importance of preserving natural resources. Campers will be taught how to recycle, preserve the environment and be taught the science of the ecology; and
5. develop a sense to enjoy the beauty of natural scenery.

The occurrence of bullying in schools is a process of group dynamics where there are role divisions among the teenagers themselves (Salmivalli et al, 2005; in Ratna Djuwita 2006, p 12). Therefore, the group situation is deemed appropriate in the delivery of interventions. The amount of changes and influences depends on the group's proper functioning and its members' dedication.

A group process that takes place stage-by-stage in an experience-based counseling model can make the counselee more aware of specific behaviors and other needs to change and how to decide, so they are more sensitive to areas of life that they want to modify. Through group activities in experience-based counseling, the counselee is expected to be able to assess how to change behavior, as required in achieving the objectives of the group. Besides individual achievements, the counselee can also see the group's achievement. As a result of this social and psychological support, counselees have opportunities to design activities in their groups.

4 CONCLUSIONS

Empirically, experience-based counseling intervention is effective in reducing the tendency of students' behavior symptoms or characteristics associated with bullying. Its intervention is also proven to effectively lower the tendency of bullying behavior related to six aspect indicators or bullying behavior symptoms.

5 RECOMMENDATIONS

Recommendations are addressed to the school counselors/tutors as well as to future researchers.

5.1 For school counselor/tutor

The role as a school counselor or a tutor requires the ability to help students develop their potential as optimally as possible. In line with the problems related to bullying, school counselors need to have the skills to deal with pupils who engage in bullying situations, either as perpetrators, victims or witnesses. School counselors or tutors can use the experience-based counseling model to reduce bullying behavior tendencies of learners as a way of handling bullies at school.

5.2 Future research

To enhance the various limitations of this study, future researchers are expected to be able to design more comprehensive research design. Related to the development of research instruments, researchers need to prepare more specific, valid and reliable instruments. It is expected that the future study is conducted not only for measuring reduction in behavioral symptoms, but also measuring the decrease of behavior intensity that occurs after the administration of the intervention. Therefore, future work needs to take into consideration time for counseling interventions.

REFERENCES

Adair, V. (1999). No bullies at this school: Creating safe schools. *Childrenz Issues: Journal of the Children's Issues Centre*, 3(1), 32–37.

Djuwita, R. (2006). *Kekerasan Tersembunyi di Sekolah: Aspek-aspek Psikososial dari Bullying*. Retrieved from www.didplb.or.id.

Drosopoulos, J.D. (2008). *Minimizing bullying behavior of middle school students through behavioral intervention and instruction*. Illinois: San Xavier University & Pearson Achievement Solution, Inc.

Gillis, H.L. & Gass, M.A. (2004). Adventure therapy with groups. In J.L. DeLucia-Waack, D.A. Gerrity, C.R. Kalodner, & M.T. Riva (Eds.). *Handbook of Group Counseling and Psychotherapy* (pp. 593–605). NY: Norton-Sage.

Mitchell, K.J., Finkelhor, D., & Wolak, J. (2004). Victimization of youths on the Internet. *Journal of Aggression, Maltreatment & Trauma*, 8(1–2), 1–39.

Newman-Carlson, D., & Horne, A.M. (2004). Bully busters: A psychoeducational intervention for reducing bullying behavior in middle school students. *Journal of Counseling & Development*, 82(3), 259–267.

Novianti, I. (2008). Fenomena kekerasan di lingkungan pendidikan. *Insania*, 13(2), 324–338.

Olweus, D. (1993). *Bullying at school: What we know and what we can do*. Malde, MA: Blackwell Publishers.

Salmivalli, C., Kaukiainen, A., & Voeten, M. (2005). Anti-bullying intervention: Implementation and outcome. *British journal of educational psychology*, 75(3), 465–487.

The team of Psychology Faculty of Indonesia University. (2008). *Peranan faktor personal dan situasional terhadap perilaku bullying di tiga kota besar*. Jakarta: FPUI.

Ideas for 21st Century Education – Abdullah et al. (Eds)
© *2017 Taylor & Francis Group, London, ISBN 978-1-138-05343-4*

The enhancement of self-regulated learning and achievement of open distance learning students through online tutorials

U. Rahayu
Universitas Terbuka, Tangerang Selatan, Banten, Indonesia

A. Widodo & S. Redjeki
Universitas Pendidikan Indonesia, Bandung, Indonesia

ABSTRACT: Self-Regulated Learning (SRL) skill is an important aspect to be successful in studying at Open and Distance Learning (ODL). Independent learning skills can be introduced and be trained to students in various ways. The aim of this study was to discuss the benefits of SRL training in enhancing SRL skills and learning achievement of ODL students through an integration of learning strategies into the process of online tutorials. The design of this study used one group pretest and posttest design. Samples of this study were students who were taking the online tutorial of Biology Learning Strategies Course (n = 16). The results showed that the self-regulated learning ability of students before and after the treatment increased, though not significantly. However, self-monitoring aspect showed significantly increased. The students' achievement improved significantly. This study recommends that ODL tutors of online tutorials should train SRL skills to their students along with the online tutorial process so that the students' independent learning skills can improve gradually.

1 INTRODUCTION

Students of distance education are autonomous learners. It means that they have the authority in controlling their learning method and determining where to learn and how to achieve their goals (Moore 2005). The behavior of autonomous learners is strongly associated with the term of self-regulated learning (SRL). Distance education students should be able to take advantage of their authority when they study. The authorization can be carried out properly if they have skills in self-regulated learning. Zimmerman and Martinez-Pons (1990) emphasize that there is a correlation between student achievement and the use of self-regulated learning strategies (1990). Furthermore, according to Schunk and Ertmer (2000), students' independent learning skills can be improved through systematic interventions. Students' independent learning can be developed through the developing time management skills, understanding and summarizing the text, noting ideas, anticipating and writing test (Zimmermann et al. 1996); the use of tools matrix note taking and prompt self-monitoring (Kauffman et al. 2011), the use of time management tools (Puspitasari 2012), and etc. Along with advances in technology, the use of ICT-based networking in education, including distance education, have increased (Malik et al. 2005, Hu & Gramling 2009), especially when they are used as students learning services (Fozdar & Kumar 2007, Jung 2005).

Online tutorial began to be used as student learning services in distance education in Indonesia in 2009. It was developed with the intention to bridge the characteristic of distance education, which is "the physical separation between teacher and student". The online tutorial process is a mean of interaction between student, teacher, and student. Students' independent learning skills are possible to be trained through online tutorials (Rahayu et al. 2015). In this study, we will integrate learning strategies of SRL including cognitive and metacognitive strategies, such as self-monitoring, self-reflection, mind mapping, and question & answer into the online tutorial process. Then, we will conduct a study that aims are to identify the difference of students' self-regulated learning skills and students' learning outcomes (pretest and posttest) between before and after online tutorials that integrate learning strategies.

2 LITERATURE REVIEW

Self-regulated learning (SRL) is a process that involves students in managing and organizing complex learning activities to achieve their academic goals (Zimmerman & Schunk 2001). In line with these opinions, Pintrich (2000) states that SRL is an active and constructive process where learners specify learning goals and monitor, manage and control their cognition, motivation, and behavior to achieve the goal.

Self-regulatory is a cyclical process which consists of three phases that are, forethought phase, performance or volitional control phase, and self-reflection phase (Zimmerman 2000). According to Shuy (2010), SRL consists of 3 aspects, those are, cognition, metacognition, and motivation. Cognition includes skills and habits in coding, remembering, recalling and critical thinking. Metacognition includes skills in which students are able to understand and monitor cognition process. Motivation includes the development of cognition and metacognition.

Each student has a different level of independent learning. It depends on their motivation to learn, the method used in the study, the utilization of existing performance, and the utilization of social resources and learning environment. Students whose level of independence of learning is low are much more likely to have lower educational achievement in schools (Zimmerman 2000).

3 RESEARCH METHODS

This research applied mixed methods. This study used one group pretest and posttest design. The independent variable of this research was the training of SRL integrated into the online tutorial. The dependent variables of the study were students' SRL skills and students' learning outcomes. The experiment was conducted during the academic year of 2016.

The populations of the study were the students who accessed the online tutorial of Biology Learning Strategy (BLS) courses in the first semester in 2016. The samples of the study were 16 students taking the online tutorial of BLS in Biology course and receiving the students' learning guide "Cerdas Strategy". The students' guide described how (1) to select and to use learning strategies, to manage and to motivate the learning process; (2) to use study time effectively; (3) to plan learning process realistically; (4) to set goals that can be achieved; (5) to plan learning targets accurately and with measurement; and (6) to plan learning objective specifically (Rahayu et al. 2016).

Data were collected using questionnaire, test, and observation sheets. Participants filled out 40 valid items modified MLSQ (motivation for Learning Strategy Questionnaire) in order to measure students' SRL skills; four-point Likert-type scale (1 = never; 2 = rarely; 3 = often; and 4 = very often), with the level of reliability (r) = 8.6 at α = 1%. Participants filled out this questionnaire before and after the online tutorials integrating learning strategies. Participants took the online pretest and posttest to measure students' achievement, which was a standard test taken from the item bank. The online pretest was conducted before the online tutorial, and posttest was conducted after the online tutorial. During the online tutorial, participants were observed how often they had practiced self-monitoring (SM), self-reflection (SR), mind mapping (MM), and questions and answers (QA) writing. Besides that, participants also did three online tutorial tasks in order to measure concepts mastery. The collected data were analyzed using SPSS 17 for windows.

4 RESULTS AND DISCUSSION

We offered two primary sets of findings. The first set of analyses focused on the difference of students' SRL skills before and after the online tutorial, including 11 aspects of SRL. Then, it was correlated with the finding during the SRL training. Second, we analyzed the difference in the students' learning outcome before and after tutoring. It was correlated with the finding during the training of SRL and doing the tasks.

4.1 *The difference of mean score on students' SRL skills before and after the online tutorial*

The data by means of students' SRL before and after online tutorial were normal (p = 0.200) and homogeneous (p = .085). Then, the data was analyzed by t-test and the result was that the mean score of students' SRL before and after online tutorial were not significantly different (p = 0.165). It means that even though the mean score of SRL was increased (SRL' mean score before the online tutorial = 3.050 and SRL' mean score after online tutorial = 3.184), but the increase was not significant. The results of this study were consistent with other research. Darmayanti (2004) found that the intervention of learning strategies to open distance learning (ODL) students did not have a significant influence on the improvement of student' learning independence. Likewise, the results study of Puspitasari (2012), who found that the intervention time management and learning strategies did not significantly influence on the improvement of learning independence. Moreover, Barnard-Brak et al. (2010) suggested that the online lectures could not improve self-regulated learning. It seems to us that the SRL skills training given to the students might need more time. It is relevant to Darmayanti' study that student independent learning, increased after 5 years when students had implemented a learning strategy (Darmayanti 2012). Then, the data were analyzed further in order to see the difference of each SRL aspect. Summary of the p-value of mean difference of each SRL aspect, before and after online tutorial can be shown in Table 1.

Table 1 showed that the mean score of self-monitoring aspect was significantly different. Self-monitoring is one of the processes that occurs in the performance phase. It is conducted during the learning process

Table 1. Summary of p-value of mean score difference of each SRL aspect before and after online tutorial.

Dimension	Aspect	P
Motivation	Intrinsic & extrinsic	0.287
	Self-efficacy	0.519
Learning strategy	Goal setting	0.910
	Strategic planning	.236
	Scientific goal accomplishment	.452
	Self-monitoring	.00**
	Resource management	.452
	Time management	.859
	Regulation effort	.105
	Self-evaluation	.207
	Self -reaction	.094

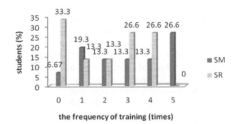

Figure 1. Frequency of student' self-regulated learning training for self-monitoring (SM) and content self-reflection (SR) at the online tutorial of BSL course.

(Schunk & Ertmer 2005). Self-monitoring is a meta-cognitive skill which is necessary for students to control their learning process. According to Kaufman (2004), self-monitoring is an important aspect of the SRL because with this ability student gain feedback related to their own performance. Without self-monitoring, effective control to the cognitive system might be very limited. In line with these results, Arsal (2009), Sungur & Tekaya (2006) emphasized that there was a significant effect of the intervention on students' metacognitive ability.

During the online tutorial process, the students were trained metacognitive SRL (SM and SR) and cognitive SRL (MM and QA). There were 27% of students performed SM five times. Students did SM four times, three times and two times were 13%. Students did not practice SM were 19%.

In addition, during the online tutorial, 26.5% of students did the content of SR four times and three times. Students did the content of SR once and two times was 13.3%. Students did not do SR were 33.3%. Figure 1 showed the training activities of regulated learning regarding SM and content of SR.

4.2 The differences of mean score on learning outcomes before and after the online tutorial

The data of mean score of students' achievement before (pretest) and after online tutorial (posttest) were

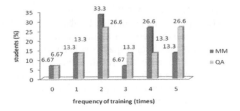

Figure 2. The frequency of student' self-regulated learning training for mind mapping and writing question and answer at the online tutorial of BSL course.

normal (p = 0.200) and homogeneous (p = .0853). Then, the data was analyzed by t-test. The t-test result was that the mean score of students' achievement before and after online tutorial were significantly different (p = 0.005). It means that there was an increase mean score of student achievement. The significant difference between pretest and posttest indicated that students' training on cognitive SRL, which were making mind mapping (MM) and writing question & answer (Q & A), had an effect on students' achievement.

During the online tutorial process, students' activities during the SRL training were; 13.3% of the students composed MM five times. Students composed MM four times were 26.6%. Students composed MM three times were 67%. Students composed MM two times were 33.3%. The rest of the students (13.3%) did not compose MM.

In addition, during the online tutorial process, 26.6% of students wrote Q & A five times. Students wrote Q & A four times and three times were 13%. Students wrote Q & A once were 27%. The students did not practice Q & A were 6.67% (see Figure 2). Then, when we analyzed further, the students who did the task for more SRL training had better posttest score than those who did not.

A significant difference between the pretest and posttest might also be the impact of metacognitive besides cognitive training undertaken by these students. The student who did more self-monitoring and self-reflection training, they arranged more mind mapping and question and answer. There was correlation between SM training and MM training (correlation coefficient (r) = 0.893, p<0.05 with α = 0.01). There was correlation between SM training and QA training (r = 0.866, p<0.05 with α = 0.01). There was correlation between SR training and MM (r = 0.808, p<0.05 with α = 0.01). There was correlation between SM training and QA training (r = 0.905, p<0.05 with α = 0.01).

Training on cognitive process encourages the students to monitor activities during their learning. (Kaufman 2011). It might occur in this case. Training of MM and QA encouraged students to monitor and do self-reflection during their learning. Information which was obtained during monitoring, it was possibly used by students to adjust their learning strategy (Zimmerman & Paulsen 1995). There was a synergism between metacognitive and cognitive training and its impact on learning outcome. It was in line with the result study of Kaufman (2011) dan Cazan's (2013). Cazan (2013) showed that the training combination of cognitive SRL and metacognitive strategies were able to improve learning outcomes. Kaufman (2011) showed that students who take notes using an online taking note matrix tool and self-monitoring prompt had better achievement than students who did not.

In addition, during the process of online tutorial, students did tasks 1, 2, and 3. The average score achieved by the students at the first task was 83.79, its second task was 92.5, and its third task was 81.4. So, the average score was about 86. It showed that students' achievement was quite high. It indicated that tasks' score influenced on the post test score.

5 CONCLUSIONS

SRL training through online tutorial integrating learning strategy has significantly improved students' achievement and self-monitoring SRL. Therefore, SRL training should be continuously conducted by ODL instructor so that ODL students' SRL skills improve gradually.

ACKNOWLEDGEMENTS

The author would like to thank the Ministry of Research, Technology and Higher Education of Indonesia for the scholarship of BPPDN while doing a Post-Graduate education in Universitashu Pendidikan Indonesia.

REFERENCES

Arsal, Z. 2009. The effects of diaries on self-regulation strategies of preservice science teachers. *International Journal of Environmental & Science Education.* 5(1): 85–103.

Barnard-Brak, L, Paton, V.O & Lan, W.Y. 2010. Self-regulation across time of first-generation online learners. ALT-J. *Research in Learning Technology.* 18(1): 61–70.

Cazan, A.M. 2013. Teaching self-regulated learning strategies for psychology students. *Procedia Social and Behavioral Sciences.* 78: 743–747.

Darmayanti, T. 2004. Efektivitas intervensi keterampilan self-regulated learning dan keteladanan dalam meningkatkan kemampuan belajar mandiri dan prestasi belajar mahasiswa pendidikan jarak jauh' [effectivity intervention self-regulated learning skills and model to enhance independent learning and students' achievement at Distance Education]. *Jurnal PJJ.* 9(2): 68–82.

Fozdar, B.I. & Kumar, L.S. 2007. Mobile learning & student retention. *International Review of Research in Open Distance Learning.* 8(2): 1–18.

Hu, H. & Gramling, J. 2009. Learning strategies for success in a web course: A descriptive exploration. *The Quarterly Review of Distance Education.* 10(2): 123–134.

Jung. 2007. Changing faces of open and distance learning in Asia. *International Review of Research in Open & Distance Learning.* 8(1): 1–6.

Kauffman, D.F. 2004. Self-regulated learning in web-based environments: instructional tools designed to facilitate cognitive strategy use, metacognitive processing, and motivational beliefs. *J. Educational Computing Research.* 30 (1&2): 139–161.

Kauffman, D.F., Zhao, R. & Yang, Y. 2011. Effect of online note taking formats and self-monitoring prompts on learning from online text: using technology to enhance self-regulated learning. *Contemporary Educational Psychology.* 36: 313–322.

Malik, N.A., Belawati, T. & Baggaley. 2005. *Framework of collaboration research and development on distance learning technology for Asia.* Paper presented at the 19th AAOU Annual Conference Jakarta.

Moore, M.G. 2005. *Learner autonomy: the second dimension of independent learning.* Retrieved 21/10/2016, from http://192.107.92.31/Corsi_2005/bibliografia%20e-learning/learner_autonomy.pdf

Pintrich, P.R. 2000. *The role of goal orientation in self-regulated learning.* In M. Boekarts, P. Pintrich, & M. Zeidner (Eds.) Handbook of self-regulation (p. 451–502). San Diego, CA: Academic Press.

Puspitasari, K. 2012. *The effect of learning strategy intervention and study time management intervention on students' self-regulated learning, achievement, and course completion in a distance education learning environment.* Ph. D thesis. The Florida State University College Education.

Rahayu, U. Widodo, U., Redjeki, S. & Darmayanti, T. 2015. *Pembiasaan keterampilan self-regulated learning: mungkinkan melalui tutorial online.* Paper presented at Temu Ilmiah Nasional Guru VII (TING VII), Universitas Terbuka, 25 th November.

Schunk, D.H. & Ertmer, P.A. 2005. *Self-regulation and academic learning: Self-efficacy enhancing intervention.* In M. Boekarts, P. Pintrich, & M. Zeidner (Eds.) Handbook of self-regulation (p. 631–646), San Diego, CA: Academic Press.

Sungur, S. & Tekkaya, C. 2006. Effect of problem-based learning and traditional instruction on self-regulated learning. *The Journal of Educational Research.* 99(5): 307–319.

Zimmerman, B.J. & Martinez-Pons, M. 1990. Students differences in self-regulated learning: relating grade, sex, and giftedness to self-efficacy and strategy use. *Journal of Educational Psychology.* 82:51–59.

Zimmerman, B.J. & Schunk, D.H. 2001. *Self-regulated learning and academic achievement: theoretical perspectives.* Hisdale, N.J: Erbaum.

Zimmermann, B.J., Bonner, S. & Kovach, R. 1996. *Developing self-regulated learners: beyond achievement to self-efficacy.* Washington, D.C: American Psychological Association.

Pedagogy (PDG)

Ideas for 21st Century Education – Abdullah et al. (Eds)
© *2017 Taylor & Francis Group, London, ISBN 978-1-138-05343-4*

Promoting individually-tailored teacher development program using the dynamic model of educational effectiveness research

S.N. Azkiyah
Universitas Islam Negeri Syarif Hidayatullah, Jakarta, Indonesia

ABSTRACT: Various teacher development programs have been conducted in order to improve teachers' competences. Yet, many of the programs fail to improve teaching quality that has been regarded as the most crucial elements for teachers. One of the reasons is the failure to address individual needs of teachers. Therefore, this paper aims to propose individually-tailored teacher development program based on difficulty level analysis of teaching skills included in the Dynamic Model of Educational Effectiveness Research. The individual teaching skills in the Dynamic Model represent eight factors of teaching quality i.e. orientation, structuring, modeling, application, questioning, creating classroom as a learning environment, assessment, and time management. These factors are developed based on previous teacher effectiveness research, which concerns teacher behavior that leads to better student outcomes. In this paper, classroom observation was conducted to 59 English teachers in DKI Jakarta and Banten to gather the data of teaching quality. The Rasch model was used to examine the difficulty level of each teaching skills. The findings revealed that teaching skills belong to orientation, CLE, modeling, and application were difficult for teachers. On the other hand, teaching skills belong to questioning, structuring, and assessment were relatively easier. These findings offer two important implications. Firstly, teacher development program should pay attention to the four difficult factors. Secondly, it is important to base TPDs on teaching skills individual teachers have mastered and those they need to improve so that TPDs meet the needs of individual teachers and hence individually-tailored teacher development programs could be promoted.

1 INTRODUCTION

Studies across different countries have consistently found that teacher is the most influential factor (e.g. Harris & Muijs, 2005; Marzano, 2007; Van der Werf, Creemers, De Jong, & Klaver, 2000). Therefore, the government in many countries including in Indonesia has prioritized various Teacher Development Programs (TPDs) to improve the quality of education. Some examples of TPD in Indonesia as the context of the study are in-service teacher training [INSET], the Islamic Schools English Language Program [ISELP], and the Madrasah Education Development Program [MEDP] (ADB, 2006; Hendayana, 2007; Jazadi, 2003).

Nevertheless, these TPDs have been considered not to be successful in improving teachers' teaching quality. Some possible causes include large classroom size, heavy teaching loads, insufficient preparation time for the teachers, noisy classrooms due to a lack of soundproofing and equipment shortages (Hendayana, 2007; Thair & Treagust, 2003). Other reasons are the absence of links between TPDs and student outcomes and of teachers' ownership of the programs (Nielson, 2003). Hence, it is not surprising to see that

the teaching quality of many teachers, especially in Indonesia are not yet satisfactory (Kaluge, Setiasih, & Tjahjono, 2004; Ree, Al-Samarrai, & Iskandar, S., 2012; Utomo, 2005).

Thus, it is urgent to look for better strategies not only to improve the quality of teachers but also to address the problems found in the previous teacher development programs. This paper offers the results of educational effectiveness research particularly Teacher Effectiveness Research (TER), which deals with various factors at the teacher level found to be related to better student outcome. TER could provide a relevant basis to decide what to do when we want to improve the quality of teachers.

The findings of TER have revealed that compared to other factors in teachers such as teachers' subject knowledge and length of experience, the actions of teachers in the classroom, which are often referred to as teacher instructional roles or teachers' teaching quality, are found to play bigger role (Creemers, 1994; Creemers & Kyriakides, 2008; Muijs & Reynolds, 2010). Furthermore, in order to contribute to the development of TER, Creemers (1994) developed a model of effective classroom, which was further developed into a dynamic model of educational effectiveness (Creemers & Kyriades,

2008). This dynamic model has four levels, i.e. context/national policy, school, teacher/classroom, and student. The model emphasizes the importance of the classroom factors especially on the behavior of teachers in promoting learning and expects the context and school levels to provide necessary conditions for the effectiveness of the classroom level.

The classroom level consists of eight factors: 1) orientation, 2) structuring, 3) questioning, 4) teaching modeling, 5) application, 6) management of time, 7) the classroom as a learning environment (CLE), and 8) classroom assessment. 'Orientation' concerns the explanation of the objectives, which is expected to help the students understand the importance of their learning activities. 'Structuring' refers to the explanation of the distribution of the series of activities of the lesson. 'Questioning' entails the attempt to categorize the questions in terms of difficulty level and type (product and process) and the reaction to the students' responses. 'Modeling' includes the provision of strategies of learning or the encouragement of students to develop their own. 'Application' relates to the immediate practice of the topics taught during the lesson. 'Management of time' requires the teacher to organize his/her lesson in such a way that the students' attention spans are maximized and that they are engaged in tasks throughout the lesson. 'CLE' includes the following components: 1) teacher-student interaction, 2) student-student interaction, 3) students' treatment by the teachers, 4) competition among the students, and 5) classroom disorder. Finally, effective teachers collect information on their students' knowledge and skills to identify their learning needs.

Several studies have been conducted to test the validity of the dynamic model (Antoniou, Demetriou & Kyriakides, 2006; Antoniou, 2009). The study of Antoniou also reveals that teaching quality represented in the classroom factors of the dynamic model could be divided into five stages, ranging from easier to more difficult skills. This finding is extremely useful for the development of teacher training programs. When teachers are found to be at level one, they can focus their improvement efforts to master the skills in level two. In other words, depending on current teachers' teaching quality, different teachers may have a different focus of improvement.

Therefore, this paper is intended to investigate the difficulty level of numerous teaching skills included in the classroom factors of the Dynamic Model. This study is expected to be a significant endeavor in designing teacher professional development programs that address individual needs of the teacher in accordance to what they have achieved and what they need to achieve. In this way, the development of individually-tailored TPDs could be promoted.

2 RESEARCH METHODS

This research used a quantitative approach, in which a structured observation instrument was used to gather teachers' teaching quality. 59 (44% M, 56% F) English teachers of Islamic Junior High Schools from DKI Jakarta (45%) and Banten (55%) voluntarily participated in the study. The focus was on the teaching of reading comprehension. The observation instrument represented the eight classroom factors of the dynamic model and consisted of 52 teaching skills. In this study, time management and CLE were combined since both concern maximum engagement of students. It was modified from original high inference observation instrument of the dynamic model and arranged in 1–5 Likert Scale. Five trained observers conducted the observation and the inter-rater reliability was good (Generalized Kappa = .72). In answering the research questions, the Rasch model (Andrich, 1988) was used to analyze the data to identify the difficulty level of each observed activity.

3 RESULTS AND DISCUSSION

The findings presented in this papers focus on the difficulty level of the teaching skills included in the classroom factors of the Dynamic Model as described in Figure 1.

The left side of Figure 1 shows teachers, whereas the right side represents teaching skills in the research instruments according to the classroom factors of the Dynamic Model. Zero (0) of the left side indicates the average score, which for the case of teachers (symbolized by X) means that those who are below zero are those who do not perform well while for the case of teaching skills, those

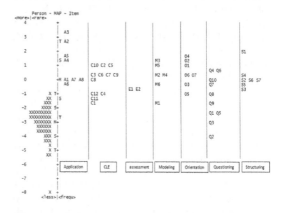

Figure 1. Map of teachers' performance according to the classroom factors of the dynamic model.

which are below zero are those which are very easy to be performed.

In other words, Figure 1 provides important information that the higher the position on the left side, the better the teaching quality of teachers is and vice versa. Regarding the teaching skill on the right side, the higher the position, the more difficult the teaching skill is. Furthermore, the teacher graph on the left side clearly illustrates that many respondents or teachers are below 0 representing the average, while the item graph on the right side illustrates that there are many items above 0. The findings mean that the teaching quality of the participants was low whereas many of the teaching skills were difficult for the participants since many of them did not perform the required skills.

These findings are not surprising since previous studies reveal similar results. As previously mentioned, Kaluge et al., (2004) find that Indonesian teachers, in general, are not yet able to deliver effective and joyful teaching and learning processes. In general, the dominant characteristic of Indonesian classrooms is a whole-class style of teaching where individual learners' needs are not sufficiently addressed and students' ideas, opinions and conceptions about their world are not fully recognized (Utomo, 2005). Furthermore, the current certification program launched by the Indonesian government has not yet improved the teaching quality and the student outcome (Ree, Al-Samarrai, & Iskandar, S., 2012). In 2012, the results of the teacher competence test showed that the national average score of teachers teaching at various levels of schooling was approximately 40–60 (out of 100) (Suharto, 2012). To add more, many teachers are also found to take a safer approach to teaching to the test to prepare the national exam (Hendayana, 2007).

Furthermore, the next interesting question is which factors are easier and which ones are more difficult. As described in Figure 1, among the factors, application, CLE, modeling, and orientation are more difficult factors because many of the items in these factors were located above 0. Orientation focuses on teachers' effort to explain the importance of learning certain topics and skills by connecting the lesson with daily life and/or the previous lesson. Knowing the importance of the materials presented, students are expected to be more motivated considering that what they learn is related to themselves and beneficial to their lives. Unfortunately, many items in this factor were above 0, which happened because teachers did not connect students' learning and students' daily lives. This finding supports Utomo (2005), who observes that most teachers only cover one topic in their lessons and no links are made to previous lessons, daily life situations or other subject domains.

Likewise, modeling, which is the provision of learning strategies and for the context of reading should happen during reading activities, was also a difficult factor. This finding is not surprising either since an old study conducted by Cooper (1986) reported that the emphasis in reading comprehension class was much more on before reading and after reading, but not during reading. Modeling concerns teachers' effort to provide learning strategies to enable students to do exercises provided by teachers in the classroom and to learn on their own outside the classroom. Furthermore, modeling is also expected to facilitate students to be self-regulated learners who are ready to learn both inside and outside the class, with and without teacher assistance. Unfortu- nately, as seen in Figure 1, more modeling items were above *logit* 0 which means that modeling is not an easy factor to implement by teachers.

The next two difficult factors were application and CLE. Most teachers provided exercises for students in the application stage, but were limited to answering questions provided in the textbooks. Concerning CLE, no grouping and collaboration were introduced, which are in line with the study of Utomo (2005), which reveals that the dominant teaching style is whole-class style.

The remaining three factors, i.e. structuring, questioning, and assessment were considerably easier for teachers since many of the items were below *logit* 0. These findings indicate that the items were easy as many participants practiced the skills. With respect to structuring, most teachers explicitly explained the topic to the students, yet they did not explain the series of activities during the lesson nor the link among different activities. Concerning questioning, a lot of teachers have engaged their students in the teaching-learning process by posing questions. Through these questions, teachers also ensured whether students understood the materials. However, the questions were limited to product questions while critical questions were not sufficiently raised. For the case of reading, product questions refer to those which answers are explicitly found in the text whereas process questions require students to think beyond the printed information in the text. Finally, questioning was the only technique used to make sure if students understood the materials.

4 CONCLUSIONS

In summary, this study aimed at investigating the difficulty level of numerous teaching skills included in the classroom factors of the Dynamic Model, the findings of which are expected to serve a basis to design individually-tailored teacher development program. Using the Rasch model, the study shows that many teaching skills included in the classroom factor of the Dynamic Model were difficult for teachers since there were more items above *logit* 0. Consequently, the teaching quality of participating teachers was low. Furthermore, concerning the factors, the analysis yields that among the factors, orientation, modeling, application, and CLE were more difficult factors whereas structuring, questioning, and assessment were considerably easier. The findings of this study imply three important aspects. Firstly, it should be noted that that in general teachers have difficulties in four factors and therefore, TPDs should carefully pay attention to the four difficult factors. Secondly, when teacher professional development program is designed, it is important to look at individual teachers by considering the teaching skills they have achieved and what they need to achieve so that improvement efforts can be more focused. In this way, individually-tailored teacher development program could be promoted. Thirdly, the five stages of teaching quality as have been found in Cyprus are not known from this study and therefore further studies should perform further analysis to investigate if the teaching skills could be divided into different stages by incorporating cluster analysis.

REFERENCES

Andrich, D. 1988. A general form of Rasch's Extended Logistic Model for partial credit scoring. *Applied Measurement in Education*. 1(4): 363–378.

Asian Development Bank. 2006. *Indonesia: madrasah education development project*. Project Number: 37475–01. Jakarta: SMEC International Pty Ltd.

Antoniou, P. 2009. *Using the dynamic model of educational effectiveness to improve teaching practice: building an evaluation model to test the impact of teacher professional development programs*. Unpublished PhD thesis. University of Cyprus: Department of Education.

Antoniou, P., Demetriou, D. & Kyriakides, L. 2006. Towards a dynamic model of educational effectiveness. A meta-analysis of studies investigating the impact of school factors on student achievement gains. *Proceedings of the IX Conference of the Cyprus Pedagogical Association. Nicosia*.

Cooper, J. D. 1986. *Improving reading comprehension*. Boston, MA: Houghton Mifflin Company.

Creemers, B. P. M. 1994. *The effective classroom*. London: Cassell.

Creemers, B. & Kyriakides, L. 2008. The dynamics of educational effectiveness. A contribution to policy, practice and theory in contemporary schools. New York & London: Routledge Taylor & Francis Group.

Harris, A. & Muijs, D. 2005. *Improving schools through teacher leadership*. London: Open University Press.

Hendayana, S. 2007. Development of INSET model for improving teacher professionalism in Indonesia. *NUE Journal of International Educational Cooperation*. 2: 97–106.

Jazadi, I. 2003. *An investigation of current constraints and potential resources for developing learner-centred curriculum frameworks for English language at high schools in Lombok, Indonesia*. Unpublished PhD thesis. Adelaide, Australia: Division of Education, Arts and Social Sciences, University of South Australia.

Kaluge, L., Setiasih & Tjahjono, H. 2004. *The Quality Improvement of Primary Children Learning through a School-Based Programme in Indonesia*. A Research Paper. East Java: Universitas Surabaya.

Marzano, R. J. 2007. *The art and science of teaching*. Alexandria, VA: Association for Supervision and Curriculum Development.

Muijs, D. & Reynolds, D. 2010. *Effective teaching: Research and practice*. London: Paul Chapman.

Nielsen, H. D. 2003. Reforms to teacher education in indonesia: does more mean better? In Beau- champ, E.R. (ed). *Comparative Education Reader*. New York: Routledge Falmer.

Ree, J. D., Al-Samarrai, S. & Iskandar, S. 2012. *Teacher certification in Indonesia: a doubling of pay, or a way to improve learning?. Policy brief*. Washington, DC: World Bank. Retrieved from http://documents.worldbank.org/curated/en/201-2/10/16843129/teacher-certification-indonesia—doubling-pay-or-way-improve-learning.

Suharto. 2012. *Refleksi 5 tahun program sertifikasi guru*. Retrieved online from http://www.m-edukasi.web.id/2012/08/refleksi-5-tahun-program-sertifikasi.html.

Thair, M. & Treagust, D.F. 2003. A brief history of a science teacher professional development initiative in Indonesia and the implications for centralized teacher development. International *Journal of Educational Development*. 23: 201–213.

Utomo, E. 2005. *Challenges of Curriculum Reform in the Context of Decentralization: The Response of Teachers to Competency Based Curriculum (CBC) and Its Implementation in Schools*. Unpublished doctoral dissertation. University of Pittsburgh.

Van der Werf, M. P. C., Creemers, B. P. M., De Jong, R. & Klaver, L. 2000. Evaluation of school improvement through an educational effectiveness model: The case of Indonesia's PEQIP project. *Comparative Education Review*. 44(3): 3.

Ideas for 21st Century Education – Abdullah et al. (Eds)
© *2017 Taylor & Francis Group, London, ISBN 978-1-138-05343-4*

Students' understanding, communication skills and anxiety of mathematics in an Islamic Junior High School using brain-based learning

T. Dahlan
Universitas Pasundan, Bandung, Indonesia

ABSTRACT: The aim of this study was to analyze the influences of brain-based learning on the enhancement of understanding, communication ability and anxiety reduction for students studying mathematics at an Islamic Junior High School (Madrasah Tsanawiyah/MTs). In addition, the relationships between students' understanding, communication skills and anxiety levels were discussed. The participants of this study were 54 MTs students at grade VII. The method used was a quasi experiment with a non-equivalent control group design. The sampling technique was purposive sampling. A pre-test, post-test, observation sheet, interview and questionnaire were used to measure each student's mathematical anxiety reduction. Data analysis was conducted in SPSS 20. The results of study were: (1) the mathematical understanding ability of students who received brain-based learning was better than students who received conventional learning; (2) the mathematical communication ability of students who received brain-based learning was better than students who received conventional learning; (3) there was no difference of mathematical anxiety of students between those who received brain-based learning and those who received conventional learning; (4) there was a positive relationship between students' mathematical understanding ability and students' mathematical communication ability; (5) there was a negative relationship between students' mathematical understanding ability and students' mathematical anxiety; and (6) there was a negative relationship between students' mathematical communication ability and students' mathematical anxiety.

1 INTRODUCTION

The ability of understanding and communication is becoming a key goal in mathematics, as stated in Ministerial Regulation 22 (Ministry of National Education, 2006), which explicitly lists some of the abilities and attitudes that must be developed during mathematics learning. Some abilities and attitudes listed are: (1) understand mathematical concepts, explain the link between concepts and apply concepts or algorithms in a flexible, accurate, efficient and precise way during problem solving; (2) using reasoning from patterns and nature, perform mathematical manipulation in making generalizations to compile evidence or explain mathematical ideas and statements; (3) solve problems that include the ability to understand the problem, devise a mathematical model, solve the model and interpret the obtained solution; (4) communicate ideas with symbols, tables, diagrams or other media to clarify the situation or problem; and (5) have respect for the usefulness of mathematics in life, be curious, attentive, and have a tenacious attitude and confidence in solving problems.

Communication skills in mathematics include the ability to interpret and explain terms and mathematical notations, both orally and in writing. Communication skills need to be the focus of attention in the study of mathematics, because through communication students can organize and consolidate mathematical thinking, express mathematical thinking coherently, analyze and evaluate strategies and mathematical thinking to one another, and explore mathematical ideas (NCTM, 2000).

Based on the description, it can be concluded that mathematical understanding and communication is needed by any person to address daily life problems. With mathematical communication, humans can find patterns, resolve problems or draw conclusions. Mathematical communication is a way to exchange opinions and ideas and develop students' mathematical understanding. Through mathematical communication, ideas become the object to be discussed. The communication process helps to build the meaning and accuracy of ideas and make it into something general. For developing students' mathematical communication skills, teachers need to expose students to various real-life problems and give opportunities to students to communicate their ideas and solve existing problems.

One strategy that can be done is to use a learning model that can maximize brain function. The learning strategy in question here is a brain-based learning approach. Brain-based learning is learning that is aligned with how the brain works naturally, designed for teaching (Jensen, 2007). Planning stages of learning with a brain-based learning approach, according to Jensen (2007), are pre-exposure phase, preparation, initiation and acquisition, elaboration phase, the incubation stage and enter the memory, the stage of verification and checking of belief and celebration stage and integration.

The main strategy that can be developed in the implementation of brain-based learning (Weiss, 2000) includes: (1) create a learning environment that challenges students' thinking skills; (2) create fun learning environment; and (3) creating a situation of active learning that is meaningful to students. This strategy must be well designed in brain-based learning for mathematics; the approach is expected to run smoothly as planned, so that learning objectives can be achieved optimally.

2 LITERATURE REVIEW

2.1 Brain-based learning

Jensen (2007) states the seven stages brain-based learning, which have been arranged in such a way that they are superficial and the teacher can still add a lot of other things into it based on the needs of each student. Accordingly, brain-based learning can be customized to different materials. The steps for implementing brain-based learning in this study were: pre-exposure, preparation, initiation and acquisitions, elaboration, incubation and inserting memory, verification and checking of belief, celebration and integration.

2.2 Mathematical understanding

Polya (Sumarmo, 2012) divides an understanding into four categories: (1) understanding the mechanical is the ability to remember and apply the formula of a simple calculation regularly; (2) understanding the inductive is the ability to apply the formula to simple issues or similar; (3) rational understanding is the ability to prove the truth of formulae or theorems; and (4) the ability to estimate the intuitive understanding of certain truth (no doubt) before further analysis.

Skemp (2006) categorize understanding into two types: instrumental and relational understanding. Instrumental understanding shows some ability to recall concepts and basic principles without knowing the links with other concepts, and the

ability to apply it to a simple calculation process with a good algorithm. Relational understanding is the ability to associate a concept or principle with other concepts.

2.3 Mathematical communication

Mathematical communication skills include how a person is able to: (1) communicate mathematical thoughts coherently and clearly to teachers or friends; (2) use mathematical language to express an idea or ideas exactly; (3) organize mathematical thoughts through communication; and (4) analyze and evaluate the mathematical thinking and strategies of others (Bell, 1978).

Widjayanti and Wahyudin (2010) define mathematical communication as a person's ability to write queries mathematically, write the reason or explanation of any mathematical arguments with which to solve a mathematical problem, use terms, tables, diagrams, notations or a mathematical formula exactly, and examine or evaluate the mathematical minds of others. With insight from these capabilities, each student can easily express mathematical ideas and the possibility to correct or express mathematical difficulty can be easily understood by discussion with others.

2.4 Relevant research

Some research related to brain-based learning, mathematical understanding and mathematical communication includes:

- Brain-based learning—Ozden and Gultekin (2008) investigated the retention of students in science subjects in Grade 5 elementary schools in Turkey in the 2004/2005 academic year. Two classes were used as experimental and control classes, respectively. Results showed that student achievement in the classroom using brain-based learning was better than that using conventional learning. Retention (memory) of students in the class using brain-based learning was also better than that in students using conventional learning.
- Comprehension and communication of mathematical ability—Results of research on improving the ability of students' mathematical understanding and communication include: Lilis (2014), found that cooperative learning called 'Two-Stay-Two Stray' can improve students' mathematical understanding in junior high school. Gardenia (2013) concluded that learning constructivism (Needham models) can improve students' mathematical understanding at vocational high school. Gumanti (2014) showed that learning to use GeoGebra can

improve students' mathematical understanding in junior high school.

3 RESEARCH METHODS

This research follows an experimental design using pre-test, post-test and control groups; brain-based learning was implemented to study the ability of understanding and communication in mathematics students. The method used was a quasi experiment, with non-equivalent control group design. The sampling technique was purposive sampling. A pre-test, post-test, observation sheet, interview and a questionnaire were used to assess student's mathematical anxiety reduction. Data analysis was conducted in SPSS 20 and Excel 2010. The research aim was to assess the improvement in the ability of understanding and communication of mathematics students after brain-based learning and conventional learning, including the assessment of mathematical anxiety in the classroom and the relationships between understanding, communication and mathematical anxiety. The subjects of this study were 54 MTs students in Grade 7. The study used a test to assess the students' abilities in understanding mathematical description, and, for mathematical communication skills and anxiety, an observation and interview sheet was used.

The data in this study include quantitative and qualitative data. In processing the quantitative data, a descriptive statistical analysis was performed to get a general idea of the improvement obtained by students in understanding mathematical ability and mathematical communication; this descriptive analysis consisted of the maximum value, minimum value, average, and standard deviation.

Then, an analysis of the mathematical understanding and communication abilities of students' was carried out using two mean difference tests. Building capacity in this study was obtained from the difference between pre-test and post-test scores. The scores of ideal mathematical understanding and communication capabilities were expressed as a score gain, normalized by the formula, Meltzer (2002):

$$\text{Gain normalized } (g) = \frac{score\ postes - score\ pretes}{score\ maximum - pretest}$$

4 RESULTS AND DISCUSSION

The results of the analysis of the pre-test, post-test, and normal gain understanding and communication ability of students are presented in Table 1. Table 1 shows no differences in pre-test

Table 1. Pre-test, post-test and n-gain understanding and communication mathematical abilities of students assessed.

			Class Brain Based Learning					
			Pretest		Postest		N-Gain	
	n	IMS	$\bar{x}$	s	$\bar{x}$	s	G	S
M.U.	27	20	3,55	1,87	13,50	0,16	0,61	0,16
M.C.	27	24	4,83	1,58	16,98	0,13	0,68	0,13

Class Konvensional					
Pretest		Postest		N-Gain	
$\bar{x}$	s	$\bar{x}$	s	G	s
2,88	1,76	11,08	2,65	0,48	0,16
4,90	1,75	14,28	2,09	0,49	0,13

Note:
M.U.: Mathematical Understanding;
M.C.: Mathematical Communication;
I.M.S.: Ideal Maximum Score.

scores in the mathematical ability and communication skills of students from the second grade, with mathematical learning classified as very low for all. But, after learning, students in the brain-based learning classes had mathematical abilities reaching 13.50/20 and gain increases (0.61) were classified as moderate and better than achievement (11.08 out of 20). The increase (0.48) for students in the conventional classes was classified as lacking in understanding mathematical abilities. Similar results were found after learning the communication skills of mathematics; students in the classroom with brain-based learning (16.98/24) achieved 'fair' and the gain increased (0.68) better than the achievement (11.08 out of 20) and gain increase (0.48) of students in the conventional learning class.

5 CONCLUSIONS

Based on the findings and discussion, this study provides the following conclusions: the ability of mathematical understanding in learners taught through brain-based learning is quite good; the ability is better than the mathematical understanding ability of students who received conventional learning (classified as moderate). Likewise, enhancements in the mathematical understanding of students subject to brain-based learning was better than the increase observed in students taught conventionally. However, the increased ability of mathematical understanding equally belong to the category of being advanced. There were advantages of brain-based learning over conventional learning in terms of the mathematical communication skills. Students in the experimental class achieved mathematical communication classified as better than that of students who received conventional learning.

Likewise, enhancements in the mathematical understanding of students taught through learning brain-based learning was better than that of the students in the conventional classroom. However, the increase in mathematical communication skills equally belonged to the category of being advanced. Based on the conclusion, we put forward some suggestions, as follows. The research was conducted on students in MTs.; the mathematical ability of students was classified as moderate. Students took longer to learn to understand mathematical concepts and their application with conventional learning. Moreover, questions of the ability of mathematical understanding and mathematical communication in this study consider the high level of mathematical ability. It is therefore recommended that study carried out in sufficient time to allow students to learn significantly.

REFERENCES

Bell, F. H. (1978). *Teaching and learning mathematics (in secondary school)*. United States: WC Brown Company Publisher.

Gardenia, N. (2013). *Improved understanding and communication abilities of vocational students through learning mathematical models constructivism needham* (Unpublished doctoral thesis). Universitas Pendidikan Indonesia, Bandung.

Gumanti, S. (2014). *Effect of GeoGebra assisted learning to increase understanding and visual thinking ability of junior high school students* (Unpublished doctoral thesis). Universitas Pendidikan Indonesia, Bandung.

Jensen, E. (2007). *Brain based learning*. Yogyakarta: Pustaka Pelajar.

Lilis, L. (2014). *Improving the ability of understanding and communication mathematical self-concept and junior high school students through cooperative learning two-stay two-stray* (Unpublished doctoral thesis). Universitas Pendidikan Indonesia, Bandung.

Meltzer, D. E. (2002). The relationship between mathematics preparation and conceptual learning gains in physics: A possible "hidden variable" in diagnostic pretest scores. *American journal of physics, 70*(12), 1259–1268.

Ministry of National Education. (2006). *Permendiknas No. 22 of 2006 on Standards for Primary and Secondary Education Unit*. Jakarta: Indonesian Ministry of National Education.

National Council of Teacher of Mathematics (NCTM). (2000). *Principles and standards for school mathematics*. USA: NCTM.

Ozden, M., & Gultekin, M. (2008). The effects of brain-based learning on academic achievement and retention of knowledge in science course. *Electronic Journal of Science Education, 12*(1), 1–17.

Skemp, R. R. (2006). National council teachers of mathematics. *Relational Understanding and Instrumental Understanding, 12*(2), 88–95.

Sumarmo, U. (2012). *Handout course evaluation in learning mathematics*. Bandung: Universitas Pendidikan Indonesia.

Weiss, R. P. (2000). Brain based learning. *Training & Development, 54*(7), 21–21.

Widjayanti, W. & Wahyudin, W. (2010). *Developed their communication skills mathematical mathematics student teachers through lectures strategy based collaborative problem*. Retrieved from: http://staff.uny.ac.id (Accessed: June 15, 2015].

Ideas for 21st Century Education – Abdullah et al. (Eds)
© 2017 Taylor & Francis Group, London, ISBN 978-1-138-05343-4

Realizing a good education in an Indonesian university context

A. Aunurrahman
IKIP-PGRI Pontianak, West Kalimantan, Indonesia

F.A. Hamied & E. Emilia
Universitas Pendidikan Indonesia, Bandung, Indonesia

ABSTRACT: This paper aims to explore Halliday's aspects of language development in order to realize a good education in an Indonesian university context. There are three aspects of language development explored in this paper. First, learning language that means a language is learned to enable students to communicate in spoken and written forms. Second, learning through language that means a language as a medium to learn other things such as critical thinking and characters. Third, learning about language that means a language as an object that is studied such as its grammatical system. At the end of the discussion, the paper suggests a genre-based approach developed under systemic functional linguistics that can be applied by lecturers of any subject to realize a good education in an Indonesian university context.

1 INTRODUCTION

People are aware of the importance of education for them. In Indonesia, adults go to the university and any equivalent higher educational institutions to learn competence they need to work. It is the task for the higher educational institutions to facilitate the adult learners' need. The institutions should provide a good education that can make them be competent workers who can work and even compete with foreign workers as Association of Southeast Asian Nations (ASEAN) economic community which is known as AEC has been established since 2015. The establishment of the AEC means that competent foreign people from ASEAN countries have equal opportunity with Indonesian people to work in Indonesia (see Jobanputra et al. 2016). Underestimating the establishment of AEC will bring the quality of the Indonesian human resources nowhere.

Due to this issue, this paper aims to explore Halliday's three aspects of language development to realise a good education in an Indonesian university context. The importance of language in education is easily seen from its use as a medium of instruction (Pinnoc & Vijayakumar 2009). In Indonesia, Indonesian language is commonly used in the classroom to communicate the teaching materials from the lecturers to the students. However, the use of language is more than just a medium of instruction. Halliday, with his systemic functional linguistics (hereafter SFL), views language into three important aspects, that is, learning language, learning through language, and learning about language. These aspects have been used to describe the language development of the children's mother tongue (Halliday 2007, Painter 2008). The writers see that these aspects are necessary to be elaborated further in order to realise a good education in the Indonesian university context.

Specifically, the aspects of the language development are elaborated in three different sections. Each section covers the nature of each language development aspect. Yet, having three sections do not mean that each aspect stands by its own. Instead, these three aspects are connected to each other (Halliday 2007). At the latter section, the paper suggests a literacy-based teaching approach that can suit the need to realise a good education in the Indonesian university context, namely the genre-based approach that is developed under SFL.

2 LEARNING LANGUAGE

Learning language means that languages are learned in order to develop students' communication in spoken and written forms. The languages here can be a first, second, third, or even a foreign language, which language is "the substance of what is being learnt" (Halliday 2007, p. 288). In a language class, the students learn specifically the four language skills, that is, listening, speaking, reading, and writing. However, the writers see that these skills are not the responsibility of the language class solely. A non-language class should also facilitate this learning language aspect.

For example, a biology lecturer assigns an academic task to the students, which is in a written form such as a short essay or a research paper. The assignment certainly is related to the biology subject. This kind of academic task requires not only specific knowledge and skills of academic writing but also the language related to the subject. Leaving the students to learn the academic writing alone in the academic writing class is not sufficient. The lecturer of the subject should also guide the students to write academically and to use the biology related language as a language depends on its social context (Halliday 1994, Martin 2014).

Accordingly, learning a language is a responsibility of any lecturers with their own subjects or disciplines. The students may learn academic terms in a language class. However, when they learn in a specific academic subject such as a biology subject, they have to learn the language related to the subject. This is where the next two aspects of language development, that is, learning through language and learning about language are involved and elaborated in the following sections.

3 LEARNING THROUGH LANGUAGE

Learning through language means that a language is used to learn other things (Halliday 2007) such as critical thinking, characters (see Aunurrahman et al., 2016) and what the earlier section has elaborated as an example, a biology subject. The key is utilizing language as a resource for making meaning (Halliday 1978). A lecturer should use the language to introduce the biology subject. A language, under SFL, can be reflected in a text as a text is the level where a meaning is negotiated (Eggins 2004).

A text, in relation to an educational context, is an important element of literacy. Specifically, literacy has varieties of reading and writing available for choice (genre), the contexts for performance (social context); and the manner (linguistic features) in which the literacy is interpreted and tested, not by experts, but by ordinary people in ordinary activities (Szwed 2001, p. 422). Here, a lecturer should provide texts or references to help the students to learn the biology subject by paying attention to the texts' genre with its context and linguistic features. By doing this will facilitate the students to develop their knowledge regarding the subject and the language that is commonly used in that subject.

At this point, language plays an important part for the education to work in a class. The lecturer should use literacy, that is, introducing relevant texts to facilitate the students to learn a particular subject by paying attention to the genre and its context and linguistic features. In accordance with the linguistic features, this leads the discussion of this paper to learning about language that is elaborated in the following section.

4 LEARNING ABOUT LANGUAGE

Learning about language means that a language as "an object in order to understand how it works" (Halliday 2007, p. 288). When students are learning a language skill (e.g. writing), they are learning modes of writing or text types (genre) to help them to communicate in a written form in a particular context, in this case, in a biology subject. At the same time, the students are learning the grammatical systems or simply the linguistic features that enable the student to fully communicate in the particular context.

In an academic writing class, the students learn the academic genre with its linguistic features (Emilia 2012). But, when the students are in a non-language class such as the biology subject, they have to learn the linguistic features and terms related to the subject in order to communicate their ideas properly. As a result, the lecturers of the non-language classes should facilitate the students to learn the language related to their subjects. Accordingly, teaching a subject is not just about teaching the content of the subject. Lecturers should also learn to facilitate the students to learn the language and its grammatical systems to enable them to communicate properly and relevant to the studied subjects. This brings the discussion further to a teaching approach that can enable lecturers to facilitate learning in their classes.

5 A GENRE-BASED APPROACH

The early three sections show that language should be developed across the curriculum. It is the task for the lecturers of language or non-language classes to learn to guide the students to develop not only the content of the subject but also how to learn the language that is required to communicate the content of the subject. The lecturers should not only assign the students, let say, a paper assignment without giving a clear instruction and a guidance of what is expected from the paper. To assist the lecturers, this paper proposes a genre-based approach (hereafter GBA), under SFL, which is commonly used to teach literacy (Christie & Derewianka 2010, Rose 2009).

The GBA is practised worldwide including in Indonesia from the secondary to tertiary level of education. The approach is also used not only to teach literacy but also to teach critical thinking and to facilitate character building (Aunurrah-man et al., 2016, Emilia & Hamied 2015, Hardini 2013) as a language can be used as a medium to learn other things (Christie 2004). The GBA that is elaborated in this paper is developed by the SFL proponents, which is known as the Sydney School in Australia (Martin 2013, Rose 2009). This section elaborates the main principles of the GBA with its teaching stages and criticisms.

There are four main principles of GBA. The first three principles are it focuses on text, purpose, meaning, and choice (Derewianka 2003). The latter principle is explicit teaching (Emilia 2005). The GBA focuses on text as meaning is constructed at the text level (Eggins 2004). This means that lecturers should use texts to introduce a certain topic. Certainly, the texts should be relevant to the topic as GBA focuses on purpose. Halliday (1993) describes that adult learners can learn something through different ways. To do so, the lecturers can provide multimodal texts where the texts are not limited to printed texts but also electronic texts and videos that can facilitate the students' learning (Walsh 2010). For example, an academic writing lecturer who wants to introduce the ethics of academic writing can provide model texts that have citations and its references. Another example is a biology lecturer who wants to introduce the human organs; the lecturer can use a descriptive text with pictures and videos to facilitate the students' learning.

The lecturers' task is not completed yet. The lecturers should also facilitate the students to learn the language related to the subject in specific. This is where the third GBA principle takes place, that is, focuses on meaning and choice. It means that lexical and grammatical choices should be learned by the students to interpret and to construct a meaning or a text. Without them, it is impossible to interpret or even to construct the meaning or the text (Halliday 1994). For example, in introducing human organs, the biology lecturer uses descriptive texts along with pictures and videos. At the same time, the lecturer introduces the human organs, the lecturer also describes the terms related to the human organs. The lecturer also then introduces the schematic structures and linguistic features of the texts. This can be a good preparation for the students before assigning a written assignment to them if the lecturer expects the students to write a descriptive text later on.

The latter principle is explicit teaching. Feez (2002) describes this explicit teaching occurs when the lecturers are helping their students in creating texts for different social contexts where they explicitly pointing out the things that have been done. As a result, the students can focus on the things that have not been done. It is also called as the explicit

pedagogy where "teacher inducts learners into the linguistic demands of genres which are important to participation in school learning and in the wider community" (Macken-Horarik 2002, p. 26).

Knowing and implementing the main principles of GBA properly should enable the lecturers to teach their subjects and to handle the language barriers during the teaching. This even will be much better if the lecturers also know the stages in the GBA. The stages that are commonly used in Indonesia consist of four stages. The stages are building knowledge of field — students are introduced to the topic that will be learned; modelling — the lecturer provides references and model texts to introduce the students to the topic and the text with its linguistic features that is expected to be constructed by them; joint construction — the students construct a text with the help of their peers and the lecturer; and independent construction — the students construct a text independently with little help from their peers and the lecturer (Emilia 2012, Feez 2002).

However, the GBA in the practice is not easy. There are three issues that should be considered by the lecturers. First, Indonesia universities have a large class. A large class creates difficulty for the lecturers to provide explicit teaching to every student (Sukyadi 2015). To cope with this issue, group work can be applied throughout the teaching stages. Further information of the group work application for the Indonesian university context can read Alwasilah & Puncochar (2016). The second and third issues were identified at the secondary education level (Sukyadi 2015). Second, the language teachers practised the GBA teaching stages in a linear sequence. Meanwhile, the stages should be conducted in a circular sequence where the lecturers can move back to any stage, depend on the students' skills and readiness (Emilia 2012, Sukyadi 2015).

Third, the language teachers tended to focus on the text's generic structure with its lexical, and grammatical choices instead of the content of the text that can facilitate the students in constructing a text (Sukyadi 2015). Yet, this paper elaborates that the three aspects of language development are important as they are connected to each other. Up to this point, lecturers of a non-language class may focus on the content of the text. At the same time, the lecturers should also facilitate the students to learn the language and terms to understand the subject. If not, the language will be a barrier for the students in learning (Pinnoc & Vijayakumar 2009).

6 CONCLUSIONS

Language plays an important part to realize a good education in the Indonesian university context.

Through the elaboration of the three aspects of language development, the lecturers should learn to facilitate not only the teaching of the content of a subject but also the language and terms that are necessary to facilitate the students in learning. To do so, a genre-based approach is proposed. Indeed, this approach is commonly used in the language teaching. Still, the lecturers can utilize the approach to suit their needs by paying attention to the issues that are elaborated in this paper.

REFERENCES

Alwasilah, A.C. & Puncochar, J. 2016. *Empowering higher education in Indonesia*. Bandung, Indonesia: Pustaka Jaya.

Aunurrahman, Hamied, F.A. & Emilia, E. 2016. Facilitating character building through an academic writing practice. *Arab World English Journal* 7(3): 146–160.

Christie, F. 2004. Systemic functional linguistics and a theory of language in education. *Ilha Do Desterro* 46: 13–40.

Christie, F. & Derewianka, B. 2010. *School discourse learning to write across the years of schooling*. New York: Continuum.

Derewianka, B. 2003. Trends and issues in genre-based approaches. *RELC Journal* 34(2): 133–154.

Eggins, S. 2004. *An introduction to systemic functional linguistics* (2nd ed.). New York: Continuum.

Emilia, E. 2005. *A critical genre-based approach to teaching academic writing in a tertiary EFL context in Indonesia* (unpublished dissertation). The University of Melbourne, Melbourne.

Emilia, E. 2012. *Pendekatan genre-based dalam pengajaran bahasa Inggris: Petunjuk untuk guru* (2nd ed.). Bandung: Rizqi Press.

Emilia, E. & Hamied, F.A. 2015. Systemic functional linguistic genre pedagogy (SFL GP) in a tertiary EFL writing context in Indonesia. *TEFLIN Journal* 26(2): 155.

Feez, S. 2002. Heritage and innovation in second language education. In A.M. Johns (Ed.), *Genre in the classroom: Multiple perspectives*: 47–68. Mahwah: Lawrence Erlbaum.

Halliday, M.A.K. 1978. *Language as social semiotic: The social interpretation of language and meaning*. London: Edward Arnold.

Halliday, M.A.K. 1993. Towards a language-based theory of learning. *Linguistics and Education* 5(2): 93–116.

Halliday, M.A.K. 1994. *An introduction to functional grammar* (2nd ed.). London: Edward Arnold.

Halliday, M.A.K. 2007. *Language and education*. (J.J. Webster, Ed.). New York: Continuum.

Hardini, S.R. 2013. *Developing character values in the teaching of narrative texts using genre based approach: A case study at a senior high school in Bandung* (unpublished postgraduate thesis). Universitas Pendidikan Indonesia, Bandung, Indonesia.

Jobanputra, S., Cole, E., Murphy, D., Skinner, M. & Trinh, Q. 2016. *The ASEAN economic community: Investment opportunities and challenges in the world's newest market*. Jones Day Publication.

Macken-Horarik, M. 2002. Something to shoot for: A systemic functional approach to teaching genre in secondary school science. In A.M. Johns (Ed.), *Genre in the classroom: Multiple perspectives*: 17–42. Mahwah: Lawrence Erlbaum.

Martin, J.R. 2013. Embedded literacy: Knowledge as meaning. *Linguistics and Education* 24(1): 23–37.

Martin, J.R. 2014. Evolving systemic functional linguistics: beyond the clause. *Functional Linguistics* 1(1).

Painter, C. 2008. Language for learning in early childhood. In F. Christie & J.R. Martin (Eds.), *Language, knowledge and pedagogy: Functional linguistic and sociological perspectives*: 131–155. New York: Continuum.

Pinnoc, H. & Vijayakumar, G. 2009. *Language and education: The missing link*. London: International Save the Children Alliance.

Rose, D. 2009. Writing as linguistic mastery: The development of genre-based literacy. In R. Beard, D. Myhill, J. Riley, & M. Nystrand (Eds.), *The SAGE Handbook of Writing Development*: 151–166. California: SAGE Publications, Inc.

Sukyadi, D. 2015. The teaching of English at secondary schools in Indonesia. In B. Spolsky & K. Sung (Eds.), *Secondary School English Education in Asia: From Policy to Practice*: 123–147. New York: Routledge.

Szwed, J. 2001. The ethnography of literacy. In E. Cushman, E.R. Kintgen, B.M. Kroll, & M. Rose (Eds.), *Literacy: A critical sourcebook*: 421–429. Boston: Bedford/St. Martin's.

Walsh, M. 2010. Multimodal literacy: What does it mean for classroom practice? *Australian Journal of Language and Literacy* 33(3): 211–239.

Ideas for 21st Century Education – Abdullah et al. (Eds)
© *2017 Taylor & Francis Group, London, ISBN 978-1-138-05343-4*

Self-criticism on the teacher-training program from the faculty of education

A. Sofyan
Universitas Islam Negeri Syarif Hidayatullah, Jakarta, Indonesia

ABSTRACT: This study aims to describe problems with institutions and teacher training programs in the Faculty of Education and to find the solutions. The issue of how to develop a high-quality teacher is vitally important to the future of education in Indonesia. There were many unresolved issues, ranging from general low quality, uneven distribution, and the lack of standardization for institutions and teachers. All of these reasons added to the problems that already exist and affected everyone, including stakeholders, in making major decisions. More than one million teachers were becoming professional through Education Faculties with Teacher Training (PLPG or PPG) activities or portfolio assessments. However, the program has not significantly increased the quality of education in Indonesia. This paper is a self-criticism and self-evaluation for the Faculty of Education, in preparing the teacher training program.

1 INTRODUCTION

Issues around teachers in Indonesia are now shifting from one of shortages in the number of teachers to an uneven distribution. The ratios of teachers to students in Indonesia are now 1:15 (school) and 1:10 (madrasah); similar to Japan (1:17). However, in terms of quality, career coaching and the competence of teachers, we are still far below a decent standard. Referring to the teacher competency test results (UKG) that were implemented in 2015 and earlier, Indonesia gained an average score of below 50, which indicates a low level of teacher competence. Similarly, the teacher competence based on early competency testing (UKA), as a pre-requirement for teachers to follow the training (PLPG), showed low scores. Even after 9–10 days of PLPG training, teacher competence did not improve.

These low quality conditions were due to many factors, including the quality of the Institute of Teachers' Education (so called LPTK), the raw input (the students entering LPTK were of low standard), and low levels of passion of LPTK alumni in being teachers. Teaching was not considered the best profession for well-educated children.

2 LITERATURE REVIEW

2.1 *Profesional teachers*

The teaching profession in Indonesia is defined as functional. This means that being a teacher requires special skills and cannot be done by just anyone. As mentioned in the Law of the Republic of Indonesia Number 14 (Year 2005, Chapter 1, Article 1, Paragraph 1), teachers are professional educators with the primary task of educating, teaching, guiding, directing, training, assessing, and evaluating students on early childhood education (ECD), formal education, primary education and secondary education. The professionalism of teachers is a necessity.

Being a professional is a job or activity carried out for the duration (generally) of a person's life that provides a source of income and requires expertise and skills, or skills that meet certain quality standards or norms and requires professional education (Chapter 1, Article 1, Paragraph 4). Associated with the teaching profession, the law on teachers and lecturers (Chapter IV) mandated a teacher to have certain qualifications and competencies, outlined in the Minister of National Education (MONE) No. 16 of 2007. In regulating the standards of academic qualifications and competence of the teachers, it mentioned that all teachers, in both ECD, primary and secondary education, should have a minimum level of academic qualification, being a Diploma IV or Strata (S1) within the appropriate field, or similar, that could suit their students' characters. In addition, teachers are required to show some competencies in their performance, across four integrated areas: pedagogical, personality, social, and professional (Cronbach et al., 1973).

Pedagogical competence shows an understanding and the application of the principles of learning and educational psychology. Personal competence shows a behavior and noble spirit with an honest and good work ethic. Social competence is the ability to socialize well with learners, colleagues, and the wider community. Professional competence is mastering of the science of teaching.

The Faculty of Education in Indonesia, which produces teachers, is the LPTK. The LPTK is a college that was given the task by the Government for the procurement program of teachers in ECD, formal education, basic education, and secondary education, as well as to organize and develop pedagogy and non-educational aspects (Chapter 1, Article 1, Paragraph 14). Thus, LPTK is formally recognized by the Government to educate prospective teachers. In Indonesia, there are more than 250 LPTK; 13 are state LPTK, 19 are FKIP State University; and 234 are private LPTK.

According to Sujanto (2007), LPTK do not get adequate resources to prepare good prospective teachers. Also, there are few professors who pursue science education, and teacher training was considered a key factor resulting in the low education quality in Indonesia.

2.2 *Tasks and competencies teacher*

The Law of National Education System No. 20 (Year 2003, Article 39, Paragraph 2) stated that educators are professionals in charge of planning and implementing the learning process, assessing the results of learning, coaching and training, and conducting research and community service, particularly for educators at a college.

Furthermore, Article 40, Paragraph 2 stated the obligations of teachers as follows: (1) to create an atmosphere of meaningful, fun, creative, dynamic and dialogical education; (2) to be a committed professional and improve the quality of education; and (3) to set an example and keep the good name of the institution, profession and position, in accordance with the trust given to him.

Operationally the statute of teachers and lecturers of Article 20 states that, in carrying out professional duties, teachers are obliged to plan learning and implement the learning process to high quality, as well as assess and evaluate learning outcomes. Thus, teachers, as one component of the subsystem of national education, which are directly in contact with the 'raw input', have an important role, especially in preparing students to not only face the future with confidence, but also build with purpose and responsibility in the face of new challenges in the era of globalization.

The roles played by teachers in the classroom to learners, as stated by Gagne (1977), are to plan for learning, manage learning and instruct. However, Petres, as quoted by Sudjana & Rivai (2002), said

there are three tasks and responsibilities of teachers: (1) the teacher as a teacher; (2) the teacher as a mentor; and (3) the teacher as class administrator. The third task of the teacher is the primary duty of the teaching profession. Teacher stresses the duty plan on implementation and improvement of the learning system. In this position, the teacher's role is very big in the development of learning, which is in charge, as well as being a source, of learning activities. Teachers must be full of initiative and creativity, because teachers know the circumstances, especially with regard to the characteristic of learners and their backgrounds.

Teachers as mentors apply pressure to show that the teacher's task is not only to teach, but also to provide assistance to learners in solving life problems they are faced with, in regard of knowledge, values, and skills towards the formation of attitudes and personality. Strictly speaking, this is an aspect of education, which is to not only convey knowledge, but also about personality development and the formation of values in students. While the task as administrator is essentially the link between the management areas of learning and management in general.

Thus, the study of the performance of a systematic effort of the profession, through an accurate description of the knowledge skills, tasks, and other critical capabilities, is meant to see a vocational profession as a teachers, especially in the classroom. According to Brophy (1992), mastery learning principles will depict skilled teachers in managing learning activities within the scope of the tasks: plan, organize, lead and evaluate. This concept, in a brief study, can be grouped into clusters of planning, implementation and evaluation of learning.

Competency comes from the word that means the ability or skill. Mulyasa (2007) suggest that teacher competence, as a descriptive of a qualitative nature of teacher behavior, appears to be entirely meaningful. The opinion gives the meaning that the competence of teachers is a qualitative description of the nature of teacher behaviors. Stated in the law of the Republic of Indonesia Number 14 (Year 2005), competence is a set of knowledge, skills and behaviors that must be owned, lived and ruled by a teacher or lecturer in performing the duties of professionalism.

3 RESEARCH METHODS

This study followed a qualitative approach by narrative. The data were analyzed by descriptive analysis, to describe the phenomena of the teacher profession program in the Faculty of Education or LPTK.

4 RESULTS AND DISCUSSION

In accordance with the mandate of law no. 14 of 2005 on Teachers and Lecturers, teachers must be qualified as professional academics, of which their qualification and competence have to be proved by a teaching certificate. For this task, the government appointed several LPTK as the organizers. Starting in 2007, teachers who met minimum academic qualifications, of Diploma IV or S1 with working experience of ten years minimum, were prioritized to obtain teacher certification using portfolio assessment, PLPG, and PPG.

After three years, the portfolio was deemed to be not objective enough, because many documents of the portfolio were fabricated. Furthermore, the consortium of teacher certification (KSG) agreed to use education and training of the teaching profession (PLPG) as a media to certify teachers. PLPG participants were invited to attend PLPG for 90 hours (8–10 days). The numbers of teachers who could follow the program were limited based on the quota for each region. By 2015, the government had a target to certify all teachers who served before 2005 through a portfolio or PLPG. However, they failed to achieve this target. Hundreds of thousands of teachers were still not certified.

Regarding the teachers competencies, some findings were uncovered as follows: based on the competency test results for prospective teachers, the average teachers' scores of mastery of content was still very low; only 44% of the teachers answered 50% of the questions correctly. It was recorded that the teachers were most poorly qualified in the subjects of physics and mathematics, while they were best at English. In addition, the percentage of teachers' pedagogic ability, based on the test results from 2015, was only 56.69%. Others findings on the teachers' competence revealed the quality of teachers was uneven amongst the Indonesian islands.

Research on the UKG showed no significant difference between the teachers at district and city areas. Teacher competence, based on the UKG scores, decreased significantly after the age of 41-years-old. Moreover, there was no significant different between the certified and non-certified teachers. The UKG scores of teachers who teach at government schools were higher than those of teachers at private schools. The higher qualifications of the teachers, the better of the UKG score.

PLPG was expected to overcome many obstacles stated previously. However, some policies were considered to slow down the process. For instance, the regulation that allows graduate students from non-teacher education institutes to become a teacher by following the PLPG program underestimated the teaching profession. This condition is very much different from the medical profession, of which only medical graduates can become a doctor to follow the physician profession education. This policy should be reconsidered since the current ratio of teachers to students, being 1 to 15, is similar to the ratio in developed countries. Therefore, it is not necessary to create teachers via PPG.

5 CONCLUSIONS

The quality of national education is based on the Faculty of Education in preparing professional teachers. The certification programs by portfolio assessment or PLPG did not significantly increase teaching skills. Therefore, the Faculty of Education must be reform the curriculum and improve the teaching skills for preparing the Teacher Profession Program (PPG) after undergraduate level.

REFERENCES

Brophy, J. (1992). Planning and Managing Learning Tasks and Activities. *Advances in Research on Teaching, 3*(2), CT 06836-1678.
Cronbach, L.J., Warthen M.B.R. & Sanders J.R. (Eds.). (1973). *Course improvement through evaluation: Educational evaluation, theory and practice*. Belmont: CA. Wadsorth.
Gagne, R.M. (1977). *The conditions of learning*. New York: Holt, Riehart and Winston, Inc.
Mulyasa, E. (2007). *Standar kompetensi dan sertifikasi guru*. Bandung: Remaja Rosdakarya.
Sudjana, N., & Rivai, A. (2002). *Media pengajaran*. Bandung: Sinar Baru Algensindo.
Sujanto, B. (2007). *Manajemen Pendidikan Berbasis Sekolah*. Jakarta: CV. Sagung Seto.

Ideas for 21st Century Education – Abdullah et al. (Eds)
© *2017 Taylor & Francis Group, London, ISBN 978-1-138-05343-4*

The implementation of asking and group competition learning strategies to improve students' creative thinking skills

D.F. Wulandari
SMPN 2 Cibalong, Tasikmalaya, Indonesia
Doctoral Program of Science Education, Universitas Pendidikan Indonesia, Bandung, Indonesia

N. Rustaman, A. Setiawan & I. Hamidah
Doctoral Program of Science Education, Universitas Pendidikan Indonesia, Bandung, Indonesia

ABSTRACT: To improve students' creative thinking skills, it had been proposed a Physics Learning Strategy base of Asking activities and group Competition (PLSAC). This learning strategy consists of learning activities together (cooperative learning), reading, asking questions, looking for possible answers, competition in the group, and ends with reflections. The aim of this research is to examine PLSAC implementation and its impact on students' creative thinking skills. The research was carried out using a quasi experimental method. Senior high school students in Tasikmalaya were chosen as the subject of this research and two classes were taken as sample on the topic of sound. This research revealed that the students that used PLSAC had significantly better creative thinking skills than those students that used the conventional method. The average N-gain of creative thinking skills for the experimental class was 0.36, considered moderate improvement, compared with the control class, which scored 0.22 (low improvement). The result was supported by the positive responses of teachers and students on the implemented learning strategies. It can be concluded that PLSAC is better than conventional learning strategies in improving the students' creative thinking skills.

1 INTRODUCTION

Commitment to promote creative thinking skills has become one of the main agendas of the education program in Indonesia as mentioned in the Secondary School Competency Standards (Permendiknas, 2006). This was in line with educational experts, who agreed on the importance of creative thinking skills possessed by students.

Fisher & Amabile (2009) stated that creative thinking is widely seen as a basic competence, as well as reading and writing. Gopinathan claimed that new needs for work and life in the 21st century include the ability of students to be independent learners, work in a group, work as a good communicator, be willing to take risks, commit to lifelong learning and be capable of creative thinking (Biggs, 2011). Tumkaya and Aybek (Aktamis & Yenice, 2010) said that the purpose of science education is to educate students to be adaptive in different conditions, think flexibly, ask questions, be creative, respect people, and be tolerant of ideas. Fogarty (1991) also warned, in an effort to integrate the science curriculum framework, that one significant outcome to be aware of in the cognitive aspects of students is creative thinking. From these descriptions, it can be concluded that the experts agree on the importance of creative thinking.

Observations of some of physics classes, and interviews with teachers and students, were conducted to reveal data about creative thinking learning. These showed that learning in physics is still not intentionally devoted to developing creative thinking skills. Based on the background and review of literature learning methods, to improve students' creative thinking skills, Wulandari, Hamidah and Setiawan (2014) proposed a Physics Learning Strategy base of Asking activities and group Competition (PLSAC). This research was conducted to implement PLSAC in a physics class on the topic of sound. The research problem was 'How is the comparison of students' creative thinking skills improved between the students who gain PLSAC and students who receive a conventional learning strategy?'

2 LITERATURE REVIEW

2.1 *The Physics Learning Strategy base of Asking activities and group Competition (PLSAC)*

Cooperative learning is most popularly used in all fields of study. Various models of cooperative learning are already known and have been widely used for many purposes, but the model of cooperative learning base of asking activities and group competition for creative thinking has never been done. This study tries to assess the use of the models in learning creative thinking.

PLSAC is a learning strategy that consists of learning activities together (cooperative learning), reading, asking questions, looking for possible answers, competition in the group, and ends with reflections (Wulandari et al., 2014). The process of reading, asking questions and looking for possible answers started with review of a text book and continued with study of the environment and real-life problems. The final activity was called a problem solving activity. In this study, students also learned using interactive multimedia and websites.

PLSAC was developed based on a literature review of various studies of creative thinking that have proven its effectiveness. Thus, the hypothesis for this study was that the PLSAC learning model would be effective in improving creative thinking skills. The indicator of this effectiveness would be a significant difference in the normalized gain between students who received PLSAC and those who received conventional learning strategy.

2.2 *Creative thinking skills*

When students learn about the environment and real-life problems, they are encouraged to ask a question of the phenomenon, formulate the problems, question the causes and effects that may occur in the future, and offer an alternative solution. The activities were inspired by the definitions of creative thinking skills given by Lawson (1979), Treffinger, Isaken, & Firestien (1982), and Torrance (1966), which define creative thinking skills as: '...the process of (1) sensing difficulties problems, gaps in information, missing element, something asked; (2) making guesses and formulating hypotheses about these deficiencies; (3) evaluating and testing these guesses and hypotheses; (4) possibly revising and retesting them; and (5) communicating the results'.

Torrance (1966) looked at creative thinking as a process that involves elements of originality, fluency, flexibility and elaboration. These capabilities represent the process to be sensitive to one's understandings, and are the main traits of creative thinking (Piaw, 2004). In the same book four characteristics of creative thinking are defined as:

1. Originality. This refers to the uniqueness of the responses that were received. Originality is indicated by an unusual, unique and rare response.
2. Elaboration. This is the ability to decipher a certain object. Elaboration is a bridge that must be passed by someone to communicate the idea of 'creativity' to the public.
3. Fluency. This is the ability to create a myriad of ideas. It is one of the most powerful indicators of creative thinking.

4. Flexibility. This is a picture of an individual's ability to change his mental device when the circumstances require it.

In this study, the indicators of elaboration are not included in the measurement.

3 RESEARCH METHODS

This study used a quasi-experimental method. The study design used a non-equivalent control group design. Data were collected through: (1) a test of creative thinking skills in the form of essay questions; (2) a questionnaire to determine students' responses; and (3) interview guidelines to reveal the teachers' responses. Data were processed by calculating the normalized gain scores and with the two mean difference test using SPSS. The questionnaire data from the students were used to determine mean score and the questionnaire data from the teachers were described.

4 RESULTS AND DISCUSSION

4.1 *The increase of creative thinking skills*

The increase of creative thinking skills were judged from preliminary and final test scores in the form of percentages. The percentage scores of creative thinking skills (averaged) achieved in preliminary tests, final tests, and the N-gain of the experimental and control classes can be seen in Table 1 and Table 2.

Table 1 and Table 2 show the results of the quantitative descriptive analysis, in the form of average values (maximum of 100) for each indicator of creative thinking skills, standard of deviation (SD), and the average percentage gain for indicators of creative thinking skills.

The analysis of Table 1 and Table 2 shows that all indicators of creative thinking skills increased, demonstrated by positive average scores of N-gain. The results show that the acquisition of N-gain in the control class was highest in the fluency indica-

Table 1. Description of the creative thinking skills score for the control group.

| Creative thinking skills indicator | Control group | | | | |
| | Pre-test | | Post-test | | <g> |
	Mean	SD	Mean	SD	(%)
Fluency	52.2	9.7	63.7	9.6	23.1
Flexibility	39.8	7.2	53.8	10.6	22.6
Originality	35.8	6.8	49.2	7.7	20.3
All indicators	42.6	7.9	55.6	9.3	22.0

Table 2. Description of creative thinking skills score for the experimental group.

Creative thinking skills indicator	Experimental group				
	Pre-test		Post-test		
	Mean	SD	Mean	SD	<g> (%)
Fluency	52.2	7.6	71.5	9.7	40.3
Flexibility	39.3	6.1	61.6	6.6	36.3
Originality	34.4	5.1	54.3	7.0	30.4
All indicators	41.9	6.3	62.5	7.7	35.7

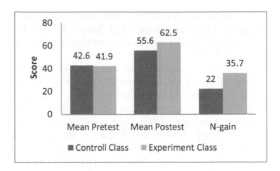

Figure 1. Comparison between the control group and experimental group.

tor, which was 0.23 (categorized in the low category) and lowest in the originality indicator, which was 0.20 (categorized in the low category); in contrast, in the experimental class, the highest N-gain in the fluency indicator was 0.40 (categorized in the medium category) and lowest occurred in the originality indicator, which was 0.30 (categorized in the low category). The average N-gain for the control class was in the low category, while the average N-gain in the experimental class was in the medium category. Thus the average N-gain of the experimental class was higher than the average N-gain in the control class. The comparison data between the control group and experimental group is shown in Figure 1.

The Shapiro–Wilk test for the N-gain data of the control class was significant ($P = 0.004$), so it can be assumed that the data are not normally distributed. For the experimental class, the test was not significant ($P = 0.872$), so it can be assumed that the data are normally distributed. The next step was to do a non-parametric statistical test (Mann-Whitney U with $\alpha = 0.005$).

Test results of the Mann-Whitney U test between the N-gain of the creative thinking control class and the experimental class were significant ($P = 0.000$), so H_0 was rejected and it can be assumed the two groups of data have a different average. The conclusion of the test is that the average N-gain creative thinking skills control class and experimental class do differ significantly.

4.2 Students' responses against implementation of PLSAC

Based on the student responses obtained, it can be concluded that the students gave positive responses to the implementation PLSAC with an average positive response of 82.96%. The main aspects of the questionnaire covered the enjoyment to learn physics with PLSAC, stimulation to ask a lot of questions, easiness to follow, encouragement to think more critically and creatively, and increasing motivation and enjoyment of group competition.

4.3 Teacher's responses against implementation of PLSAC

Based on teachers' responses obtained in the study, it can be concluded that the teachers gave a positive response to the implementation of PLSAC. The teachers said that this learning strategy has the potential to improve the quality of the physics learning process. Advantages of this strategy were: it could increase the students' motivation to read lessons before coming to class, because there was a pre-learning session and they wanted to be the best in the group competition session. This strategy can also improve students' participation/activity in the classroom.

4.4 Discussion

The positive result of this learning strategy is in accordance with the objectives when the learning strategy was proposed. The hypothesis for this study, that the PLSAC learning model will be effective in improving creative thinking skills, was supported by the results of the research. The indicators of the statement are the significant difference in the normalized gain between the students who received PLSAC and those who received a conventional learning strategy.

The asking aspect in PLSAC was intended to question reading a text book that goes into the process of understanding the subject matter, as well as the application of the concept. In the phase of application of the concept, PLSAC had an activity similar to problem-based learning (PBL). Characteristics that distinguish PLSAC and PBL are that the main feature of the group competition in PLSAC does not exist in the structure of PBL.

Other aspects in PLSAC are cooperative learning. In an effective cooperative group, the students will learn from each other because during their discussions about the content of the material, cognitive conflict will arise; poor reasoning will be discussed and an understanding of higher quality will appear (Slavin, 2010). This study outcome is supported by Qin, Johnson & Johnson (1995); Heller, Keith, & Anderson (1992); and Benckert & Pettersson (2008).

The use of cooperative learning in PLSAC was implicated by the increase in the group of the cooperative learning model. Thus, it can be said that the group of cooperative learning model now has a variety of new learning strategies, apart from learning models that known in the world of Cooperative Learning Model, such as: Student Teams Achievement Division (STAD), Teams Games Tournaments (TGT), Number Head Together (NHT) and Jigsaw.

The results of the analysis of students' responses to PLSAC reinforces the assumption of the motivation theory, that the purpose of cooperative learning provides an incentive for students to perform academic tasks that are a positive influence on cooperation among students. Constructivist theory requires students themselves, who must build their knowledge with the help of the teacher as a conduit of stairs that will take students to a higher understanding, with a record of the students themselves who must climb these stairs (Slavin, 2010). In connection with cooperative learning, the role of the teacher as a grant can be supplemented by a peer who has a knowledge level slightly higher than the other students concerned. Kuhn (1972) found that a small difference in level between a child's cognitive and social impact of a model is more conductive to the growth of the cognitive rather than the big difference between the cognitive level of learners within a social model.

A further aspect of PLSAC is the competition group. Humans basically like working together; they argue, discuss, and will always seek to compete with the competency of the opponent through debate or discussion (Sharan, 1990). This is shown in this model; learning activities of students in the classroom were dense and various. The challenge for teachers in implementing this model, is not in the effort to increase student activity, but in terms of maintaining the effectiveness and focus of student activity to keep it running in the corridor.

High student activity is due to high levels of student motivation. The study reinforces research by Lam, et al. (2004); Parrenas and Parrenas (1993); Johnson, Johnson, & Smith (1998); Wynne (1995); Attle & Baker (2007); Dyson & Grineski (2001) and Tauer and Harackiewicz (2004), who found the positive effect of competition on the results of student performance and motivation to learn in the classroom.

This research enriches that done by Chin (2000), who used learning strategies of teacher questioning in a cooperative learning environment. Although, in Chin (2000), the question of the teachers was more dominant than the portion of the students'

questions; in this study, the emphasis was on the questions raised by the students. Nevertheless, the similarity in the methods of questioning and cooperative environment are a common thread between Chin (2000) and this PLSAC research. The study also enriches the research carried out by Tiong (2002) and Miri, David & Uri (2007), who reported learning real-life problem solving can improve thinking skills.

The use of technology in this study considered the use of MMI and website in students' learning. The results of this research corroborates with those from Price et al. (2009); Wheeler, Waite, & Bromfield (2002); and Craig, Van Lehn, and Chi (2008), who reported the positive effects of using of web-based technologies to improve learning of creative thinking.

5 CONCLUSIONS

Improvement of students' creative thinking skills using PLSAC was significantly higher than that of students who received a conventional learning strategy. This showed that PLSAC learning strategy was more effective than conventional learning strategy in improving creative thinking skills. The results were supported by the positive responses of teachers and students on the implemented learning strategies. It can be concluded that PLSAC is better than conventional learning strategies in improving the students' creative thinking skills. Based on the results of this study, it is recommended that PLSAC is used in physics learning processes.

REFERENCES

Aktamis, H. & Yenice, N. (2010). Determination of the science process skill and critical thinking levels. *Procedia Social and Behavioral Sciences, 2*, 3282–3288.

Attle, S. & Baker, B. (2007). Cooperative learning in a competitive environment: classroom applications. *International Journal of Teaching and Learning in Higher Education, 19*(1), 77–83.

Benckert, S. & Pettersson, S. (2008). Learning physics in small-group discussions–three examples. *Eurasia Journal of Mathematics, Science & Technology Education, 4*(2), 121–134.

Biggs, J.B. (2011). *Teaching for quality learning at university: What the student does.* United Kingdom: McGraw-Hill Education.

Chin, C. (2000). Promoting critical and creative thinking in chemistry through teacher questioning. In L.S. Chia, H.K. Boo, S.N. Tan, & M.F. Tsoi (Eds.), *Chemistry teacher network* (pp. 7–14). Singapura: Singapore National Institute of Chemistry.

Craig, S.D., Van Lehn, K. & Chi. M.T.H. (2008). Promoting learning by observing deep-level reasoning questions on quantitative physics problem solving with andes. In K. McFerrin, R. Weber, R. Weber,

R. Carlsen, & D.A. Willis (Eds.). *Proceedings of the 19th international conference for the society for information technology & teacher education* (1065–1068). Chesapeake, VA: AACE.

Dyson, B. & Grineski, S. (2001). Using cooperative learning structures in physical education. *Journal of Physical Education, Recreation, and Dance, 72*(2), 28–31.

Fisher, C.M. & Amabile, T. (2009). Creativity, improvisation and organizations. *The Routledge companion to creativity*, 13–24.

Fogarty, R. (1991). *How to integrate the curricula*. United States: Skylight Publishing.

Heller, P., Keith, R. & Anderson, S. (1992). Teaching problem solving through cooperative grouping. Part 1: group versus individual problem solving. *American Journal of Physics, 60*(7), 627–636.

Johnson, D., Johnson, R. & Smith, K. (1998). Cooperative learning returns to college. *Change, 30*(4), 26–35.

Kuhn, D. (1972). Mechanisms of change in the development of cognitive structures. *Child development*, 833–844.

Lam, S.F., Yim, P.S., Law, J.S. & Cheung, R.W. (2004). The effects of competition on achievement motivation in Chinese classrooms. *British Journal of Educational Psychology, 74*(2), 281–296.

Lawson, A.E. (1979). *1980 AETS Yearbook the psychology of teaching for thinking and creativity. Clearing house for science, mathematics, and environmental education*. Ohio: The Ohio State University College of Education.

Miri, B., David, B.C. & Uri, Z. (2007). Purposely teaching for the promotion of higher-order thinking skills: A case of critical thinking. *Research in science education, 37*(4), 353–369.

Parrenas, C.S. & Parrenas, F.Y. (1993). Cooperative learning, multicultural functioning, and student achievement. In L.M. Malave (Ed.), *Proceedings of the national association for bilingual education conferences* (pp. 181–189). Washington, DC: ERIC.

Permendiknas. (2006). *Peraturan Mendiknas Nomor 23/2006. Standar kompetensi lulusan untuk satuan pendidikan dasar dan menengah*. Jakarta: Departemen Pendidikan Nasional.

Piaw, C.Y. (2004). *Creative and critical thinking styles*. Serdang: Universiti Putra Malaysia Press.

Price, S., Roussos, G., Falcão, T.P. & Sheridan, J.G. (2009). Technology and embodiment: Relationships and implications for knowledge, creativity and communication. *Beyond Current Horizons*, 29, 1–22.

Qin, Z., Johnson, D.W. & Johnson, R.T. (1995). Cooperative versus competitive efforts and problem solving. *Review of Educational Research, 65*(2), 129–143.

Sharan, S. (1990). *Cooperative learning: theory and research*. New York: Praeger.

Slavin, R.E. (2010). Cooperative learning, teori, riset dan praktik. Bandung: Nusa Media.

Tauer, J.M. & Harackiewicz, J.M. (2004). The effects of cooperation and competition on intrinsic motivation and performance. *Journal of Personality and Social Psychology, 86*(6), 849–861.

Tiong, H.B. (2002). Promoting thinking in science. In C.A.C. Shook, & C.Y. Mee (Eds.) Teachers' handbook in teaching thinking skills across disciplines. Singapura: Prentice Hall.

Treffinger, D.J., Isaken, S.G. & Firestien, R.L. (1982). Theoretical perspectives on creative learning and its facilitation: an overview. *The Journal of Creative Behavior, 17*(1), 9–17.

Torrance, E.P. (1966). The Torrance tests of creative thinking-norms-technical manual research edition-verbal tests, forms A and B-figural tests, forms A and B. Princeton, NJ: Personnel Press.

Wheeler, S., Waite, S.J. & Bromfield, C. (2002). Promoting creative thinking through the use of ICT. *Journal of Computer Assisted Learning, 18*, 367–378.

Wulandari, D.F., Hamidah, I. & Setiawan, A. (2014). Physics of learning strategy to train critical and creative thinking skills. *International Journal of Science and Research (IJSR), 3*(11), 2976–2981.

Wynne, E.A. (1995). Cooperation-competition: An instructional strategy. *Phi Delta Kappan Fastbacks, 387*, 7–27.

Ideas for 21st Century Education – Abdullah et al. (Eds)
© 2017 Taylor & Francis Group, London, ISBN 978-1-138-05343-4

Implementation of the government's law on the management of Islamic religious education in the community

D.F. Sjoraida, A. Asmawi, D. Mariana & R.K. Anwar
Universitas Padjadjaran, Bandung, Indonesia

ABSTRACT: This study made an in-depth systematic analysis of the implementation of the Government's law on the management of Islamic religious education in the community. The study employed a multiple case-study method. The cases were three Muslim communities, all located in the City of Bandung. There were three categories of community: traditional, modern, and combination. Interviews, observations, and document analyses were used to collect the data. Path analysis was used to understand the implementation of Islamic religious education from the highest level of the Government's law to the implementation at the society level. The findings showed that all the rules could be performed by the communities and the assisting government, especially through the Ministry of Religious Affairs, from the ministry level, to the religious affair offices in sub-districts.

1 INTRODUCTION

As well as general education in Indonesia, religious education had serious considerations, which aim to provide sufficient space to anyone who would practice it, either individually or in groups. Religious education was given significant attention and space in the National Education System. It is based on the background of Indonesia, which retains some religions, but is not a religious state. The existence of this diversity has long been perceived as a strong capital for religious believers to implement it in the form of religious educational activities, in accordance with the guidance of their respective religions.

Legally, Islamic Religious Education, which is part of Religious Education enshrined in the regulations and legislation, such as the Law on National Education System No. 20 (Year 2003) and Government Regulation No. 55 (Year 2007) on Religious Education. In the latter Law, Chapter VI, Article 26 (4) states that *majlis taklims* are part of non-formal education units held by the public and regulated by the Government in other regulations.

Islamic education in the community, such as organizing *majlis taklims* or study clubs for children, Qur'anic Educational Garden, Diniyah even Pesantren, has been regulated by the Government, as stipulated in the regulations and legislation, so that all activities are under the umbrella of Islamic Religious Education law and its legality is guaranteed by the state up to the implementation guidelines and technical instructions. But all regulation is still much unknown by the Islamic community broadly, which affects the implementation in the field.

Socialization of the Government may be not maximized and all stakeholders in the community have not contributed to the maximum they can, so regulation of Islamic Religious Education in societies is not fully known. The organizational form of public education, curriculum, facilities and infrastructure, means of evaluation, objectives of education activities and others, are directly related to the format of Islamic education in the community. In the end, it affects the forms of activity, which are irregular and unstructured, objectives that are not really clear, the results are not optimal, and it is difficult to evaluate the activity.

The above reality shows the importance of knowing and implementing government regulations on Islamic education in the community, so that all of the activities are clear, the program is focused and scalable, and maximum optimal results are achieved.

The purpose of this study was to answer the question, 'How is the implementation of government regulations on Islamic Education in *majlis taklims*?'

The above question is to fill the gap between theories and studies that exist on Islamic Religious Education, which are mostly around the field of formal education institutions. Thus, this study focuses on the implementation of government policies on Islamic education in the area of non-formal education.

This research involves sociology of education, which is 'a sociological analysis of the processes involved in educational institutions' (Stalcup, 1968). This research is limited to the aspects and matters relating to the patterns developed by *majlis taklims* in organizational action, in developing the spiritual development of society. The researched aspect is the implementation of regulations on Islamic education in the community. This involves 'various functions and structural problems of analytic functions in conjunction with the entire social system and society.' (Morrish, 1978).

2 LITERATURE REVIEW

2.1 *Implementation of public policy*

Good public policy is a policy that can be implemented in order to achieve objectives. The definition of execution (implementation) of a public policy, according to Pressman and Wildavsky (1973), is 'implementation as to carry out, accomplish, fulfill, produce, complete.' The implementation of the policy is an action of processes that bring, finish, fulfill, produce and complete the objectives of the policy. Pressman and Wildavsky (1973), furthermore, explain that the implementation of policy is basically the ability to 'build relationships' in a chain of causation so that the policy has an impact. Implementation of a policy will not be effective if the relationship between all of the agencies that implement policies produces a 'deficit of policy implementation'. In the implementation of policies, goals must be clearly defined and well understood, the resources must be provided, the chain of command must be able to unite and control the sources of such power, and the system must be able to communicate effectively and control of individuals and organizations involved in the implementation of tasks.

Howlett and Ramesh (1995) suggest that policy implementation is 'the process whereby a program of policies are carried out. The words denote the translation of plans into practice.' Jones (1984) formulates them as 'a process of getting additional resources so as to figure out what is to be done … is that set of activities directed toward putting a program into effect.'

Meanwhile, Gordon, Lewis & Young (1977) says that implementation of a policy should be done with regard to a variety of activities aimed at the realization of the program. In this case, the administrator set up a way to organize, interpret and implement policies that were selected. Organizing means managing resource units and methods for implementing the program. Interpretation with respect to translate the language or terminology program into plans and guidelines that are acceptable and feasible, while applying means using instruments, working or providing routine

services, and making the payments. In other words, the implementation stage is realizing the goals of the program. In this case, to note is the preparation of implementation, which is to think and calculate carefully the various possibilities of success and failure, including barriers and opportunities that exist and the ability of the organization entrusted with the task to implement the program. Thus, according to Pressman and Wildavsky (1973), Howlett and Ramesh (1995) and Gordon et al., (1977), the implementation of public policy has the following prerequisites:

1. The implementation of public policies contains some purposes or goals;
2. In a public policy there is an idea that underlies the emergence of a policy;
3. A public policy contains a series of activities aimed at the realization of the program.
4. In realizing the program, the task of government administrator (bureaucracy) is to interpret, organize and apply the policy.
5. Implementing a policy requires a wide range of instruments and resources.

The successful achievement of policy objectives depends on the actors participating in the implementation of the policy. Therefore, Howlett and Ramesh (1995), Jones (1984), and Gordon et al., (1977) explain that the implementation of the policy can be seen as a process of strategic interaction that exists of cooperation between a large number of actors involved in the implementation of policies, to achieve policy mandates. The participation of a group of such individuals in the implementation of the policy may have some forms of collaboration with government agencies.

2.2 *Outside school education of Islamic religion*

Majlis taklims can be understood as institutions that organize religious propaganda characterized by non-formal education, irregular time to study, and participants called pilgrims; it is specifically aimed at efforts to popularize Islam (Siregar & Shofiuddin, 2003). In simple terms it can be said that *majlis taklims* are containers or venues for teaching and learning or teaching Islamic religious knowledge, or places to carry out teaching or study of Islam.

The existence *majlis taklims* in the middle of the community aims to increase the knowledge and religious beliefs that would encourage religious teaching experience, as a means of gathering members of the public, and raising awareness and household welfare, and an environment of congregation (Alawiyah, 1997). Still in the same context, *majlis taklims* are also useful to foster and develop religious life, to form a society that fear Allah SWT, into a spiritual event, and maintain *silaturahim* (brotherhood) with fellow Muslims, and convey ideas that are beneficial for the development of the people and nation (Djaelani, 2007). Meanwhile, the purposes of holding *majlis taklims*, as defined by Chirzin (1997), are:

1. Laying the foundation of faith in the provisions and all supernatural things;
2. The spirit and the worship that permeates all activities of human life and the universe;
3. As an inspiration, motivation and stimulation to all potential members who can be developed and activated maximally and optimally with private construction activity and productive work for the common good;
4. All the activity or activities to become a solid unity and harmony.

Judging from the structure of the organization that has one, a *majlis taklim* can be categorized as an organization of outside school education, that is, a non-formal educational institution, because it is not supported by a set of rules of academic curricula or duration of study, there is no increase in class level, there are no book report cards and no diplomas, as required in formal education institutions, namely schools (Huda, 1980).

Although categorized as non-formal educational institutions of Islam, *majlis taklims* have their own position in society (Ambari & Abdullah, 1996). This is because *majlis taklims* are platforms for building and developing the religious life to create a society that fears God. In addition, *majlis taklims* are spiritual theme parks, because their maintenance is done casually. Other factors that make people interested, is that *majlis taklims* are non-formal education institutions, as well as a gathering place to make Islam alive and as a medium to deliver ideas that are beneficial to the development of the people and nation.

The scriptures written in Indonesian are usually used as a handle (Rasyid, 2004) and some other translated books. Meanwhile, according to guidelines of *majlis taklims* produced by the Islamic Propagation Coordination (KODI) (Huda, 1980), the materials presented in *majlis taklims* are:

1. Religious Knowledge Group – this includes teachings on monotheism, tafsir, fiqh, hadith, character, Islamic history, and Arabic.
2. General Knowledge Group – this is directly related to the lives of people associated with religion. That is, in presenting these descriptions the teachings based on the arguments of both verses of the Koran or hadith, or examples from the life of Prophet Muhammad.

These non-formal education institutions have long grown and developed in the middle of the Muslim community as propaganda plus education institutions and have become institutions most wanted by the Muslim community, in developing religious insight (Siregar & Shofiuddin, 2003).

As a part of the National Education System, *majlis taklims* perform their functions at the level of non-formal, more flexible, and open education, and are one solution that should provide opportunities for the public to add and complement the knowledge that they acquire in formal education, especially in the religious aspect. *Majlis taklims* have gained support from the community, which can be seen from the increasing numbers of *majlis taklims* from year to year.

3 RESEARCH METHODS

3.1 *Research design*

The method used in this study is a qualitative research method, with a descriptive approach. Qualitative research is a method of research that aims to gain an understanding of reality through the process of inductive thinking. Through qualitative research researchers can identify the subject and feel what is experienced in everyday life. The descriptive approach is a method in researching the status of human groups, an object, a condition, a system of thought, or an event in the present (Basrowi, 2005).

This method is characterized by obtaining data, considering it, describing it, and then analyzing it. Basically, qualitative research is seen as a procedure that can produce descriptive data, in the form of words written or spoken, of people and observed behavior (Kahmad, 2000). In this method, the researchers not only collect data, but also interpret the data and look for things behind the data along with their supporting theories (Punch, 1998).

This research is a case study with a phenomenological approach. Taylor and Bogdan (1998) affirm that the task of the phenomenological users of qualitative methodology is to capture the process of interpretation. Phenomenology assumes that human beings are creatures or individuals who have an awareness of something, that is, that something is meaningful. Phenomenology aims at describing human or individual understanding of something.

3.2 *Subject characteristics*

This research was located in South Bandung, of Bandung Regency, with the reason that in this location there are many *majlis taklims* that are very interesting to study. The population is quite dense. Trading conditions in the area are quite good, while the territory and population is quite dynamic, within the informal sector of life. Thus, the area is quite busy, so if *majlis taklims* can develop, this can be considered as an achievement here.

3.3 *Data collection process*

Data collection, to further facilitate the course of the research process, required accurate techniques in this study, including: observation, interviews, and documentation.

a. Observation

Observation is defined as a systematic inspection and recording of phenomena being investigated (Hadi, 1997). Observations made here include a thorough look at the reality of *majlis taklim* activity in South Bandung.

Observation data collection methods used to collect research data about social phenomena, by observing, searching for answers, and looking for evidence of the phenomena, take place through the five senses. Observation in the collection of data necessary for the purposes of this study is seen or taken directly from *majlis taklim* members in South Bandung.

b. Interview

Interviewing is a technique of data collection done through conversation by directly interviewing representative people, in accordance with their knowledge and experience. Interviews conducted in this study were carried out with members and administrators of *majlis taklims*, the authorities, and public figures in South Bandung.

c. Documentary Study

Documentary study is a technique of collecting data related to the problems being studied; in this study it was done by reviewing the documents related to the management of *majlis taklims*, especially the aspects of the implementation of the Government's regulation.

d. Data analysis

The data analysis used here is a descriptive data analysis, in the form of data reduction, data display and conclusions or data verification. Based on the principle of qualitative research, data analysis was carried out in the field and in conjunction with the data collection process. Data reduction and data presentation are two components of data analysis (Bamberger, 2000). Inferences are made if data collection is considered complete and adequate. If there is a conclusion that is considered inadequate, necessary verification activity would be done with more focused goals. Those three activities are interacted with each other to obtain a strong conclusion and were done in a cycle. So if there was a shortage, the researchers could return to the field to cover the shortage. The data analysis

process is called *Interactive Analysis Model*, which was developed by Miles and Huberman (Punch, 1998).

4 RESULTS AND DISCUSSION

The curriculum can be interpreted as a plan or design of teaching (*majlis taklim*), created and implemented to achieve the objectives that have been set by the *majlis taklim*. In this case, the curriculum contains material that guide a mu'allim (teacher) in presenting the material. *Majlis taklim* materials are listed in order or stages from one meeting to the next meeting, so as to facilitate the clerics in teaching and make the material easier to understand by the *majlis taklim* pilgrims.

4.1 *Goal of curricula*

In structuring the curriculum to support the successful development of the *majlis taklim*, there are some things to note: the purpose of the curriculum (standards of competence), materials (teaching modules), activities (methods, tools, and location of study), and evaluation.

Especially with regard to the curriculum of *majlis taklim*, it can be seen in Article 23 paragraph (2) of the Regulation, saying: 'The curriculum of *majlis taklim* is open with reference to the understanding of the Qur'an and the Hadith as a base to increase faith and piety to Allah SWT, and noble character.' (Penamas, 2010).

From the findings in the field, it was found that there was a lack of determination to standardize the competence as learning goals of *majlis taklim*. On average, the purpose of education was already integrated (built-in) with the establishment and performance of lectures on those *majlis taklims*. The Chairman of *majlis taklim* Miftahul Huda, for example, said that the goal to continue what already exists or has been will not to do, in this case, by the previous management (interview with Muttaqin, October 2, 2015). This was confirmed by the Chairman of *majlis taklim* Al-USROH, who stated that the study has already become a family tradition (interview with Nana Heryana, October 1, 2015). Here it appears that the purpose of education was not clearly documented, although in terms of religious intuition, was probably already imprinted on each self of the members of the *majlis taklims*.

As with the *majlis taklim* of Persistri, Southern Bandung, which clearly states the purpose of the activity, the Section of Development and Human Resources Development of Persistri, for example, organizes such activities as follows:

1. *Halaqoh* or coaching members in May 2015, with the goal, 'Realization of the quality and

quantity of members who are committed to one goal in the Organization', which was attended by 100 members.
2. The practice of reading and writing the Koran every week, with the goal of 'Increasing literacy Qur'an', which was attended by 36 participants.
3. Training become MC/Moderator in June 2015, with the goal of, 'Availability of a professional Master of Ceremony', which was attended by 16 participants.

Or the Section of Family Consultation, in conducting Sakinah Family Development on January 3, 2015, which was aimed at 'Increasing knowledge of harmonious family', which was attended by 50 participants.

4.2 Module

Teaching materials or teaching modules given in the *majlis taklim*s in Southern Bandung included six objectives, namely: reading of the Koran, the sciences of the Qur'an, Hadith, faith, sharia, character and history of Islam. Mostly, these materials are from the preachers or clerics who gave the material. But it was also written down minimally, in organizing the themes. This is done, for example, in the *majlis taklim of* Persistri, Southern Bandung.

These materials were given in the form of a fixed curriculum so that the materials were continuous in a certain direction. Curriculum instruction can be tailored to the needs of the pilgrims' study material.

Inferred from the three *majlis taklims* sampled in the sub-district of South Bandung, the following are among the study materials of *majlis taklim*:

I. On belief:
  1. The meaning of faith and its influence on life.
  2. Characteristics of Tawheed and Aqeedah of Islam.
  3. The hazard of polytheism.
  4. Deviant streams in Islam.
II. On study of worship:
  1. Definition of fiqh of worship and its aspects.
  2. Taharah and its aspects.
  3. Salat and its aspects.
  4. Fasting and its aspects.
  5. Zakat and its aspects.
  6. Haji and its aspects.
III. On fiqh munakahat:
  1. Khitbah dan aspeknya.
  2. The goals of marriage.
  3. Hadhanah and its urgency.

4. Marriage of different religions.
5. Hidden marriage and its aspects.
6. Divorces/divorce and its effects.
7. Tips on building harmonious family.
IV. On Islamic economics:
  1. Characteristics of Islamic economics.
  2. The behavior of Islamic economics.
  3. Purchase and its aspects.
  4. Endowments and its aspects.
  5. Grants and prizes.
  6. Banking.
  7. Riba (usury) and its implications on the economy.
V. On ethics:
  1. Moral and scope.
  2. Benefits of morality in life.
  3. Tips on building human morality.
VI. On Social Problems
VII. Etc.

4.3 Activities of learning

4.3.1 Method of learning

The teaching system applied in *majlis taklim* consists of a variety of methods. In general, the various methods used in *majlis taklim* are:

1. Lecturing – It is illumination with a narrative spoken by a cleric. This is the method most often used by *majlis taklims* in South Bandung.
2. Questions and answers – This method makes a more active participant. Activity is stimulated by the questions presented, either by the cleric or by the participants (an interview with Ani Kurniasari, October 1, 2015).
3. Exercise – This method increases training for skills and dexterity.
4. Discussion – This method is used if there is a question that should be discussed.
5. Demonstration and practice.
6. Study tour (field trips).
7. Etc.

4.3.2 Location of learning

Public education of Islam was mostly done in a mosque, although it was possible for the study to be conducted at the home of one of the congregation members, in the hall, or elsewhere. The mosque is a good place to carry out study, because it is a place of worship for Muslims and doing hospitality. The use of the mosque as a place of execution of *majlis taklim* activities restores its function as a center of Islamic propaganda.

In the regulation about the locations of *majlis taklims*, Article 23, Paragraph (3) of the Regulation reads 'The *majlis* taklim held in mosques, prayer rooms, or anywhere else that qualifies.' (Penamas, 2010).

4.4 Evaluation of learning

Evaluation in *majlis taklims* seems to have never existed. This is because of its characteristics as a non-formal system of education. However, when observed, evaluation was done on teaching activities in *majlis taklim*.

For study participants, for example, evaluation was seen from the increasing knowledge of ladies on reading and writing of the Qur'an. Likewise, when there was a quiz competition among the participants, it appears that there was an increased knowledge of the participants study (an interview with Yoyoh, October 12, 2015).

For the cleric, evaluation from study participants and managers often occurred when dispersed after recitation. They provided comments, such as the cleric was nice, a lot of jokes; or that cleric is not good, too many 'dirty words', and so on. The consequence was to ascertain whether the cleric should be invited back or not (an interview with Pupun, October 10, 2015).

Likewise, the board, the chairperson and his/her staffs could be replaced by ways of evaluation in everyday life, which was not visible. This evaluation was intuitive and not yet documented. Therefore, the related institutions, in this case, the Ministry of Religious Affairs, could provide instruments of evaluation that would be given to the *majlis taklims* for the development of their progress.

5 CONCLUSIONS

The existence *majlis taklims*, which became a place of education and instruction of Islamic teaching was one of the implementations of Government Regulation of 2007, which is still in the finishing touches. *Majlis taklim* is one of the religious educational institutions that serves to form Indonesian humans having faith and fear of God Almighty, as well as noble as aspired to in the Act. The Government, in this case the Ministry of Religion, from top to bottom, that is, up to the level of the Office of Religious Affairs at sub-districts, has been guiding the implementation of the regulation.

In terms of management of *majlis taklim*, communities have implemented the regulation at all costs as much as possible. Basically, these efforts are self-supporting, but the involvement of the Ministry of Religion, by sending field educators was highly appreciated in the *majlis taklims*.

In terms of the curriculum of the *majlis taklims*, communities have implemented the regulation by holding their curriculum for drafting and upon request by the pilgrims. The involvement of the Ministry of Religion that gives examples of curriculum for teaching materials is very helpful for organizing activities of the *majlis taklims*.

REFERENCES

Alawiyah, T. (1997). *Strategi dakwah di lingkungan majlis taklim.* Bandung: Mizan. Cetakan Pertama.

Ambari, H.M., & Abdullah, T. (1996). *Ensiklopedi Islam.* Jakarta: PT. Ikhtiar Baru Van Hoeve.

Bamberger, M. (Ed.) (2000). *Integrating quantitative and qualitative research in development projects. Directions in Development.* Washington DC: The World Bank.

Basrowi, M.S. (2005). *Pengantar sosiologi.* Bogor: Ghalia Indonesia.

Chirzin, M.H. (1997). *Pesantren dan pembaharuan.* Jakarta: LP3ES. Cetakan Ketiga.

Djaelani, B.M. (2007). *Ensiklopedi Islam.* Yogyakarta: Panji Pustaka Yogyakarta.

Gordon, I., Lewis, J. & Young, K. (1977). Perspectives on policy analysis. In *Public Administration Bulletin, 25*, 26–35.

Hadi, S. (1997). *Research.* Yogyakarta: Andi Offset.

Howlett, M. & Ramesh, M. (1995). *Studying public policy: Policy cycles and policy subsystem.* New York: Oxford University Press.

Huda, N. (1980). *Pedoman majlis majlis taklim, proyek penerangan, bimbingan dakwah khutbah agama Islam pusat.* Jakarta: LP3ES.

Jones, C.O. (1984). *An introduction to the study of public policy.* Monterey, CA: Brooks/Cole Publishing Company.

Kahmad, D. (2000). *Sosiologi Agama.* Bandung: Rosdakarya.

Morrish, I. (1978). *The sociology of education: An introduction.* London: George Allen & Unwin.

Penamas (2010). *Himpunan peraturan bidang pendidikan agama Islam pada masyarakat & pemberdayaan masjid.* Bandung: Kementerian Agama Kantor Wilayah Jawa Barat.

Pressman, J. & Wildavsky, A. (1973). *Implementation.* Berkeley, CA: University of California Press.

Punch, K.F. (1998). *Introduction to social research: Qualitative and quantitative approaches.* London: Sage Publications.

Rasyid, S. (2004). Fiqh Islam, cet. Ke-37. *Bandung: Sinar Baru Algesindo.*

Siregar, H.I. & Shofiuddin, M. (2003). *Pendidikan agama luar sekolah (studi tentang majlis taklim).* Jakarta: Puslitbang Pendidikan Agama dan Keagamaan Badan Litbang Agama dan Diklat Keagamaan Departemen Agama RI.

Stalcup, R.J. (1968). *Sociology and education.* Columbus, OH: C.E. Merrill.

Taylor, S.J. & Bogdan, R. (1998). *Introduction to qualitative research methods.* Hoboken, NJ: John Wiley and Sons Inc.

Ideas for 21st Century Education – Abdullah et al. (Eds)
© 2017 Taylor & Francis Group, London, ISBN 978-1-138-05343-4

Education of cultural and national characteristics based on local wisdom through social studies at SMP Negeri 1 Singaraja school, Bali

I.W. Kertih

Universitas Pendidikan Ganesha, Bali, Indonesia

ABSTRACT: This study aimed to explain the effect of sociocultural community context, whether in terms of local, national or global community, in relation to the implementation of the cultural educational program and the development of national characteristics at SMP Negeri 1 Singaraja school. In particular, this study aimed to explain the practices of instilling national characteristic values through the school educational program and in the specific subject of social studies. This study used an educational ethnographic research design. The data analysis in this study was done by adopting a qualitative approach. The results showed: (1) how the sociocultural contexts of the Balinese, at the level of local, national and global community, provide the foundations for implementing an educational program that is oriented to the development of national characteristic values; (2) that social studies provides important strategies for cultural education and the development of national characteristics; (3) how the creation of an educational climate in the school and social education that is based on the wisdoms and values of the local Balinese community can produce a young generation of Balinese with Indonesian national character.

1 INTRODUCTION

The phenomena of modernization and globalization have, nowadays, spread all around the world. Indonesian community and culture are not free of the effects of global liberalization. The influx of foreign values caused by globalization and the strengthening of local culture cause various clashes of values and interests in the community (Tilaar, 1999). This has an effect on the emergence of various problems, such as: (1) the disruption of the values of the national philosophy and ideology of Pancasila; (2) the shift of ethical values in the life of the community, and the nation; (3) a decrease in the awareness of the national philosophy and ideology; (4) a disintegrative threat to the nation; (5) the weakening of national autonomy (Kemendiknas, 2010).

In these conditions, awareness and commitment emerges from various groups in the community – from educational scholars, bureaucrats and practitioners, to the general public – about the importance of cultural and national character education, and how its implementation should be prioritized and consolidated in all types, paths and levels of education in Indonesia (Kemendiknas, 2010, p. 2). Cultural and national character education is empowering, humanizing and civilizing and, as such, is surely an important key to fully developing students' potential. In these processes, education helps to develop knowledge, values, systems of belief, norms, traditions or habits, rules, and skills that are coherent and useful for the individual, schools, families and the community, as well as for the wider nation.

As an integral part of the educational curriculum in school, social studies is, of course, also responsible for conducting character education in an effort to advance a dignified Indonesian national civilization and for the integrity of the unitary state of the Republic of Indonesia. To achieve this objective, social studies education has to become an integrative study that can empower the entire potential of students in giving them the capability to overcome problems in life in various dimensions of space, time, activity, and values in the environment (Hasan, 2010). By its nature as an integrative study of social science, humanity, religion, and culture, social studies education has to be able to bring the students to realize truths, and also achieve happiness in life, to produce good citizens who have strong knowledge, and are able to lead a meaningful life (Somantri, 2001).

The problem is that social studies teaching in schools has, thus far, been oriented toward the mastery of certain subjects which has not enabled teachers to integrate the mission of character education and national character education in an optimal fashion. There are two constraints faced by the teacher, especially in social studies. First, the school curriculum, especially for social studies, in developing the concepts of character and nation building is dominated by the concept of nationalism (Widja, 1993). Second, the social studies curriculum still has the concept of social sciences education, which has smaller scope than social studies, making it difficult for the teacher to integrate the mission of character education and the instilling of national cultural values. This is because the scientific values that are the basis for the social studies curriculum are oriented toward the mastery of scientific materials in the area of study and tend to be singular, affording less respect for the real-life pluralistic and multi-cultural community at the heart of the students' environment.

To be able to integrate character education and national cultural values into social studies teaching, there is a need to reconstruct the policies and implementation practice of social studies education at school. In relation to this effort, one needs to start from the construction of contextual input (policies and school social environment), educational process, and students' achievement in social studies education, which is the main reason for conducting this study.

2 RESEARCH METHODS

This study was primarily conducted in a school and classroom setting, supported by the community background surrounding the school. The school chosen was SMP Negeri 1 Singaraja, which is located at Gajah Mada 109, Singaraja, Banjar Bali village in the Buleleng district, in the Buleleng regency of Bali. The subjects of the study were: the Buleleng government, committee leaders, principles and co-principles, the teachers that teach social studies, and students of the academic year 2012/2013.

Based on the problems identified and the aims of this study, it used educational ethnography design. The use of this design was based on the rationale that the focus of this study was to describe and give a detailed account of cultural phenomena that can be reconstructed, according to the perspective of the participants in relation to a unity of systems, in the form of programs, activities and events (Nasution, 1988). On the basis of the collection of these types of data, this study employed some major methods/techniques of data collection, including (1) observational method, (2) interviews, (3) document study, and (4) discussion method. The process of data analysis in this study was done simultaneously and cyclically using the process often called interactive cyclical data analysis. The data analysis technique was, therefore, qualitative. All of the data were analyzed through a range of activities comprising: (1) data reduction, (2) data display, (3) data interpretation, (4) data verification, (5) conclusion drawing (Miles & Huberman, 1992).

3 RESULTS AND DISCUSSION

3.1 Cultural context and values

In relation to the local culture of the community, the wisdom and values that need to be developed in the context of changes in the globalization era, and that are believed to have the capacity for strengthening the national character values of the students, were revealed as: (1) the importance of maintaining existing values, attitudes, and religious practices; (2) familiar values, such as togetherness, that have at their core the ideal value of *pawongan* (humanity), creating harmonious social relations among individuals, family members, and citizens; (3) the need to apply the concepts of *menyama braya* and *Tri Kaya Parisudha* (everybody is equal) as the foundation for regulating interactions in the school community and developing tolerance, responsibility, and harmony in the life of a school community that is increasingly more heterogeneous and is oriented toward multicultural life; (4) the unity and harmony between human beings and nature; (5) the ethical and aesthetic values of the living arts of the Balinese.

Religion is seen not only as the controller of the behavior and attitudes of humans, but also acts as a motivation, self-discipline, and guide to behavior in achieving life goals in the community with the principles of *suka duka, paras paros sarpanaya, segilik seguluk selunglung sebayantaka,* and *saling asah saling asuh* as the foundations for creating a harmonious relationship among citizens. The values that become the foundation of these principles is the study of *Tat Tvam Asi*. In this case, Balinese, in general, tends to appreciate a balance in material and spiritual well-being, or *sekala* and *niskala*. In doing an activity, work has to contain values such as: *metaksu, nyalanang jengah*, and the spirit of *puputan*. All of the values above are, in principle, the elaboration of the values of *Tri Hita Karana*, which forms the central values (core values) of the foundation for the Balinese in their relationship with their creator, their fellow human beings, and the natural environment.

For the Balinese, these ideal values should be passed from one generation to the next to preserve Balinese community and culture. As has been explained by Giroux (1981), there is a connection between social culture and education. Moreover, Pai (1990) states that education is influenced by society's prevailing perceptions and values. Education as a part of human life cannot be separated from the sociocultural context and processes that surround it, including the values involved. This means that education used to build the behavior, spread knowledge, emphasize thinking process, promote values, cognitive skills, social skill, and also the values of truth will be determined by how the community sees the world and their values, even the conceptual ones.

It is these values that need to be taught to students as the foundation for strengthening the development of the 18 national character values, namely: 'religious, honest, tolerant, disciplined, hardworking, creative, democratic, curious, having national spirit, loving homeland, appreciating achievement, communicative, peace-loving, diligent, taking care of the environment, and showing great responsibility'. Here we can see the uniqueness available to SMP Negeri 1 Singaraja in implementing its educational program. In terms of the implementation of cultural and national character education, such uniqueness will make it more likely that an educational climate with a multicultural nuance can be created.

3.2 Cultural and character education through social studies teaching

In terms of SMP Negeri 1 Singaraja, this study found that social studies as a subject taught at the school plays a strategic role in the implementation of cultural and national character education. From the perspective of the school's principles, social studies have the same position as other subjects in the implementation of cultural and national character education at the school.

In social studies, it appears that the teachers have made great efforts to integrate national character values into every basic competence (KD) in the teaching programs that they develop. From the point of view of the teachers in SMP Negeri 1 Singaraja, social studies is a study that has as its main purpose helping students to become good citizens. Even though other studies also have a similar purpose, the development of the citizen has become the main concern of social studies.

This opinion is similar to that of Barr et al. (1978, pp. 22–111), who state that social studies has developed in three main traditions, which are: 'social studies taught as Citizenship Transmission, social studies taught as Social Science, and social studies taught as Reflective Inquiry'. The tradition of 'social studies taught as Citizenship Transmission' is concerned with a social learning modus which has the purpose of developing a good citizen based on the values and norms that have been accepted in a country. The tradition of 'social studies taught as Social Science' is the social study modus that also

develops the character education of a good citizen, shown by the ability to see and solve social problems and develop personality using the visions and methods of the social scientists. The tradition of 'social studies taught as Reflective Inquiry' is the social study modus that is concerned with the development of the good character of the citizen with the main characteristic that they can make decisions.

On this basis, social study is essentially education in the national characteristics, so social study cannot be separated in terms of the principles of the program and the educational process. Social study cannot be separated by the relationship between education and the social activities of human life. When these things are related to one another, it gives rise to the need for social study that has a clear philosophical foundation that explains the relationship between an educational program and the citizen's activities in everyday life. Given a clear philosophical foundation, the vision, mission, purpose, program, content, learning, and values of social studies can later be directed to achieve the goal of social studies as the place for character education (Hasan, 2010; Winataputra, 2010).

Social studies do not merely focus on concepts, principles, and theories in relation to social facts yet they should be implemented by the students in their real life. As the place for education in the national characteristics, social studies must develop the life skills of the citizen in completing subjects, and be meaningful and authentic, whether it is in relation to personal skills, intellectual, academic or even vocational skills. These skills develop only if social studies is able to reach and completely integrate all the individual competencies of the students, whether these involve the social knowledge, social disposition, social skill, social confidence, social commitment, or social competence that are needed in order to build the civic culture that contains the values and morality of Indonesia (Sukadi, 2006).

In order to reach this goal, students are forced to not only believe in themselves as a result of what happens in the classroom, but identify and be capable of searching out the wider range of learning sources required. The variation in the things that are important to people in their life and kindness means that they can be regarded as learning resources for schools, including in the social studies (Schuncke, 1988). The use of citizens as learning sources can be a fun educational tool and is able to enrich the variety of media in the application of social studies in Elementary school. Community resources are convenient and valuable assets for the enrichment of elementary social studies. The variety and convenience of such resources will vary from community to community.

4 CONCLUSIONS

This study produced a number of findings as follows:

1. *Tri Hita Karana, Tri Kaya Parisudha, Tat Tvam Asi, Asah Asih Asuh, Segilik Seguluk, Menyama Braya, Nyalanang Jengah,* and *Kerja Metaksu* are local wisdoms of the Balinese community that need to be preserved and transformed into a foundation to strengthen national character values.
2. The creation of an educational climate in the school and social studies education based on local wisdoms and values can be used as media for civilizing national character values to develop a young, modern Balinese generation with Indonesian national character.
3. Social studies education as a means for cultural and national character education should be interpreted as a process of student empowerment that enables them to develop knowledge and insights, values and attitudes, as well as social skills, through participation in the teaching of the sociocultural life of the community.

The findings of this study have the theoretical implication that social studies education in school needs to be developed by using an approach of social reconstruction in order that students have the ability to think globally, act locally, and respect nationalism. Thus, social studies education as a means to instill the national character needs to be treated and developed in the context of local culture, without ignoring the commitment to national life, and the development of global thinking ability.

REFERENCES

Barr, R.D., Barth, J.L. & Shermis, S.S. (1978). *The nature of the social studies*. Palm Springs, CA: ETS Publications.

Giroux, H.A. (1981). *Ideology, culture, and the process of schooling*. Philadelphia, PA: Temple University Press.

Hasan, S.H. (2010). *Inovasi pembelajaran IPS dalam membangun karakter peserta didik*. Bandung, Indonesia: FPIPS-UPI.

Kemendiknas. (2010). *Pengembangan pendidikan budaya dan karakter bangsa; Bahan pelatihan penguatan metodologi pembelajaran berdasarkan nilai-nilai budaya untuk membentuk daya saing dan karakter bangsa*. Jakarta, Indonesia: Badan Penelitian dan Pengembangan Pusat Kurikulum Kemendiknas.

Miles, M.B. & Huberman, A.M. (1992). *Analisis data kualitatif: buku sumber tentang metode-metode baru* (Translated by Tjetjep Rohendi Rohidi). Jakarta, Indonesia: UI Press.

Nasution, S. (1988). *Metode penelitian naturalistik kualitatif*. Bandung, Indonesia: Tarsito.

Pai, Y. (1990). *Cultural foundations of education*. New York, NY: Macmillan.

Schuncke, G.M. (1988). *Elementary social studies: Knowing, doing, caring*. New York, NY: MacMillan.

Somantri, M.N. (2001). *Menggagas pembaharuan pendidikan IPS*. Bandung, Indonesia: PT Remaja Rosdakarya.

Sukadi, S. (2006). *Pendidikan IPS sebagai rekonstruksi pengalaman budaya berbasis idiologi Tri Hita Karana pada SMU Negeri 1 Ubud Gianyar Bali*. Bandung, Indonesia: Sekolah Pascasarjana UPI.

Tilaar, H.A.R. (1999). *Beberapa agenda reformasi pendidikan nasional dalam perspektif abad 21*. Magelang, Indonesia: Tera Indonesia.

Widja, I.G. (1993). Pelestarian budaya: Makna dan implikasinya dalam proses regenerasi bangsa. In T.R. Sudhartha, et al. (Eds.), *Kebudayaan dan kepribadian bangsa* (pp. 45–63). Denpasar, Indonesia: Upada Sastra.

Winataputra, U.S. (2010). Peran pendidikan Ilmu Perngetahuan Sosial (PIPS) dalam konteks pembangunan karakter bangsa; Kebijakan, konsep, dan implementasi.

Ideas for 21st Century Education – Abdullah et al. (Eds)
© 2017 Taylor & Francis Group, London, ISBN 978-1-138-05343-4

The impact of a STEM project-based learning approach in vocational high school on students' mathematical communication ability

A. Ismayani & Y.S. Kusumah
Universitas Pendidikan Indonesia, Bandung, Indonesia

ABSTRACT: This study aimed to analyze the influence of a STEM project-based learning approach on students' mathematical communication abilities. The research involved 36 students from a vocational high school at Cianjur, West Java, Indonesia. Based on prior mathematical ability, the students were divided into high, medium, and low groups. A set of mathematical communication ability tests was used as an instrument. The results showed that STEM project-based learning provided a significant enhancement of students' mathematical communication ability. The improvement in mathematical communication ability of students who had high prior mathematical ability was significantly better than that of students who were in the medium and low groups, where the impact was not significant.

1 INTRODUCTION

Mathematics has become an integral part of our life. Its application in other scientific fields evolves in accordance with the development of science and mathematics itself. The importance of mathematics makes it an integral part of education. Mathematics levels in schooling range from elementary to middle and on to higher education.

Mathematics in schools is aimed at developing a variety of student mathematical abilities. Mathematical communication skills are among these. It is stated in the content standards of the curriculum for mathematics courses for primary and secondary education, issued by the Ministry of National Education (2006), that the objectives of mathematics courses at school include enabling students to communicate ideas with symbols, tables, diagrams, or other media to clarify situations or problems.

Along with this, the National Council of Teachers of Mathematics (NCTM, 2000) argues that the general objectives of learning mathematics in schools are: (1) learning to communicate (mathematical communication); (2) learning to reason (mathematical reasoning); (3) learning to solve problems (mathematical problem-solving); (4) learning to associate ideas of mathematics (mathematical connection); (5) forming a positive attitude toward mathematics. Through mathematical communication, students can organize their mathematical thinking both orally and in writing within the learning process.

At the same time, the myth of mathematics as a difficult and frightening subject is still common in schools, including *Sekolah Menengah Kejuruan*

(SMK) or vocational high school. In SMK, mathematics is treated as an adaptive group lesson. Although classified as a compulsory subject, students often take less notice of it than subjects seen as more relevant to their respective areas of expertise. The effect is a lack of student mathematical ability. All mathematical communication ability expectations which can be observed from the results of a study on the response of the students' answers in Programme Internationale for Student Assesment (PISA) and Trends in International Mathematics and Science Study (TIMMS) for middle school students in Indonesia (Wardhani & Rumiati, 2011).

The importance of mathematical communication skills is based on the reality that the need to communicate mathematics often arises in the real world, including in the world of work. When there is a need to use mathematics and communicate it to a non-mathematical person in a workplace interaction, mathematical communication ability is definitely required. Vocational school students who are generally faced with the workplace as soon as they graduate need to be prepared for this challenge, despite the fact that even college students are often not prepared for this need (Wood, 2011).

Given this phenomenon, efforts should be made toward improvement. In line with the reform of education in the field of mathematics which requires teachers to be able to create a learning environment that stimulates the intellectual development of students (Fennema et al., 1996), lessons should allow students to explore mathematical ideas and make connections to fields outside mathematics.

One of the things that can be done is to apply techniques that can take students to higher levels of activity and creativity, by giving them freedom to solve problems and communicate mathematical ideas during the learning process. One such possible technique involves application of the project-based learning model.

New York City Department of Education (2009) depicts project-based learning as a series of instructional designs which are devised to help students learn. The learning model is dynamic because in its activities learners are empowered to pursue content knowledge that they will learn for themselves and then asked to demonstrate their understanding in the various models of work method.

Yaron (2003), Alacapinar (2008), and Koparan and Guven (2014) all found indications that the implementation of project-based learning can foster motivation and self-actualization of students at all levels and the results are significant for effective learning, so that student learning outcomes become better than with conventional learning. Rambely et al. (2013) confirmed that project-based learning can stimulate students' interest to think of an investigation and encourage them to conduct it. From various studies, it is expected that this learning model can be widely applied in teaching mathematics at a variety of schooling levels.

Meanwhile, the need to integrate mathematics with other fields of science has, today, become a necessity. The progress of science and information technology and communication has also affected the development of mathematics significantly. Therefore, an approach to project-based learning that accommodates such a need is the integration of STEM (Science, Technology, Engineering and Mathematics) into the learning process, hereinafter referred to as STEM project-based learning.

In the context of learning mathematics, STEM project-based learning has very good potential to provide meaningful learning, training a student's ability to problem-solve through a project integrated with one or more STEM fields and practice communicating mathematical ideas, in addition to providing experience for the student that mathematics brings real benefits to life and the world around them. Daugherty (2013) states that the ultimate goal of learning in STEM project-based learning is a cognitive activity of students in learning, which includes the learning of content that the students are expected to know.

2 LITERATURE REVIEW

2.1 Mathematical communication ability

Communication can, in general, be interpreted as a way to convey a message from the messenger to the message recipient to inform opinions or good behaviors, both verbally and indirectly, through a given medium. In this communication, consideration should be given as to how to ensure that the message is clear and can be understood by others. To enable the ability to communicate, people can do so using a variety of languages, including mathematics.

Kluver (2012) states that communication is at the core of social reality and social reality itself emerged as a result of communicative interaction. In the context of mathematics learning in the classroom, communication has been recognized as a process by which ideas become objects of reflection and discussion, and as the development of mathematical understanding in an individual who in the process of public expression obtains validation and justification (Sáenz-Ludlow, 2006). Barton, Poisard, & Domite (2006), in *The Language of Mathematics*, examines the relationship between language and mathematics. He concludes that mathematics is created by communication; specifically, that mathematics is formed in communicative actions, and learning and doing mathematics involves talking about mathematics.

The Indonesia mathematics curriculum standards – *Kurikulum Tingkat Satuan Pendidikan* (KTSP) or School-Based Curriculum (SBC) – explained that mathematics as a means of communication can: (1) disclose and explain thinking about mathematical ideas and their relationship; (2) establish mathematical definitions and make generalizations obtained through investigation; (3) express mathematical ideas orally and in writing; (4) present mathematical content that has been read or written with its understanding; (5) explain and ask questions related to mathematics that has been read or heard; (6) appreciate the economic value, power, and beauty of mathematical notation, as well as its role in developing mathematical ideas (Departemen Pendidikan Nasional, 2006).

The NCTM (2000) argue that instructional programs for the learning of mathematics, ranging from preschool to 12th-grade level, should enable students to: (1) organize and consolidate their mathematical thinking through communication; (2) communicate (deliver) their mathematical thinking clearly and directly to friends, teachers, or others; (3) analyze and evaluate the mathematical thinking and strategies created by others; (4) use mathematical language to express mathematical ideas precisely.

Correspondingly, Sumarmo and Hendriana (2014) describe a variety of activities under mathematical communication, among which are: (1) illustrating or representing real objects, pictures and diagrams in the form of an idea or mathematical symbols; (2) explaining mathematical ideas, situations, and relations orally or in writing by using real objects, pictures, graphs and algebraic expressions; (3) representing a daily occurrence in mathematical language or symbols or constructing a mathematical model of an event; (4) listening, discussing, and writing about mathematics; (5) reading and understanding a mathematical representation; (6) developing a conjecture, building an argument, or formulating definitions and generalizations; (7) revisiting a mathematical description or paragraph in their own language. These abilities can be classified into low- or high-level mathematical thinking skills depending on the complexity of the communication involved.

2.2 STEM project-based learning

Integrated STEM education programs combine two or more disciplines contained in STEM— science, technology, engineering, or mathematics— and give equal attention to their standards and objectives (Laboy-Rush, 2010). At the center of the various activities in such programs is the involvement of students in defining and formulating a solution to an authentic real-world problem.

Ritz and Fan (2014) reveal that the application of STEM education has taken place in several countries, and each has a variety of forms in terms of implementation. In Indonesia, STEM integration, as a learning approach, is not yet popular. Nevertheless, the concept of integration between the fields of STEM has begun to be voiced in the Indonesian educational curriculum, including in 'Kurikulum 2013'. Although not explicitly coined in terms of 'STEM', the concept of 'thematic integration' that appears in Kurikulum 2013 indicates the need for the integration of various disciplines within the learning of particular fields of study, and this is in line with the concept of integration in STEM.

Table 1 outlines the definition of STEM literacy according to the National Governor's Association Center for Best Practices (Asmuniv, 2015).

In the project-based learning that is designed in this study, the integration of STEM includes the three areas of mathematics, technology, and engineering. The technology considered in this study is the use of various Information and Communication Technology (ICT) devices, specifically, a computer and internet media. From an engineering perspective, the research is associated with a practical subject, which is design and web programming, and in terms of mathematics it is associated with the topic of statistics.

The learning steps carried out in STEM project-based learning follow the terminology of project-based learning in general, namely: (1) defining fundamental questions, (2) planning the project, (3) scheduling, (4) monitoring, (5) testing results, and (6) evaluating experience.

Table 1. Definition of STEM literacy.

Science
Scientific literacy: The ability to use scientific knowledge and processes to understand the world, nature and the ability to participate making decisions to influence it.
Technology
Technology literacy: Knowledge of how to use new technology, understanding of how new technology is developed, and the ability to analyze how new technologies affect the individual community, nation, and world.
Engineering
Design literacy: Understanding of how technology can be developed through process engineering/design using the theme of project-based learning by integrating multiple subjects (interdisciplinary).
Mathematics
Mathematical literacy: The ability to analyze, argue, and communicate ideas effectively and how to recognize, formulate, solve, and interpret the solution of a mathematical problem when applied to a variety of different situations

3 RESEARCH METHODS

3.1 Research design

This research is a quasi-experimental design with one group and pre- and post-tests. In terms of the quasi-experimental research study, the subjects were not selected randomly, but rather used the existing classes. This was done to maintain the class in accordance with its normal conditions, because if there was adjustment through random selection of class participants then it could affect the normal dynamic of the class and thereby affect the experimental results. This is in line with Cohen et al. (2007) who suggest that in educational research, random sample selection is almost certainly impossible.

3.2 Instruments

The instruments used in this study were test and non-test instruments. The test instrument was a set of essay-type tests of mathematical communication ability, and the non-test instrument was a sheet of observation and interview guides. The instruments (test and non-test) were assessed by an expert. In addition to expert validation, the test instruments were also tested with students outside the classroom research. The latter activity was carried out to determine the validity of the whole test, for each of its items, its reliability, and its level of difficulty. Data processing of the tests was carried out using the Rasch model with Winsteps® software.

4 RESULTS AND DISCUSSION

4.1 *Results*

The purpose of this study was to observe the effect of implementation of STEM project-based learning on the mathematical communication ability of students overall and based on Prior Mathematical Ability (PMA) – high, medium or low. For this purpose, the data were collected in the form of pre- and post-test scores, and the results of observation of student activity. For mathematical communication skills, normalized gain (n-gain) was also calculated to assess the degree of improvement.

Further analysis was carried out to observe the differences in the achievement of mathematical communication skills before and after the STEM project-based learning, which was done by testing the average difference for paired samples. The differences test is conducted using paired-sample t-tests because the data was normally distributed and had homogeneous variance. The results are shown in Table 2.

The hypotheses for the statistical tests conducted were:

H_0: Implementation of STEM project-based learning does not have a significant effect on students' mathematical communication ability.

H_1: Implementation of STEM project-based learning has a significant effect on students' mathematical communication ability.

A summary of the test results of students' mathematical communication ability before and after STEM learning using one test sample t-test is shown in Table 3.

Based on the analysis of the data in Table 3, we obtained a significance value Sig. = 0.000 < 0.05.

Therefore H_0 is rejected, which means that the application of STEM project-based learning in mathematics had a significant influence on the achievement of students' mathematical communication ability. The average increase in students' mathematical communication ability was 0.63 (Table 2). This improvement value is moderate.

Further analysis was conducted on the n-gain to see whether there were significant differences in the increase of students' mathematical communication ability according to PMA (high, medium, and low). The difference test was performed using Analysis Of Variance (ANOVA) because the data was distributed normally and the variances were homogeneous. A summary of the results of the one-way ANOVA of the n-gain data is shown in Table 4.

The one-way ANOVA shows a significance level of 5%, resulting in significance = 0.001 < 0.05 (Table 4), which indicates that there are significant differences in the increase of students' mathematical communication ability based on PMA (high, medium, and low). Because there are differences in mathematical communication ability among students at the level of PMA, the ANOVA was followed by a post hoc Scheffé test to observe any difference between the three PMA groups of high, medium, and low. The test results are shown in Table 5.

Table 5 shows that some pairings of PMA levels had differences and others did not. For the pairs of high and medium PMA, and medium and low PMA, there were no significant differences but for the pair of high and low PMA there were significant differences in improvement. This difference indicates that the average increase in mathematical

Table 2. Descriptive statistics of data.

	Paired-sample statistics		
	N	Mean	Std. deviation
Pre-test	36	15.6944	10.36152
Post-test	36	67.9167	16.44797
N-gain	36	0.63	0.17

Table 3. Analysis of paired-sample t-test regarding students' mathematical communication before and after STEM project-based learning.

	Paired-sample test		
	df	F	Sig. (2-tailed)
Pre-test/Post-test	−26.42236	35	.000

Table 4. Test result of one-way ANOVA of n-gain data according to PMA (high, medium, and low).

	Paired samples test		
	df	F	Sig.
Between group	2	8.300	.001
Within group	33		
Total	35		

Table 5. The average difference of improvement of mathematical communication ability for each PMA.

	Paired-sample test	
	Mean difference	Sig.
High/Medium	.14575	.060
High/Low	.28625	.001
Medium/Low	−.14575	.072

Table 6. Summary results of student activity.

Student activity	Mean	(%)
Following learning with enthusiasm, paying attention to the subject matter and mathematical problems posed by the teacher	2.67	88.89
Actively conducting question and answer sessions between the students and teachers of the subject matter and the mathematical problems posed	2.50	83.33
Putting forward ideas to solve the problems posed by the teacher in a group or class	2.50	83.33
Giving comments, feedback, questions, suggestions, or criticisms of the proposed solutions of other students	2.50	83.33
Conducting argument, giving feedback, or answering questions during the discussion and presentation	2.67	88.89

communication ability in the high PMA group differed significantly from the increase in ability of the low PMA group.

In addition to quantitative data analysis, qualitative data were collected to support the research, including observational data. Table 6 presents a summary of data from the observation of student activity.

4.2 Discussion

The data analysis indicated that the average achievement in students' mathematical communication abilities after learning with STEM project-based learning is significantly improved. Observation of the student learning activity showed positive results, associated with students' mathematical communication ability. This is because in the STEM project-based learning students are encouraged to make understanding a concept meaningful. Students are invited to explore mathematical ideas through a project activity, so that they are actively involved in the process. It fosters critical, creative, and analytical thinking in students and thus increases high-level thinking skills (Capraro & Slough, 2013). STEM project-based learning also fosters cooperation, communication between peers and teachers, and problem-solving skills, as well as self-discipline.

STEM project-based learning helps students to bridge the mathematical knowledge learned in school and the real world. Integration between several disciplines (mathematics with technology and engineering) in STEM project-based learning helps students to appreciate that mathematics is closely related to other fields. This is consistent with the culture of vocational high school, where

students are required to put into practice the theoretical lessons gained from their classroom.

The data analysis based on the level of PMA revealed that students' mathematical communication ability increase in all the three groups performing as respondents of this study. The ability is categorized into "high" and "medium". These findings concur with those of Han et al. (2015), who state that the implementation of STEM project-based learning can improve student achievement in different ability groups (high, medium, and low), although Han et al. also stated that the benefits of STEM project-based learning are more beneficial to low-ability groups than others. In our study, the implementation of STEM project-based learning showed better effects for the students with higher PMA levels, findings that concur with those of Yoon (2009), who stated that students of high ability derive better benefits in a self-regulated, student-directed learning environment, because student-centered learning is a key feature in STEM project-based learning.

Observations of student learning activities in the STEM project-based learning class showed positive results: student activity was considered high. The students' enthusiasm is seen when they are happily involved in various learning activities. This result is in line with the findings of Tseng et al. (2011) in a study that showed that the students' attitude, especially toward engineering, changes after project-based learning with STEM.

5 CONCLUSIONS

Based on the research findings and data analysis results, it can be concluded that the implementation of STEM project-based learning in mathematics education at vocational high school has a positive effect on students' mathematical communication skills. Integrating STEM in project-based learning provides a better effect on students' mathematical communication skills for those with higher levels of PMA than others; at the levels of medium and low PMA, the effect was not significant.

REFERENCES

Alacapinar, F. (2008). Effectiveness of project-based learning. *Eurasian Journal of Education Research, 32,* 17–34.
Asmuniv. (2015). *Pendekatan Terpadu Pendidikan STEM Upaya Mempersiapkan Sumber Daya Manusia Indonesia Yang Memiliki Pengetahuan Interdisipliner Dalam Menyosong Kebutuhan Bidang Karir Pekerjaan Masyarakat Ekonomi ASEAN (MEA)*. Malang, Indonesia: PPPPTK BOE Malang. Retrieved from

http://www.vedcmalang.com/ppppt kboemlg/index.php/menuutama/listrik-electro/1507-asv9

Barton, B., Poisard, C., & Domite, M. D. C. (2006). Cultural connections and mathematical manipulations. *For the Learning of Mathematics, 26*(2), 21–24.

Capraro, R.M. & Slough, W.S. (2013). *STEM project-based learning: An integrated science, technology, engineering, and mathematics (STEM) approach.* Rotterdam, The Netherlands: Sense Publishers.

Cohen, L., Manion, L. & Morrison, K. (2007). *Research methods in education* (6th ed.). London, UK: Routledge.

Daugherty M.K. (2013). The prospect of an "A" in STEM education. *Journal of STEM Education, 14*(2), 10–15.

Departemen Pendidikan Nasional. (2006). *Panduan penyusunan kurikulum tingkat satuan pendidikan.* Jakarta, Indonesia: BSNP.

Fennema, E., Carpenter, T.P., Franke, M.L., Levi, L., Jacobs, V.R. & Empson, S.B. (1996). A longitudinal study of learning to use children's thinking in mathematics instruction. *Journal for Research in Mathematics Education, 27*(4), 403–434.

Han, S., Capraro, R.M. & Capraro, M.M. (2015). How science, technology, engineering, and mathematics (STEM) project-based learning (PBL) affects high, middle, and low achievers differently: The impact of student factors on achievement. *International Journal of Science and Mathematics Education, 13*(5), 1089–1113.

Kluver, J. (2012). Communication and understanding mathematical foundation and practical applications. *Computational and Mathematical Organization Theory, 18*(2), 221–231.

Koparan T. & Guven, D. (2014). The effect on the 8th grade students' attitude toward statistics of project-based learning. *European Journal of Educational Research, 3*(2), 73–85.

Laboy-Rush, D. (2010). *Integrated STEM education through project-based learning.* New York, NY: Learning.com.

NCTM. (2000). *Principles and standards for school mathematics.* Reston, VA: National Council of Teachers of Mathematics.

NYC Department of Education. (2009). *Project-based learning: Inspiring middle school students to engage in deep and active learning.* New York, NY: NYC Department of Education.

Rambely, A.S., Ahmad, R.R., Majid, N., M-Suradi, N.R., Din, U.K.S., A-Rahman, I., ... Abu-Hanifah, S. (2013). Project-based activity: Root of research and creative thinking. *International Education Studies, 6*(6), 66–71.

Ritz, J.M. & Fan, S. (2014). STEM and technology education: International state-of-the-art. *International Journal of Technology and Design Education, 25*(4), 429–451.

Sáenz-Ludlow, A. (2006). Classroom interpreting games with an illustration. *Educational Studies in Mathematics, 61*, 183–218.

Sumarmo, U. & Hendriana, H. (2014). *Penilaian dan pembelajaran matematika.* Bandung, Indonesia: Refika Aditama.

Tseng, K.H., Chang, C.C., Lou, S.J. & Chen, W.P. (2011). Attitudes towards science, technology, engineering and mathematics (STEM) in a project-based learning (PjBL) environment. *International Journal of Technology and Design Education, 23*(1), 87–102.

Wardhani, S. & Rumiati. (2011). *Instrumen penilaian hasil helajar matematika SMP: Belajar dari PISA dan TIMSS.* Yogyakarta, Indonesia: PPPPTK Matematika.

Wood, N.L. (2011). Practice and conception: Communicating mathematics in the workplace. *Educational Studies in Mathematics, 70*(1), 109–125.

Yaron, D. (2003). Implementation and assesment of project-based learning in a flexible environment. *International Journal of Technology and Design Education, 13*, 255–272.

Yoon, C. (2009). Self-regulated learning and instructional factors in the scientific inquiry of scientifically gifted Korean middle school students. *Gifted Children Quarterly, 53*(3), 203–216.

Ideas for 21st Century Education – Abdullah et al. (Eds)
© *2017 Taylor & Francis Group, London, ISBN 978-1-138-05343-4*

Students' attitude to biodiversity in Ciptagelar indigenous village

H.W. Kelana, T. Hidayat & A. Widodo
Universitas Pendidikan Indonesia, Bandung, Indonesia

ABSTRACT: This study aimed to investigate students' attitude to biodiversity as representatives of the young generation of the indigenous people of Kasepuhan Ciptagelar. The rationale for this study was that this population is accustomed to living with nature in the surrounding forest with great wisdom and care. This study was descriptive research, in which data about students' attitudes to biodiversity was gathered through a questionnaire consisting of 25 questions rated on a Likert-type scale. The participants were 30 students, from grades 7 to 9, of SMP Negeri 4 Cisolok Satu Atap junior high school. The results showed that the students' average score for the biodiversity aspect of sensitivity toward, and positive values for, the prevention and remediation of biodiversity problems and issues was 78.27%, which was categorized as excellent. Likewise, the students' average score in the context of belief in personal and societal efficacy in relation to biodiversity was 79.77%, also categorized as excellent.

1 INTRODUCTION

Mount Halimun-Salak National Park (MHSNP) is a National Park in West Java, Indonesia. Administratively, it is located in three districts: Bogor and Sukabumi in West Java, and Lebak in the province of Banten (Priyadi et al., 2010). With altitudes between 500 and 2000 meters above sea level, it consists of rain forest and contributes to important habitat for many species.

More than 700 species of flowering plants belonging to the natural forest, covering 390 genera and 119 families, have been identified in the park (Priyadi et al., 2010). Endangered orchids that can be found in this park are *Bulbophyllum binnendijkii, B. angustifolium, Cymbidium ensifolium*, and *Dendrobium macrophyllum*.

MHSNP also provides habitat for a number of protected and endangered animals. Previous studies have recorded the existence of 11 important mammals, such as the Javan gibbon (*Hylobates moloch*), Javan leaf monkey (*Presbytis comata*), ebony leaf monkey (*Trachypitheus auratus*), leopard cat (Felis bengalensis), Javan leopard (*Panthera pardus melas*), Asian palm civet (*Paradoxurus hermaphroditus*), black giant squirrel (*Ratufa bicolor*), wild boar (*Sus scrofa*), and barking deer (*Muntiacus muntjak*). Some 52 important birds can also be found in the park, including endemic birds such as the crested serpent eagle (*Spilornis cheela*), chestnut-bellied partridge (*Arborophila javanica*), and the Javan hawk-eagle (*Nisaetus bartelsi*), Indonesia's national bird and also known as Garuda (Cahayadi, 2003; Harahap et al., 2004; Prawiradilagal & Marakarman, 2004; GHSNPMP-JICA, 2005).

Sundanese traditional communities form the indigenous peoples who live inside the MHSNP area. They form a community called 'Kasepuhan Ciptagelar'. These indigenous peoples live wisely, managing the environment with great local wisdom and care. Local wisdom refers to knowledge or values, principles, advice or thoughts that serve as the basis of ethical and moral assessment in forming an intelligent world view (Idris et al., 2015). The indigenous peoples believe that rice is a symbol of life, and they only grow rice with a local heritage from their ancestors. There are 168 local rice varieties, and it is local taboo for them to be sold or taken out of the village.

The indigenous peoples treat the earth like a mother, not to be exploited, but utilized wisely to maintain natural balance. In managing the forest, they divide it into *leuweung garapan* (open forest), *leuweung titipan* (protected forest) and *leuweung tutupan* (closed forest). This demonstrates that local knowledge plays a role in the conversation of biodiversity all the way from gene to ecosystem. Each indigenous people has a diversity of traditional knowledge associated with the use and management of biodiversity, as a source of food, medicinal raw materials and other materials needed for life and livelihood (BAPPENAS, 2016).

Besides the role of local knowledge in environmental and biodiversity conservation, formal education in schools can be a means of dissemination of environmental awareness and preservation of biodiversity. In accordance with the demands of the curriculum, education is not only intended to build students' knowledge about biodiversity but also to form students' attitudes to the subject.

In forming students' attitudes, awareness and concrete action in relation to conservation becomes more important (Rustaman, 2013). Therefore, this study aimed to investigate the Biodiversity Attitude (BA) of students in junior high school *Sekolah Menengah Pertama* (SMP) Negeri 4 Cisolok Satu Atap as representative of the young generation of the indigenous people of Kasepuhan Ciptagelar.

2 RESEARCH METHODS

2.1 *Study site*

This study was conducted in Ciptagelar indigenous village, in the Cisolok sub-district of Sukabumi district, which is part of the MHSNP area. Generally, Ciptagelar has a cool and pleasant climate with an average temperature of 21–28°C. Specifically, the data were collected at SMP Negeri 4 Cisolok Satu Atap local secondary school. The study site can be seen in Figure 1.

2.2 *Methods*

A preliminary study was conducted prior to the research, to determine the profile of the community and the way they manage their natural environment. The respondents to be included in the main study were taken from this preliminary study. This type of study was qualitative, using a descriptive method. In this study, the subjects were students of SMP Negeri 4 Cisolok Satu Atap ($n = 30$), in seventh, eighth and ninth grades, acting as representatives of the young generation of the indigenous people of Kasepuhan Ciptagelar. To complete the data related to the students' BAs, in-depth interviews were conducted.

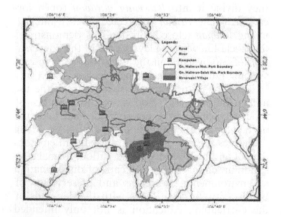

Figure 1. Map of MHSNP and Ciptagelar village (adapted from Suganda, 2009).

Table 1. Interpretation of BA score.

Percentage (%)	Category
$BA \geq 70$	Excellent
$70 > BA \geq 30$	Good
$BA < 30$	Weak

2.3 *Data analysis*

The BA framework used in this study was adapted from the attitude to biodiversity assessment instrument developed by the World Wildlife Fund (1996), including sensitivity toward, and positive values for, the prevention and remediation of biodiversity problems and issues, as well as beliefs about personal and societal efficacy relating to biodiversity. The data about students' BAs was gathered through a questionnaire which consisted of 25 questions rated on a Likert-type scale. To derive percentages for students' BAs, Equation 1 was used:

$$BA = \frac{f}{N \times I \times R} \times 100\% \qquad (1)$$

where
  BA = Biodiversity Attitude
  F = Total score of observation frequency
  N = Maximum score
  I = Items number
  R = Respondent number.

To interpret these percentages, the guidance shown in Table 1 was used.

3 RESULTS AND DISCUSSION

3.1 *Student profile*

Of the total of 30 students selected at SMP Negeri 4 Cisolok Satu Atap, 12 were in grade 7, 13 were in grade 8, and six were in grade 9. In terms of gender, the numbers of male and female students were 12 and 18, respectively (see Figure 2). The small number of young people in Kasepuhan Ciptagelar who receive formal education was due to a lack of parental awareness in sending their children to school, together with an economic motive. From the point of view of the parents, who are generally subsistence farmers, children were directly encouraged to move as fast as possible into the network of the family economy (Adimihardja, 1992).

3.2 *Sources of Students' biodiversity attitudes*

Students' BAs can be influenced by various sources. Based on Figure 3, the most influential

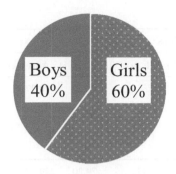

Figure 2. Split of boy and girl students in SMP Negeri 4 Cisolok Satu Atap research subjects.

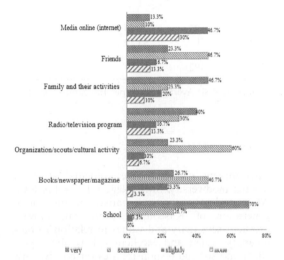

Figure 3. Sources and levels of influence on students' BAs.

source in relation to a student's BA is school, with 70% of students identifying school as very influential, and a further 26.7% and 3.3% of students indicating school to be somewhat or slightly influential, respectively. This finding demonstrates that a majority of students felt that schools had a role in forming their attitudes toward biodiversity. As well as school, organization/scouts/cultural activity, friends, and family and their activities are all environments in which students socialize.

The source that did not play much of a role was the internet, with 30% indicating that it did not have any role and a further 46% assigning it only slight influence; just 13.3% identified the internet as very influential and 10% as somewhat so. Most students felt that the internet did not contribute to forming their attitude to biodiversity. Based on interviews, affordability was the main reason for minimal internet access. In line with previous

studies, cultural transmission, including knowledge and attitude, can occur through parents, social activity such as cultural activity, and friends of the same age (Reyes-Garcia et al., 2009). Nowadays, schools could be as important in transmitting student attitudes as media information.

3.3 Students' biodiversity attitudes

3.3.1 Student's sensitivity toward, and positive values for, the prevention and remediation of biodiversity problems and issues

In terms of student attitude to the overall aspect of sensitivity toward, and positive values for, the prevention and remediation of biodiversity problems and issues, the Average Score per Aspect (ASA) was 78.27%, categorized as excellent. For specific indicators, student sensitivity had the lower score (66.70%, categorized as good), and the holding of positive values scored 79.20%, categorized as excellent (see Table 2).

Most students did not realize that a lot of rice varieties showed biodiversity at gene level and they were unaware that there were so many protected and endangered species around, such as Oa (Javan gibbon) and Maung tutul (Javan leopard). They simply chose rice as a plant that must be maintained on the mandate of their ancestors, while Javan gibbon and leopard were not so important as to be protected.

Biodiversity issues are important because human activities are the main cause of environmental quality decline and biodiversity loss. Attitudes can function to encourage individuals and communities to value the environment and consider it important, in order to inspire participation in the process of improving and protecting the environment for the betterment of their own livelihoods (Shrestha, 2005).

3.3.2 Student's belief about personal and societal efficacy relating to biodiversity

In terms of student's belief about personal and societal efficacy in relation to biodiversity, the ASA was 79.77%, categorized as excellent. More specifically, with regard to the student's locus of control, such as political action, persuasion, eco management, consumerism, and legal action, the Average Score per Indicator (ASI) reached 79.20%, indicating that students believe they have an excellent locus of control. For the indicator of student hopefulness vs. hopelessness, the ASI was 77.10%, also categorized as excellent. Based on the associated statement, most students believed in the sustainability of biodiversity as long as they could manage their environment in a fair way.

The ASI for students' assumptions of personal responsibility for biodiversity was 83.80%, greater than their assumptions of societal responsibility,

Table 2. Students' attitudes toward aspects of biodiversity.

Aspect	Indicator	Data analysis				
		EF ($n = 30$)	OF ($n = 30$)	ASI (%)	ASA (%)	Category
Sensitivity toward, and positive values for, the prevention and remediation of biodiversity problems and issues	1. Values	1560	1235	79.20	78.27	Excellent
	2. Sensitivity	120	80	66.70		

Belief about personal and societal efficacy relating to biodiversity (predictors of behavior)	1. Locus of control
	   a. Political action
	   b. Persuasion
	   c. Eco management
	   d. Consumerism
	   e. Legal action
		1200	1047	79.20	79.77	Excellent

	2. Hopefulness/hopelessness	240	185	77.10		
	3. Assumptions of personal responsibility	240	201	83.80		
	4. Societal responsibility	240	192	80.00		

EF: Expectation Frequency ($N \times I \times R$); OF: Observation Frequency; ASI: Average Score per Indicator (%); ASA: Average Score per Aspect (%).

indicating an excellent attitude. In practice, holding to this principle means that natural resource management must be based on self-awareness. So the most important belief is that humans do not have the right to decide the fate of all other things (Suganda, 2009).

Students have excellent awareness that biodiversity is the responsibility of human beings. This is in line with their wider beliefs. With the concept of *Jagat Leutik, Jagat Gede—Jagat Leutik Sanubari, Jagat Gede Bumi Langit* (awareness of the microcosm and macrocosm—the small world of consciousness and the immensity of the natural universe), they believe that humans are only one of many living creatures in the universe (Suganda, 2009).

In education, understanding students' attitudes is important in planning curricula and instruction because the biodiversity issue is an abstract and complex one, and it is difficult for students to understand and learn (Huang & Lin, 2014). Therefore, it is necessary for educators to understand students' attitudes before developing biodiversity learning on a contextual basis for better understanding. This is expected to increase students' awareness of the prevention and remediation of biodiversity problems; not only to respect their ancestors but so that they have belief and understanding that what they do locally has an impact globally. Indonesian biodiversity, at the level of genes, species and ecosystems, plays an important role in the global life of the biosphere (Gadgil, Berkes, & Folke, 1993).

4 CONCLUSIONS

Environment can influence students' attitudes toward biodiversity. SMP Negeri 4 Cisolok Satu Atap students, as representatives of the young generation of Kasepuhan Ciptagelar, showed excellent sensitivity and values in relation to biodiversity issues, and excellent beliefs about personal and societal responsibilities for biodiversity. Understanding students' attitudes in this area is important in increasing students' awareness of the prevention and remediation of biodiversity problems.

REFERENCES

Adimihardja, K. (1992). *The Traditional Agricultural Rituals and Practices of the Kasepuhan Community of West Java'*. The Heritage of Traditional Agriculture among the Western Austronesians, Canberra: The Australian National University, 33–46.

BAPPENAS. (2016). *Indonesia Biodiversity Strategy and Action Plan (IBSAP) 2015–2016.* Jakarta, Indonesia: Badan Perencanaan Pembangunan Nasional.

Cahayadi, I. (2003). *Analisis spasial struktur dan fungsi koridor hutan antara Taman Nasional Gunung Halimun dengan Hutan Lindung Gunung Salak.* Bogor, Indonesia: Tesis Program Pascasarjana.

Gadgil, M., Berkes, F., & Folke, C. (1993). Indigenous knowledge for biodiversity conservation. *Ambio,* 151–156.

GHSNPMP-JICA. (2005). *Ecological study, Halimun-Salak corridor, Mount-Salak National Park* (Unpublished).

Retrieved from http://www.sci.kagoshima-u.ac.jp/~itp/student/siryo/Corridor%20Study-Eng2005.pdf

Harahap, S.A., Ikbal, W., Rachmady, R. & Sutrisno, E. (2004). *Corridor assessment on leopard to traverse from Gunung Halimun to Gunung Salak.* Bogor, Indonesia: Biodiversity Conservation Indonesia.

Huang, H & Lin, Y.K. (2014). Undergraduate student's attitude toward biodiversity. *Universal Journal of Educational Research, 2*(4), 379–386.

Idris, M., Ramli, M., Agustang, A. & Kesuma, A.I. (2015). Bureaucracy ethics based in public service local wisdom in Gowa. *Mediterranean Journal of Social Sciences, 6*(6), 419–424.

Prawiradilagal, D. & Marakarman, A. (2004). *Kominitas burung pada koridor Halimun-Salak.* Laporan Teknik, Teknik 2003, Proyek Investarisasi dan Karakterisasi. Bogor, Indonesia: Sumberdaya Pusat Penelitian Biologi LIPI.

Priyadi, H., Takao, G., Rahmawati, I., Supriyanto, B., Nursal, W.I. & Rahman, I. (2010). *Five hundred plant species in Gunung Halimun-Salak National Park, West Java (A checklist including Sundanese names, distribution and use).* Bogor, Indonesia: Center for International Forestry Research (CIFOR).

Reyes-Garcia, V., Broesch, J., Mir, L.C., Pelaez, N. F., McDade, T.W., Parsa, S., … Rodriguez, M.R.M. (2009). Cultural transmission of ethnobotany knowledge and skills: An empirical analysis from an Amerindian society. *Evolution and Human Behavior, 30*(4), 274–285.

Rustaman, N.Y. (2013). Trend penelitian pendidikan biologi dalam rangka mengembangkan perilaku konservasi. In *Prosiding Seminar Nasional Pendidikan dan Penelitian Biologi 2013, Universitas Pendidikan Indonesia, Bandung.*

Shrestha, N. (2005). *Education and communication for biodiversity conservation in Nepal.* Kathmandu, Nepal: World Wildlife Fund Nepal Program.

Suganda, U. (2009). *The Ciptagelar Kasepuhan indigenous community, West Java (Forest for the future: Indigenous forest management in a changing world).* Jakarta, Indonesia: AMAN & DTE.

World Wildlife Fund. (1996). *The development of a biodiversity literacy assessment instrument (Report to the National Environmental Education Training Foundation).* Stevens Point, WI: Wisconsin Center for Environmental Education.

Ideas for 21st Century Education – Abdullah et al. (Eds)
© *2017 Taylor & Francis Group, London, ISBN 978-1-138-05343-4*

Determinate factors of mathematics problem solving ability toward spatial, verbal and mathematical logic intelligence aspects

K. Kusaeri & B. Sholeh
Universitas Islam Negeri Sunan Ampel, Surabaya, Indonesia

ABSTRACT: This research aimed to know and to analyze the direct and indirect effects of three multiple intelligences (spatial, verbal, mathematical-logic intelligences) towards students' mathematical problem solving ability. The subjects were 280 of 9th grade students of SMP Negeri 37 Surabaya. The data were collected by mathematical logic intelligence, problem solving ability, and psychological tests. The psychological test was used to measure psychological constructs of spatial and verbal intelligences. The data analysis used inferential statistics, path analysis. The results showed that: (1) the mathematical logic intelligence was the first determinant factor in the students' ability in mathematical problem solving (42.7%), and the second was the verbal intelligence (29.2%); (2) the verbal and mathematical logic intelligences had a direct effect to the ability of mathematical problem solving, simultaneously. The spatial intelligence affected indirectly the ability of mathematical problem solving. Individually or simultaneously, it appears that verbal and mathematical logic intelligences had a direct effect to the ability of mathematical problem solving. It implies that mathematics teachers of SMP/MTS should give more opportunities and stimulus of mathematical logic problems and verbal abilities to their students.

1 INTRODUCTION

Gardner's Multiple Intelligence Theory has contributed a lot in the world of education (Klcin 1997). It especially has enriched and inspired teachers to create various kinds of teaching methods and techniques which help stimulate student's potentials (Murtadlo 2012). However, of all nine types of intelligence, only some of them are relevant and effective in helping students to solve mathematical problems because mathematical problem solving process needs not only one's understanding of the problem itself, but also high imagination and ability to visualize and configure spatial knowledge as well as ability to observe and analyze numbers (Tambunan 2006).

A student's language competence has great impact to his ability to solve problems since the ability will help him in understanding mathematical problems, which are normally stated in written form using mathematic abstract symbols. He will eventually be better able to describe phenomenon, generalize concepts, and to draw conclusions than his peers who are strong in other types of intelligence (Nugraha 2012). This ability relates closely to what Gardner mentioned as verbal intelligence.

Meanwhile, the ability of the child to visually imagine spatial configurations requires a special skill which needs practice and proper regular repetitions. This is related to spatial intelligence. Children with this intelligence are more likely to dabble with visual objects than with abstract symbols. They are also relatively easy to learn to use visual images and have advantages in terms of visual imagination (Tambunan 2006).

The ability to observe and analyze numbers requires logical thinking. This intelligence is a combination between numeracy ability and logic. Children who have logical-mathematical intelligence tend to be able to understand a problem, and analyze and solve them appropriately (Suhendri 2011). With this intelligences, children are able to think and devise solutions (exit) from the logical sequence (reasonable), are able to understand the pattern of relations as well as the process of deductive and inductive thinking (Susanti & Werdiningsih 2009).

The role of those three kinds of intelligences towards students' problem-solving ability has been supported by some previous studies. A study conducted by Landau (1984) and Campbell et al. (1995), for example, found a relationship between the spatial and mathematical problem-solving ability. Battista (1990) and Fennema & Tartre (1985) found an interaction between verbal intelligence with problem-solving abilities.

In Indonesian context, similar research was conducted with the focus on students 'achievement, not on their mathematical problem-solving ability. One of the examples is a research conducted by Tambunan (2006) which found out a positive relationship between spatial ability and academic achievement. Suhendri (2011) in his study found a significant influence of logical-mathematical intelligence on students' mathematics learning outcomes.

Based on the above explanation, it appears that verbal intelligence, spatial and mathematical logic does not always have a contribution to mathematical problem solving ability. This situation is thought-provoking and very much an open possibility of further research. Therefore, this study specifically attempted to test this situation.

2 LITERATURE REVIEW

Learning is expected to provide enough space for children to develop their full potentials (Jayantika et al., 2013). In this context, learning does not only focus on the cognitive development of students alone, but it is also directed to develop their talents and potential.

Every child has different talents; sports, art and some are gifted in the field of processing numbers (numerical). A child who is gifted in specific areas is more likely to achieve better in the field than other children. That potential intelligence is unique for each child (Murtadlo 2012).

Problem solving activity is one of the ways to develop children's intelligence. Through problem-solving experiences given to a child, a schema on facts and experiences will be established and can be used to solve the next problem (Tri Hariastuti & Saman 2007). In solving a mathematical problem, in particular, it takes specific understanding, analysis, calculation and imagination as well as verbal intelligence, logical mathematical and visual-spatial (Indragini 2010).

Studies which examine the contribution of three kinds of intelligences on learning outcomes or mathematical problem solving skills continue to emerge. One of these studies was conducted by Jayantika et al. (2013). The results indicated that spatial intelligence and logical mathematical intelligence both simultaneously and significantly contributed to mathematics achievement. These results show that spatial intelligence and logical mathematical intelligence are important factors that determine the mathematics achievement. Therefore, this study proposed that mathematics achievement can be improved by increasing the students' spatial and logical mathematical intelligence.

Another study conducted by Foster (2012) indicated that the relationships between the students' verbal skills and between their spatial skills were not as strong as would be expected. In fact, each of the two skills appeared to have stronger relationships with the other problem type. Additionally, although no strong relationships were observed

among the students' cognitive skills (i.e., verbal, spatial, and analytical skills), the strongest pairwise relationship was shown to exist between their verbal and analytical skills.

Although Campbell et al. (1995) found that vividness of visual imagery had no efect on students' problem solving success, Diezmann & Watters (2000), in Brisbane, Australia, found that spatial intelligence has a significant contribution to the students' achievement on mathematics subject. This supports the argument that logical reasoning was a greater contributor to mathematical succes than vivid visual images.

From studies that have been described above, it appears that there is a lack of agreement in the role of spatial, verbal, mathematical-logic intelligences to a child's mathematical problem solving abilities. It may be caused by the different researchers' perspectives to problem-solving ability, or that there is a definition on the spatial, verbal, mathematical-logical intelligences themselves and how each intelligence correlated to each other. Therefore, this study has the possibility to provide the initial framework in interpreting the studies that have been conducted before and to find the link of each intelligence (spatial, verbal, mathematical-logical intelligences) or how each intelligence contributes to a child's ability of mathematical problem solving.

3 RESEARCH METHODS

3.1 Research design

The study began with a theoretical study of multiple intelligences affecting mathematics. The results would be used to build a structural model (lines) of the theoretical relationship between the variables being discussed. In addition to quantitative data, qualitative data is also used to strengthen the quantitative data interpretation. The participants of the study were 280 of class IX students of SMP Negeri 37 Surabaya, the academic year of 2016/2017.

3.2 Research variables

The variables of this study were divided into two parts, namely the independent variable (exogenous) and the dependent variable (endogenous). The independent variable is the spatial intelligence (X1), verbal intelligence (X2) and logical-mathematical intelligence (X3), while the dependent variable is the mathematical problem solving ability (Y).

3.3 Research instruments

Two types of research instruments were used in this research: (1) tests—to collect data on logical mathematical intelligence, verbal intelligence problem-solving ability, and (2) psychological test—to measure psychological constructs of spatial intelligence.

3.4 Data analysis

Data collected for this research were analyzed both quantitatively and qualitatively. Quantitative analysis was done using inferential statistics which included path analysis and regression analysis. The qualitative analysis was used to supplement the information obtained from the quantitative analysis.

4 RESULTS AND DISCUSSION

In this research report, the variable of spatial intelligence is symbolized with X1, verbal intelligence with X2, logical mathematical intelligence with X3 while mathematical problem solving ability with Y. The results summary of the calculation of the four variables is presented in the Table 1.

The correlation between variables was calculated using the Pearson product moment correlation. The results are set forth in the form of a diagram as the representation of the theoretical model which is built in this study (Figure 1).

Testing of the model was done using path analysis. Trimming model of path analysis was applied in this research with an expectation to improve the structural model of path analysis itself. This was done by excluding the exogenous variables with insignificant path coefficient.

Testing is done to look at the contribution of each variable (X_1 to Y, X_2 to Y, X_3 to Y, X_1 and X_2 to Y, X_2 and X_3 to Y, X_1 and X_3 to Y, and X_1, X_2, and X_3 towards Y). The results showed a relationship between variables X_1, X_2, and X_3 towards Y as follows.

Figure 2 shows that verbal intelligence contributes to a student's mathematical problem solving ability in the amount of approximately 29.2%. Likewise, logical mathematical intelligence

Table 1. Result summary of the calculation of the four variables of the research.

	X_1	X_2	X_3	Y
Mean	4,63	5,27	13,33	16,22
Median	5	5	13	15,5
Modus	4	6	11	15
Standard Deviation	1,33	1,57	3,56	3,86
Variance	1,78	2,49	12,68	14,92
Total	167	190	480	584

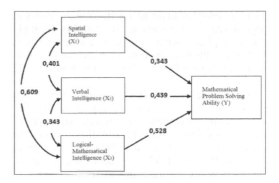

Figure 1. The coefficient correlation between variables X_1, X_2, X_3 and Y on a theoretical.

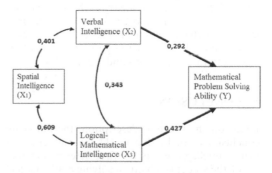

Figure 2. Line Model along with the path coefficient linking between the three kinds of intelligences and mathematics problem-solving skills.

contributes to the mathematical problem solving abilities as big as 42.2%. However, this figure also showed that spatial intelligence does not directly affect a student's ability of solving mathematical problems. Spatial intelligence contributes to the student's logical thinking ability and verbal thinking skills, which in turn affects the ability of solving mathematical problems.

These findings suggest that the stronger one's ability to think logically and the higher his verbal thinking skills, then his math problem-solving abilities would also increase. This statement indicates that a child with logical mathematical intelligence and verbal skills has the ability to manage logic and understand the context of the problems by applying logical thinking, counting, arranging pattern of relationships, and solving problems. The result is in line with Foster's (2012), who found that children with high verbal ability and logical-mathematical skills outperform other children.

A child with strong verbal intelligence and logic, according to Gunawan (2003), is able to think and devise solutions in logical sequence, have good numeracy skills and be able to think logically and orderly. On the other hand, children who have weaker ability in math and arranging solutions in logical sequence, results in the inability to apply the concepts, principles, formulas, and units, which will eventually have an impact on their lack of ability in solving mathematical problems (Fatoke et al., 2013).

Meanwhile, the result of this research, which shows the absence of a direct impact on the ability of spatial intelligence research in mathematical problem solving, is different from the results of the study conducted Landau (1984) and Campbell, Collis & Watson (1995). The differences occurred most probably because the instrument used to measure the spatial intelligence in this research is a psychological test. This means that the spatial abilities measured did not focus dimensional spatial ability and spatial geometry as the material that is usually taught and is often found when children learn geometry in junior high school, but on how to recognize the geometrical pattern changes that exist both in 2 dimensions and 3 dimensions.

5 CONCLUSIONS

Based on the previous description, the following conclusions can be drawn: (1) the mathematical logical intelligence was the first determinant factor of the students' ability in solving mathematical problems (42.7%), while the second was the verbal intelligence (29.2%); and (2) when combined, the verbal and mathematical logic intelligences had direct effects to the students' ability of mathematical problem solving, while the spatial intelligence had indirect effect on the ability of mathematical problem solving.

REFERENCES

Battista, M.T. 1990. Spatial visualization and gender differences in high school geometry. *Journal for Research in Mathematics Education* 21(1): 47–60.

Campbell, K.J., Collis, K.F. & Watson, J.M., 1995. Visual processing during mathematical problem solving. *Educational Studies in Mathematics* 28: 177–194.

Diezmann, C.M. & Watters, J.J. 2000. Identifying and supporting spatial intelligence in young children. *Contemporary Issues in Early Childhood* 1(3): 299–313.

Fatoke, A.O., Ogunlade, T.O. & Ibidiran, V.O. 2013. The effect of problem-solving instructional strategy and numerical ability on student's learning outcomes. *The international Journal of Engineering and Science (IJES)* 2(10): 97–102.

Fennema, E. & Tartre, L.A. 1985. The using of spatial visualization in mathematics by girls' and boys'. *Journal for Research in Mathematics Education* 16(3): 184–206.

Foster, A.C. 2012. *The contribution of spatial, verbal, and analytical skills to problem solving performance.* Unpublished dissertation. Chicago: Illinois Institute of Technology.

Gunawan, 2003. *Genius learning strategy.* Jakarta: Gramedia Pustaka Utama.

Indragiri, A. 2010. *Kecerdasan optimal: Cara ampuh memaksimalkan kecerdasan anak.* Jogjakarta: Starbooks.

Jayantika, T., Ardana, I.M. & Sudiarta, I.G.P.S. 2013. Konstribusi bakat numerik, kecerdasan spesial, dan kecerdasan logis matematis terhadap prestasi belajar matematika siswa SD Negeri di Kabupaten Buleleng. *e-Journal Program Pascasarjana Universitas Pendidikan Ganesha Program Studi Matematika* 2(1): 1–12.

Klein, P.D. 1997. Multiplying the problems of intelligence by eight: A critique of Gardner's Theory. *Canadian Journal of Education* 22(4): 337–394.

Murtadlo, A. 2012. Kecerdasan dalam pembelajaran matematika. *Edu-Math* 3: 18–31.

Suhendri, H. 2011. Pengaruh kecerdasan matematis-logis dan kemandirian terhadap hasil belajar matematika. *Jurnal Formatif* 1(1): 29–39.

Tambunan, S.M. 2006. Hubungan antara kemampuan spasial dengan prestasi belajar matematika. *Jurnal Makara, Sosial Humaniora* 10(1): 27–32.

Tri Hariastuti, R. & Saman, A. 2007. Mengembangkan kecerdasan emosional anak. *Jurnal Pendidikan Dasar* 8(1): 101−110.

Relationship between factors that improve student achievement in primary teacher education institutes

S. Ratnaningsih

Universitas Islam Negeri Syarif Hidayatullah, Jakarta, Indonesia

ABSTRACT: The aim of this study was to investigate the relationship between factors that affect students' achievement in the primary teacher education institutes of state universities in Indonesia. The method used in this research was a survey method with correlation testing. From a total population of 1008 university students, data were collected from a sample of 205. Data were collected through inquiry and then analyzed using the Structural Equation Modeling (SEM) analysis technique with the AMOS 4.0 software package. The results showed that the perception levels of students with regard to the profession of educator, the role of the lecturer, classroom climate, the library service, and student motivation were all in the high category. Moreover, among all of these variables, there was either a direct or an indirect correlation.

1 INTRODUCTION

The quality of education at university level is still low, even though many efforts, including curriculum improvement, renovation of the lecture hall, provision of learning facilities, addition of books and literature, and improvement of lecture quality, have been delivered by government and other agencies. These programs have not significantly improved the quality of education as yet.

Improving quality of education should be a concern of the government and other communities, including parents (Tilaar, 2002). Parents should motivate students, which in turn can improve students' educational achievement. The same thing was also articulated by Tirtarahardja and Lasulo (1994), who stated that the quality of the output of an education system was affected by the raw input (student), the instruments input (faculty, curriculum, facilities exist), and the environmental input (social, cultural, security, political). Skaalvik and Skaalvik (2013) also found that learning structures were a predictor of the perception of teachers as emotionally supportive, and also predicted that student motivation has positive results.

The research presented here involved students of institutes of teacher education at the three state universities in Java, Indonesia, consisting of the Universitas Negeri Malang (UNM), Universitas Negeri Jakarta (UNJ), and Universitas Pendidikan Indonesia (UPI). The assumption made in this study was that the areas of East and Central Java were represented by UNM, West Java was represented by UPI, and other areas of Java were represented by UNJ. The aims of this study were to measure the relationship between student perceptions of the profession of educators and how they relate to the role of the lecturer in the classroom, classroom climate, library services, learning motivation and the learning achievements the students have earned.

2 RESEARCH METHODS

This study aimed to explain the presence or absence of a relationship between the various variables based on the size of the correlation coefficient, using a survey of correlation type. The relationships were investigated between student perceptions and the variables of: the profession of educator (X_1); the role of the lecturer in class (X_2); classroom climate (X_3); library services (X_4); motivation to learn (X_5); learning achievement (Y).

The research was conducted at the state universities on the island of Java: UNM, UNJ, and UPI. To find out what was happening in the field, researchers used sampling techniques from a population which was considered representative. From a total population of 1008 university students, a sample of 205 was used.

Simultaneous Structural Equation Modeling (SEM) was used to analyze the data. SEM is able to explain relationships and instead of designing a theory, its use is intended to examine and justify a model. The main requirement to use SEM is to build a hypothetical model that consists of structural models and measurement models in the form of the path diagram-based justification theory. SEM is a set of statistical techniques that allow the simultaneous testing of a series of relationships. These relationship are built between one or more of the independent variables.

2.1 Hypothesis testing of overall model

The overall model or models in this SEM analysis study consisted of a structural model and a measurement model.

2.2 Hypothesis testing of goodness of fit through evaluation model

Analysis of data to test the research hypotheses used a procedural model of SEM, which consists of seven steps (Hair et al., 1998; Solimun, 2002; Ferdinand, 2002): 1) the development of model-based concepts and theories; 2) the development of a flow diagram (path diagram); 3) the conversion of flowcharts into a structural model (equation statistics); 4) selection of an input matrix and estimation models; 5) assessment of problem identification; 6) goodness-of-fit evaluation; 7) interpretation and modification of the model.

2.3 Suitability index and test statistics

2.3.1 Chi-squared statistic (χ^2)
The most fundamental statistical test to measure the overall fit is the likelihood ratio test or chi-squared statistic. The chi-squared statistic is very sensitive to the size of the sample used. Therefore, if the sample size is such that there are more than 200 samples, the chi-squared statistic must be accompanied by other test tools (Hair et al., 1998). The model we tested will be considered good or satisfactory when the chi-squared value is low. The smaller the value, the better ^{2} capital (shown by the chi-square results, which shows no difference at all) i.e. null hypothesis, H_0, applies and gets accepted by the probability of the cut-off value of $p > 0.05$ or 0.10.

2.3.2 The measure of significance probability
A significance test for measuring the differences of covariance matrix was carried out. A significance probability value greater than 0.05 indicates that the model is acceptable.

2.3.3 RMSEA (Root Mean Square Error of Approximation)
RMSEA is an index that can be used to compensate for the chi-squared statistic in a large sample (Baumgartner & Homburg, 1996). The RMSEA value shows the goodness of fit that can be expected if the capital is estimated in the population (Hair et al., 1998). An RMSEA value greater than or equal to 0.08 is an index of the inadmissibility of the capital which shows a close fit of the capital by degrees of freedom (Browne & Cudeck, 1993).

2.3.4 GFI (Goodness-of-Fit Index)
The suitability index (goodness-of-fit index) will calculate the weighted proportion of the variance in the sample covariance matrix, described by the covariance matrix of the population estimate (Tanaka, 1993). GFI is a non-statistical measure that has a range of values between 0 (poor fit) and 1.0 (perfect fit). A high value in this index indicates a better fit.

2.3.5 AGFI (Adjusted Goodness-of-Fit Index)
GFI is an analog of the R^2 of the regression. AGFI can adjust the degrees of freedom that are available to test (Arbuckle, 1999). Values below 2.0, or even sometimes below 3.0, are indicative of an acceptable fit between the model and the data (Arbuckle, 1999).

2.3.6 TLI (Tucker–Lewis Index)
TLI is an alternative incremental fit index which compares the model being tested against a baseline model (Baumgartner & Homburg, 1996). The recommended value as a reference for the adoption of a model is a value > 0.95 (Hair et al., 1998), and a value close to 1 indicates a very good fit (Arbuckle, 1999).

2.3.7 CFI (Comparative Fit Index)
The magnitude of this index is in the range of 0 to 1, with values approaching 1 indicating the highest degree of fit (a very good fit) (Arbuckle, 1999). CPU recommended value is > 0.95. The advantage of this index is that it is not influenced by the size of the sample because it was not to measure the level of acceptance of a model.

2.4 Interpretation and modification of model

The final step is to interpret any model that does not fit and modify or refine it. Based on the covariance estimation parameters of the SEM, the model was a structural one. The model was then used to explain the phenomenon studied.

3 RESULTS AND DISCUSSION

It was revealed that the average perception level of university students regarding the profession of educator, the lecturer's role, the classroom climate, the library service, and students' motivation were in the high category. Moreover, the model was well-developed, meaning that the model can be used to explain the relationship among the variables.

The results showed that there was a direct relation between: a) the lecturer's role and student motivation to study; b) the lecturer's role and the classroom climate; c) the lecturer's role and the library service; d) the library service and student motivation to study; e) the library service and learning achievement; f) the classroom climate and student motivation to study; g) student motivation to study and learning achievement.

An indirect relation was also seen between: a) university students' perception of the educator profession and study achievement through study motivation; b) the lecturer's role and study achievement through study motivation; c) the classroom climate and study achievement through study motivation.

The findings of the relationship between students' perceptions of the educator profession and students' motivation to learn is complementary to the theory put forward by Brown and Phelps (1961). These finding suggests that a positive outlook in primary school teachers to the profession of educator will have an impact on students' learning motivation. (Yuan & Zhan, 2016). The relationship between the role of the lecturer and student motivation to study the theory stated by Brown and Phelps (1961). So this finding is expected to give the ability to strengthen students' motivation to learn the course material and practice in the field.

There was a relationship between the library services and learning motivation (Hardjoprakoso, 1992). Thus, the campus should make the library as comfortable and representative as possible, so students might gain knowledge to support the teaching and learning process. The existence of a relationship between study motivation and learning achievement was in line with Willerman's (1979) theories.

The finding of an indirect relationship between classroom climate and learning achievement through study motivation was complementary to the theory of Sceerens and Bosker (1997). The implication of this finding is that the faculty and its leaders can create a classroom climate that supports learning, so that the classroom atmosphere will both make students enthusiastic about learning and deliver learning outcomes as expected.

The result also showed that there was no relationship between learning perceptions and motivation (Brown & Phelps, 1961). With these findings, leaders and lecturers should be able to establish a positive perception in students by providing a picture of the future prospects for professional educators. That there was no direct relationship between the role of the lecturer in the classroom and student achievement through study motivation means corroboration of the theories of Sonhaji (1989) and Usman (2002). So, the implication of this research is that lecturers should be able to create

an atmosphere of enjoyable learning for students to follow world developments, including mastering English and technology.

Based on the findings, the suggestions are as follows: 1) there should be a high standard of selection of students who enter the institutes of teacher education (Paulick et al., 2013; Azigwe et al., 2016); 2) the lecturer should pay attention to the activities and facilities in the classroom to create a conducive environment for the learning process; 3) the lecturer should intensify the interactions among students; 4) the lecturer should create rules to raise the level of discipline in classes; 5) the lecturer should give the students intrinsic motivation through reinforcement and the support of their achievements; 6) the lecturer should encourage students to study in the library.

4 CONCLUSIONS

It is concluded that the students' levels of perception regarding the profession of educator, the role of the lecturer, classroom climate, the library service, and student motivation were in the high category. Moreover, there were direct or indirect relationships between the perception levels of students with regard to the profession of educator, the role of the lecturer, classroom climate, the library service, and student motivation.

REFERENCES

Arbuckle, J.L. (1999). *Amos 4.0 user guide*. Chicago, IL: SmallWaters Corporation.

Azigwe, J.B., Kyriakides, L., Panayiotou, A. & Creemers, B.P.M. (2016). The impact of effective teaching characteristics in promoting student achievement in Ghana. *International Journal of Educational Development, 51*, 51–61.

Baumgartner, H., & Homburg, C. (1996). Applications of structural equation modeling in marketing and consumer research: A review. *International journal of Research in Marketing, 13*(2), 139–161.

Brown, E.J. & Phelps, A.T. (1961). *Managing the classroom* (2nd ed.). New York, NY: The Ronald Press.

Browne, M.W. & Cudeck, R. (1993). Alternative Ways of Assessing Model Fit. In K.A. Bollen & J.S. Long (Eds.), *Testing structural equation models* (pp. 136–192). Newbury Park, CA: Sage.

Ferdinand, F.A. (2002). *Structural equation modelling dalam penelitian manajemen*. Semarang, Indonesia: BP Undip.

Hair, J.F, Anderson, R.E., Tatham, R.L. & Black, W. (1998). *Multivariate data analysis* (5th ed.). Upper Saddle River, NJ: Prentice Hall.

Hardjoprakoso. M. (1992). *Perpustakaan Sekolah: Petunjuk untuk membina, Memakai dan memelihara Perpustakaan Sekolah*. Jakarta: Perpusnas RI.

Paulick, I., Retelsdorf, J. & Möller, J. (2013). Motivation for choosing teacher education: Associations with teachers' achievement goals and instructional practice. *International Journal of Educational Research, 61*, 60–70.

Sceerens, J. & Bosker, R. (1997). *The foundations of educational effectiveness.* New York, NY: Elsevier Science.

Skaalvik, E.M. & Skaalvik, S. (2013). School goal structure: Associations with students' perceptions of their teachers as emotionally supportive, academic self-concept, intrinsic motivation, effort, and help seeking behavior. *International Journal of Educational Research, 61*, 5–14.

Solimun, M.S. (2002). *Structural equation modelling, lisrel dan amos.* Makalah disajikan dalam Diklat Angkatan II, Program Pascasarjana Unibraw, Malang, 11–13 Maret 2002.

Sonhaji, A. (1989). Akuntansi sosial: Peranannya dalam Mengukur Tanggung Jawab sosial Perusahaan, Suatu tinjauan Analitis. *Majalah Akuntansi, 10*(1), 25–33.

Tanaka, J. S. (1993). Multifaceted conceptions of fit in structural equation models. In K. A. Bollen & J. S. Long (Eds.). *Testing structural equation models.* Newbury Park, CA. Sage.

Tilaar, H. A. R. (2002). *Membedah pendidikan nasional.* Jakarta: Refika Cipta.

Tirtarahardja, U. & Lasulo, L. 1994, *Pengantar Pendidikan,* Jakarta: P3MTK—Direktorat Jenderal Pendidikan Tinggi.

Usman, M. U. (2002). *Menjadi guru profesional.* Bandung, Indonesia: PT Remaja Rosda Karya.

Willerman, L. (1979). *The psychology of individual and group differences.* San Francisco, CA: W.F. Freeman.

Yuan, R. & Zhan, L.J. (2016). Exploring student teachers' motivation change in initial teacher education: A Chinese perspective. *Teaching and Teacher Education Journal, 61*, 142–152.

Ubiquitous Learning (UBL)

Ideas for 21st Century Education – Abdullah et al. (Eds)
© *2017 Taylor & Francis Group, London, ISBN 978-1-138-05343-4*

Developing dynamic instructional media to promote explorative activities in geometry lessons

S. Sariyasa
Universitas Pendidikan Ganesha, Bali, Indonesia

ABSTRACT: The main purpose of this study was to develop instructional media to accompany teaching material for selected topics in plane geometry for junior high school in Bali. To do so, a development study was conducted following Plomp's developmental model. Based on the results of needs assessment and analysis of instructional processes in the classroom, it was found that one problem for teachers is the difficulty in delivering geometry instruction. This is partly due to the unavailability of appropriate learning media. Computer—aided learning, together with dynamic mathematics software, offers a promising solution to this problem. To achieve this, a number of instructional media have been developed, based on GeoGebra. The developed media is dynamic and explorative in nature. The use of such media is expected to bring the geometry classroom to life in the sense that students can actively engage in the process of learning achievement both physically and mentally. With the aid of these media, students are able to explore geometrical objects by dragging them with a mouse, to observe what happens during this dragging process, to discuss with their counterparts, to formulate conjectures, and—facilitated by teachers—draw conclusions. This learning process is in line with the scientific approach currently being emphasized in the curriculum. In addition, these activities provide opportunities for students to develop mathematical competencies and higher-order thinking skills. In this way, the internalization of geometrical concepts occurs more constructively and, thus, persists longer in students' cognitive constructs.

1 INTRODUCTION

Traditionally, geometry has been taught with pencil and paper. Accordingly, a diagram in traditional geometry instruction may pose difficulties in visually describing dynamic processes in constructing geometrical concepts. However, students are required to build dynamic visualizations of a geometrical object in their cognitive studies. This may lead to students' failure to understand geometrical concepts, with the consequence that many, if not most, students might be demotivated when it comes to learning geometry.

King and Schattschneider (1997) have stated that, in a traditional geometry course, students are told definitions and theorems and are assigned problems and proofs; they do not experience the discovery of geometric relationships for themselves, nor invent any mathematics as a result. Reys et al. (in Furner & Marinas, 2007) have suggested that geometry 'is best learned in a hands-on active manner, one that should not rely on learning about geometry by reading from a textbook'. A study by the author revealed that most teachers have difficulties in delivering some geometrical concepts (Sariyasa, 2015). One finding of the study is that teachers do not fully utilize media in their geometry instruction, which was partly due to the lack of availability of appropriate instructional media.

One solution to this problem is to introduce technology-aided learning environments. Some studies have shown that technology can be used in instruction and gives positive impacts on the quality of the instruction (Crompton, 2015; McPherson & Tyson, 2006; Keong et al., 2005). Dede (2000) has indicated that technology can be used to strengthen student learning and enhance pedagogy. The use of technology in student instruction has other benefits, such as providing students with more chances to study (Roberts, 2012).

Preiner (2008) proposed two forms of technology usage in mathematics instruction, that is, virtual manipulators and mathematical software tools. Since virtual manipulators have limited capability to conduct mathematical experiments, many teachers use educational software packages that enable both teachers and students to visualize and explore mathematical concepts in their own creative ways (Barzel in Preiner, 2008, p. 29). It should be noted, however, that technology is intended to support the learning and teaching of mathematics. Empirical results obtained from the use of technology give us some hints and better points of view in establishing deductive arguments. Furthermore, the use of technology in mathematics teaching should support and facilitate conceptual development, exploration, reasoning, and problem-solving (Garofalo et al., 2000). In addition, the National Council of Teachers of

Mathematics (NCTM) has stressed that teachers should use technology to enhance their students' learning opportunities by selecting or creating mathematical tasks that take advantage of what technology can do efficiently and well—graphing, visualizing, and computing (NCTM, 2000, p. 25).

Therefore, this research was aimed at developing geometry teaching material for *Sekolah Menengah Pertama* (SMP), or junior high school, and creating GeoGebra-based instructional media. To achieve this, a Plomp development model was used. The model consisted of five phases, namely: (a) preliminary investigation; (b) design; (c) realization/construction; (d) test, evaluation, and revision; (e) implementation.

2 LITERATURE REVIEW

Geometry is an important part of school mathematics for which a number of technological tools have been developed and implemented. Among these tools, dynamic programs (such as GSP, Cabri, and GeoGebra) play a significant role in geometry teaching and learning (Yanik, 2013). With the aid of Dynamic Geometry Software (DGS), a Dynamic Learning Environment (DLE) can be established to facilitate geometry instruction.

In a DLE, students can quickly and accurately explore geometrical objects, discover patterns, formulate a conjecture, test an idea, and explore mathematical relationships (Yanik & Flores, 2009), in a way that cannot be done with just pencil and paper (Jones, 2000). A DLE provides features that enable extensive and instant manipulation through the power of drag mode. This feature allows students to instantly experiment with as many examples as possible and get immediate feedback.

Geometry instruction that encourages exploration and stimulates active participation is expected to improve the quality of students' learning. Giving students the opportunity to discuss and share their ideas will certainly improve their mathematical competencies. The use of a DGS is intended to provide a learning environment that allows students to explore geometry objects and investigate their attributes and properties while preserving the rigorous nature of mathematics. One such DGS is GeoGebra (originally created by Markus Hohenwarter); it is free, is easy to operate, is widely used, and has features to support interactive exploration. GeoGebra enables construction and animation of geometrical objects such that exploration and investigation are easily carried out interactively by both teachers and students. This feature enables the possibility of studying geometrical objects through manipulation, analysis, conjecture, and testing. These activities help students to better and more easily understand and master geometrical concepts. Doğan and Içel (2011) have found that

GeoGebra positively affects students' learning and achievement, and also improves motivation. Study by Haciomeroglu and Andreasen (2013) has shown that students' understanding of mathematics improved after using a DGS.

To optimize DLE-oriented geometry instruction aided by GeoGebra, it is necessary to develop appropriate geometry teaching material. Such teaching material is equipped with GeoGebra-based instructional media, so that students themselves can carry out visualization, exploration, and investigation in their efforts to construct geometrical concepts. As stated by Keşan and Calişkan (2013), 'through the use of dynamic geometry software, students can create geometric drawings or do interactive investigations on dynamic geometric shapes prepared by the teacher'. Visualization helps students to imagine or even to manipulate abstract concepts, enabling them to grasp a better understanding of those concepts. Therefore, geometry teaching material incorporating GeoGebra-based media will bring more meaningful instruction for students and, consequently, internalization of concepts is expected to occur more smoothly. This is supported by the opinion of Falcade et al. (2007) that the use of technology as a tool 'may foster [the] internalization process'.

3 RESEARCH METHODS

This research was aimed at developing geometry teaching material for junior high school (SMP) and creating GeoGebra-based instructional media. To achieve this, we used the plomp development model consisting of five phases, that is, (a) preliminary investigation, (b) design, (c) realization/construction, (d) test, evaluation, and revision, and (e) implementation. The first four phases were completed in the first and second years of the research (2015 and 2016). Over this two-year period, geometry teaching material and GeoGebra-based instructional media were developed. These two product types were then validated by content experts, media experts, and teachers.

A total of 36 mathematics teachers were chosen as follows. The teachers were selected from 12 SMPs in the regencies of Buleleng, Tabanan, Jembrana, and Karangasem in the province of Bali. From each regency, three SMPs were chosen and three mathematics teachers (teachers of grades 7, 8 and 9) were selected from each SMP.

The sources of data were relevant documents, mathematics teachers, students, school principals, and experts. Data were gathered using the techniques of documentation, interview, and questionnaire. The instruments used were questionnaire, interview guidance, and documentation guidance. Data were then analyzed descriptively. The implementation phase will be conducted in the final year of the research (2017).

4 RESULTS AND DISCUSSION

Needs assessment was carried out qualitatively by analyzing grade 7 and 8 syllabi for geometry, analyzing textbooks and sampling lesson plans. A set of questionnaires was distributed to teachers.

The questionnaires revealed that one main problem concerns the availability of media that can enable the visualization of concepts. Teachers are aware of the importance of media, but they do not fully utilize media in their geometry instruction. If they use media, then their media is static, and is not capable of being manipulated and explored. The idea of incorporating computers into geometry instruction was welcomed enthusiastically by teachers. Teachers wanted media that was dynamic and enabled manipulation so that demonstration of a concept could be done by just altering a figure on the computer screen. It should not be necessary to create the figure again and again. According to teachers, such media can save time and are attractive to students.

Based on these results, geometry teaching materials were developed and GeoGebra-based instructional media were created. The geometry teaching materials were based on the current curriculum and included elementary topics such as the basic concepts of geometry, triangles, quadrilaterals, circles, and transformations. GeoGebra-based instructional media were created such that they were practical and effective. Further, GeoGebra-based media were developed by considering students as the primary users while having regard to teachers as key users too. This consideration followed an approach proposed by Garofalo et al. (2000).

In the following sections, some examples of GeoGebra-based media and samples of investigative activities using these media are presented (taken from Sariyasa, 2016).

4.1 Median of a triangle

The following example is adapted from Johnston-Wilder and Mason (2005, pp. 39–43).

In the GeoGebra graphic window, draw the triangle *ABC*. Construct a point on one of the sides, say side *AB*. Call this point *D*. Draw a line through *D* and parallel to *BC*. This line intercepts side *AC* at point *E*. Draw segments *BE* and *CD* and name the intersection of these two segments *M*. Finally, draw the line *AM* and let it cut the side *BC* at point *N* (see Figure 1). Ask students to measure and write down the lengths of *BN* and *CN*. Students are then requested to move point *D* with the mouse along the side *AB* and note down the lengths of *BN* and *CN*. Now, change the shape of the triangle by dragging one of its vertices with the mouse. One possible result of this action is shown in Figure 2. Guide students to conclude that the point *N* is the midpoint of the side *BC*, regardless of both the position of *D* on the side *AB* and the shape of the triangle. Hence,

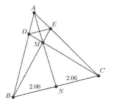

Figure 1.   Segment *AN* is a median.

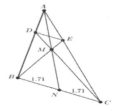

Figure 2.   Different shape of triangle; *AN* is still a median.

the segment *AN* is always a median of the triangle. The question is 'how to prove that *AN* is always a median?' Teachers can invite students to engage in a discussion to mathematically prove this result.

The investigation can be continued further. Teachers may ask students to explore the properties of the three medians of a triangle (such as that the three medians of any triangle are concurrent; the centroid of a triangle divides each median in the ratio 1:2). We notice that each median divides the triangle into two smaller triangles. Using GeoGebra, students easily verify that the area of these two triangles is the same. The three medians divide the triangle into six smaller triangles. The area of these six triangles is always equal regardless of the type of the original triangle. Using GeoGebra, this property can be reinvented by students just by dragging the vertices of the triangle and observing the changes that happen. Teachers can motivate students to move further and investigate the shape of the triangle for which the six smaller triangles are congruent with each other. An assignment can be given to students to formally prove the results obtained in this investigation.

4.2 Midpoint quadrilateral

Consider an arbitrary convex quadrilateral. Determine the midpoints of the sides of the quadrilateral. If we connect these midpoints to form a new quadrilateral, we know that the resulting quadrilateral is a parallelogram (see Figure 3). Teachers can create learning activities together with GeoGebra-based instructional media to investigate this fact. In GeoGebra, start with four points and construct a convex quadrilateral with these four points as vertices. Mark the midpoint of each side and join them to form a new quadrilateral.

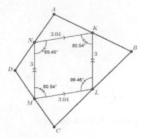

Figure 3.   Midpoint quadrilateral is a parallelogram.

Ask students to explore the original quadrilateral by dragging one of the vertices. Notice that as the shape of the original quadrilateral changes, the shape of the resulting quadrilateral also changes. Students are instructed to observe the resulting quadrilateral and record the measurements of all angles, the length of each side, and the positions of two opposite sides. At this stage, students can have a group discussion to conjecture as to whether the resulting quadrilateral is a parallelogram. Ask students to replace the original quadrilateral with a square, a rectangle, a rhombus, a trapezium, and a kite. Ask students to observe and discuss what happens to the midpoint quadrilateral. Further, let students investigate the conditions for which the midpoint quadrilateral is a rectangle, a square, a rhombus, or a kite.

5   CONCLUSIONS

By employing dynamic geometry software, a dynamic learning environment can be established. GeoGebra-based instructional media has been created to open up opportunities for students to dynamically visualize, explore, and investigate geometrical concepts and their properties. This will encourage students to engage in the learning process more deeply and actively. Thus, it is expected that such a learning environment will enhance students' internalization and understanding of geometrical concepts.

REFERENCES

Crompton, H. (2015). Using context-aware ubiquitous learning to support students' understanding of geometry. *Journal of Interactive Media in Education, 1*, 1–11.
Dede, C. (2000). Emerging influences of information technology on school curriculum. *Journal of Curriculum Studies, 32*, 281–303.
Doğan, M. & Içel, R. (2011). The role of dynamic geometry software in the process of learning: GeoGebra example about triangles. *International Journal of Human Sciences, 8*, 1441–1458.
Falcade, R., Laborde, C. & Mariotti, M.A. (2007). Approaching functions: Cabri tools as instruments of semiotic mediation. *Educ Stud Math, 66*, 317–333.

Furner, J.M. & Marinas, C.A. (2007). Geometry sketching software for elementary children: Easy as 1, 2, 3. *Eurasia Journal of Mathematics, Science & Technology Education, 3*, 83–91.
Garofalo, J., Drier, H., Harper, S., Timmerman, M.A. & Shockey, T. (2000). Promoting appropriate uses of technology in mathematics teacher preparation. *Contemporary Issues in Technology and Teacher Education, 1*, 66–88.
Haciomeroglu, E.S. & Andreasen, J.B. (2013). Exploring calculus with dynamic mathematics software. *Mathematics and Computer Education, 47*, 6–18.
Johnston-Wilder, S. & Mason, J. (2005). *Developing thinking in geometry.* London, UK: Paul Chapman Publishing.
Jones, K. (2000). Providing a foundation for deductive reasoning: Students' interpretation when using dynamic geometry software and their evolving mathematical explanations. *Educational Studies in Mathematics, 44*, 55–85.
Keong, C.C., Horani, S. & Daniel, J. (2005). A study on the use of ICT in mathematics teaching. *Malaysian Online Journal of Instructional Technology (MOJIT), 2*, 43–51.
Keşan, C. & Calişkan, S. (2013). The effect of learning geometry topics of 7th grade in primary education with dynamic geometer's sketchpad to success and retention. *TOJET, 12*, 131–138.
King, J.R. & Schattschneider, D. (1997). *Geometry turned on!* Washington, DC: Mathematical Association of America.
McPherson, R.F. & Tyson, V. (2006). Creating your own interactive computer-based algebra teaching tools: A no programming zone. *Contemporary Issues in Technology and Teacher Education, 6*, 293–301.
NCTM. (2000). *Principles and standards for school mathematics.* Reston, VA: National Council of Teachers of Mathematics.
Preiner, J. (2008). *Introducing dynamic mathematics software to mathematics teachers: The case of GeoGebra* (PhD thesis). University of Salzburg, Austria.
Roberts, G.R. (2012). Technology and learning expectations of the net generation. In D.G. Oblinger & J.L. Oblinger (Eds.), *Educating the Net Generation.* Boulder, CO: EDUCAUSE.
Sariyasa. (2015). Techno geometry: Enriching learning of geometry with technology. In K. Setemen, K.S. Mahedy, I.G.P. Sindu & P.H. Suputra (Eds.), *Innovative research across disciplines: 1st International Conference, ICIRAD 2015 proceedings, Denpasar, Bali, 18–19 November 2015* (pp. 32–35).
Sariyasa. (2016). *Creating dynamic learning environment to enhance students' engagement in learning geometry.* Paper presented at the 3rd International Conference on Mathematics, Science, and Education, Semarang State University, Semarang, 3rd September.
Yanik, H.B. (2013). Learning geometric translations in a dynamic geometry environment. *Education and Science, 38*, 272–287.
Yanik, H.B. & Flores, A. (2009). Understanding rigid geometric transformations: Jeff's learning path for translation. *The Journal of Mathematical Behavior, 28*, 41–57.

Ideas for 21st Century Education – Abdullah et al. (Eds)
© *2017 Taylor & Francis Group, London, ISBN 978-1-138-05343-4*

Internship information system availability on vocational high school websites

N. Amelia, A.G. Abdullah, M. Somantri & A.A. Danuwijaya
Universitas Pendidikan Indonesia, Bandung, Indonesia

ABSTRACT: The information systems of Vocational High Schools (VHS) should be complete with information about internships. This research was conducted to analyze the level of internship information system availability on VHS websites in West Java, Indonesia. The methods used were content analysis and survey. A survey was conducted on the websites of 209 public schools, specializing in the field of technology and engineering, in West Java, Indonesia. The findings from this research are that most such schools pay limited attention to the availability of their website, especially in relation to the dissemination of information about internships. The expected result can be input to the school about the availability of information of internships on the school website.

1 INTRODUCTION

Internship is an opportunity for students to combine work-related experience and knowledge with formal education in schools (Ahmad & Shariff, 2013). The main purpose of the practice is that students are given the opportunity to apply previously acquired knowledge to an industry by integrating theory and practice to improve their learning (Mercader-Trejo et al., 2016). Industrial working practice also aims to encourage learning experiences that are authentic to enable the integration of knowledge and practical experience in a particular context (Chanlin & Hung, 2015).

During the internship process, a partnership between schools and industries is built in accordance with the students' knowledge and skills acquired during the internship period. Internship training involves three main parties, that is, the students who are conducting the internship, school facilitators, and industrial institutions (Shariff, 2008). As the program involves three parties, information system facilities that will accommodate all of the parties, particularly the students, in the implementation of the internship are needed.

An internship information system is an alternative to the traditional process of internship implementation because the system is financially affordable and students do not need to travel long distances to meet with their teacher supervisor (Jeske & Axtell, 2016). An internship information system has been developed to automate the existing processes of internship which are still performed manually. The system allows a monitoring function into the feasibility of internship, online registration, student assignment by a teacher supervisor, a schedule of visits, the use of an online logbook, and regular monitoring (Sarlan et al., 2008).

The purpose of the internship information system is to connect students with academic institutions and companies in an effort to manage the ongoing process of internship (Fuerstner et al., 2015). In addition, the system is also used as a medium of communication for all those involved in the internship program. The system can easily be used as a support for the program (Sarlan et al., 2008).

Internship information systems have features in common with some of the work associated with the virtual world. These systems are more likely to operate in support of activities that are temporary and project-based. The systems also provide the facility to conduct all interactions, but the systems are also very dependent on technology, software, and good communication skills in all of the individuals involved. This is because a student doing an internship might not always be able to have face-to-face meetings with their supervisor nor parties from industry (Jeske & Axtell, 2016). Therefore, this study was conducted to analyze the availability of such internship information systems on the websites of Vocational High Schools (VHS) in West Java, Indonesia.

2 RESEARCH METHODS

This study used a qualitative approach, and the methods used in this research were content analysis, to examine the documents in the form of general categorization (Sung & Ho, 2012), and a survey. The survey was carried out on 209 websites of public secondary vocational schools in technology and engineering in West Java, Indonesia. The samples studied were internship information systems on the school websites. In general, this research has involved the stages of preliminary study, the development of models and hypotheses for the study, data collection and analysis, and eventually concludes with the results of the interpretation of the analysis.

3 RESULTS AND DISCUSSION

The data were gathered from direct observation of the official web pages of VHS, taken from the *Data Pokok Sekolah Menengah Kejuruan* (SMK) website (http://datapokok.ditpsmk.net/datapokoksmk/). The gathered data concerned the ownership of the official website, the complete information about partners and internships, updated internship information and the availability of online information system applications on the websites of VHS specializing in technology and engineering in West Java, Indonesia.

The results showed that the total number of vocational high schools in technology and engineering in West Java was 209. Figure 1 indicates that the number of these schools with official websites was 151 (72.25%), while 58 schools (27.75%) did not have official websites. The findings on the classification of the domains used by each school's website can be seen in Figure 2.

Figure 2 shows that most of the schools already used commercial domains, such as *sch.id* used by 88 schools, *.org* used by six schools, and *.net* by two schools. It has become mandatory for schools to have an official website with a commercial domain name. However, some schools used free domains, such as *blogspot*, used by 40 schools, *wordpress*, as used by nine schools, and others (*ktp.info, ucoz. com, hol.es, weebly.com*) used by six schools.

The mapping of school website ownership is quite prevalent in every town and district in West Java. In the use of the domain, the majority

of schools have been using commercial website domains, such as *sch.id*, *.net* and *.org*. In terms of the several schools that still use non-commercial domains, this should be a concern for the schools because as formal institutions they should employ good websites, and even greater concern attaches to those schools that do not have a website at all. Basically, the ownership of the website that a school uses is very important, because the school's website is one of the key internet-based media that serves to share information and data between users of the school and the students, parents, and other parties associated with education in these schools (Noureddine, 2010).

Of the 151 schools which had an official website, Figure 3 shows that there were 40 schools (26.49%) that provided complete information on their websites for their partner companies, while there were 111 schools that did not have any relevant information on their websites. The information on the websites consisted of a list of school partnerships with various companies in accordance with the areas of expertise of each school. Only six schools out of the 40 were not updating the website information on their partnerships.

The study found that internship information provided on school websites was limited. Only a few schools included such information, while most of the school websites did not. This is very worrying because this information is very important for the continuity of the internship process. It is during the internship process that partnerships

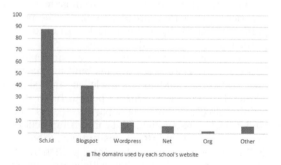

Figure 2.   The domain classification of school websites.

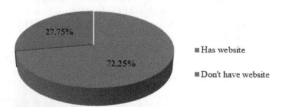

Figure 1.   The proportions of VHS having ownership of school websites in West Java.

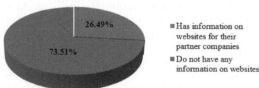

Figure 3.   A comparison of VHS based on the level of completeness of school website information provided for school partners.

between schools and industries are built (Shariff, 2008).

Figure 4 indicates that 20 schools provided internship information on their websites, while 131 schools did not provide such information. The internship information consisted of historical data on the internship program based on area of expertise, students' names, internship processes, detailed internship activities, placements, final presentation, and writing a report. Of these 20, there were eight schools that were not updating such information on their websites.

The internship information provided on websites is limited. A number of schools listed very limited information on internships by comparison with the information on partner companies. This should be a concern for the schools because disseminating clear internship information is very crucial in supporting the continuity of the process. In the perception of the user, a website must be easy to use and easy to understand (Mebrate, 2010).

Overall, there were 20 schools that provided complete internship information on their official website. Figure 5 shows that there were only 12 schools that regularly updated the internship information on their websites, while 139 schools did not update their information. The process of updating occurred over several different timescales. There were five schools that had most recently updated their information in 2016, three schools in 2015, one school in 2014, two schools in 2012, and one school in 2008.

The process of updating internship information was very limited as only a few websites were regularly updated. Information systems must be frequently updated so that the information provided

Figure 6. Online application in internship information systems at VHS.

is up to date and valid. Schools need to pay more attention to the availability of websites that can be accessed anytime needed. Then they must ensure that they update information regularly and provide the latest information related to academic and non-academic matters (Dwi et al., 2015).

As indicated in Figure 6, there were only two schools that used online application to provide internship information on websites, and 149 schools did not have such an online application system. The two schools were SMK 1 Patrol Indramayu and SMK 13 Bandung. SMK 1 Patrol has a specific information system about public and industry relations. This system has a menu that provides a list of company names, information about internship implementation, internship schedules, and internship guides. However, the information was not updated regularly. Meanwhile, SMK 13 Bandung has an online application based on Android, named SINDROID 13. Apart from teaching and learning media, the application can also be used for monitoring the implementation of internships online.

4 CONCLUSIONS

Based on the findings and discussion above, several conclusions can be drawn. The mapping of school website ownership is evenly distributed in schools in West Java. The information provided on school websites about partner companies, internship programs, and updates of such information is not sufficient and has limited management. This includes online application systems for internships which are extremely limited.

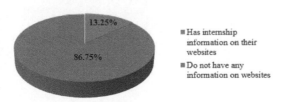

Figure 4. Level of completeness of internship information on websites in VHS.

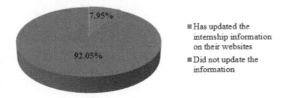

Figure 5. Updating process for school websites that provide internship information.

REFERENCES

Ahmad, N. & Shariff, S.M. (2013). Students' practicum performance of industrial internship program. *Procedia —Social and Behavioral Sciences*, *90*, 513–521.

Chanlin, L. & Hung, W. (2015). Evaluation of an online internship journal system for interns. *Procedia— Social and Behavioral Sciences*, *191*, 1024–1027.

Dwi, D., Suwawi, J. & Darwiyanto, E. (2015). Evaluation of academic website using ISO/IEC 9126. In *3rd*

International Conference on Information and Communication Technology (ICoICT) (pp. 222–227).

Fuerstner, I., Nadj, A. & Anisic, Z. (2015). Prototyping a career center online internship- and job-offer module. In *IEEE 13th International Symposium on Intelligent Systems and Informatics (SISY)* (pp. 89–94).

Jeske, D. & Axtell, C.M. (2016). Going global in small steps: E-internships in SMEs. *Organizational Dynamics*, *45*(1), 55–63.

Mebrate, T.W. (2010). *A framework for evaluating academic website's quality from student's perspective* (Master's thesis). Delft University, The Netherlands.

Mercader-Trejo, F., López, A.R. & Granada, G.L. (2016). Technical internships as a means of acquiring professional skills for future metrologists. *Measurement*, *84*, 1–6.

Noureddine, A. (2010). TKVM : A method and tool for school web sites knowledge cartography. *2010 International Conference on Machine and Web Intelligence (ICMWI)* (pp. 489–496).

Sarlan, A.B., Fatimah, W. & Wan, B. (2008). Student industrial internship web portal. *2008 International Symposium on Information Technology*, *1*, 1–10.

Shariff, S. (2008). *Industrial practicum training among operations management students: A case study in Malaysia* (Unpublished doctoral dissertation). Universiti Putra Malaysia, Malaysia.

Sung, H.T. & Ho, Y.C. (2012). Analysis of vocational high school websites in Taiwan. In *Proceedings of the 2012 International Symposium on Computer, Consumer and Control* (pp. 431–434). Washington, DC: IEEE Computer Society.

High school students' perceptions of the application of Edmodo to English language learning

P. Purnawarman, A.A. Danuwijaya & A.R. Ningrum
Universitas Pendidikan Indonesia, Bandung, Indonesia

ABSTRACT: The progressive improvement in technology offers the opportunity for innovation in learning environments. Edmodo, which presents as a social networking platform, promotes solutions for the development of teaching and learning activities, including English language learning. This current study investigated high school students' activities in Edmodo and their perceptions in relation to the application of Edmodo to English language learning. The study used a mixed-method approach, and data were collected from 112 tenth-grade students in one of the senior high schools in Bandung. The instruments were questionnaire and interview. The findings showed that the students mostly used Edmodo for submitting their tasks and tests. The students showed various perceptions, from agreement to disagreement, on the use of Edmodo for English language learning. The highest positive response showed that students agreed that Edmodo could save them time and effort in getting information related to their English learning. Problems in the use of Edmodo acknowledged by the students in this research were unstable data when submitting multiple-choice tests, and the difficulty of opening web links in Edmodo when using the mobile application.

1 INTRODUCTION

The rapid development in technology, especially in the 21st century, has promoted the expansion of the way people learn and teach. With technology, people can learn and teach without barriers of time or place. It also has created a new teaching and learning atmosphere. Robyler (2003, p. 2) states that computers and other technology resources have improved in capabilities and user-friendliness for educators. The way students and teachers share information through technologies establishes new standards in learning and teaching activities. Harmer (2007, p. 193) explains that the application of technological inventions such as computers and the internet extend the opportunities for students and teachers in composing materials in an alternative way to using pencils, pens and paper. Nomass (2013) claims that for effective learning and teaching of language, the modern technological ways should be followed. Cennamo et al. (2010, p. 10) add that the new technologies make it easier to integrate the new learning theories and pedagogies such as active learning, knowledge construction, cooperative learning, and guided discovery into the classroom.

A number of researchers have investigated the integration of technology in assisting teaching and learning activities (Al-Said, 2015; Enriquez, 2014; Nomass, 2013; Purnawarman et al., 2016; Shyamlee, 2012). One of the focuses in researching the integration of technology with education relates to its usefulness in solving several problems in teaching

and learning. The ineffectiveness of communication in learning is one of the problems in teaching and learning. With the concern about improving education, technology makes a significant impact through its innovation on teaching and learning. One of the new technology products that is focusing on education is Edmodo. Edmodo, which was developed as a social and educational platform for students and educators, has attracted several researchers to study its effectiveness (Al-Kathiri, 2015; Al-Said, 2015; Bayburtsyan, 2016; Enriquez, 2014; McClain et al., 2015; Purnawarman et al., 2016; Pop, 2013). For example, Al-Said (2015, p. 178) shows that the use of Edmodo as an application for a mobile learning environment can help the learning process, especially in the communication of learning.

However, only a few studies have been conducted to investigate the use of Edmodo in a high school context, especially in relation to English language learning in senior high school. Therefore, this research was conducted to establish students' perceptions of the application of Edmodo to English language learning in one of the senior high schools in Bandung.

2 LITERATURE REVIEW

2.1 *The integration of technology in education*

The function of technology in education has been developing since technology was first introduced. With the rapid development in technology, it has become almost inseparable from human life. To make life easier, technology can be found almost everywhere in helping human jobs, including in education. The role of computers and technology in education has been transformed since Apple introduced its first desktop in 1977 (Cennamo et al., 2010, p. 11): it transformed the position of technology in the field of education.

Moreover, Cennamo et al. (2010, p. 10) declare that the integration of technology as integral parts of the teaching or learning environment represents the most up-to-date phase in the development of the use of digital technologies (computers, software, the internet, etc.). It means that teachers and students move from being consumers of online content to becoming creators, and technology plays a significant role in solving educational problems (Cennamo et al., 2010).

2.2 The rationale for selecting technology for education

The decision to use technology in education is determined according to several considerations. Robyler (2003, p. 11) declares several elements of a rationale for using technology in education. The first consideration in choosing suitable technology for education can be seen in students' motivation. As Bauer (2010, p. 179) states, people will participate actively in the learning process when they are motivated to learn. The capabilities and productivities that derive from using technological tools in education need to be investigated, especially in relation to students' motivation or engagement in learning. A technology that is suitable for learning can engage students' attention in participating during teaching and learning activities.

Another consideration in choosing technology is related to its uniqueness and capabilities in learning instruction (Robyler, 2003). In education, the technology is supposed to accommodate the informational and educational journey of the learner. Moreover, technology for education also should facilitate the data record of the learners' progress.

Cennamo et al. (2010, p. 10) state that the inclusion of relevant technologies as integral and natural contributors to the entire educational process is the current goal of technology integration in education. With suitable technological tools, the educational process for attaining teaching and learning goals can be facilitated. With technology, the teacher can also provide student-friendly tasks: students can more easily do their tasks with the support of technology.

With the rapid development of technology, students are required to have technology literacy skills. Technology literacy is described as the ability to use, manage, and comprehend technology (Dugger, 1997, p. 98). It is important when technology becomes an integral part of human life.

2.3 Edmodo as an English language learning platform

Edmodo is a free social networking and microblogging service designed specifically for education (Corbell & Corbell, 2011, p. 24). Edmodo has been developed as a social and education platform for students and educators. With multiple functional features, Edmodo can assist teacher and students in teaching–learning activities. According to the U.S. Congress Office of Technology Assessment (1995), many teachers' jobs can be achieved more easily and efficiently by using technology. These include: preparing materials, developing lessons, evaluating student progress, enlisting parent participation, keeping up with advances in pedagogy and content, and participating in the professional community. These teachers' jobs can be accomplished with the use of Edmodo. Carlson and Raphael (2015) explain that Edmodo offers a single organized repository and documentation for a teacher's class, practice, and class interaction. As a result, Edmodo is used for teaching and learning activities in many schools.

Previous studies have shown that Edmodo has contributed to improvements in teaching and learning (Enriquez, 2014; Al-Said, 2015; Purnawarman et al., 2016). Enriquez (2014) declares that, as a supplementary tool, Edmodo provided some benefits, such as improving active participation in online activities, providing uncomplicated features for submitting online tasks (quiz, students' feedback, etc.), effortless access to reference materials and improving students' motivation as a result of online activities and discussions. However, he also described some drawbacks in its employment, such as being time-consuming, struggle in following Edmodo procedures, plagiarism, and problems with internet access for some students.

Another study claims that Edmodo increases effective communication of learning and saves time (Al-Said, 2015), because Edmodo allows more opportunities for students and teachers to get and give information outside their limited time together in the classroom. Meanwhile, a study by Purnawarman et al. (2016, p. 249) showed that Edmodo, as a blended learning platform for teaching writing, can facilitate student development through interactivity and meaningful writing tasks via the *Note* menu in Edmodo, which contributes critical-thinking skills through cognitive processing.

3 RESEARCH METHODS

3.1 *Design and participants*

The current research used mixed methods as the research design to discover students' activities in Edmodo and their perspectives toward the application of Edmodo for English language learning. Creswell (2003, p. 15) points out that mixed-method research combines multiple approaches to data collection in a study. The mixed-method approach was used to obtain comprehensive data for this research. One of the mixed-method approaches that were chosen for this research was triangulation, used as a means of seeking convergence across qualitative and quantitative methods (Creswell, 2009).

The target population for this research was around 156 students who were students in digital classrooms in the tenth grade in one of the senior high schools in Bandung. The researcher chose the sample participants of this study randomly from four different classes (two science classes, one social class, and one language class) using the Slovin formula. Based on this calculation, 112 students actually participated in the study.

3.2 *Data collection instruments*

The data in this research were collected through a triangulation approach. Creswell (2012, p. 629) states that triangulation is 'the process of corroborating evidence from different individuals, types of data, or methods of data collection'. This research used a triangulation approach that consisted of literature research and interviews (qualitative data), which were combined with surveys or questionnaires (quantitative data).

In the literature research, the data were obtained through collective information from various texts and other resources related to the topic of research in Edmodo. The literature research became not only the reference for generating the interviews and the questionnaire, but also the reference for discussion of the findings. In this research, the literature research was focused on the topic of technology, especially the use of Edmodo for language learning.

The questionnaires were used in this research to collect the students' responses and investigate attitudes, perceptions, and opinions (Cohen et al., 2007). This research used a closed questionnaire in relation to the students' perceptions of the application of Edmodo for English language learning. Closed questionnaires were used in this research to collect a range of responses based on what respondents might choose (Cohen et al., 2007, p. 321). The data accumulated from the question-naire were about the students' scale of perception in their experience in using Edmodo for English language learning. The questionnaire itself was a closed type of questionnaire. There were 32 statements that were divided into three dimensions or topics: measurement and academic achievement, mobile communication and interaction, and information access. In the questionnaires, there were 26 positive statements and six negative statements, each with four choices of response: strongly agree, agree, disagree, and strongly disagree.

This research used interviews to enhance the questionnaire results. One-on-one type interviews, with one participant in the study at a time (Creswell, 2012, p. 218), were used in this study. The interviews were undertaken with 12 sample participants who were selected randomly from four tenth-grade classes in the senior high school. Furthermore, open-ended questions were used to gather the participants' honest experiences and provide them with options for responding according to their open-ended responses (Creswell, 2012, p. 218). Five questions were used in the interview for this research. In order to enable the students to answer the questions as easily as possible, the questions for both questionnaire and interview were delivered in *Bahasa Indonesia*.

3.3 *Data analysis*

The data analysis was divided into two parts. First, students' activities in using Edmodo were identified through interviews aimed at gathering the students' experiences in using Edmodo. Second, the students' perceptions toward Edmodo for learning English language were sought using the closed questionnaire. Information was accumulated using a four-point Likert scale, from strongly agree to strongly disagree. The students checked one box based on their experience in relation to each question. The assigned scores ranged from four for positive statements to one for negative statements. The questionnaire results were then analyzed using descriptive statistical analysis that was coded using SPSS version 23. Descriptive statistical analysis is used to describe information contained in many scores with just a few indicators, for example, mean and median (Fraenkel et al., 2012, p. 187). The indicators that were used in this research were mean, Standard Deviation (SD) and percentage.

The total scores in the questionnaire were calculated on a scale ranging from 32 to 128 and, as a result, the mean scores of perception were elucidated in three levels, ranged as low, average, and high (see Table 1).

Table 1. Three levels of mean score for perception.

Perception level	Mean score of perception
Low	32–64
Average	65–97
High	98–128

4 RESULTS AND DISCUSSION

4.1 Students' perceptions toward Edmodo for English learning

The findings from the questionnaire data showed that the overall student perception of the application of Edmodo to English language learning was of an 'average' level. This was based on the overall mean of all perception scores, which was 88.09, placing it in the average-level range (65–97).

This result showed the perceptions of students of Edmodo for English language learning varied between positive and negative responses. This result differs from the study by Al-Said (2015), which reported students' perception levels of Edmodo at a 'high' level.

The analysis data were divided into three dimensions which include measurement and academic achievement, mobile communication and interaction, and information access. In the first dimension, related to measurement and academic achievement, the students reported positive responses in their perceptions of the use of Edmodo for English language learning, visible in the high percentage of students' scores for the positive statements. The highest mean and percentage scores in this dimension were for the statements '*Edmodo can facilitate English learning*' and '*I agree with the possibility of carrying out English quizzes through Edmodo*', with respective percentages of agreement of 82.14% and 71.43%, means of 2.90 and 2.93, and SDs of 0.44 and 0.58. This implied that students agreed that Edmodo can assist English learning method (66.07%), especially in terms of quizzes. The students also believed that Edmodo could help in improving both their English academic achievement and their self-learning skills (statements 3 and 7), where the same high percentage of agreement was seen for both (74.11%), with slightly different respective means (2.78 and 2.86) and SDs (0.50 and 0.52).

At lower levels, the students believed that Edmodo could increase the effectiveness of English learning and the learning environment. This could be seen from the differing responses related to effectiveness in which 60.71% students agreed with a positive statement (no. 4) while 53.57% of students disagreed with a negative statement (no. 11), with different respective means (2.76

and 2.56) but the same SD (0.60). These results reflect the view of Farr and Murray (2016, p. 1), who explain that technology-enhanced practices have transformed teaching and learning methods. These results are also confirmed by studies such as Al-Kathiri (2015), Al-Said (2015), Enriquez (2014), Purnawarman et al. (2016), Pop (2013) and Shams-Abadi et al. (2015), which confirm Edmodo can be used to improve teaching and learning activities.

Furthermore, in statements 1 and 12, the students gave positive responses in respect of their motivation (62.50%) and improvement in their English (66.07%) after using Edmodo. It was the result of the study regarding the students' positive response in the usefulness of Edmodo for getting immediate feedback (70.54%), enriching English skills, continuous English course evaluation (62.50%), and improving students' critical-thinking skill (52.68%). These results are confirmed by Cennamo et al. (2010, p. 12), who state that technology tools spare the resources and capacities of students and teachers from mundane tasks so that they can focus on activities that promote greater collaboration, more in-depth study, and critical-thinking skills.

In the second dimension, related to mobile communication and interaction, students replied with both positive and negative perspectives. The students agreed that Edmodo could help them learn English anytime (69.64% with mean 2.99 and SD 0.58). The highest mean in this dimension attached to the statement '*Edmodo fosters collective cooperation among students*', with 71.43% agreement, a mean of 3.04, and an SD of 0.56. On the other hand, the lowest mean score in this dimension attached to the negative statement '*I think that Edmodo cannot be used to facilitate communication between students*', which had a mean of 2.46, an SD of 0.67, and a percentage disagreement of 50%. This showed that half of the students believed that Edmodo could help their interaction with other students. Furthermore, they agreed that Edmodo could encourage isolated or shy students to interact (66.07%), and could also facilitate the interaction and communication between teacher and learner (68.75% agreement). Although the students reported a positive perspective on the use of Edmodo for interaction, they demonstrated a negative perspective and significant disagreement with statement 19, '*I believe that learning using Edmodo can strengthen social relations between students*', with a percentage of disagreement of 47.32%, mean of 2.51, and SD of 0.62. It can be concluded that, even though the students felt positively about the use of Edmodo for interaction and communication, they still believed that their social relations could not be strengthened through Edmodo. This result contradicts the results of the study conducted by Al-Said (2015), which showed a high and positive mean score (3.93) in relation to a statement on the stronger social relations between students as a result of using Edmodo.

In the third dimension, the students revealed a positive perspective toward information access in Edmodo. The highest mean score in this dimension belonged to statement 23, *'By using Edmodo, I save time and effort in getting information about English material'*, with a percentage agreement of 76.79%, a mean of 3.06, and SD of 0.51. The students also agreed that they could get quicker and easier access to the material and information of the English course at any time with Edmodo, based on statements 20 (66.96% agreement), 24 (76.79% agreement), and 28 (72.32% agreement). Thus, based on statement 21, it was not surprising that the students believed that learning English with Edmodo was easier than traditional learning approaches, with 54.46% in agreement, a mean of 2.73, and an SD of 0.64. The students also enjoyed learning English with Edmodo (68.75% agreement) and would like to learn more Edmodo skills (75.89% agreement), especially because it contributed to development of their searching skills (78.57%). These positive responses in relation to the efficiency and accessibility of using Edmodo for learning activities were similar to the findings of Al-Kathiri (2015), Al-Said (2015), Bayburtsyan (2016), Enriquez (2014), Purnawarman et al. (2016), Pop (2013) and Shams-Abadi et al. (2015).

In terms of high scores for disagreement related to Edmodo statements, this could be seen in statements 27 (60.71% disagreement), 29 (61.61% disagreement), 30 (47.32% disagreement), and 32 (65.18% disagreement). Thus, the students thought that using Edmodo as a mobile learning application was not difficult. Although the students did not experience the problem of poor communication networks when accessing Edmodo, they still had difficulty in conducting discussions using Edmodo (53.57% agreement). Furthermore, the students disagreed with contributing to discussions about Edmodo (60.71%). They also showed disagreement in relation to a higher number of negative Edmodo experiences in comparison to positive ones (65.18% disagreement).

4.2 Students' activities in Edmodo

Based on the interview results, students shared their experiences related to English language learning activities in Edmodo. In the first question, the students' answers revealed some similar activities. The 12 students in four different classes (two science classes, one social class, and one language class) described some common and some distinct English learning activities. These were answering quizzes set by the teacher, handing in assignments (book-sourced tasks and tasks set by teachers), completing tests, reading learning material posted by teachers, opening and submitting video material into Edmodo, participating in discussions, and providing analysis feedback.

Such Edmodo-enabled learning activities were also found in a study by Enriquez (2014), which found that, as a supplementary tool, Edmodo contributes several benefits to students' perceptions, such as active participation in online activities, uncomplicated facilities for submitting online tasks (quizzes, students' feedback, etc.), effortless access to reference materials, and encouragement of student motivation as a result of online activities and discussions.

One of the students in the social class (L1) stated that it was common for the students to complete assignments, quizzes, and homework using Edmodo. Another student in the social class (L2) described some activities for which Edmodo was used, such as questioning practice, remedial tests, and sending in tasks.

In the language and science classes, the students described more Edmodo activities related to their English learning. In the science class, one of the students (L4) responded that Edmodo was mostly used for completing works and sharing them with teachers:

The activities that we did were doing tasks, quizzes, then sharing web material from the teacher, submitting video, and during the lesson the answers were sent into Edmodo [Excerpt 1].

We do a lot activities through Edmodo, Miss. Several of these are doing essays, tests, quizzes, sometimes reading learning material from the files in Edmodo, group discussions, sending in video tasks, and analyzing the feedback from the teacher [Excerpt 2].

Based on the interviews, it can be concluded that the Edmodo-based English language learning activities in the language class were more varied than in the other two classes, science and social. The language class students did more activities in Edmodo, such as analysis and group discussion, that were useful for intensifying their English ability, which aligned with their majoring in language. On the other hand, the students in the science and social classes replied that they used Edmodo mostly for sending in completed tasks set by the teacher.

According to the interviews, most of the students (seven out of 12) only opened Edmodo around two to four times in a week, when they received tasks from their teacher, while the others opened it almost daily but only for checking. One of the students, in the meantime, opened Edmodo when answering the question about times.

Students also described the positive and negative value of their experience in using Edmodo. Based on the interviews, there were some positive experiences for the students when using Edmodo, such as convenience and efficiency in terms of time, location and effort, an improvement in student self-study, a help to enhancing student knowledge, and easier access to information.

One of the students responded on the topic of positive experience in using Edmodo:

The positive values from my experiences are more efficiency and speed when completing a task, and achieving additional knowledge [Excerpt 3].

However, some students had issues in using Edmodo, such as unstable data when submitting a multiple-choice answer sheet, time-consuming opening of web links in the mobile application, limitations in teachers' explanations, and inconvenience in using the mobile application.

The reply from one of the students in the language class (L10) related to such problems in Edmodo:

Because the link must be opened through the web, it is sometimes difficult, especially using a handphone. Sometimes the answers of multiple choices can be different after submission, especially during tests [Excerpt 4].

5 CONCLUSIONS

This research investigated high school students' activities in Edmodo for English language learning and their perceptions toward the application of Edmodo to English language learning. Several points can be summarized according to the findings and discussion of this research.

First, the students in senior high school used Edmodo for learning English language both inside and outside classes. The students usually used Edmodo for handing in their assignments, both from book-sourced tasks and teacher-set tasks. With Edmodo, they could also easily read the material for the English course from the files and links that were posted by the teacher. Moreover, Edmodo facilitated some learning activities that could be done outside the class: reading learning material posted by teachers, analyzing feedback from teachers, completing tests and quizzes, participating in discussions, and opening and submitting video material into Edmodo. With the Edmodo mobile application, students could learn and study English language with other students and their teacher in any place and at any time.

Furthermore, the findings from the student questionnaires revealed a perception of using Edmodo for English language learning at an 'average' level, with a mean score of 88.09 (average-level range was 65–97), which placed it in between positive and negative responses. The students' response as to whether Edmodo can facilitate English learning and its environment was positive. They also believed that Edmodo helped them in communications and interactions with other students and their teacher. On the other hand, they indicated that Edmodo could not help strengthen social relations between students. Based on the interview findings, the students said that they only opened and used Edmodo for doing tasks and other activities related to study. The students also found difficulty in conducting discussions through Edmodo. However, the students gave the highest percentage of agreement to the notion that Edmodo could save them time and effort in getting information related to their English language learning material.

Lastly, based on the interview findings, the students had some problems when using Edmodo that need further attention. The students said that sometimes there were unstable data when submitting multiple-choice tests; they reported that their answers were different from what they had submitted and, as a result, their English scores had been affected by the problem. Moreover, the students also found difficulty in opening web links in Edmodo when using the mobile application; they said that it took longer than when using a laptop.

As to recommendations, it is suggested that more educational institutions give consideration to using Edmodo in learning activities. With many positive responses associated with Edmodo, its use is recommended for improving learning methods and the learning environment. In terms of further research, it is hoped that there will be new focuses related to the use of Edmodo, for example, focusing on speaking skills, or teachers' perceptions, among others.

REFERENCES

Al-Kathiri, F. (2015). Beyond the classroom walls: Edmodo in Saudi secondary school EFL instruction, attitudes and challenges. *English Language Teaching*, *8*(1), 189–204.

Al-Said, K.M. (2015). Students' perceptions of Edmodo and mobile learning and their real barriers towards them. *TOJET: The Turkish Online Journal of Educational Technology*, *14*(2), 167–180.

Bauer, C. (2010). *Promotive activities in technology-enhanced learning: The impact of media selection on peer review, active listening and motivational aspects.* Frankfurt, Germany: Peter Lang.

Bayburtsyan, K. (2016). The use of Edmodo, virtual learning management platform, in the context of promoting mobile learning. *Journal of Teaching English for Specific and Academic Purposes*, 4(1), 75–84.

Carlson, G. & Raphael, R. (2015). *Let's get social: The educator's guide to Edmodo*. Eugene, OR: International Society for Technology in Education.

Cennamo, K.S., Ross, J.D. & Ertmer, P.A. (2010). *Technology integration for meaningful classroom use: A standards-based approach* (International ed.). Belmont, CA: Wadsworth Cengage Learning.

Corbell, J.R. & Corbell, M.E. (2011). The birth of a social networking phenomenon. In C. Wankel (Ed.), *Educating educators with social media* (pp. 13–32). Bingley, UK: Emerald Group Publishing.

Creswell, J.W. (2009). *Research design: Qualitative, quantitative, and mixed methods approaches (3rd ed.)*. Los Angeles, CA: SAGE Publications.

Creswell, J.W. (2012). *Educational research: Planning, conducting, and evaluating quantitative and qualitative research* (4th ed.). Harlow, UK: Pearson Education.

Dugger, W.E. (1997). Technology for all Americans: Providing a vision for technological literacy. *Educational Horizons*, 75(2), 97–100.

Enriquez, M.A.S. (2014). Students' perceptions on the effectiveness of the use of Edmodo as a supplementary tool for learning. In *DLSU Research Congress, De La Salle University, Manila, Philippines*. Retrieved from http://www.dlsu.edu.ph/conferences/dlsu_research_congress/2014/_pdf/proceedings/LLI-II-010-FT.pdf

Farr, F. & Murray, L. (2016). *The Routledge handbook of language learning and* technology. New York, NY: Routledge.

Fraenkel, J.R., Wallen, N.E. & Hyun, H.H. (2012). *How to design and evaluate research in education* (8th ed.). New York, NY: McGraw Hill.

Harmer, J. (2007). *The practice of English language teaching* (4th ed.). Harlow, UK: Pearson Education.

McClain, C., Brown, A. & Price, G. (2015). Characterization personified: Using Edmodo to strengthen student interaction with literature. *Meridian*, 18. Retrieved from https://ced.ncsu.edu/meridian/index.php/meridian/article/view/74

Nomass, B.B. (2013). The impact of using technology in teaching English as a second language. *English Language and Literature Studies*, 3(1), 111–116.

Pop, A. (2013). Edmodo e-portfolios in EFL—A case study. In *8th International Conference on Virtual Learning* (pp. 25–26).

Purnawarman, P., Susilawati, S. & Sundayana, W. (2016). The use of Edmodo in teaching writing in a blended learning setting. *Indonesian Journal of Applied Linguistics*, 5(2), 242–252.

Robyler, M.D. (2003). *Integrating educational technology into teaching* (3rd ed.). Upper Saddle River, NJ: Pearson Education.

Shams-Abadi, B.B., Ahmadi, S.D. & Mehrdad, A.G. (2015). The effect of Edmodo on EFL learners' writing performance. *International Journal of Educational Investigations*, 2(2), 88–97.

Shyamlee, S. (2012). Use of technology in English language teaching and learning: An analysis. In *2012 International Conference on Language, Medias and Culture, Singapore* (pp. 150–156). Singapore: IACSIT Press.

U.S. Congress Office of Technology Assessment. (1995). *Teachers and technology: Making the connection*. OTA-HER-616 (database online). Retrieved from https://www.princeton.edu/ota/disk1/1995/9541/9541.PDF

A web-based model to enhance competency in the interconnection of multiple levels of representation for pre-service teachers

I. Farida
Universitas Islam Negeri Sunan Gunung Djati, Bandung, Indonesia

L. Liliasari & W. Sopandi
Universitas Pendidikan Indonesia, Bandung, Indonesia

D.H. Widyantoro
Institut Teknologi Bandung, Bandung, Indonesia

ABSTRACT: This study aimed to design a web-based learning model to enhance pre-service teachers' competencies in the Interconnection of Multiple Levels of Representation (IMLR). The model contains multi-modal representations with assignments and probing questions; it creates social engagement through online discussion forums and online assessment as feedback on learning performance. The validity of the model was evaluated by expert judgment, while the feasibilty of the model was explored through a limited test with students using the quasi-experimental method. The results showed that the implementation of a web-based model increased the pre-service teachers' abilities in IMLR on each subtopic of chemical equilibrium in aqueous solution. The pre-service students also showed good abilities to resolve problems with interconnection patterns that progressed from macroscopic to submicroscopic and symbolic, rather than starting from submicroscopic and moving to symbolic and macroscopic. It can be concluded that the web-based learning model enhanced the pre-service teachers' understanding of the submicroscopic level, changing existing problem-solving ability patterns from macroscopic–symbolic into six interconnection patterns, and improving student learning patterns.

1 INTRODUCTION

As stated by Johnstone (in Treagust, 2008), the characteristics of chemistry involve three levels of chemical representation, that is, macroscopic, submicroscopic, and symbolic. These three levels of chemical representation contain inter-connectedness information. While macroscopic observable chemical phenomena are the basis of chemistry, explanations of these phenomena usually rely on the symbolic and submicroscopic levels of representation (Treagust, 2008). Consequently, the ability of learners to understand the role of each level of chemical representation and the ability to transfer from one level to another is an important aspect of generating understandable explanations. This ability is also referred to as Interconnection of Multiple Levels of Representation (IMLR) competence.

In general, teaching and learning have been restricted to the level of macroscopic and symbolic representations. Many high school teachers tend only to use these two levels. They often do not integrate the three representation levels in their teaching but move among them without highlighting their inter-connectedness. Teachers often assume that students are able to connect symbolic to submicroscopic representations on their own

(Tasker & Dalton, 2006). Students' ability to solve mathematical problems became the criterion by which they were deemed to have understood chemical concepts. However, such a view could hinder students in achieving representational competence (Chittleborough & Treagust, 2007).

Reviews of various empirical studies supported these statements. One previous study found that first-year students had difficulty in describing the scheme and transfer of symbolic representation to submicroscopic representation in acid–base equilibrium (Devetak et al., 2006). Another study showed that high school students had difficulty in representing submicroscopic levels of ionic equilibrium in weak acids, weak bases, salt hydrolysis, and buffer solutions (Murniati & Sopandi, 2007). Such problems are caused by the lack of ability of teachers to use a variety of modes of submicroscopic representation and connect them to other levels of representation (Savec et al., 2006; Weerawardhana & Ferry, 2006; Akselaa & Lundell, 2008).

Students' representational competence is tied to the learning process in classrooms, the practical laboratory and textbooks. Chemistry teachers and pre-service chemistry teachers must achieve their own internal connection of the three levels of representation, as well as re-representing the three levels in their teaching

(Farida et al., 2010). Based on the consideration that the effectiveness of teaching and learning in school depends on teachers' competence, we should endeavor to increase the professional competence of chemistry teachers through the provisioning of IMLR competence. This paper discusses the results of research on the characteristics of a web-based learning model designed to increase the IMLR competence of student chemistry teachers, and the results of its testing.

2 RESEARCH METHODS

This research aims to engender IMLR competence using a web-based model of learning, and to analyze its impact on increasing the ability of pre-service chemistry teachers. The model was developed through three phases: a preliminary study phase, model design, and model validation. The design model was validated by expert judgment, and limited tryout on 31 students, and a model revision was then performed. The revised model was implemented to 37 students using the quasi-experimental method: a one-group pre-test/post-test design. IMLR competence was measured using an online test (two-tier multiple-choice test). An online questionnaire was used to determine student opinion of the model.

3 RESULTS AND DISCUSSION

Theoretical studies were reviewed to analyze the characteristics of chemical concepts, the mapping relationship among those concepts and the levels of chemical representation. For this purpose, we chose the concept of chemical equilibrium in aqueous solution for study. Contextually, this concept plays a crucial role in many biological and environmental processes (McMurry & Fay, 2006).

The results of the conceptual analysis showed that there are three main types of concept of chemical equilibriums in aqueous solution: 1) abstract concepts with concrete examples; 2) concepts in the form of processes; 3) concepts in the form of principles. The representation levels included macroscopic, submicroscopic and symbolic.

The prerequisite concepts involved the three levels of representation too, and were: 1) proton transfer reaction (Brønsted–Lowry acid–base concepts); 2) weak acids, weak bases and water dissociation; 3) the strength of acids and bases, and pH; 4) solubility. These prerequisite concepts must be understandable to appreciate three main concepts: 1) salt hydrolysis; 2) buffer solutions; 3) solubility equilibrium.

An analysis of levels of representation is required to determine the modes of representation that are appropriate for these concepts. The modes of representation available were pictorial mode, graphics, animation, simulation and ChemSense Animators (http://chemsense.sourceforge.net/).

The ChemSense Animator tool provides shape templates for modeling techniques and processes from the simple to the complex to help visualize and explain chemical phenomena (Toplis, 2008). An animation of a chemical process can be constructed frame by frame using a simple toolbox, so as to support reasoning and discussion about the submicroscopic changes occurring during the reaction. The use of modes of representation is integrated in the learning, as suggested by Tasker and Dalton (2006), and by Mayer (in Kozma & Russell, 2005).

The characteristics of web-based learning model found in this study were:

1. Content- and activity-based learning (problem-based and question-driven).
2. Problem-solving is done through periodic tasks for each topic in order to develop competence in IMLR. Based on these tasks, students' IMLR competence can be explored and elaborated.
3. Learning materials for each topic are multimodal representations in the form of text, images, graphics, animation, simulation and tools aimed at facilitating the development of IMLR ability in any topic. Their use is integrated into learning activities via the web.
4. A social atmosphere is created through an online discussion forum to allow students to construct meaning and reflect their abilities.
5. Online assessment was conducted independently as progress feedback on individual learning performance.

These characteristics included three main elements associated with web-based learning (Garrison & Vaughan, 2008): 1) cognitive presence; 2) social presence; 3) the role of instructors in creating and facilitating the cognitive and social climate (teacher presence).

Learning phases were developed according to these characteristics as follows: orientation, exploration, elaboration, reflection, confirmation, and evaluation. The orientation phase is performed offline or through a face-to-face meeting, while the subsequent phases are implemented in a web-based learning environment. The courseware for the web-based learning is installed into a learning management system (Moodle 2.0). It is in accordance with every teaching phase. Dynamic features of Moodle 2.0 apply the principles and learning strategies based on social pedagogy constructivism (Stocker, 2010), thus enabling: interactive learning materials management; periodic content upload; multimedia integration, which facilitates multiple levels of representation; communication forums; and online assessment (Gudimetla & Mahalinga, 2006).

Thus, the Moodle features that are activated are:

1. Lesson activity: an interactive web page with questions in card format, which present the topic of

acid–base equilibrium. This feature is integrated with animation, slideshows and images related to the topic. Lesson activity can enable students to get feedback so that they can assess and reflect on their own knowledge and then play it back to really understand the learning content. Lesson activity features act as a triggering event and can be adapted.

2. Assignment involving advanced uploading of files that is used to load worksheet and send tasks. Assisted tasks are accomplished using ChemSense Animator tools, animation and simulation. For completion of acid–base, salt hydrolysis, and buffer solution tasks the Animator tool can be used, and can be supplemented with animations and PhET science simulations (http://phet.colorado.edu).

3. Discussion forum: for reflection and sharing of knowledge between students and discussion of the problems they face in relation to the study topics. From this discussion forum, how students make use of IMLR to solve problems can be traced.

4. Online assessment tools are used to test the ability of students' IMLR competence before and after the implementation of learning activities. Online assessment is carried out to obtain feedback on the progress of students' learning performance.

5. Chat room: used to communicate synchronously, so that students can directly exchange ideas, ask questions about various aspects of the material or technical terms. This feature was not enabled at the time of the online test/quiz.

6. Feedback feature: includes an online questionnaire to obtain feedback from students regarding the learning model. This feature can be accessed after the student has completed the entire learning package.

Each feature has a sequenced arrangement of access and time limit, so that student access to learning content is more structured. The continuity of learning of each student can be traced by using block completion tracking settings, as advised by Stocker (2010) and Gillani (2003).

Thus, instructional design has to accommodate the characteristics of effective online learning according to Horton (2006) and Dawley (2007). Learning tools were developed by incorporating the principles of dynamic linking, as suggested by Kaput (in Snelson, 2005), and according to Mayer's multimedia learning principles (in Kozma & Russell, 2004). For each topic, eight indicators were developed that interconnect each level of representation to a relational understanding, as described by Treagust et al. (2003).

Overall, the assessment involves expert judgment of the design of the courseware material, and testing of the validity of its content. A limited test was conducted with 31 students and was intended to establish the design feasibility of the learning steps through the use of the web and supporting devices.

On the basis of the limited testing, there were several components that needed repair, including the presentation of text and images, instructions for using the application, and the strategy for using discussion forums and features, as well as consideration of the allocation of timings to access assignments. Through trial, it is found through the reliability test using Cronbach's Alpha is at 0.86 with index of discrimination, difficulty level, and internal validity.

According to Table 1, the highest increase observed in student IMLR competence was on the topic of acid–base equilibrium, while the lowest increase was in relation to buffer solutions. The overall improvement in capability was moderate (N-gain = 0.5). A non-parametric Wilcoxon test was conducted on the pre-test/post-test data to determine significance, for all topics: the value of asymptote using 2-tailed significance level = 0.00 < 0.05. This indicates that the IMLR abilities of the students increased significantly after the learning intervention for all topics.

Data on the activity and quality of student writing was obtained from the discussion forum and traced from the data recordings of when students logged in to each forum. There was a tendency for discussion activity on a topic to decrease from topic I to IV. Students who did not actively follow these discussions averaged 37%. These students were recorded only as observers on the discussion forum: they were seen reading posts on the forum, but did not leave comments or ask questions.

Compared with the results of the preliminary phase of the study, these findings showed that the overall pattern of interaction occurring between students was strengthened and the ability of students to solve problems through IMLR was improved. These findings indicate the importance of the role of online discussion forums as a process of interaction (Snelson, 2005), which encourages the formation of knowledge (Stocker, 2010) and a deeper learning experience (Kozma & Russell, 2005).

The obligation for every student to post and answer at least one question on the discussion forum was intended as an encouragement to take an active

Table 1. Data of pre-test/post-test analysis of student IMLR competence on all topics.

Topic	Mean score (%)		Mean gain	N-gain
	Pre-test	Post-test		
I. Acid–base equilibrium	10	68	58	0.7
II. Hydrolysis of salts	21	58	37	0.5
III. Buffer solutions	17	52	35	0.4
IV. Solubility equilibrium	16	58	42	0.5
Mean	20	58	38	0.5

role in discussions. For every post or answer, students get an assessment based on set criteria. Based on the recordings of the discussions, the higher-performing students developed more ideas and answered more questions. They were more active and motivated to answer questions posed by other students. When they had a doubt about something, they tended to ask the lecturer for confirmation. In any discussion forum, the lecturer only responded if there was an indication of conceptual error. However, most of the medium- and low-performing students remained at zero levels of participation (not following the discussion, posting an idea or answering a question). Most of them only entered the discussion forums at the end of the time limit, when they simply read a post in the forums without making any comment. Such decreases in general discussion activity are a problem that often occurs in online learning. Garrison and Vaughan (2008) have explained that the motivation of students decreases in online discussions, because their expectation is that they will always get a response to the problems they face. However, differences in online schedules cause such responses to be delayed, so that the feedback becomes less focused. This is one of the weaknesses of an asynchronous communication system. Justifications given manually can effect decreases in motivation too. Therefore, consideration should be given to an automatic assessment system for forum discussions.

Overall, most of the students responded positively to the web-based model. They felt that the learning was more interactive, motivating and structured, even though previously they had never learned through the web. They expected web-based learning to be extended to other chemical topics.

4 CONCLUSIONS

The web-based model enhanced the pre-service teachers' competencies through its comprehensive features, which enabled them to experience many multimodal representations. Through periodic problem tasks given on each topic, students' IMLR competence could be explored and elaborated. Multimodal representation, in the form of text, images, graphics, animation, simulation and tools, facilitated the development of students' IMLR abilities. The online discussion forum feature in the web-based model allowed students to construct meaning and reflect their abilities. Thanks to these characteristics, the web-based model designed here can be used as an alternative for improving pre-service teachers' competencies in IMLR.

REFERENCES

Akselaa, M. & Lundell, J. (2008). Computer-based molecular modeling: Finnish schoolteachers' experiences and views. *Chemistry Education Research and Practice, 9*(4), 301–308.

Chittleborough, G.D. & Treagust, D.F. (2007). The modelling ability of non-major chemistry students and their understanding of the sub-microscopic level. *Chemistry Education Research and Practice, 8*, 274–292.

Dawley, L. (2007). *The tools for successful online teaching.* London, UK: Information Science Publishing.

Devetak, I., Urbančič, M., Wissiak Grm, K.S., Krnel, D. & Glažar, S.A. (2006). Sub microscopic representations as a tool for evaluating students' chemical conceptions. *Acta Chimica Slovenica, 51*(4), 799–814.

Farida, I., Liliasari, Widyantoro, D.H. & Sopandi, W. (2010). Representational competence's profile of pre-service chemistry teachers in chemical problem solving. In *Proceedings 4th International Seminar of Science Education.*

Garrison, D.R. & Vaughan, N. (2008). *Blended learning in higher education.* San Francisco, CA: Jossey-Bass.

Gillani, B.B. (2003). *Learning theories and the design of e-learning environments.* Lanham, MD: University Press of America.

Gudimetla, P. & Mahalinga. (2006). The role for e-learning in engineering education: Creating quality support structures to complement traditional learning. In *Proceedings 17th Annual Conference of the Australasian Association for Engineering Education, Auckland, New Zealand.*

Horton, W.K. (2006). *E-learning by design.* San Francisco, CA: Wiley.

Kozma, R. & Russell, J. (2005). Modeling students becoming chemists: Developing representational competence. In J. Gilbert (Ed.), *Visualization in science education* (pp. 121–145). Dordrecht, The Netherlands: Springer.

McMurry, J. & Fay, R. (2006). *Chemistry* (4th ed.). New York, NY: Prentice Hall.

Savec, V.F., Vrtačnik, M., Gilbert, J.K. & Peklaj, C. (2006). In-service and pre-service teachers' opinion on the use of models in teaching chemistry. *Acta Chimica Slovenica, 53*, 381–390.

Snelson, C. (2005). Designing dynamic online lessons with multimedia representations. *Journal of Educators Online, 2*(1), 1–12.

Sopandi, W. & Murniati. (2007). Microscopic level misconceptions on topic acid base, salt, buffer, and hydrolysis: A case study at a state senior high school. In *Proceedings of 1st International Seminar on Science Education.* Bandung, Indonesia: SPS UPI.

Stocker, V.L. (2010). *Science teaching with Moodle 2.0.* Birmingham, UK: Packt Publishing.

Tasker, R. & Dalton, R. (2006). Research into practice: Visualization of the molecular world using animations. *Chemistry Education Research and Practice, 7*, 141–159.

Toplis, R. (2008). Probing student teacher's subject content knowledge in chemistry: Case studies using dynamic computer models. *Chemistry Education Research and Practice, 9*, 11–17.

Treagust. D.F. (2008). The role of multiple representations in learning science: Enhancing students' conceptual understanding and motivation. In Y.J. Lee & A.L. Tan (Eds.), *Science education at the nexus of theory and practice* (pp. 7–23). Rotterdam, The Netherlands: Sense Publishers.

Treagust, D.F., Chittleborough & Mamiala. (2003). The role of submicroscopic and symbolic representations in chemical explanations. *International Journal of Science Education, 25*(11), 1353–1368.

Weerawardhana, A. & Ferry, B. (2006). Use of visualization software to support understanding of chemical equilibrium: The importance of appropriate teaching strategies. In *Proceedings of the 23rd Annual Ascilite Conference: Who's learning? Whose technology? 3–6 December 2006, University of Sydney, Australia.*

Author index